Lecture Notes in Computer Science 13930

Founding Editors

Gerhard Goos
Juris Hartmanis

Editorial Board Members

The series Lecture Notes in Computer Science (LNCS), including its subseries Lecture Notes in Artificial Intelligence (LNAI) and Lecture Notes in Bioinformatics (LNBI), has established itself as a medium for the publication of new developments in computer science and information technology research, teaching, and education.

LNCS enjoys close cooperation with the computer science R & D community, the series counts many renowned academics among its volume editors and paper authors, and collaborates with prestigious societies. Its mission is to serve this international community by providing an invaluable service, mainly focused on the publication of conference and workshop proceedings and postproceedings. LNCS commenced publication in 1973.

Michael Khachay · Yury Kochetov ·
Anton Eremeev · Oleg Khamisov ·
Vladimir Mazalov · Panos Pardalos
Editors

Mathematical Optimization Theory and Operations Research

22nd International Conference, MOTOR 2023
Ekaterinburg, Russia, July 2–8, 2023
Proceedings

 Springer

Editors
Michael Khachay 🆔
Krasovsky Institute of Mathematics
and Mechanics
Ekaterinburg, Russia

Anton Eremeev 🆔
Sobolev Institute of Mathematics
Omsk, Russia

Vladimir Mazalov 🆔
Institute of Applied Mathematical Research
Petrozavodsk, Russia

Yury Kochetov 🆔
Sobolev Institute of Mathematics
Novosibirsk, Russia

Oleg Khamisov 🆔
Melentiev Energy Systems Institute
Irkutsk, Russia

Panos Pardalos 🆔
University of Florida
Gainesville, FL, USA

ISSN 0302-9743   ISSN 1611-3349 (electronic)
Lecture Notes in Computer Science
ISBN 978-3-031-35304-8   ISBN 978-3-031-35305-5 (eBook)
https://doi.org/10.1007/978-3-031-35305-5

This Springer imprint is published by the registered company Springer Nature Switzerland AG
The registered company address is: Gewerbestrasse 11, 6330 Cham, Switzerland

Preface

This volume contains the refereed proceedings of the 22nd International Conference on Mathematical Optimization Theory and Operations Research (MOTOR 2023)[1] held during July 2–8, 2023, in Ekaterinburg, capital of the Urals, Russia.

This year, we celebrate the 90th anniversary of academician Ivan Ivanovich Eremin (1933–2013). Academician I. I. Eremin was a widely known Soviet and Russian mathematician, a specialist in mathematical programming and operations research. He introduced the well-known Eremin-Zangwill exact penalty functions, established a brilliant theory of improper (singular) linear and convex programs having valuable applications in mathematical economics, and put forward a family of highly efficient iterative algorithms, called by him Fejér methods. In the early 1960s, he established the Ural scientific school on optimization theory and methods. MOTOR 2023 is devoted to his blessed memory.

MOTOR 2023 was the fifth joint scientific event unifying a number of well-known conferences held in Ural, Siberia, and the Far East of Russia for a long time

- The Baikal International Triennial School Seminar on Methods of Optimization and Their Applications (BITSS MOPT), established in 1969 by academician N. N. Moiseev, with 17 events held up to 2017,
- The All-Russian Conference on Mathematical Programming and Applications (MPA), established in 1972 by I. I. Eremin, with 15 events held up to 2015,
- The International Conference on Discrete Optimization and Operations Research (DOOR), which was organized nine times between 1996 and 2016,
- The International Conference on Optimization Problems and Their Applications (OPTA), which was organized seven time in Omsk between 1997 and 2018.

---

[1] http://motor2023.uran.ru

The first four events of this series, MOTOR 2019[2], MOTOR 2020[3], MOTOR 2021[4], and MOTOR 2022[5] were held in Ekaterinburg, Novosibirsk, Irkutsk, and Petrozavodsk, Russia, respectively. As per tradition, the main conference scope included, but was not limited to, mathematical programming, bi-level and global optimization, integer programming and combinatorial optimization, approximation algorithms with theoretical guarantees and approximation schemes, heuristics and meta-heuristics, game theory, optimal control, optimization in machine learning and data analysis, and their valuable applications in operations research and economics.

In response to the call for papers, MOTOR 2023 received 189 submissions. Out of 89 full papers considered for review (100 abstracts and short communications were excluded for formal reasons) only 29 papers were selected by the Program Committee (PC) for publication in this volume. Each submission was single-blind reviewed by at least three PC members or invited reviewers, experts in their fields, in order to supply detailed and helpful comments. In addition, the PC recommended the inclusion of 29 papers in the supplementary volume after their presentation and discussion during the conference and subsequent revision with respect to the reviewers' comments.

The conference featured ten invited lectures:

- Kamil Aida-Zade (Institute of Control Systems, Azerbaijan), "Feedback control on the class of zonal control actions"
- Mario R. Guarracino (Higher School of Economics, Russia), "Semi-supervised Learning with Depth Functions"
- Milojica Jaćimović (University of Montenegro, Montenegro), "Strong convergence of extragradient-like methods for solving quasi-variational inequalities"
- Pinyan Lu (Shanghai University of Finance and Economics, China), "Algorithms for Solvers: Ideas from CS and OR"
- Panos Pardalos (University of Florida, USA), "Artificial Intelligence, Smart Energy Systems, and Sustainability"
- Eugene Semenkin (Reshetnev Siberian State University of Science and Technology, Russia), "Hybrid evolutionary optimization: how self-adapted algorithms can automatically generate applied AI tools"
- Yaroslav D. Sergeev (University of Calabria, Italy), "Numerical Infinities and Infinitesimals in Optimization"
- Alexander A. Shananin (Lomonosov Moscow State University, Russia), "General equilibrium models in production networks with substitution of inputs"
- Predrag S. Stanimirović (University of Niš, Serbia), "Optimization methods in gradient and zeroing neural networks"
- Vladimir V. Vasin (Krasovsky Institute of Mathematics and Mechanics, Russia), "Fejér type iterative processes for quadratic minimization problems".

We thank the authors for their submissions, members of the Program Committee, and all the external reviewers for their efforts in providing exhaustive reviews. We thank

---

[2] http://motor2019.uran.ru

[3] http://math.nsc.ru/conference/motor/2020/

[4] https://conference.icc.ru/event/3/

[5] http://motor2022.krc.karelia.ru/en/section/1

our sponsors and partners: Krasovsky Institute of Mathematics and Mechanics, the Ural Mathematical Center, the Ural Branch of the Russian Academy of Sciences, the Sobolev Institute of Mathematics and Mathematical Center in Akademgorodok, the Center for Research and Education in Mathematics, and the Higher School of Economics (Nizhny Novgorod). We are grateful to the colleagues from the Springer LNCS and CCIS editorial boards for their kind and helpful support.

July 2023

Michael Khachay
Yury Kochetov
Anton Eremeev
Oleg Khamisov
Vladimir Mazalov
Panos Pardalos

Organization

Honorary Chairs

Yury Evtushenko   Dorodnicyn Computing Centre, Russia
Panos Pardalos   University of Florida Gainesville, USA

General Chair

Michael Khachay   Krasovsky Institute of Mathematics and
     Mechanics, Russia

Program Committee Chairs

Anton Eremeev   Sobolev Institute of Mathematics, Russia
Oleg Khamisov   Melentiev Energy Systems Institute, Russia
Yury Kochetov   Sobolev Institute of Mathematics, Russia
Vladimir Mazalov   Institute of Applied Mathematical Research,
     Russia

Program Committee Members

Alexander Afanasiev   Institute for Information Transmission Problems,
     Russia
Vladimir Beresnev   Sobolev Institute of Mathematics, Russia
George Bolotashvili   Georgian Technical University, Georgia
Igor Bykadorov   Sobolev Institute of Mathematics, Russia
Tatiana Davidović   Mathematical Institute of the Serbian Academy of
     Sciences and Arts, Serbia
Alexander Dolgui   IMT Atlantique, France
Adil Erzin   Sobolev Institute of Mathematics, Russia
Alexander Gasnikov   Moscow Institute of Physics and Technology,
     Russia
Milojica Jaćimović   University of Montenegro, Montenegro
Valeriy Kalyagin   National Research University Higher School of
     Economics, Russia

Vadim Kartak	Ufa State Aviation Technical University, Russia
Alexander Kazakov	Matrosov Institute for System Dynamics and Control Theory, Russia
Andrey Kibzun	Moscow Aviation Institute, Russia
Alexnader Kononov	Sobolev Institute of Mathematics, Russia
Dmitri Kvasov	University of Calabria, Italy
Bertrand Lin	National Yang Ming Chiao Tung University, Taiwan
Vittorio Maniezzo	University of Bologna, Italy
Nevena Mijajlovic	University of Montenegro, Montenegro
Evgeni Nurminski	Fast East Federal University, Russia
Nicholas Olenev	Doronicyn Computing Centre, Russia
Mikhail Posypkin	Federal Research Center on Computer Science and Control, Russia
Oleg Prokopyev	University of Pittsburg, USA
Artem Pyatkin	Sobolev Institute of Mathematics, Russia
Soumyendu Raha	Indian Institute of Science, Bengaluru, India
Anna Rettieva	Institute of Applied Mathematical Research, Russia
Eugene Semenkin	Siberian State Aerospace University, Russia
Yaroslav Sergeyev	Universita della Calabria, Italy
Angelo Sifaleras	University of Macedonia, Greece
Alexander Strekalovsky	Matrosov Institute for System Dynamics and Control Theory, Russia
Tatiana Tchemisova	University of Aveiro, Portugal
Raca Todosijević	Université Polytechnique Hauts-de-France, France
Igor Vasilyev	Matrosov Institute for System Dynamics and Control Theory, Russia
Alexander Vasin	Lomonosov Moscow State University, Russia
Vladimir Vasin	Krasovsky Institute of Mathematics and Mechanics, Russia

Industry Session Chair

Igor Vasilyev	Matrosov Institute for System Dynamics and Control Theory, Russia

Organizing Committee

Alexey Makarov (chair)	Ural Branch of RAS, Ekaterinburg, Russia
Igor Kandoba (co-chair)	Krasovsky Institute of Mathematics and Mechanics, Ekaterinburg, Russia
Alexander Petunin (co-chair)	Ural Federal University, Ekaterinburg, Russia
Boris Digas	Krasovsky Institute of Mathematics and Mechanics, Ekaterinburg, Russia
Artem Firstkov	Krasovsky Institute of Mathematics and Mechanics, Ekaterinburg, Russia
Majid Forghani	Krasovsky Institute of Mathematics and Mechanics, Ekaterinburg, Russia
Nina Kochetova	Sobolev Institute of Mathematics, Novosibirsk, Russia
Polina Kononova	Sobolev Institute of Mathematics, Novosibirsk, Russia
Timur Medvedev	Higher School of Economics, Nizhny Novgorod, Russia
Ekaterina Neznakhina	Krasovsky Institute of Mathematics and Mechanics, Ekaterinburg, Russia
Yuri Ogorodnikov	Krasovsky Institute of Mathematics and Mechanics, Ekaterinburg, Russia
Roman Rudakov	Krasovsky Institute of Mathematics and Mechanics, Ekaterinburg, Russia
Ksenia Ryzhenko	Krasovsky Institute of Mathematics and Mechanics, Ekaterinburg, Russia

Organizers

Krasovsky Institute of Mathematics and Mechanics, Russia
Ural Mathematical Center, Russia
Ural Branch of RAS, Russia
Sobolev Institute of Mathematics, Russia
Mathematical Center in Akademgorodok, Novosibirsk, Russia
Higher School of Economics, Nizhny Novgorod, Russia
Melentiev Energy Systems Institute, Russia
Ural Federal University, Russia

Additional Reviewers

Aizenberg, Natalia
Antipin, Anatoly
Aroslonkin, Artem
Berikov, Vladimir
van Bevern, René
Davydov, Ivan
Dong, Zhang
Enkhbat, Rentsen
Erokhin, Vladimir
Forghani, Majid
Gomoyunov, Mikhail
Gruzdeva, Tatiana
Iliev, Victor
Ivanov, Sergey
Izmest'ev, Igor
Kazakovtsev, Lev
Khachai, Daniil
Khandeev, Vladimir
Khlopin, Dmitry
Khoroshilova, Elena
Kotov, Vladimir
Kumkov, Sergey
Kulachenko, Igor
Lavlinskii, Sergey
Lempert, Anna
Makarovskikh, Tatiana

Melnikov, Andrey
Neznakhina, Katherine
Obrosova, Natalia
Ogorodnikov, Yuri
Orlov, Andrey
Panasenko, Anna
Panin, Artem
Parilina, Elena
Plyasunov, Alexander
Popov, Leonid
Ratushnyi, Alexey
Rettieva, Anna
Rozenberg, Valeriy
Servakh, Vladimir
Simanchev, Ruslan
Sotnikova, Margarita
Stonyakin, Fedor
Sukhoroslov, Oleg
Tarasyev, Alexander
Ushakov, Anton
Yarullin, Rashid
Zabotin, Igor
Zakharova, Yulia
Zubov, Vladimir
Zyryanov, Alexander

Abstracts of Invited Talks

Feedback Control on the Class of Zonal Control Actions

Kamil Aida-Zade (iD) and Samir Guliev

Institute of Control Systems, Baku, Azerbaijan
kamil_aydazade@rambler.ru

Abstract. We propose an approach to feedback control for both nonlinear objects with lumped parameters and point sources of objects with distributed parameters. The approach lies in the fact that the entire set of possible phase states of an object is divided into a nite number of zones (subsets), and the synthesized controls are determined not by the measured state values themselves, but by the zonal values of the zone parameters to which belong the current measured object states. We have obtained necessary optimality conditions for the zonal values of the feedback parameters, and carried out computational (computer) experiments on some test problems of feedback control both for objects with lumped parameters and for distributed parameters.

Keywords: Feedback control · Optimality conditions

Semi-Supervised Learning with Depth Functions

Mario R. Guarracino [iD]

Higher School of Economics, Moscow, Russia
mario.guarracino@gmail.com

Abstract. Depth functions have been exploited in supervised learning for many years. Given that the depth of a point is somehow a distribution-free measure of its distance from the center of a distribution, their use in supervised learning arose naturally and it has seen a certain degree of success. Particularly, DD-classifers and their extensions have been extensively studied and applied in many applied elds and statistical settings. What has not been investigated so far is their use within a semisupervised learning framework. That is, in case some labeled data are available along with some unlabeled data within the same training set. This case arises in many applications and it has been proved that combining information from labeled and unlabeled data can improve the overall performance of a classifier. For this reason, this work aims at introducing semi-supervised learning techniques in association with DD-classifiers and at investigating to what extent such a technique is able to improve DD-classifier performances. Performances will be evaluated by means of an extensive simulation study and illustrated on some real data sets.

Keywords: Semi-supervised learning techniques · Depth functions · Performance evaluation

Strong Convergence of Extragradient-like Methods for Solving Quasi-variational Inequalities

Milojica Jaćimović [iD]

University of Montenegro, Podrogica, Montenegro
milojica@jacimovic.me

Abstract. The goal of this talk is to study convergence of approximation methods for quasi-variational inequalities with the moving set. First, we propose the extragradient dynamical system and under strong monotonicity we show strong convergence with exponential rate of generated trajectory to the unique solution of quasi-variational inequality. Further, the explicit time discretization of this dynamical system leads to an extragradient algorithm with relaxation parameters. We prove the convergence of the generated iterative sequence to the unique solution of the quasi-variational inequality and derive the linear convergence rate under strong monotonicity.

Keywords: Quasi-variational inequalities · Extragradient dynamical system · Convergence rate

Algorithms for Solvers: Ideas from CS and OR

Pinyan Lu(iD)

Institute for Theoretical Computer Science, Shanghai University of Finance and Economics, China
lu.pinyan@mail.shufe.edu.cn

Abstract. The MIP/LP solvers are primarily developed by the operations research community while SAT/SMT solvers are primarily developed by the computer science community. However, these problems are closely related with each other. In recent years, there have been many attempts to combine the algorithmic techniques of both sides to develop better solvers. In this talk, I will discuss these and our attempts and try to provide a unified perspective and framework.

Keywords: MIP and SAT solvers · Efficient algorithms · General framework

Artificial Intelligence, Smart Energy Systems, and Sustainability

Panos Pardalos (iD)

University of Florida, Gainesville, USA
p.m.pardalos@gmail.com

Abstract. Distribution systems face significant changes due to the growing number of distributed and variable energy generation resources and the smart grid implementation. The traditional design paradigm can no longer meet the need for greater resilience, power quality, and customer participation. On the other hand, smart grid implementation brings a large amount of data that can be used to better plan a distribution system. Growing energy demand and limited investment capital make distribution system planners look to these advances in smart grid technology to identify new approaches to achieve load reliability. When planning a distribution system, the main goal is to meet the most economically and reliably timed demand growth. The planning methodology must ensure that every opportunity for savings or power quality improvement is exploited. This is not a straightforward task, even in traditional systems, since the distribution networks are usually large in extent, with a large amount of data to be analyzed. In addition, new regulations from authorities and the modernization of power systems highlight the importance of a constant update and improvement of methodologies and planning techniques. The ongoing changes bring enormous opportunities and challenges to traditional and new players and require huge planning and operation methods changes. With more and innovative players entering the sector, artificial intelligence-based approaches can be the key to dealing with the new challenges and ensuring the systems and the respective players' sustainability, both in economic and environmental terms. The drive to make utilities more efficient through AI, machine learning, and data science has resulted in major benefits for every actor in the energy sector, including generators, distributors, the environment, taxpayers, and consumers.

Keywords: Sustainability · Smart energy and distribution systems · Artificial intelligence

Hybrid Evolutionary Optimization: How Self-Adapted Algorithms can Automatically Generate Applied AI Tools

Eugene Semenkin ⓘ

Reshetnev Siberian State University of Science and Technology, Krasnoyarsk, Russia
eugenesemenkin@yandex.ru

Abstract. When designing AI tools or machine-learning models, one must make multiple choices: which approach should be used, which structure of model is more appropriate for the problem in hand and which settings and parameters must be applied. All of these choices are mathematically reduced to some kind of optimization problem. In fact, one has to solve multi-criteria multi-dimensional multi-scale "black box" optimization problems with algorithmically-given objectives and/or constraints. Bio-inspired algorithms, e.g. evolutionary algorithms (EAs), could be used for solving the described problems. However, the effectiveness and efficiency of EAs depends essentially on the choice of their settings and the tuning of their parameters, which is a separate and very complicated decision-making problem. An EA with appropriate settings and parameters can be very effective, but in the opposite case, this algorithm can fail. Moreover, such a useful property of EAs as their universality (their independence from the properties of the problem), which allows them to be used in solving the widest class of optimization problems, means that it is impossible to use properties of problems convenient for optimization (such as convexity, monotony and unimodality) in cases where such properties exist in the problem being solved. The approach that will be discussed in the lecture allows us to simplify the use of EAs and other bio-inspired algorithms by means of the self-configuration, coevolution and hybridization of EAs with problem-specific algorithms. Details of the approach will be described, and examples of approach deployment in the automated design of AI-tools will be given.

Keywords: Artificial intelligence tools · Evolutionary algorithms · Blackbox optimization

Numerical Infinities and Infinitesimals in Optimization

Yaroslav D. Sergeyev

University of Calabria, Rende, Italy
yaro@dimes.unical.it

Abstract. In this talk, a recent computational methodology is described (see [1, 2]). It has been introduced with the intention to allow one to work with infinities and infinitesimals numerically in a unique computational framework. It is based on the principle 'The part is less than the whole' applied to all quantities (finite, infinite, and infinitesimal) and to all sets and processes (finite and infinite). The methodology uses as a computational device the Infinity Computer (a new kind of supercomputer patented in several countries) working numerically with infinite and infinitesimal numbers that can be written in a positional system with an infinite radix. On a number of examples (numerical differentiation, divergent series, ordinary differential equations, fractals, set theory, etc.) it is shown that the new approach can be useful from both theoretical and computational points of view. The main attention is dedicated to applications in optimization (local, global, and multi-objective)(see [1–7]). The accuracy of the obtained results is continuously compared with results obtained by traditional tools used to work with mathematical objects involving infinity. The Infinity Calculator working with infinities and infinitesimals numerically is shown during the lecture. For more information see the dedicated web page http://www.theinfinitycomputer.com and this survey: The web page developed at the University of East Anglia, UK is dedicated to teaching the methodology: https://www.numericalinf inities.com/.

Keywords: Numerical infinities and infinitesimals · Optimization techniques · Infinity Computer

References

1. Yaroslav, D.S., Leone R.D.: Numerical infinities and infinitesimals in Optimization. Springer, Cham (2022). https://doi.org/10.1007/978-3-030-93642-6
2. Yaroslav, D.S.: Numerical infinities and infinitesimals: Methodology, applications, and repercussions on two Hilbert problems. EMS Sur. Math. Sci. **4**(2), 219–320 (2017)
3. Leone R.D., Egidi, N., Fatone, L.: The use of grossone in elastic net regularization and sparse support vector machines. Soft Computing. **24**, 17669–17677 (2020)
4. Astorino, A., Fuduli, A.: Spherical separation with innitely far center. Soft Comput. **24**, 17751–17759 (2020)

5. Leone, R.D., Fasano, G., Roma, M., Yaroslav, D.S.: Iterative grossone-based computation of negative curvature directions in large-scale optimization. J. Optimization Theory Appl. **186**(2), 554–589 (2020)
6. Leone R.D, Fasano, G., Yaroslav, D.S.: Planar methods and grossone for the conjugate gradient breakdown in nonlinear programming. Comput. Optim. Appl. **71**(1), 73–93 (2018)
7. Cococcioni, M., Pappalardo, M., Yaroslav, D.S.: Lexicographic multiobjective linear programming using grossone methodology: theory and algorithm. Appl. Math. Comput. **318**, 298–311 (2018)

General Equilibrium Models in Production Networks with Substitution of Inputs

Alexander Shananin(iD) and Natalia Obrosova(iD)

Lomonosov Moscow State University, Russia
alexshan@yandex.ru

Abstract. The developed countries made a transition from intensive to extensive growth in 1980s, that led to goods and services range extension. Related changes in the technologies during the economic globalization caused supply chains sophistication and complex production network development in local and global economies. The problem of sustainable development of regional economies while supply chains were localized was raised by recent shocks due to the pandemic or to economic sanctions. Proper tools are needed to describe modern production networks taking into account the substitution of inputs. The traditional interindustry balance method based on the Leontief linear model and nonnegative matrices theory should not be used due to its base assumption of the constancy of direct requirement coefficients in a supply network. Modern methods demand new mathematical tools that reflect substitution of inputs in complex production networks. The tools presented by the authors were developed as a result of analysis of the resource allocation problem with positively homogeneous neoclassical production functions and the dual problem whose solutions are the price indexes of intermediate inputs. Our way of studying the problem is based on the construction of the Young dual transform for equilibrium price indexes. It is proved that in the case of CES technologies with constant elasticity of substitution (Cobb-Douglas in particular) the resource allocation problem has an explicit solution. In this case the comparative statics method based on the official national accounts statistics is developed by the authors. The method allows forecasting of intersectorial links in the medium term under given scenarios of the internal or external shocks taking into account the substitution of inputs. The point of the method is solving the inverse problem of non-linear balance identification and further verification of the model on the basis of the offical input-output tables statistics. In this way the elasticities of substitution of intermediate inputs are the verification parameters of the model. The method was successfully tested on the input-output tables statistics of Russia and Kazakhstan. Applications to the analysis of the intersectoral links of several countries with different levels of market relations maturity demonstrated that in the case of a decentralized economy

the nonlinear balance with Cobb-Douglas production functions gives a
more precise forecast than the traditional linear Leontief model approach.

Keywords: Production networks · Equilibrium models

Optimization Methods in Gradient and Zeroing Neural Networks

Predrag S. Stanimirović ⓘ

University of Niš, Serbia
predrag.stanimirovic@pmf.edu.rs

Abstract. The topic of our lecture is a class of recurrent neural networks (RNN) dedicated to finding zeros of equations or minimizing nonlinear functions. Optimization RNN models are divided into two global classes: Gradient Neural Networks (GNN) and Zhang Neural Networks (ZNN). GNN models are aimed at solving time-invariant problems, while ZNN models are able to solve time-varying problems. The design of GNN and ZNN models arises from the choice of an appropriate error function. ZNN dynamics is a certain dynamical system whose states evolve over a state space continuously based on the time derivative of the error function. Some new error functions resulting from nonlinear gradient-descent and quasi-Newton optimization methods are presented. A modification of ZNN dynamical evolution based on higher-order hyperpower iterative methods is described. We discuss the problems of nondifferentiability and division by zero (DBZ) which appear relatively frequently in timevarying dynamical systems. The GNN design is defined as a movement along the negative gradient of the Frobenius norm of the error function inside the time interval. Modifications of the GNN flow based on gradient-descent and conjugate-gradient optimization methods are considered. In general, dynamical systems are defined as continuous-time analogies of known nonlinear optimization algorithms, such as the class of gradient-descent algorithms or various quasi-Newton methods for solving nonlinear optimization problems. The convergence of various modified dynamical systems aimed at solving time-varying matrix equations are investigated.

Keywords: Recurrent, gradient and Zhang neural networks · Gradient-descent and quasi-Newton methods

Iterative Processes of Fejér Type for Quadratic Optimization Problem

Vladimir V. Vasin

Krasovsky Institute of Mathematics and Mechanics, Ekaterinburg, Russia
vasin@imm.uran.ru

Abstract. This lecture presents a short overview of iterative solution methods of Fejér type for the well-known quadratic constrained optimization problem, which was introduced and widely studied by acad. I.I. Eremin— the founder of the Ural mathematical programming school. Along with common quadratic programs, we consider several variants of the basic problem, which have numerous applications. We direct attention to some special settings of the problem in question including metric projections, linear programming, etc., which are of separate interest. Fejér-type methods attract the interest of specialists in numerical optimization, since, along with convergence, one can prove conditions of their stability to small perturbations of the input data. Thus, these methods induce self-regularizing algorithms, unlike some well-known primal numerical optimization techniques.

Keywords: Fejér iterative processes · Quadratic optimization · Stability

Contents

Stochastic Optimization

Scheduling

Game Theory

Optimal Control and Mathematical Economics

Invited Papers

General Equilibrium Models in Production Networks with Substitution of Inputs

Natalia Obrosova[1,2,4]($\boxtimes$) (iD) and Alexander Shananin[1,2,3] (iD)

[1] Federal Research Center Computer Science and Control of the Russian Academy of Sciences, 40 Vavilov st., Moscow 119333, Russia
{nobrosova,alexshan}@ya.ru

[2] Moscow Center for Fundamental and Applied Mathematics, Lomonosov Moscow State University, GSP-1, Leninskie Gory, Moscow 119991, Russia

[3] Moscow Institute of Physics and Technology (State University), 9 Institutskiy per., Dolgoprudny, Moscow Region 141701, Russia

[4] Federal State Budgetary Institution "All-Russian Research Institute of Labor" of the Ministry of Labor and Social Protection of the Russian Federation, Parkovaya st. 29, Moscow 105043, Russia

Abstract. The diversification of the inter-industry connections of modern economies seriously hindered the input-output analysis, that traditionally was made by the Leontief model. We suggest the new methodology for scenario analysis of the input-output linkages in complex production networks, taking into account the substitution of inputs. This article reviews briefly our recent results in this field. The methodology is based on the resource allocation problem with positively homogeneous neoclassical production functions and the dual problem which solutions are the price indexes of intermediate inputs. The way of problem studying is based on the construction of the Young dual transform for equilibrium price indexes. In the case of CES technologies with constant elasticity of substitution (Cobb-Douglas in particular) the comparative statics method based on the official national accounts statistics is developed by authors. Method allows in a middle-term to forecast intersectoral linkages under given scenarios of the internal or external shocks taking into account the substitution of inputs. The method was successfully tested on the input-output tables statistics of Russia, Kazakhstan and other large economies with different stages of market relations maturity.

Keywords: input-output model · Young duality · convex optimization problem · production networks · CES function · Cobb-Douglas function · equilibrium · elasticity of substitution

1 Introduction

The developed countries made a transition from extensive to intensive growth in 1980s, that led to goods and services range extension. Related changes in

Supported by grant RSF 23-21-00429.

M. Khachay et al. (Eds.): MOTOR 2023, LNCS 13930, pp. 3–22, 2023.
https://doi.org/10.1007/978-3-031-35305-5_1

the technologies during the economic globalization caused increasing of supply chains sophistication degree and complex production network development in local and global economies. The problem of sustainable development of regional economies while supply chains was localized raised due to latest pandemic and sanction economic shocks. Proper tools are needed to describe the intersectoral flows taking into account the substitution of inputs which is an important feature of the modern production networks.

A long tradition in economics and mathematical modeling has the Leontief linear model of intersectoral balance and the nonnegative matrices theory [1–5]. The methods based on the Leontief-type models were successfully used during the period of extensional recovery growth (about thirty years from the middle of the last century) for the intersectoral connections analysis, economic multipliers estimation and growth drivers detection. Let

$X = (X_1, \ldots, X_m)^\top$ be a vector of sector outputs

$W = (W_1, \ldots, W_m)^\top$ be a vector of final consumption

$A = \left\| a_i^j \right\|$ - $m \times m$ -Leontief matrix, where a_i^j mean the norm of material cost of good of industry i for a unit of output of industry j. Then the Leontief balance model has the form

$$X = AX + W.$$

The basic hypothesis of Leontief model is the stability of the technological coefficients a_i^j, i.e. the Leontief matrix A is assumed to be a constant over the considered years.

If the Leontief matrix A is productive [5,6] then the Leontief inverse $(E - A)^{-1} \geq 0$ exists which coefficients are interpreted as economic multipliers

$$X = (E - A)^{-1}W.$$

In this case the Leontief inverse can be calculated as a sum of the convergent series [6]

$$(E - A)^{-1} = \sum_{k=0}^{+\infty} A^k.$$

Leontief Input-Output (IO) analysis is actively used by developed countries up to the present in particular for many applications, for ex. energy and environmental input-output models, labour market analysis, etc. The construction of the symmetric Input-Output tables that are the initial data for IO methods has been supported for a long time within the framework of large international projects (for ex., the WIOD Database, http://www.wiod.org/new_site/database/niots. htm).

In modern production networks the growth of substitutability of inputs lead to the violation of the basic hypothesis of the linear Leontief method. The technological coefficients a_i^j lose stability. So the traditional linear IO model can not be correctly used.

Against the background of the financial crises of the late 20th and early 21st centuries, scientific publications began to increasingly discuss new methods for analyzing of the structural imbalances and the spread of shocks in network economies, taking into account the substitution of inputs in supply chains. (for ex., [7–9]). In the cycle of empirical and theoretical studies [10–12] the insufficiency of Leontiev's method is discussed. In order to analyze the aggregated fluctuations as a result of a random microshocks through the production network, these papers propose to describe the intersectoral interactions based on nonlinear dependencies. A model of perfect competitive equilibrium in a supply chain with a Cobb-Douglas production function and a utility function with a constant consumption proportion is considered. By the number of examples, the dependences of the economic multiplier of industry shocks on the topology and dominant sectors of the production network are studied. The ideas are continued in [13], where the problems of assessing the stability of financial networks under shocks conditions are considered by examples of networks topologies. Approaches to the analysis of the macroconsequences of microeconomic shocks in production networks of different topologies in terms of more general network models of competitive equilibrium with production technologies with constant elasticity of substitution (CES functions), as well as examples of models of imperfect competition, taking into account transaction costs, are considered in [14–16].

Despite the interesting theoretical and empirical results, the mathematical methods of the mentioned above articles are not applicable to analysis of the intersectoral linkages and forecasts of economic multipliers, taking into account the substitution of production factors on the base of actual data of the official input-output statistics of the state.

Our approach is motivated by the ideas from [10–16]. However it bases on new mathematical methods for developing the class of non-linear intersectoral balance models which are interpretable in terms of the official statistics of regions. In contrast to the Leontief model, our approach comes from the more general production technology definition that provides substitution of inputs in the supply network. That's the main advantage of our models. Moreover we suggest a clear and applicable methodology for identifying and verifying of such intersectoral balance models using the official national accounts statistics of states.

The main purpose of the research is the development of the methodology for scenario forecasting of the input-output balance, taking into account the input substitution resulted from shifts of external economic parameters.

This article reviews briefly our recent theoretical and practical results in this field [17–24].

We set the resource allocation problem with positively homogeneous neoclassical production functions on the production network with external primary factors and final consumers. We show that the solution of the problem corresponds to a competitive market equilibrium, where the equilibrium prices of intermediates are derived from the special form of the dual problem that we call the Young dual problem.

It is proved that in the case of CES technologies with constant elasticity of substitution (Cobb-Douglas in particular) the resource allocation problem has an explicit solution. In this case we suggest the applicable comparative statics method based on the official national accounts statistics. The method allows us to forecast the intersectoral flows in a middle-term scale under given scenarios of the internal or external shocks taking into account the substitution of inputs. The point of the method is solving the inverse problem of nonlinear balance identification and further verification of the model on the base of the official input-output tables data sets. By that the elasticities of substitution of intermediate inputs are the verification parameters of the model. The method was successfully tested on input-output tables statistics of Russia and Kazakhstan. Applications of our method to the analysis of the intersectoral flows in several economies that differs by the level of market relations maturity demonstrated that in the case of decentralized economy the nonlinear balance with Cobb-Douglas production functions gives more precise forecast than more traditional model with Leontief technologies.

2   The Baseline Non-linear Inter-industry Balance Model

Optimal Recourse Allocation Problem. Consider a network economy with m pure industries. Each industry $i = 1, .., m$ produces a distinct product that can be either consumed by several final consumers in the total amount $X^0 = (X_1^0, \ldots, X_m^0) \geq 0$ or used as the intermediate input $X^j = (X_1^j, \ldots, X_m^j)$ of industry $j = 1, .., m$. Besides that, the input of each industry j, $j = 1, .., m$ includes n types of primary factors amounting by vector $l^j = (l_1^j, \ldots, l_n^j) \geq 0$ which are not produced by the production network. Assume that primary factors are limited by $l = (l_1, \ldots, l_n) \geq 0$, i.e.

$$\sum_{j=1}^{m} l^j \leq l.$$

Denote by Φ_k the class of concave, monotonically nondecreasing, continuous and positively homogeneous of degree one functions on $R_{\geq 0}^k$, that vanish at the origin. Each industry j employs a unique production technology $F_j(X^j, l^j) \in \Phi_{m+n}$ from the class Φ_{m+n} to transform the intermediate input to output. Finally, preferences of the final consumer are described by the utility function $F_0(X^0) \in \Phi_m$. Note that

1. $F_j(0,0) = 0$, $j = 1, .., m$ and $F_0(0) = 0$,
2. we measure the above mentioned amounts in a constant prices for a once fixed base year.

Set a problem of optimal resources allocation [17]: for a given values $l = (l_1, \ldots, l_n) \geq 0$ of primary factors limits evaluate X^j, X^0, l^j as a solution of the following problem

$$F_0(X^0) \to \max \tag{1}$$

$$F_j(X^j, l^j) \geq \sum_{i=0}^{m} X_j^i, j = 1, \ldots, m \tag{2}$$

$$\sum_{j=1}^{m} l^j \leq l \tag{3}$$

$$X^0 \geq 0, \ X^1 \geq 0, \ldots, X^m \geq 0, \ l^1 \geq 0, \ldots, l^m \geq 0. \tag{4}$$

Assumption 1. Productivity condition. The supply chain is productive, i.e., there exists $\hat{X}^1 \geq 0, \ldots, \hat{X}^m \geq 0, \ \hat{l}^1 \geq 0, \ldots, \hat{l}^m \geq 0$ such that

$$F_j\left(\hat{X}^j, \hat{l}^j\right) > \sum_{i=1}^{m} \hat{X}^i_j, \ j = 1, \ldots, m.$$

Corollary 1 [17]. If the set of sectors is productive and $l = (l_1, \ldots, l_n) > 0$, then the optimization problem (1)–(4) satisfies to the Slater condition.

Assumption 2. Denote

$$A(l) = \left\{ X^0 = (X^0_1, \ldots, X^0_m) \geq 0 \,\middle|\, X^0_j \leq F_j\left(X^j, l^j\right) - \sum_{i=1}^{m} X^i_j, \ j = 1\ldots m; \right.$$
$$\left. \sum_{j=1}^{m} l^j \leq l, \ X^1 \geq 0, \ldots, X^m \geq 0, \ l^1 \geq 0, \ldots, l^m \geq 0 \right\}.$$

There exists $\hat{l} \in R^n_{\geq 0}$ such that the set $A\left(\hat{l}\right)$ is bounded.

Corollary 2 [19]. The set $A(l)$ is bounded, convex and closed for any $l \in R^n_{\geq 0}$.

Theorem 1 [17]. *The set of vectors* $\left\{\hat{X}^0, \hat{X}^1, \ldots, \hat{X}^m, \hat{l}^1, \ldots, \hat{l}^m\right\}$, *which satisfy to the constraints (2)–(4) is the solution of the optimization problem (1)–(4) if and only if there exist Lagrange multipliers* $p_0 > 0$, $p = (p_1, \ldots, p_m) \geq 0$ *and* $s = (s_1, \ldots, s_n) \geq 0$ *such that*

$$\left(\hat{X}^j, \hat{l}^j\right) \in Arg\max\{p_j F_j\left(X^j, l^j\right) - pX^j - sl^j \mid X^j \geq 0, l^j \geq 0\}, \ j = 1, \ldots, m \tag{5}$$

$$p_j \left(F_j\left(\hat{X}^j, \hat{l}^j\right) - \hat{X}^0_j - \sum_{i=1}^{m} \hat{X}^i_j \right) = 0, \ j = 1, \ldots, m \tag{6}$$

$$s_k \left(l_k - \sum_{j=1}^{m} \hat{l}^j_k \right) = 0, \ k = 1, \ldots, n \tag{7}$$

$$\hat{X}^0 \in Arg\max\{p_0 F_0(X^0) - pX^0 \mid X^0 \geq 0\}. \tag{8}$$

We interpret Lagrange multipliers $p = (p_1, \ldots, p_m)$ to constraint (2) as prices of products and Lagrange multipliers $s = (s_1, \ldots, s_n)$ to constraint (3) as prices on primary production factors.

The Theorem 1 provides that the optimal resources allocation corresponds to a market equilibrium mechanisms, i.e., if the price p_j of the j-th product is positive, then demand for j-th product equals to its supply and the price $p_j > 0$ is equilibrium price for the product j in the production network.

Cost Functions and Consumer Price Index. Young Transform. Let $q_j(p, s)$, $j = 1, .., m$ and $q_0(q)$ be the functions defined by

$$q_j(p, s) = \inf \left\{ \frac{pX^j + sl^j}{F_j(X^j, l^j)} \, \Big| \, X^j \geq 0, l^j \geq 0, F_j(X^j, l^j) > 0 \right\}, \tag{9}$$

$$q_0(q) = \inf \left\{ \frac{qX^0}{F_0(X^0)} \, \Big| \, X^0 \geq 0, F_0(X^0) > 0 \right\}$$

That definitions provide the Young transform of the corresponding production functions $F_j(X^j, l^j)$ and utility function $F_0(X^0)$ [26,27]. It can be proved that the Young transform of the function from the class Φ_k is the function from the class Φ_k too (see [26]). Therefore $q_j(p, s) \in \Phi_{m+n}$, $q_0(q) \in \Phi_m$. Moreover, the Young transform is an involution in the class Φ_k, i.e.

$$F_0(X^0) = \inf \left\{ \frac{qX^0}{q_0(q)} \, \Big| \, q \geq 0, q_0(q) > 0 \right\},$$

$$F_j(X^j, l^j) = \inf \left\{ \frac{pX^j + sl^j}{q_j(p, s)} \, \Big| \, p \geq 0, s \geq 0, q_j(p, s) > 0. \right\}.$$

We call the function $q_j(p, s)$ a cost function of the industry j and the function $q_0(q)$ a consumer price index [26].

Remark 1 [17]. Consider the constant elasticity of substitution function (CES function) with the elasticity $\sigma = \frac{1}{1+\rho}$.

$$f(x_1, \ldots, x_n) = \left(\left(\frac{x_1}{\omega_1} \right)^{-\rho} + \left(\frac{x_2}{\omega_2} \right)^{-\rho} + \ldots + \left(\frac{x_n}{\omega_n} \right)^{-\rho} \right)^{-\frac{1}{\rho}}, \tag{10}$$

where $\rho \in (-1, 0) \cup (0, +\infty)$, $\omega_1 > 0, \ldots, \omega_n > 0$. The Young transform of the CES function (10) is the CES function $g(\lambda_1, \ldots, \lambda_n)$, $(\lambda_1, \ldots, \lambda_n) \in R_{>0}^n$

$$g(\lambda_1, \ldots, \lambda_n) = \left((\lambda_1 \omega_1)^{\frac{\rho}{1+\rho}} + (\lambda_2 \omega_2)^{\frac{\rho}{1+\rho}} + (\lambda_n \omega_n)^{\frac{\rho}{1+\rho}} \right)^{\frac{1+\rho}{\rho}}. \tag{11}$$

with elasticity of substitution $(1 + \rho)$.

The case of $\rho \to 0$ in (10) corresponds to the Cobb-Douglas function (with elasticity equals to 1)

$$f_{CD}(X_1, .., X_n) = AX_1^{a_1}..X_n^{a_n} \tag{12}$$

where $A > 0$, $a_i > 0$, $\sum_{i=1}^n a_i = 1$, $i = 1..n$. In this case the Young transform of (12) is the Cobb-Douglas function too

$$q_{CD}(p_1, .., p_n) = \frac{1}{f_{CD}(a_1, .., a_n)} p_1^{a_1} ... p_n^{a_n}. \tag{13}$$

The optimal value of (1) depending on $l = (l_1, .., l_n)$ (right part of (3)) in the problem (1)–(4) we call the aggregate production function $F^A(l)$. It can be proved that $F^A(l) \in \Phi_n$. The Young transform $q_A(s)$ of the function $F^A(l)$ we call the aggregate cost function, i.e.

$$q_A(s) = \inf\left\{\frac{sl}{F^A(l)} \,\middle|\, l \geq 0, F^A(l) > 0.\right\} \in \Phi_n,$$

$$F^A(l) = \inf\left\{\frac{sl}{q_A(s)} \,\middle|\, s \geq 0, q_A(s) > 0\right\}.$$

Young Dual Problem for Price Indexes. We couple the initial problem (1)–(4) with the dual problem by the following theorem.

Theorem 2 [17,18]. *If Lagrange multipliers* $\hat{p} = (\hat{p}_1, \ldots, \hat{p}_m) \geq 0$, $\hat{s} = (\hat{s}_1, \ldots, \hat{s}_n) \geq 0$ *to the problem (1)–(4) satisfy to (5)–(8) then* $\hat{p} = (\hat{p}_1, \ldots, \hat{p}_m) \geq 0$ *is the solution of the following problem*

$$q_0(p) \rightarrow \max_p \tag{14}$$

$$q_j(\hat{s}, p) \geq p_j, \, j = 1, \ldots, m \tag{15}$$

$$p = (p_1, \ldots, p_m) \geq 0. \tag{16}$$

Moreover, the aggregate cost function $q_A(\hat{s}) = q_0(\hat{p}(\hat{s}))$.

The convex programming problem (14)–(16) we call the Young dual problem to the problem (1)–(4).

Solution of the primal problem of resource allocation (1)–(4) and solution of the dual problem (14)–(16) provide the market equilibrium point $\left\{\hat{X}^0, \hat{X}^1, ..., \hat{X}^m, \hat{l}^1, ..., \hat{l}^m, \hat{p}_1, \ldots, \hat{p}_m, \hat{s}_1, \ldots, \hat{s}_n\right\}$ of the production network. The shifts in final consumption $\hat{X}_0$ and in prices of resources $\hat{s}$ result in equilibrium shifting. Thus, if the model is identified and calibrated on the base of the official input-output statistics we obtain a useful tool for evaluating and forecasting macroeconomic responses of the production network to the external or internal shocks (pandemics, sanctions, etc.).

3 Inverse Problem of Identification. Input Data Set Requirements and Notation

To identify the model on the base of actual data we should solve the inverse problem of identification. The statement of the problem provides that the solution of the developed nonlinear inter-industry balance model should reproduce the actual intersectoral financial flows data set that is known from the official statistics of the region.

The initial data for identification and calibration of the model are the symmetric Input-Output (IO) tables $Z(t)$ for the series of years. IO tables are published periodically in yearbooks of national accounts statistics. The frequency of these publications varies across countries. The standard [30] of the Input-Output tables include the set of tables which collect the detailed information about production of the economy sectors and distribution of the output among the intermediate and final consumption.

One of the main effects from contemporary economic shocks (pandemics, sanctions) on regional economics is violation of the logistics of foreign trade and drops in labor demand and supply (for ex., resulted from pandemic lockdowns). Therefore in papers [19–24]) we focus on two primary factors - import and labor for model evaluations of regional production networks.

The required data set for the model identification can be obviously obtained from that IO standard. We need the three quadrants of the standard symmetric IO table of domestic products at basic prices coupled with the row of intermediate input of imported goods that is available from the standard symmetric IO table of imports. So the initial input data set $Z(t)$ has three quadrants (I, II, III):

Table 1. Input-Output table structure

$$
\begin{pmatrix}
 & & & products & final\ consumption \\
 & & 1 & 1 \dots m & \\
products & & \dots & Z_i^j & Z_i^0 \\
 & & m & & \\
gross\ value\ added & & & Z_{m+1}^1 \dots Z_{m+1}^m & \\
imported\ intermediates & & & Z_{m+2}^1 \dots Z_{m+2}^m &
\end{pmatrix}
$$

Quadrant I: Z_i^j, $i,j = 1..m$ is the amount of money that industry i received from industry j for the intermediate inputs produced by i and consumed by j.

Quadrant II: a unique column vector $Z^0 = \left(Z_1^0, .., Z_m^0\right)^\top$ of the total final consumption (households, government, export etc.) in the economy.

Quadrant III includes the two rows ($n = 2$) that display the intermediate use of primary factors: gross value added as the measure of labor employed by pure industries $Z_{m+1}^j = l_1^j$ and imported intermediate inputs of industries $Z_{m+2}^j = l_2^j$, $j = 1..m$.

Values

$$
A_j = \sum_{i=1}^{m+2} Z_i^j, \quad j = 1..m. \tag{17}
$$

give the total inputs (of intermediate and primary factors), consumed by the pure industry $j = 1, \ldots, m$. Due to the symmetry of the IO table the value A_j equals to the total consumption of product $j = 1, \ldots, m$, i.e.

$$
A_j = \sum_{i=1}^{m} Z_j^i + Z_j^0, \quad j = 1, \ldots, m,
$$

$$\sum_{j=1}^{m} \left(Z_{m+1}^{j} + Z_{m+2}^{j} \right) = \sum_{i=1}^{m} Z_{i}^{0} = A_{0}. \tag{18}$$

The sections below (except the Sect. 8) use the following notation

$$a_{ij} = \frac{Z_{i}^{j}}{\sum_{k=1}^{m+n} Z_{k}^{j}}, \quad b_{kj} = \frac{Z_{m+k}^{j}}{\sum_{k=1}^{m+n} Z_{k}^{j}}, \quad a_{i0} = \frac{Z_{i}^{0}}{\sum_{i=1}^{m} Z_{i}^{0}} \quad i,j = 1..m, \; k = 1..n.$$
$$\tag{19}$$

The notation above obviously implies that

$$\sum_{i=1}^{m} a_{i0} = 1, \quad \sum_{i=1}^{m} a_{ij} + \sum_{k=1}^{n} b_{kj} = 1, \quad \sum_{k=1}^{n} b_{kj} > 0, \; i,j = 1..m, \; k = 1..n.$$

Since $\sum_{i=1}^{m} a_{i}^{j} < 1$, $j = 1..m$ the Leontief $(m \times m)$ matrix $A = \|a_{ij}\| \geq 0$ is productive. Therefore the Leontief inverse exists

$$(E - A)^{-1} = \|\omega_{ij}\|_{i,j=1,..,m} \geq 0, \tag{20}$$

where E -$(m \times m)$ is identity matrix.

Denote $B = \|b_{kj}\| \geq 0$ -$(n \times m)$ matrix.

The inverse identification problem lies in how to calculate the parameters of the optimal resources allocation task (1)–(4) so that its solution reproduces the given IO table $Z(t)$ for the fixed base year t.

In our previous papers we solve the inverse problem of identification and verify our model for the actual national account statistics of several regions in the class of

- Cobb-Douglas (CD) technologies and utility functions [19–21],
- Leontief technologies [20],
- Constant Elasticity of Substitution (CES) technologies and utility [22–24],
- Nested CES technologies and CES utility function [25].

As we will see later, the identification of the model with CES technologies keeps free parameters of elasticities of substitution of factors (except the limit cases of CD and Leontief technologies where the elasticity equals to 1 and 0, correspondingly). Evaluation of elasticities is a solution of model verification problem. The quality of verification depends on availability of official statistics of IO tables for several years. We develop a unique method for each case driven by the research objectives and regional economic features [22–25]. If the model is identified and verified correctly then we obtain a useful tool for scenario IO analysis. In the next section we explain the base concept of this tool.

Schemes of identification and verification of the model as well as the results of evaluation on the base of actual national IO statistics in each of above mentioned cases we present briefly in Sects. 5–7.

4 Concept of Scenario IO Analysis with the Nonlinear Intersectoral Balance Model

Fix the base year with the symmetric IO table Z, which is known from the official statistics and evaluate the coefficients of production and utility functions in the model as the solution of inverse problem of identification. Then we calibrate the model on the base of national IO data set for a several years to evaluate the parameters of elasticities of substitution of factors (except the cases of CD and Leontief technologies).

The base assumption for the scenario IO analysis is the stability over several years of the technologies and utility parameters.

Thus, once identified and calibrated on the base of actual national IO statistics the developed nonlinear IO model allows to evaluate the scenario response of the production network to various shocks (pandemics, sanctions) as well as to construct consistent scenario forecasts of the I-st and the III-rd quadrants of IO table.

We define the scenario conditions by the two groups of values:

- price indexes on primary inputs (in relation to the base year) $s = (s_1, s_2)^\top$, where $s_1 = s_L$ is evaluated as a consumer price index and $s_2 = s_I$ equals to the foreign currency exchange rate index;
- the vector of total spending of the aggregate final consumer (in current prices of the considered period) $\hat{Z}^0 = \left(\hat{Z}_1^0, \ldots, \hat{Z}_m^0 \right)^\top$

The shock may appear from external (for ex., international trade shock) or internal causes (for ex., shift in households demand) and expresses in shifting of vector $s = (s_1, s_2)$ components and/or vector $\hat{Z}^0$ components. As well as in the case of IO table scenario forecasting we should set the values $s = (s_1, s_2)^\top$ and $\hat{Z}^0$ for the forecast year.

If we set the scenario conditions for the target year then we resolve the Young dual problem (14)–(16) with scenario input $s = (s_1, s_2)^\top$ and evaluate the equilibrium price indices $p = (p_1, .., p_m)^\top$ of intermediate domestic inputs. Then we solve the resource allocation problem (1)–(4) with the scenario input $\hat{Z}^0$. The solution gives the scenario values of input-output flows that correspond to the shifted equilibrium state in terms of the model. Thus, we can evaluate the following scenario macroeconomic characteristics of the production network:

- equilibrium price indices $p = (p_1, .., p_m)^\top$ of domestic inputs;
- total output of industries $\hat{Y}_j = p_j F_j \left(X_1^j, \ldots, X_m^j, l_1^j, l_2^j \right)$, $j = 1, .., m$;
- quadrant I of the symmetric IO table $\hat{Z}_i^j = p_j X_i^j$ $i, j = 1..m$;
- quadrant III of the symmetric IO table (two rows):
  - gross value added (GVA) of industries (characteristic of equilibrium labor supply) $\hat{Z}_{m+1}^j = s_1 X_{m+1}^j$, $j = 1, .., m$;
  - imported intermediates of industries $\hat{Z}_{m+2}^j = s_2 X_{m+1}^j$, $j = 1, .., m$;

5   The Case of Leontief Technologies [20]

Fix the official IO table Z of the considered region for the base year and follow the notation from the Sect. 3.

The Leontief technologies with the fixed proportions of material costs have the form

$$F_j\left(X^j, l^j\right) = min\left(\frac{X_1^j}{a_{1j}}, ..., \frac{X_m^j}{a_{mj}}, \frac{l_1^j}{b_{1j}}, ..., \frac{l_n^j}{b_{nj}}\right), \ j = 1..m.$$

For further scenario calculations the Leontief technologies $F_j\left(X^j, l^j\right)$ are assumed to be fixed after we identify them according to the symmetric IO table Z for a base year (see Sect. 3). The Young transform (9) of Leontief production function corresponds to the Leontief cost function

$$q_j\left(p, s\right) = \sum_{i=1}^{m} p_i a_{ij} + \sum_{k=1}^{n} s_k b_{kj}, \ j = 1..m.$$

In accordance to the concept from the Sect. 4 we can now evaluate the scenario characteristics of the production network for a target year if we set scenario conditions s and $\hat{Z}^0$ for that year.

The Young dual problem (14)–(16) with the Leontief cost function has an explicit solution

$$p = \left(E - A^T\right)^{-1} B^T s$$

for the given vector $s = (s_1, s_2)^\top$ of price indexes on primary factors for a target year. Note that we put the unity price index vector (p, s) for the base year.

Denote $D\left(\hat{p}\right) = \left(d_i^j\right)_{i,j=1,..,m}$ the diagonal matrix, where $d_i^i = p_i > 0$. For the given scenario total consumption vector $\hat{Z}^0$ we evaluate the final consumption vector in material flows (for a target year) as $W = \left(D\left(p\right)\right)^{-1} \hat{Z}^0$ and the total output vector as

$$\hat{Y} = \left(E - A\right)^{-1} W = \left(E - A\right)^{-1}\left(D\left(p\right)\right)^{-1} \hat{Z}^0$$

Then the IO table for a target year (I and III quadrant) are evaluated as follows

$$I: \ D\left(p\right) A D\left(\hat{Y}\right) \quad III: \ D\left(s\right) B D\left(\hat{Y}\right)$$

We discuss the quality of scenario IO analisys on the base of Leontief technologies in comparison to our results for Cobb-Douglas technologies in [20]. The main results we give briefly at the last section of the paper.

6   The Case of Cobb-Douglas Technologies and Utility [19–21]

Solution of Inverse Problem of Identification. Recall the notation and assumptions from the Sect. 3. Let production technologies and utility function in (1)–(4) be Cobb-Douglas (CD) functions, i.e.

$$F_0(X) = X_1^{a_{10}}...X_m^{a_{m0}}, \quad F_j(X,l) = \alpha_j X_1^{a_{1j}}...X_m^{a_{mj}} l_1^{b_{1j}}...l_n^{b_{nj}}, \quad j = 1..m, \quad (21)$$

where $\alpha_j = \dfrac{A_j}{\prod_{i=1}^{m}(Z_i^j)^{a_{ij}} \prod_{k=1}^{n}(Z_{m+k}^j)^{b_{kj}}} > 0$. Denote

$$a^0 = \left(a_1^0, \ldots, a_m^0\right)^T, a^j = (a_{1j}, \ldots, a_{mj})^T, b^j = (b_{1j}, \ldots, b_{nj})^T, \quad j - 1..m.$$

It follows from Remark 1 that the Young transform of utility function and j-th technology are Cobb-Douglas consumer price index and cost function

$$q_0(p) = \frac{1}{F_0(a^0)} p_1^{a_{10}}...p_m^{a_{m0}}, \tag{22}$$

$$q_j(p,s) = \frac{1}{F_j(a^j, b^j)} p_1^{a_{1j}}...p_m^{a_{mj}} s_1^{b_{1j}}...s_n^{b_{nj}}. \tag{23}$$

The following proposition holds.

Proposition 1 [21]. Fix the supply vector of primary inputs in the problem (1)–(4) as $l = (l_1, ..., l_n)$, where $l_k = \sum_{j=1}^{m} Z_{m+k}^j$, $k = 1..n$ (Table 1). Then $\{\hat{X}_i^0 = Z_i^0, \hat{X}_i^j = Z_i^j, \hat{l}_k^j = Z_{m+k}^j \ i, j = 1..m, \ k = 1..n\}$ is a solution of convex programming problem (1)–(4) with Cobb-Douglas technologies (21).

Thus, the problem (1)–(4) explains the official symmetric IO table Z for once fixed base year (Table 1). Recall that we consider two primary factors: labor (gross value added) and imported intermediates. We format the quadrant III of the base IO table Z accordingly (Table 1). Therefore $n = 2$ above.

Young Dual Problem Solution. In [21] it is proved that for the given Cobb-Douglas consumer price index (22) and cost functions (23) the Young dual problem (14)–(16) has an explicit solution

$$s_1^{c_{j1}}...s_n^{c_{jn}}, \quad j = 1..m, \tag{24}$$

where $C = \|c_{kj}\| = \left(E - A^T\right)^{-1} B^T$ - $m \times n$-matrix, E is identity matrix.

The aggregate cost function can be written as $q_A(s) = s_1^{\gamma_1} ... s_n^{\gamma_n}$, where $\gamma = (\gamma_1, ..., \gamma_n)^T = C^T a^0$, $\gamma_1 + ... + \gamma_n = 1$.

Industry Centrality Measure. In terms of the nonlinear IO balance model we can introduce the measure of industry centrality [19]. We can measure the centrality v_j^{CPI} of industry j by the elasticity of consumer price index on the productivity α_j of industry j. It is proved in [19] that

$$v_j^{CPI} = -\frac{\alpha_j}{q_A}\frac{\partial q_A}{\partial \alpha_j} = \sum_{k=1}^{n} a_k^0 \omega_{kj},$$

where ω_{kj} are the elements of the Leontief inverse (20). The centrality v_j^{CPI} evaluates how the change of productivity in the industry affects the price level in the economy. In [19] we demonstrate the advantage of v_j^{CPI} centrality ranking in comparison to the Leontief centrality of the first $d_i = \sum_{j=1, j\neq i}^{m} a_{ij}$, $i = 1..m$ and the second $dd_i = \sum_{j=1}^{m}\sum_{k=1}^{m} a_{ik}a_{kj}$, $i = 1..m$ degrees on example of symmetric IO tables 2016–2019 of Kazakhstan. Evaluation shows that the rating by v_j^{CPI} is more stable for the considered statistics. Moreover the centrality v_j^{CPI} takes into account the structure of final demand in the economy in contrast with others centrality measures. For example, it follows from this fact that the impact of export volumes of industry is significant for the centrality value. This is a natural economic result for an export-oriented economy.

Scenario Evaluation. In the case of Cobb-Douglas technologies the Theorem 1 implies that the vector of total output equals to

$$\hat{Y} = (E - A)^{-1}\hat{Z}^0.$$

Then the quadrants I and III of the symmetric IO table for a target year are evaluated as follows

$$\hat{Z}_i^j = p_i X_i^j = a_{ij}\hat{Y}^j, \ \hat{Z}_{m+t}^j = s_t X_{m+t}^j = b_{tj}\hat{Y}^j, \ i,j = 1..m, \ t = 1..n,$$

where p_i is the solution (24) of the dual problem (14)–(16), $n = 2$.

7   The Case of CES Technologies [22–24]

Fix the base year with the official IO table Z of the region (Table 1). Let production technologies and utility function in (1)–(4) be CES functions

$$F_j\left(X^j, l^j\right) = \left(\sum_{i=1}^{m}\left(\frac{X_i^j}{w_i^j}\right)^{-\rho_j} + \sum_{k=1}^{n}\left(\frac{l_k^j}{w_{m+k}^j}\right)^{-\rho_j}\right)^{-\frac{1}{\rho_j}} \quad j = 1,..,n \quad (25)$$

$$F_0\left(X^0\right) = \left(\sum_{i=1}^{m}\left(\frac{X_i^0}{w_i^0}\right)^{-\rho_0}\right)^{-\frac{1}{\rho_0}} \quad\quad (26)$$

where $\rho_j, \rho_0 \in (-1, 0) \cup (0, +\infty)$, $w_1^j > 0, \ldots, w_{m+n}^j > 0$, $j = 0,..,m$. Constant elasticity of substitution of the industry j equals to $\sigma_j = \frac{1}{1+\rho_j}$, $j = 0, 1,..,m$.

Solution of Inverse Problem of Identification. Following the notation of the Sect. 3 fix the parameters in (25) and (26) for the given IO table Z of the base year as follows

$$w_i^j = (a_{ij})^{\frac{1+\rho_j}{\rho_j}}, \quad w_{m+k}^j = (b_{kj})^{\frac{1+\rho_j}{\rho_j}}, \quad w_i^0 = \left(Z_i^0\right)^{\frac{1+\rho_0}{\rho_0}} \left(\sum_{t=1}^m Z_t^0\right)^{-\frac{1+\rho_0}{\rho_0}},$$

where $i, j = 1, .., m$, $k = 1, 2$.

Proposition 2 [22]. Fix the vector of supply of primary inputs as follows $l = (l_1, \ldots, l_n)$, $l_t = \sum_{j=1}^m Z_{m+t}^j$, $t = 1..n$. Then the set of values $\left\{ \hat{X}_i^0 = Z_i^0, \hat{X}_i^j = Z_i^j, \hat{l}_t^j = Z_{m+t}^j, i, j = 1..m, t = 1..n \right\}$ is the solution of the convex programming problem (1)–(4) with CES technologies (25).

Thus, the problem (1)–(4) explains the symmetric IO table Z in the base year. Note that the parameters of elasticity of substitution stay free parameters at this stage.

Young Dual Problem Solution. The Young transform of the CES production function and utility is the CES cost function and CES consumer price index (see Remark 1) as follows

$$q_j(p, s) = \left(\sum_{i=1}^m \left(w_i^j p_i \right)^{\frac{\rho_j}{1+\rho_j}} + \sum_{k=1}^n \left(w_{m+k}^j s_k \right)^{\frac{\rho_j}{1+\rho_j}} \right)^{\frac{1+\rho_j}{\rho_j}}, \quad j = 1, .., m. \quad (27)$$

$$q_0(p) = \left(\sum_{i=1}^m \left(w_i^0 p_i \right)^{\frac{\rho_0}{1+\rho_0}} \right)^{\frac{1+\rho_0}{\rho_0}}. \quad (28)$$

The Theorem 2 implies that the solution of the Young dual problem (14)–(16) is the price indexes vector $p = (p_1, \ldots, p_m) \geq 0$ that satisfies to the following system

$$\left(\sum_{i=1}^m a_{ij}(p_i)^{\frac{\rho_j}{1+\rho_j}} + \sum_{k=1}^n b_{kj}(s_k)^{\frac{\rho_j}{1+\rho_j}} \right)^{\frac{1+\rho_j}{\rho_j}} = p_j, \quad j = 1, \ldots, m$$

for the given vector $s = (s_1, \ldots, s_n)$ and fixed values of $\rho_1, .., \rho_m$.

Scenario Evaluation. In accordance to the Sect. 4, given the initial data set for the target year we resolve the resource allocation problem (1)–(4) and the Young dual problem (14)–(16). As a result we obtain the new values of IO table. Denote $\Lambda = \|\lambda_{ij}\|$ - $(m \times m)$-matrix (we assume that it is productive), where

$$\lambda_{ij} = a_{ij} \left(\frac{p_i(s)}{p_j(s)} \right)^{\frac{\rho_j}{1+\rho_j}}, \quad i, j = 1..m.$$

Then for the new scenario conditions s and $\hat{Z}^0$ the vector of total output $\hat{Y} = \left(\hat{Y}_1, .., \hat{Y}_m\right)$ equals to $\hat{Y} = (E - \Lambda)^{-1}\hat{Z}^0$, where E is the identity $(m \times m)$-matrix. The coefficients of the first and the third quadrants of the IO table equal to

$$\hat{Z}_i^j = \hat{Y}_j \left(w_i^j \frac{p_i(s)}{p_j(s)}\right)^{\frac{\rho_j}{1+\rho_j}} = \lambda_{ij}\hat{Y}_j, \ i,j = 1..m.$$

$$\hat{Z}_{m+k}^j = \hat{Y}_j \left(w_{m+k}^j \frac{s_k}{p_j(s)}\right)^{\frac{\rho_j}{1+\rho_j}} = b_{kj}\hat{Y}_j \left(\frac{s_k}{p_j(s)}\right)^{\frac{\rho_j}{1+\rho_j}} \quad k = 1..n, j = 1..m.$$

Here p_i is the solution of the dual problem (14)–(16) for the new value of s and for the fixed value of ρ_j, $j = 1..m$. Recall that elasticity parameters ρ_j we evaluate as a result of model calibration with the actual IO data base of the region.

8   The Case of Nested CES Technologies [25]

Driven by the international trade shocks of the past few years we consider the nested-CES technologies for studying the substitution of imported and domestic intermediates of economy sectors. In this case the j-th technology in (1)–(4) takes the form

$$F_j\left(X^j, l^j, I^j\right) = \hat{F}_j\left(G_j\left(X^j, l^j\right), I^j\right) \in \Phi_{m+2}, \qquad (29)$$

where $\hat{F}_j\left(G_j, I^j\right) \in \Phi_2$, $G_j\left(X^j, l^j\right) \in \Phi_{m+1}$. Given the initial IO table Z we use further the following denotation

$$
\begin{aligned}
a_{ij} &= \frac{Z_i^j}{\sum_{k=1}^{m+1} Z_k^j} & a_{i0} &= \frac{Z_i^0}{\sum_{i=1}^{m} Z_i^0} \\
b_{L,j} &= \frac{Z_{m+1}^j}{\sum_{k=1}^{m+1} Z_k^j} & b_{G,j} = \frac{\sum_{k=1}^{m+1} Z_k^j}{\sum_{k=1}^{m+2} Z_k^j}, & b_{I,j} &= \frac{Z_{m+2}^j}{\sum_{k=1}^{m+2} Z_k^j}, \ i,j = 1..m.
\end{aligned}
\qquad (30)
$$

Solution of Inverse Problem of Identification. Let the output of industry j be the CES-nested structure as follows

$$\hat{F}_j\left(G_j, I^j\right) = \left(\left(\frac{G_j}{w_G^j}\right)^{-\rho_{2j}} + \left(\frac{I^j}{w_I^j}\right)^{-\rho_{2j}}\right)^{-\frac{1}{\rho_{2j}}}, \qquad (31)$$

where

$$G_j\left(X^j, l^j\right) = \left(\sum_{i=1}^{m}\left(\frac{X_i^j}{w_i^j}\right)^{-\rho_{1j}} + \left(\frac{l^j}{w_L^j}\right)^{-\rho_{1j}}\right)^{-\frac{1}{\rho_{1j}}}. \qquad (32)$$

Here the following restrictions are assumed $\rho_{1j}, \rho_{2j} \in (-1,0) \cup (0, +\infty)$, $w_1^j > 0, .., w_m^j > 0$, $w_L^j > 0$, $w_I^j > 0$, $w_G^j > 0$, $j = 1, .., m$. The utility function of the final consumer $F_0\left(X^0\right)$ is still given by CES function (26). Given the IO table Z in the base year (Table 1) fix the parameters of the functions $\hat{F}_j, G_j, F_0$ as follows $(i, j = 1, .., m)$

$$w_i^j = (a_{ij})^{\frac{1+\rho_{1j}}{\rho_{1j}}}, \quad w_L^j = (b_{Lj})^{\frac{1+\rho_{1j}}{\rho_{1j}}}, \quad w_I^j = (b_{Ij})^{\frac{1+\rho_{2j}}{\rho_{2j}}}, \quad w_G^j = (b_{Gj})^{\frac{1+\rho_{2j}}{\rho_{2j}}},$$

$$w_i^0 = (a_{i0})^{\frac{1+\rho_0}{\rho_0}},$$

Theorem 3. *Given the IO table Z (Table 1) for the base year fix the vector of supply of primary inputs as follows $l = (l_L, l_I) = \left(\sum_{j=1}^m Z_{m+1}^j, \sum_{j=1}^m Z_{m+2}^j\right)$. Then the set of values*

$$\left\{\hat{X}_i^0 = Z_i^0, \hat{X}_i^j = Z_i^j, \hat{l}^j = Z_{m+1}^j, \hat{I}^j = Z_{m+2}^j, i = 1, \ldots, m; \; j = 1, \ldots, m\right\}$$

is the solution of the convex programming problem (1)–(4) with nested CES technologies (29) and utility (26).

As in previous cases the Theorem 3 explains the initial IO table Z of the fixed base year for any values of elasticities.

Young Dual Problem Solution. The Young transform of the CES consumer price index is given by (28) in previous section. We prove in [25] that the cost function $q_j\left(\hat{s}, p\right)$ from the left part of (15) with $\hat{s} = (s_L, s_I)$ takes the nested form

$$q_j\left(s_L, s_I, p_1, .., p_m\right) =$$

$$\left(b_{Gj}\left(\left(\sum_{i=1}^m a_{ij}\left(p_i\right)^{\frac{\rho_{1j}}{1+\rho_{1j}}} + b_{Lj}\left(s_L\right)^{\frac{\rho_{1j}}{1+\rho_{1j}}}\right)^{\frac{1+\rho_{1j}}{\rho_{1j}} \cdot \frac{\rho_{2j}}{1+\rho_{2j}}} + b_{Ij}\left(s_I\right)^{\frac{\rho_{2j}}{1+\rho_{2j}}}\right)^{\frac{1+\rho_{2j}}{\rho_{2j}}}\right)$$

$$(33)$$

The Theorem 2 implies that the vector of price indexes $p = (p_1, \ldots, p_m) \geq 0$ is the solution of the Young dual problem (14)–(16) with the cost function $q_j\left(\hat{s}, p\right)$ (33) for the given vector of price indexes of resources $s = (s_L, s_I)$ and fixed parameters $\rho_1 = (\rho_{11}, \ldots, \rho_{1m})$, $\rho_2 = (\rho_{21}, \ldots, \rho_{2m})$, $j = 1..m$.

Remark 1. It is obviously that if for a certain product k the strict inequality in (15) with (33) holds at the equilibrium point then the cost of product k is greater than its price p_k. The production in this case is unprofitably, i.e. $X_k = 0$ at the equilibrium point. However in applications, if we consider high aggregated production complexes of the economy, the case of zero output is impossible and (15) turns to equality [25].

9 Applications. Model Calibration

We are interesting in scenario IO analysis in terms of developed framework to evaluate the responses of regional economies to external or internal shocks. Present world trends of deglobalization and structural changes in markets and prices caused by sanctions and covid-19 pandemic lead to production and supply chains disruptions. The substitution of inputs in the model makes it possible to evaluate the middle-term (1–5 years) macroeconomic consequences of such processes.

Up to the time we approve the developed IO framework on the high aggregated level of production network for several regional economies [19,20,23–25]. The good evaluation accuracy of the main macroeconomic characteristics confirms the quality of our model.

In [20] we compare the accuracy of evaluations on the base of the developed IO framework with Leontief technologies (L-model, without substitution of intermediate inputs) and the more general IO framework with Cobb-Douglas technologies (CD-model) that fix the financial structure of production costs. We use the World IO Database (http://www.wiod.org/) to identify the model (we fix 2007 as the base year) for three groups of economies: well developed (USA, Germany, Poland); catching-up, more centralized economies (Greece, Russia); fast growing economies with significant state regulation (China, Brazil). We forecast the IO tables (56 sectors) of the selected economies to 2008–2014 by the CD-model and L-model. Evaluations show that the L-model forecast is further consistent to essentially administrative regulated economies. The CD-model forecast gives is more adequate for economies without a strong state control. The example of Imported Intermediates forecasts and statistics we show in Fig. 1. In [22] we apply the IO framework with CES technologies to analyze the dominant macroeconomic response to covid-19 pandemic of Kazakhstan economy. We aggregate the production network of Kazakhstan to six large industrial complexes up to their relation with international trade. We identify the model in accordance to national IO statistics 2013 and calibrate the model on the base

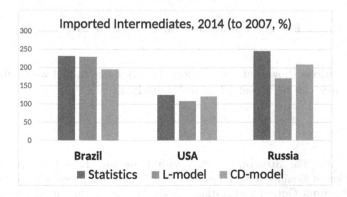

Fig. 1. Imported Intermediates. Statistics and evaluations.

of corresponding IO statistics 2013–2019. Elasticities of substitution of inputs for each of six industrial complexes are the calibration parameters. We evaluate the parameters of elasticities $\rho = (\rho_1, .., \rho_6)$ by minimizing the residual sum of squares of Total output, Gross Value Added (GVA) and Imported Intermediates (Import used) values, i.e.,

$$\sum_{t=2013}^{2019} \left[\left(\hat{Y}(t)(\rho) - Y(t) \right)^2 + \left(\hat{V}(t)(\rho) - V(t) \right)^2 + \left(\hat{I}(t)(\rho) - I(t) \right)^2 \right] \to \min_{\rho}, \tag{34}$$

where $Y(t), I(t), V(t)$ are the statistics of Total output, GVA and Import used values, and $\hat{Y}(t)(\rho), \hat{V}(t)(\rho), \hat{I}(t)(\rho)$ are the corresponding evaluated values. Evaluations of 2014–2019 IO tables with the calibrated model show the good prediction quality of our method. The main totals of the economy (evaluation and statistics) are shown on the Fig. 2. Scenario evaluations with the model show [22] that the external shock (drop in export flows) of Covid-19 pandemic was more significant for macro indicators of the Kazakhstan economy than the internal shock (drop in domestic final consumption).

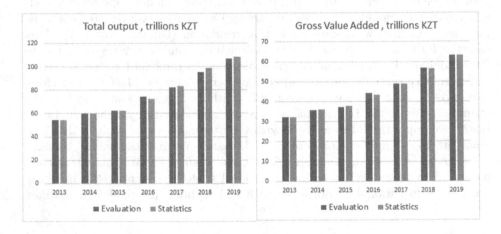

Fig. 2. Statistics and evaluation, 2013–2019

The evaluations show that the developed IO balance model with substitution of intermediates can be a useful tool of IO analysis.

References

1. Leontief, W.: The Structure of American Economy, 1919–1939: An Empirical Application of Equilibrium Analysis. Oxford University Press, New York (1951)
2. Stone, R.: Input-Output and National Accounts. Organisation for European Economic Co-operation (1961)

3. Hulten, C.: Growth accounting with intermediate inputs. Rev. Econ. Stud. **45**(3), 511–518 (1978)
4. Miller, R.: Input-Output Analysis Foundations and Extensions, 2nd edn. Cambridge University Press, New York (2009)
5. Nikaido, H.: Convex Structures and Economic Theory. Mathematics in Science and Engineering, vol. 51. Academic Press, New York (1968)
6. Ashmanov, S.: Introduction to Mathematical Economics. Fizmatlit, Moscow (1984)
7. Atalay, E.: How important are sectoral shocks? Am. Econ. J. Macroecon. **9**(4), 254–280 (2017)
8. Acemoglu, D., Ufuk, A., William K.: Networks and the macroeconomy: an empirical exploration. In: Eichenbaum, M., Parker, J. (eds.) National Bureau of Economic Research Macroeconomics Annual, vol. 30, pp. 276–335 (2016)
9. Shea, J.: Complementarities and comovements. J. Money Credit Bank. **34**, 412–433 (2002)
10. Acemoglu, D., Ozdaglar, A., Tahbaz-Salehi, A.: Cascades in Networks and Aggregate Volatility. NBER Working Papers 16516, National Bureau of Economic Research, Inc. (2010)
11. Acemoglu, D., Ozdaglar, A., Tahbaz-Salehi, A.: Microeconomic origins of macroeconomic tail risks. Am. Econ. Rev. **107**(1), 54–108 (2017)
12. Acemoglu, D., Carvalho, V., Ozdaglar, A., Tahbaz-Salehi, A.: The network origins of aggregate fluctuations. Econometrica **80**(5), 1977–2016 (2012)
13. Acemoglu, D., Ozdaglar, A., Tahbaz-Salehi, A.: Systemic risk and stability in financial networks. Am. Econ. Rev. **105**(2), 564–608 (2015)
14. Baqaee, D.: Cascading failures in production networks. Econometrica **86**(5), 1819–1838 (2018)
15. Carvalho, V.: From micro to macro via production networks. J. Econ. Perspect. **28**(4), 23–48 (2014)
16. Carvalho, V., Tahbaz-Salehi, A.: Production Networks. A Primer, Cambridge Working Papers in Economics 1856, Faculty of Economics, University of Cambridge (2018)
17. Shananin, A.A.: Young duality and aggregation of balances. Dokl. Math. **102**(1), 330–333 (2020). https://doi.org/10.1134/S1064562420040171
18. Shananin, A.: Problem of aggregating of an input-output model and duality. Comput. Math. Math. Phys. **61**(1), 153–166 (2021)
19. Boranbayev, S., Obrosova, N., Shananin, A.: Production network centrality in connection to economic development by the case of Kazakhstan statistics. In: Olenev, N.N., Evtushenko, Y.G., Jaćimović, M., Khachay, M., Malkova, V. (eds.) OPTIMA 2021. LNCS, vol. 13078, pp. 321–335. Springer, Cham (2021). https://doi.org/10.1007/978-3-030-91059-4_23
20. Obrosova, N., Shananin, A., Spiridonov, A.: On the comparison of two approaches to intersectoral balance analysis. In: Journal of Physics: Conference Series, vol. 2131, no. 2 (2021)
21. Rassokha, A.V., Shananin, A.A.: Inverse problems of the analysis of input-output balances. Math. Models Comput. Simul. **13**(6), 943–954 (2021). https://doi.org/10.1134/S2070048221060193
22. Shananin, A., Rassokha, A.: Inverse problems in analysis of input-output model in the class of CES functions. J. Inverse Ill-Posed Probl. **29**(2), 305–316 (2021)
23. Kerimkhulle, S., Obrosova, N., Shananin, A., Azieva, G.: The nonlinear model of intersectoral linkages of Kazakhstan for macroeconomic decision-making processes in sustainable supply chain management. Sustainability **14**(21) (2022)

24. Boranbayev, A., Obrosova, N., Shananin, A.: Nonlinear input-output balance and Young duality: analysis of Covid-19 macroeconomic impact on Kazakhstan. Siberian Electron. Math. Rep. **19**(2), 835–851 (2022)
25. Obrosova, N., Shananin, A., Spiridonov, A.: Nonlinear input-output model with nested CES technologies for the analysis of macroeconomic effects of a foreign trade shock. Lobachevskii J. Math. **44**(1), 401–417 (2023)
26. Shananin, A.: Non-parametric method for the analysis of industry technological structure. Matem. Mod **11**(9), 116–122 (1999)
27. Shananin, A.: Integrability problem and the generalized nonparametric method for the consumer demand analysis. Proc. Moscow Inst. Phys. Technol. **4**(1), 84–98 (2009)
28. Agaltsov, A.D., Molchanov, E.G., Shananin, A.A.: Inverse problems in models of resource distribution. J. Geom. Anal. **28**(1), 726–765 (2017). https://doi.org/10.1007/s12220-017-9840-1
29. Vratenkov, S., Shananin, A.: Analiz strukturi potrebitelskogo sprosa s pomoschyu economicheskih indeksov. Soobscheniya po prikladnoy matematike, VC RAN SSSR, Moscow (1991)
30. System of National Accounts 2008, European Commission, International Monetary Fund, Organisation for Economic Co-operation and Development, United Nations and World Bank, New York (2009)

Mathematical Programming
and Applications

On Decentralized Nonsmooth Optimization

Savelii Chezhegov[1][ID], Alexander Rogozin[1(✉)][ID], and Alexander Gasnikov[1,2,3][ID]

[1] Moscow Institute of Physics and Technology, Moscow, Russia
aleksandr.rogozin@phystech.edu
[2] Institute for Information Transportation Problems, Moscow, Russia
[3] Caucasus Mathematic Center of Adygh State University, Moscow, Russia

Abstract. In decentralized optimization, several nodes connected by a network collaboratively minimize some objective function. For minimization of Lipschitz functions lower bounds are known along with optimal algorithms. We study a specific class of problems: linear models with nonsmooth loss functions. Our algorithm combines regularization and dual reformulation to get an effective optimization method with complexity better than the lower bounds.

Keywords: convex optimization · distributed optimization

1 Introduction

The focus of this work is a particular class of problems in decentralized nonsmooth optimization. We assume that each of computational agents, or nodes, holds a part of a common optimization problem and the agents are connected by a network. Each node may communicate with its immediate neighbors, and the agents aim to collaboratively solve an optimization problem.

On the class of smooth (strongly) convex functions endowed with a first-order oracle, decentralized optimization can be called a theoretically well-developed area of research. For this setting, [13] proposed lower bounds and optimal dual algorithms. After that, optimal gradient methods with primal oracle were developed in [8]. Even if the network is allowed to change, lower bounds and optimal algorithms are known and established in a series of works [7,9,11].

However, the case when local functions are non-smooth is not that well studied. Algorithm proposed in [14] uses a gradient approximation via Gaussian smoothing. Such a technique results in additional factor of dimension. Distributed subgradient methods [12] are not optimal and only converge to a neighborhood of the solution if used with a constant step-size. In other words, development of an optimal decentralized algorithm for networks is an open research question.

As noted below, we restrict our attention to a particular class of decentralized non-smooth optimization problems. Namely, we study linear models with nonsmooth loss functions and an entropic regularizer. Problems of such type arise in

The research is supported by the Ministry of Science and Higher Education of the Russian Federation (Goszadaniye) 075-00337-20-03, project No. 0714-2020-0005.

traffic demands matrix calculation [1,15], optimal transport [10] and distributed training with decomposition over features [2].

Traffic Problems. Following the arguments in [1], for example, one seeks to minimize $g(x)$ subject to constraints $Ax = b$. Here function $g(x)$ may be interpreted as some similarity measure between x and a supposed solution. Moving the constraint $Ax = b$ as a penalty into the objective, we obtain a problem of type

$$\min_{x \in \mathbb{R}^d} g(x) + \lambda \|Ax - b\|,$$

where $\|Ax - b\|$ is a norm of vector. If the g represents a similarity measure given by KL divergence, we obtain an optimization problem for linear model with entropic regularizer.

Optimal Transport. Another example is entropy-regularized optimal transport [10]. In paper [10] the authors show that an optimal transportation problem can be rewritten as

$$\min_{\mathbf{x} \in \Delta_n^n} \min_{\mathbf{y} \in \Delta_2^n} \langle \mathbf{x}, \mathbf{a} \rangle - \langle \mathbf{y}, \mathbf{b} \rangle + \langle \mathbf{Ax}, \mathbf{y} \rangle + \lambda_{\mathbf{x}} \langle \mathbf{x}, \log \mathbf{x} \rangle - \lambda_{\mathbf{y}} \langle \mathbf{y}, \log \mathbf{y} \rangle,$$

where Δ_n denotes a unit simplex of dimension n. This illustrated that entropy-regularized linear models can arise in saddle-point optimization, as well.

Distributed ML. In distributed statistical inference and machine learning one may want to train a model in a distributed way [2]. Consider a dataset with a moderate number of training examples and a large number of features. Let the dataset be split between the nodes not by samples but by features. Let ℓ be the common loss function, and for each agent i introduce its local dataset (A_i, b_i) and the corresponding regularizer $r_i(x)$. That leads to a fitting problem

$$\min_{x_1, \ldots, x_m} \ell \left(\sum_{i=1}^{m} A_i x_i - b_i \right) + \sum_{i=1}^{m} r_i(x_i).$$

Our Contribution. In our work we propose a dual algorithm for non-smooth decentralized optimization. The dual problem is smooth although the initial one is non-smooth, but also subject to constraints. The constraints can be equivalently rewritten as a regularizer. We show that a resulting regularized problem can be solved by an accelerated proximal primal-dual gradient method.

We study a specific class of problems, and our approach allows to break the lower bounds in [14]. Omitting problem parameters, the iteration and communication complexities of our algorithm are $O(\sqrt{1/\varepsilon})$, while lower bounds suggest that at least $\Omega(1/\varepsilon)$ communication rounds and at least $\Omega(1/\varepsilon^2)$ local computations at each node are required.

1.1 Notation

Let $\otimes$ is the Kronecker product. Let $\Delta_d = \{x \in \mathbb{R}^d : \sum_{i=1}^m x_i = 1,\ x_i \geq 0,\ i = 1,\ldots,m\}$ be a unit simplex in $\mathbb{R}^d$. By Δ_d^m we understand a product of m simplices that is a set in $\mathbb{R}^{md}$. For $p \geq 0$, let $\|x\|_p = \left(\sum_{i=1}^d |x_i|^p\right)^{1/p}$ denote the p-norm in $\mathbb{R}^d$. By $\langle a, b \rangle$ we denote a scalar product of vectors. Also let $\mathbf{x} = \mathrm{col}[x_1,\ldots,x_m] = (x_1^\top \ldots x_m^\top)^\top \in \mathbb{R}^{md}$ denote a column vector stacked of $x_1,\ldots,x_m \in \mathbb{R}^d$. Similarly for matrices $C_1,\ldots,C_m \in \mathbb{R}^{n \times d}$, introduce $\mathrm{col}[C_1,\ldots,C_m] = (C_1^\top \ldots C_m^\top)^\top \in \mathbb{R}^{mn \times d}$. Moreover, let $\mathrm{diag}[C_1,\ldots,C_m]$ denote a block matrix with blocks $C_1,\ldots,C_m$ at the diagonal. For $x \in \mathbb{R}^d$, let $\log x$ denote a natural logarithm function applied component-wise. We define $\mathbf{1}_n$ to be a vector of all ones of length n and $\mathbf{I}_n$ to be an identity matrix of size $n \times n$. Also denote the i-th coordinate vector of $\mathbb{R}^n$ as $e_i^{(n)}$. Let $\lambda_{\max}(C)$ and $\lambda_{\min}^+(C)$ denote the maximal and minimal nonzero eigenvalue of matrix C. Let $\sigma_{\max}(C)$ and $\sigma_{\min}^+(C)$ denote the maximal and minimal nonzero singular values of C.

Given a convex closed set Q, let Π_Q denote a projection operator on it and denote its interior int Q. For a closed proper function $h(x): Q \to \mathbb{R}$ and a scalar $\gamma > 0$, define proximal operator as

$$\mathrm{prox}_{\gamma h}(x) = \arg\min_{y \in Q} \left(h(x) + \frac{1}{2\gamma} \|y - x\|_2^2 \right).$$

2 Problem and Assumptions

Consider m independent computational entities, or agents. Agent i locally holds a dataset consisting of matrix A_i and labels b_i. Let $A = \mathrm{col}[A_1,\ldots,A_m] \in \mathbb{R}^{n \times d}$ be the training samples and $\mathbf{b} = \mathrm{col}[b_1,\ldots,b_m]$ be the labels. The whole training dataset $(A, \mathbf{b})$ is distributed between m different machines. We consider p-norm minimization over unit simplex with entropy regularizer.

$$\min_{x \in \Delta_d} \frac{1}{m} \|Ax - \mathbf{b}\|_p + \theta \langle x, \log x \rangle, \tag{1}$$

where $\theta > 0$ is a regularization parameter.

The agents can communicate information through a communication network. We assume that each machine is a node in the network that is represented by a connected undirected graph $\mathcal{G} = (\mathcal{V}, \mathcal{E})$. The nodes can communicate if and only if they are connected by an edge.

With graph $\mathcal{G}$ we associate a communication matrix W that has the following properties.

Assumption 1
1. *(Network compatibility)* $[W]_{ij} = 0$ if $i \neq j$ and $(i, j) \notin \mathcal{E}$.
2. *(Positive semi-definiteness and symmetry)* $W \succeq 0$, $W^\top = W$.
3. *(Kernel property)* $Wx = 0$ if and only if $x_1 = \ldots = x_m$.

We also introduce the condition number of the communication matrix.

$$\chi = \frac{\lambda_{\max}(W)}{\lambda_{\min}^+(W)}. \tag{2}$$

In order to get a distributed formulation, assign each agent i in the network a local copy of the solution vector x_i. Define $\mathbf{x} = \mathrm{col}[x_1, \ldots, x_m]$, $\mathbf{A} = \mathrm{diag}[A_1, \ldots, A_m]$ and introduce $\mathbf{y} = \mathbf{A}\mathbf{x}$.

$$\min_{\mathbf{x} \in \Delta_m^d} \|\mathbf{y} - \mathbf{b}\|_p + \theta \langle \mathbf{x}, \log \mathbf{x} \rangle \tag{3}$$

$$\text{s.t. } \mathbf{W}\mathbf{x} = 0$$

$$\mathbf{y} = \mathbf{A}\mathbf{x}$$

The complexity of distributed methods typically depends on the condition number of the communication matrix (it is χ defined in (2)) and on condition numbers of objective functions. For brevity we introduce

$$\sigma_{\max}(\mathcal{A}) = \max_{i=1,\ldots,m} (\sigma_{\max}(A_i)), \ \sigma_{\min}^+(\mathcal{A}) = \min_{i=1,\ldots,m} \sigma_{\min}^+(A_i). \tag{4}$$

3   Dual Problem

Let us derive a dual problem to (3). It is convenient to introduce $F(\mathbf{y}) = \|\mathbf{y} - \mathbf{b}\|_p$, $G(\mathbf{x}) = \theta \langle \mathbf{x}, \log \mathbf{x} \rangle$ for the next estimations.

3.1   Conjugate Functions

Let us derive the conjugate functions F^* and G^*. Let $q \geq 1$ be such that $\frac{1}{p} + \frac{1}{q} = 1$.

$$F^*(\mathbf{t}) = \sup_{\mathbf{y} \in \mathbb{R}^{mn}} (\langle \mathbf{t}, \mathbf{y} \rangle - F(\mathbf{y})) = \sup_{\mathbf{y} \in \mathbb{R}^{mn}} (\langle \mathbf{t}, \mathbf{y} - \mathbf{b} \rangle - \|\mathbf{y} - \mathbf{b}\|_p) + \langle \mathbf{t}, \mathbf{b} \rangle$$

$$= \sup_{\mathbf{r} \in \mathbb{R}^{mn}} (\langle \mathbf{t}, \mathbf{r} \rangle - \|\mathbf{r}\|_p) + \langle \mathbf{t}, \mathbf{b} \rangle$$

$$= \begin{cases} \langle \mathbf{t}, \mathbf{b} \rangle, & \|\mathbf{t}\|_q \leq 1 \\ +\infty, & \text{otherwise} \end{cases}$$

Last equation is a result of conjugate function for $\|x\|_p$, which is taken from a classical book by Boyd [3], Chap. 5.

In order to compute G^*, introduce $g(x) = \theta \langle x, \log x \rangle : \mathbb{R}^d \to \mathbb{R}$ and note that $G(\mathbf{x}) = \sum_{i=1}^m g_i(x_i)$.

$$g^*(t) = \sup_{x \in \Delta_d} (\langle t, x \rangle - \theta \langle x, \log(x) \rangle).$$

Writing a Lagrange function:

$$L(t, x) = \langle t, x \rangle - \theta \langle x, \log(x) \rangle + \lambda \left(1_d^\top x - 1 \right)$$

$$\nabla_x L(t, x) = t - \theta \log x - \theta 1_d + \lambda 1_d = 0 \Rightarrow x = \exp\left(\frac{t}{\theta} + 1_d \left(\frac{\lambda}{\theta} - 1 \right) \right)$$

$$1_d^\top x = 1 \Rightarrow \exp\left(\frac{\lambda}{\theta} - 1 \right) 1_d^\top \exp\left(\frac{t}{\theta} \right) = 1 \Rightarrow \exp\left(\frac{\lambda}{\theta} - 1 \right) = \frac{1}{1_d^\top \exp\left(\frac{t}{\theta} \right)}.$$

As a consequence

$$x = \frac{\exp\left(\frac{t}{\theta}\right)}{1_d^\top \exp\left(\frac{t}{\theta}\right)}.$$

Using equation to x,

$$g^*(t) = \theta \log\left(1_d^\top \exp\left(\frac{t}{\theta}\right)\right).$$

As noted above, $G(\mathbf{x})$ is separable, i.e. $G(\mathbf{x}) = \sum_{i=1}^m g(x_i)$. Therefore,

$$G^*(\mathbf{t}) = \sup_{\mathbf{x}\in\Delta_d^m}\left(\langle \mathbf{t}, \mathbf{x}\rangle - \sum_{i=1}^m g(x_i)\right) = \sum_{i=1}^m \sup_{x\in\Delta_d}\left(\langle t_i, x\rangle - g(x)\right) = \sum_{i=1}^m g^*(t_i).$$

It is convenient to express t_i through $\mathbf{t}$. Introduce matrix

$$\mathbf{E}_i = \left(e_i^{(m)}\right)^\top \otimes \mathbf{I} = [0\ldots0\ \mathbf{I}\ 0\ldots0]. \tag{5}$$

Then $t_i = \mathbf{E}_i\mathbf{t}$. It holds

$$G^*(\mathbf{t}) = \sum_{i=1}^m g^*(\mathbf{E}_i\mathbf{t}).$$

3.2   Dual Problem Formulation

Let us derive a dual problem to (3). It is convenient to denote $F(\mathbf{y}) = \|\mathbf{y} - \mathbf{b}\|_p$, $G(\mathbf{x}) = \theta\langle\mathbf{x}, \log\mathbf{x}\rangle$. Introduce dual function

$$\begin{aligned}
\Phi(\mathbf{z}, \mathbf{s}) &= \inf_{\mathbf{x}\in\Delta_d^m, \mathbf{y}\in\mathbb{R}^n}\left[F(\mathbf{y}) + G(\mathbf{x}) + \langle\mathbf{z}, \mathbf{W}\mathbf{x}\rangle + \langle\mathbf{s}, \mathbf{A}\mathbf{x} - \mathbf{y}\rangle\right] \\
&= \inf_{\mathbf{y}\in\mathbb{R}^{mn}}\left[F(\mathbf{y}) - \langle\mathbf{s}, \mathbf{y}\rangle\right] + \inf_{\mathbf{x}\in\Delta_d^m}\left[G(\mathbf{x}) + \langle\mathbf{W}\mathbf{z} + \mathbf{A}^\top\mathbf{s}, \mathbf{x}\rangle\right] \\
&= -\sup_{\mathbf{y}\in\mathbb{R}^{mn}}\left[\langle\mathbf{s}, \mathbf{y}\rangle - F(\mathbf{y})\right] - \sup_{\mathbf{x}\in\Delta_d^m}\left[\langle-\mathbf{W}\mathbf{z} - \mathbf{A}^\top\mathbf{s}, \mathbf{x}\rangle - G(\mathbf{x})\right] \\
&= -F^*(\mathbf{s}) - G^*(-\mathbf{W}\mathbf{z} - \mathbf{A}^\top\mathbf{s}).
\end{aligned}$$

As a consequence, dual problem can be formulated as

$$\min_{\mathbf{z}\in\mathbb{R}^{md}, \mathbf{s}\in\mathbb{R}^{mn}} F^*(\mathbf{s}) + G^*(-\mathbf{W}\mathbf{z} - \mathbf{A}^\top\mathbf{s}).$$

Results from 3.1 and 3.2 leads us to final dual problem formulation

$$\min_{\mathbf{z}, \mathbf{s}:\|\mathbf{s}\|_q\le1} \langle\mathbf{s}, \mathbf{b}\rangle + \sum_{i=1}^m \theta\log\left(1_d^\top \exp\left(-\frac{1}{\theta}\mathbf{E}_i\left(\mathbf{W}\mathbf{z} + \mathbf{A}^\top\mathbf{s}\right)\right)\right). \tag{6}$$

The constrained problem above is equivalent to a regularized problem

$$\min_{\mathbf{z},\mathbf{s}} \ \langle \mathbf{s}, \mathbf{b} \rangle + \sum_{i=1}^{m} \theta \log \left(\mathbf{1}_d^\top \exp \left(-\frac{1}{\theta} \mathbf{E}_i (\mathbf{W}\mathbf{z} + \mathbf{A}^\top \mathbf{s}) \right) \right) + \nu \|\mathbf{s}\|_q^q, \qquad (7)$$

where $\nu > 0$ is a scalar.

As a result, the dual problem writes as

$$\min_{\mathbf{q}} \ H(\mathbf{z},\mathbf{s}) + R(\mathbf{z},\mathbf{s}) \qquad (8)$$

$$H(\mathbf{z},\mathbf{s}) = \langle \mathbf{s}, \mathbf{b} \rangle + \sum_{i=1}^{m} \theta \log \left(\mathbf{1}_d^\top \exp \left(-\frac{1}{\theta} \mathbf{E}_i (\mathbf{W}\mathbf{z} + \mathbf{A}^\top \mathbf{s}) \right) \right).$$

$$R(\mathbf{z},\mathbf{s}) = \nu \|\mathbf{s}\|_q^q.$$

Recall problem (8) and denote $\mathbf{B} = (-\mathbf{W} \ -\mathbf{A}^\top)$, $\mathbf{q} = \mathrm{col}[\mathbf{z},\mathbf{s}]$, $\mathbf{p} = \mathrm{col}[0,\mathbf{b}]$. With slight abuse of notation we write $H(\mathbf{q}) = H(\mathbf{z},\mathbf{s})$ and $R(\mathbf{q}) = R(\mathbf{z},\mathbf{s})$. Problem (8) takes the form

$$\min_{\mathbf{q}} \ H(\mathbf{q}) + R(\mathbf{q}).$$

Here H is a differentiable function and R is a regularizer, or composite term. Problems of such type are typically solved by proximal optimization methods.

4 Algorithms and Their Complexity Analysis

4.1 Similar Triangles Method

We apply an accelerated primal-dual algorithm called Similar Triangles Method (STM) [4].

Algorithm 1. Similar triangles method (STM)

Require: $A_0 = \alpha_0 = 0$, $\mathbf{q}^0 = \mathbf{u}^0 = \mathbf{y}^0$.

1: **for** $k = 0, \ldots, N-1$ **do**

2:    Find α_{k+1} from equality $(A_k + \alpha_k)(1 + A_k \mu) = L\alpha_{k+1}$ and put $A_{k+1} = A_k + \alpha_{k+1}$.

3:    Introduce

$$\phi_{k+1}(\mathbf{x}) = \alpha_{k+1}\left(\left\langle \nabla H(\mathbf{y}^{k+1}), \mathbf{x} \right\rangle + R(\mathbf{x}) \right) + \frac{1 + A_k \mu}{2} \left\| \mathbf{x} - \mathbf{u}^k \right\|_2^2 + \frac{\mu \alpha_{k+1}}{2} \left\| \mathbf{x} - \mathbf{y}^{k+1} \right\|_2^2$$

4:    $\mathbf{y}^{k+1} = \frac{\alpha_{k+1}\mathbf{u}^k + A_k \mathbf{q}^k}{A_{k+1}}$

5:    $\mathbf{u}^{k+1} = \arg\min_{\mathbf{q}} [\phi_{k+1}(\mathbf{x})]$

6:    $\mathbf{q}^{k+1} = \frac{\alpha_{k+1}\mathbf{u}^{k+1} + A_k \mathbf{q}^k}{A_{k+1}}$

7: **end for**

First, note that line 5 of Algorithm 1 can be decomposed into a gradient step and computation of proximal operator of R.

$$\mathbf{u}^{k+1} = \arg\min_{\mathbf{q}} [\phi_{k+1}(\mathbf{x})] = \text{prox}_{\gamma_k R} [\mu\gamma_k \mathbf{y}^{k+1} + (1-\mu\gamma_k)\mathbf{u}^k - \gamma_k \nabla F(\mathbf{y}^{k+1})],$$

where $\gamma_k = \frac{\alpha_{k+1}}{1+\mu A_{k+1}}$. Let us show that this operator can be easily computed. Let $\mathbf{t} = \text{col}[\mathbf{t_z}, \mathbf{t_s}]$. By definition of proximal operator we have

$$\text{prox}_{\gamma_k R}(\mathbf{t}) = \arg\min_{\mathbf{s}} \left(\frac{1}{2\gamma_k} \|\mathbf{t} - \mathbf{s}\|_2^2 + R(\mathbf{q}) \right)$$

$$= \arg\min_{\mathbf{z},\mathbf{s}} \left(\frac{1}{2\gamma_k}(\|\mathbf{t_s} - \mathbf{s}\|_2^2 + \|\mathbf{t_z} - \mathbf{z}\|_2^2) + \nu \|\mathbf{s}\|_q^q \right).$$

Let $\tilde{\mathbf{q}} = \text{col}[\tilde{\mathbf{z}}, \tilde{\mathbf{s}}] = \text{prox}_{\gamma_k R}(\mathbf{t})$. We have $\tilde{\mathbf{z}} = \mathbf{t_z}$. Let $\tilde{s}_i$ denote the i-th component of $\tilde{\mathbf{s}}$ and t_i denote the i-th component of $\mathbf{t_s}$; then $\tilde{s}_i$ can be found from equation

$$t_i - \tilde{s}_i + \gamma_k q\nu |\tilde{s}_i|^{q-1} = 0.$$

We assume that the equation above can be efficiently numerically solved w.r.t. $\tilde{s}_i$. For example, it can be done by solution localization methods such as binary search. As a result, we see that the proximal operator of R can be computed cheaply.

Let us formulate the result on convergence of Algorithm 1 for the problem (8).

Theorem 1. *Algorithm 1 requires*

$$O\left(\left(\frac{\theta m \left(\sigma_{\max}^2(\mathcal{A}) + \sigma_{\max}^2(W)\right) \|\log x^* + \mathbf{1}_d\|_2^2}{\min((\sigma_{min}^+(\mathcal{A}))^2, (\lambda_{min}^+(W))^2)\varepsilon} \right)^{1/2} \right)$$

iterations to reach ε-accuracy for the solution of the problem (8).

Before we prove the above result, we need to formulate some observations. We need to find Lipschitz constant for dual problem. Namely, let us find the Lipschitz constant for function $G^*(-\mathbf{W}\mathbf{z} - \mathbf{A}^\top\mathbf{s})$ as a function of $\mathbf{q} = \text{col}[\mathbf{z},\mathbf{s}]$.

Lemma 1. *Function $H(\mathbf{q})$ has a Lipschitz gradient with constant*

$$L_H = \frac{m \left(\sigma_{\max}^2(\mathcal{A}) + \sigma_{\max}^2(W)\right)}{\theta}.$$

Proof. According to [6], if a function is μ-strongly convex in norm $\|\cdot\|_2$, then its conjugate function $h^*(y)$ has a $\frac{1}{\mu}$-Lipschitz gradient in $\|\cdot\|_2$.
Using the fact from [3], Chap. 3, we obtain that the conjugate function of

$$h(x) = \log\left(\sum_{i=1}^{d} \exp(x_i) \right).$$

is

$$h^*(y) = \begin{cases} \langle y, \log y \rangle, & y \in \Delta_d \\ \infty, & \text{otherwise} \end{cases}$$

To have a constant of strongly convexity, we can find a minimal eigenvalue of Hessian of $h^*(y)$

$$\nabla^2 h^*(y) = \text{diag}\left(\frac{1}{y_1}, \ldots, \frac{1}{y_d}\right).$$

For any $y \in \text{int}\Delta_d$, we have that $1/y_i \geq 1$, $i = 1, \ldots, d$. Therefore, we have $\lambda_{\min}(\nabla^2 h^*(y)) \geq 1$, i.e. $\mu_{h^*} \geq 1$.

As a consequence, for function

$$h(x) = \log\left(\sum_{i=1}^d \exp(x_i)\right).$$

Lipschitz constant is equal to $L_h = 1$.

Therefore, for a function

$$g^*(x) = \theta h\left(\frac{x}{\theta}\right).$$

Lipschitz constant is equal to $L_g = 1/\theta$.

Introduce $\mathbf{B} = \left(-\mathbf{W}, -\mathbf{A}^\top\right)$. We have

$$H(\mathbf{q}) = G^*(\mathbf{Bq}) + \langle \mathbf{s}, \mathbf{b} \rangle = \sum_{i=1}^m g^*(\mathbf{E}_i \mathbf{Bq}) + \langle \mathbf{s}, \mathbf{b} \rangle.$$

It holds

$$\|\nabla H(\mathbf{q}_2) - \nabla H(\mathbf{q}_2)\|_2 = \|\mathbf{B}^\top \nabla G^*(\mathbf{Bq}_2) - \mathbf{B}^\top \nabla G^*(\mathbf{Bq}_2)\|_2$$

$$\leq \sigma_{\max}(\mathbf{B}) \|\nabla G^*(\mathbf{Bq}_2) - \nabla G^*(\mathbf{Bq}_1)\| \leq \sigma_{\max}(\mathbf{B}) \sum_{i=1}^m \|\nabla g^*(\mathbf{E}_i \mathbf{Bq}_2) - \nabla g^*(\mathbf{E}_i \mathbf{Bq}_1)\|_2$$

$$\leq \sigma_{\max}(\mathbf{B}) \sum_{i=1}^m \frac{\sigma_{\max}(\mathbf{E}_i \mathbf{B})}{\theta} \|\mathbf{q}_2 - \mathbf{q}_1\|_2 \leq \frac{m\sigma_{\max}^2(\mathbf{B})}{\theta} \|\mathbf{q}_2 - \mathbf{q}_1\|_2$$

$$= \frac{m(\sigma_{\max}^2(\mathbf{A}) + \sigma_{\max}^2(\mathbf{W}))}{\theta} \|\mathbf{q}_2 - \mathbf{q}_1\|_2$$

$$\leq \frac{m\left(\max_{i=1,\ldots,m}(\sigma_{\max}^2(A_1), \ldots, \sigma_{\max}^2(A_m)) + \sigma_{\max}^2(W)\right)}{\theta} \|\mathbf{q}_2 - \mathbf{q}_1\|_2$$

$$= L_H \|\mathbf{q}_2 - \mathbf{q}_1\|_2,$$

which finishes the proof of lemma.

For writing a complexity of solver for our problem, we also need to bound the dual distance.

Lemma 2. *Let* $\mathbf{q}^* = \mathrm{col}[\mathbf{z}^*, \mathbf{s}^*]$ *be the solution of dual problem* (8) *and let* x^* *be a solution of* (1). *It holds*

$$\|\mathbf{q}^*\|_2^2 \leq R_{dual}^2 = \frac{\theta^2 \, m \| \log x^* + \mathbf{1}_d\|_2^2}{\min((\sigma_{min}^+(\mathcal{A}))^2, (\lambda_{min}^+(W))^2)}.$$

Proof. Let $(\mathbf{x}^*, \mathbf{y}^*)$ be a solution to primal problem (3). In particular, we have $\mathbf{x}^* = \mathbf{1}_m \otimes x^*$. Then $(\mathbf{x}^*, \mathbf{y}^*, \mathbf{z}^*, \mathbf{s}^*)$ is a saddle point of Lagrange function. For any $\mathbf{x} \in \Delta_m^d$, $\mathbf{y} \in \mathbb{R}^{md}$, $\mathbf{z} \in \mathbb{R}^{md}, \mathbf{s} \in \mathbb{R}^{mn}$ it holds

$$F(\mathbf{y}^*) + G(\mathbf{x}^*) + \langle \mathbf{z}, \mathbf{W}\mathbf{x}^* \rangle + \langle \mathbf{s}, \mathbf{A}\mathbf{x}^* - \mathbf{y}^* \rangle$$
$$\leq F(\mathbf{y}^*) + G(\mathbf{x}^*) + \langle \mathbf{z}^*, \mathbf{W}\mathbf{x}^* \rangle + \langle \mathbf{s}^*, \mathbf{A}\mathbf{x}^* - \mathbf{y}^* \rangle$$
$$\leq F(\mathbf{y}) + G(\mathbf{x}) + \langle \mathbf{z}^*, \mathbf{W}\mathbf{x} \rangle + \langle \mathbf{s}^*, \mathbf{A}\mathbf{x} - \mathbf{y} \rangle.$$

Substituting $\mathbf{y} = \mathbf{y}^*$ we obtain

$$G(\mathbf{x}) \geq G(\mathbf{x}^*) + \langle -\mathbf{W}\mathbf{z}^* - \mathbf{A}^\top \mathbf{s}^*, \mathbf{x} - \mathbf{x}^* \rangle.$$
$$-\mathbf{W}\mathbf{z}^* - \mathbf{A}^\top \mathbf{s}^* = \nabla G(\mathbf{x}^*).$$

Recalling that $\mathbf{B} = (-\mathbf{W}, -\mathbf{A}^\top)$ we derive

$$\mathbf{B}\mathbf{q}^* = \nabla G(\mathbf{x}^*).$$
$$\langle \mathbf{B}^\top \mathbf{B}\mathbf{q}^*, \mathbf{q}^* \rangle = \|\nabla G(\mathbf{x}^*)\|_2^2.$$
$$\lambda_{min}^+(\mathbf{B}^\top \mathbf{B}) \|\mathbf{q}^*\|_2^2 \leq \|\nabla G(\mathbf{x}^*)\|_2^2.$$
$$\|\mathbf{q}^*\|_2^2 \leq \frac{\|\nabla G(\mathbf{x}^*)\|_2^2}{\lambda_{min}^+(\mathbf{B}^\top \mathbf{B})}.$$

We have

$$\lambda_{min}^+(\mathbf{B}^\top \mathbf{B}) = \lambda_{min}^+(\mathbf{B}\mathbf{B}^\top) = \lambda_{min}^+(\mathbf{W}^2 + \mathbf{A}^\top \mathbf{A}) = \lambda_{min}^+(\mathbf{W}^2 \otimes \mathbf{I}_d + \mathbf{I}_m \otimes \mathbf{A}^\top \mathbf{A})$$
$$= \min((\lambda_{min}^+(\mathbf{W}))^2, (\sigma_{min}^+(\mathbf{A}))^2) = \min\left((\lambda_{min}^+(W))^2, (\sigma_{min}^+(\mathcal{A}))^2\right).$$

and

$$\|\nabla G(\mathbf{x}^*)\|_2^2 = \theta^2 \|\log \mathbf{x}^* + \mathbf{1}_{md}\|_2^2 = \theta^2 m \|\log x^* + \mathbf{1}_d\|_2^2.$$

As a result, we obtain

$$R_{dual}^2 = \frac{\theta^2 \, m \| \log x^* + \mathbf{1}_d\|_2^2}{\min((\sigma_{min}^+(\mathcal{A}))^2, (\lambda_{min}^+(W))^2)}.$$

Now we prove the theorem about complexity of Similar Triangles Method.

Proof of Theorem 2

Proof. First, note that solution accuracy ε for problem (1) is equivalent to accuracy $m\varepsilon$ for problem (8). STM requires $O((L_H R_{dual}^2/(m\varepsilon))^{1/2})$ iterations to reach ε-accuracy. Combining the results from Lemmas 1 and 2 we obtain the final complexity.

4.2   Accelerated Block-Coordinate Method

In previous section, our approach was based on a way where we apply a first-order method without separating the variables. But we can treat variable blocks **z** and **s** separately and get a better convergence bound. We apply an accelerated method ACRCD (Accelerated by Coupling Randomized Coordinate Descent) from [5]. We describe the result only for the case $p = 1$. In this case, we apply ACRCD not to regularized dual problem (8), but to constrained version of dual problem (6). We also note that ACRCD is primal-dual, so solving the dual problem with accuracy ε is sufficient to restore the solution of the primal with accuracy ε.

Algorithm 2. ACRCD

Require: Define coefficients $\alpha_{k+1} = \frac{k+2}{8}$, $\tau_k = \frac{2}{k+2}$. Choose stepsizes $L_{\mathbf{z}}$, $L_{\mathbf{s}}$. Put
$\overline{\mathbf{z}}^0 = \underline{\mathbf{z}}^0 = \mathbf{z}^0$, $\overline{\mathbf{s}}^0 = \underline{\mathbf{s}}^0 = \mathbf{s}^0$.

1: **for** $k = 0, 1, \ldots, N-1$ **do**
2:     $\mathbf{z}^{k+1} = \tau_k \underline{\mathbf{z}}^k + (1 - \tau_k)\overline{\mathbf{z}}^k$
3:     $\mathbf{s}^{k+1} = \tau_k \underline{\mathbf{s}}^k + (1 - \tau_k)\overline{\mathbf{s}}^k$
4:     Put $\xi_i = 1$ with probability η and $\xi = 0$ with probability $(1 - \eta)$, where $\eta = \frac{\lambda_{\max}(W)}{\lambda_{\max}(W) + \sigma_{\max}(\mathcal{A})}$
5:     **if** $\xi_i = 1$ **then**
6:         $\overline{\mathbf{z}}^{k+1} = \mathbf{z}^{k+1} - \frac{1}{L_{\mathbf{z}}}\nabla H_{\mathbf{z}}(\mathbf{z}^{k+1}, \mathbf{s}^{k+1})$
7:         $\underline{\mathbf{z}}^{k+1} = \underline{\mathbf{z}}^k - \frac{2\alpha_{k+1}}{L_{\mathbf{z}}}\nabla H_{\mathbf{z}}(\mathbf{z}^{k+1}, \mathbf{s}^{k+1})$
8:     **else**
9:         $\overline{\mathbf{s}}^{k+1} = \Pi_{[-1,1]^{mn}}\left[\mathbf{s}^{k+1} - \frac{1}{L_{\mathbf{s}}}\nabla H_{\mathbf{s}}(\mathbf{z}^{k+1}, \mathbf{s}^{k+1})\right]$
10:        $\underline{\mathbf{s}}^{k+1} = \Pi_{[-1,1]^{mn}}\left[\underline{\mathbf{s}}^k - \frac{2\alpha_{k+1}}{L_{\mathbf{s}}}\nabla H_{\mathbf{s}}(\mathbf{z}^{k+1}, \mathbf{s}^{k+1})\right]$
11:    **end if**
12: **end for**

Theorem 2. *To reach accuracy ε of the solution with probability at least $(1-\delta)$, Algorithm 2 requires N_{comm} communication rounds and N_{comp} local computations, where*

$$N_{comm} = \frac{m^{1/4}}{\sqrt{\theta\varepsilon}}\frac{\lambda_{\max}(W)}{\lambda_{\min}^+(W)}\left(2\theta^2\left\|\log \mathbf{x}^* + \mathbf{1}_d\right\|_2^2 + 2n\sigma_{\max}^2(\mathcal{A}) + n(\lambda_{\min}^+(W))^2\right)^{1/2}\log\left(\frac{1}{\delta}\right),$$

$$N_{comp} = \frac{m^{1/4}}{\sqrt{\theta\varepsilon}}\frac{\sigma_{\max}(\mathcal{A})}{\lambda_{\min}^+(W)}\left(2\theta^2\left\|\log \mathbf{x}^* + \mathbf{1}_d\right\|_2^2 + 2n\sigma_{\max}^2(\mathcal{A}) + n(\lambda_{\min}^+(W))^2\right)^{1/2}\log\left(\frac{1}{\delta}\right).$$

First, we need to estimate Lipschitz constants for gradients of each block of variables. If we consider the function H as a function of two blocks of variables, the next result follows.

Lemma 3. *Function $H(\mathbf{z}, \mathbf{s})$ has a $L_{\mathbf{z}}$-Lipschitz gradient w.r.t. $\mathbf{z}$ and $L_{\mathbf{s}}$-Lipschitz gradient w.r.t. $\mathbf{s}$, where*

$$L_{\mathbf{z}} = \frac{\sqrt{m}\sigma_{\max}^2(W)}{\theta}, \quad L_{\mathbf{s}} = \frac{\sqrt{m}\sigma_{\max}^2(\mathcal{A})}{\theta}.$$

Proof. Recall that we denoted $\mathbf{x} = \mathrm{col}[x_1, \ldots, x_m]$ and consider $\mathbf{s}_1, \mathbf{s}_2 \in \mathbb{R}^{nm}$. Also denote $[\mathbf{x}]_i = E_i \mathbf{x} = x_i$.

$$
\begin{aligned}
&\|\nabla_{\mathbf{s}} H(\mathbf{z}, \mathbf{s}_2) - \nabla_{\mathbf{s}} H(\mathbf{z}, \mathbf{s}_1)\|_2 \\
&= \|\mathbf{A}\nabla G^*(-\mathbf{Wz} - \mathbf{A}^\top \mathbf{s}_2) - \mathbf{A}\nabla G^*(-\mathbf{Wz} - \mathbf{A}^\top \mathbf{s}_1)\|_2 \\
&\leq \sum_{i=1}^m \left\| A_i \nabla g^* \left(-[\mathbf{Wz}]_i - [\mathbf{A}^\top \mathbf{s}_2]_i\right) - A_i \nabla g^* \left(-[\mathbf{Wz}]_i - [\mathbf{A}^\top \mathbf{s}_1]_i\right) \right\|_2 \\
&\overset{①}{=} \sum_{i=1}^m \left\| A_i \nabla g^* \left(-[\mathbf{Wz}]_i - A_i^\top [\mathbf{s}_2]_i\right) - A_i \nabla g^* \left(-[\mathbf{Wz}]_i - A_i^\top [\mathbf{s}_1]_i\right) \right\|_2 \\
&\leq \frac{\sigma_{\max}(\mathcal{A})}{\theta} \sum_{i=1}^m \left\| A_i^\top [\mathbf{s}_2]_i - A_i^\top [\mathbf{s}_1]_i \right\|_2 \leq \frac{\sigma_{\max}^2(\mathcal{A})}{\theta} \sum_{i=1}^m \left\| [\mathbf{s}_2]_i - [\mathbf{s}_1]_i \right\|_2 \\
&\overset{②}{\leq} \frac{\sqrt{m} \sigma_{\max}^2(\mathcal{A})}{\theta} \|\mathbf{s}_2 - \mathbf{s}_1\|_2,
\end{aligned}
$$

where ① holds due to the structure of $\mathbf{A} = \mathrm{diag}[A_1, \ldots, A_m]$ and ② holds by convexity of the 2-norm.

Now consider the gradient w.r.t. $\mathbf{z}$. Let $[\mathbf{x}]^{(i)} = [x_1^{(i)} \ldots x_m^{(i)}]^\top$ denote a vector consisting of i-th components of $x_1, \ldots, x_m$. We have $[\mathbf{Wx}]_i = W[\mathbf{x}]^{(i)}$ due to the structure of $\mathbf{W} = W \otimes \mathbf{I}_d$.

$$
\begin{aligned}
&\|\nabla H_{\mathbf{z}}^*(\mathbf{z}_2, \mathbf{s}) - \nabla H_{\mathbf{z}}^*(\mathbf{z}_1, \mathbf{s})\|_2 \\
&= \|\mathbf{W}\nabla G^*(-\mathbf{Wz}_2 - \mathbf{A}^\top \mathbf{s}) - \mathbf{W}\nabla G^*(-\mathbf{Wz}_1 - \mathbf{A}^\top \mathbf{s})\|_2 \\
&\leq \sum_{i=1}^m \left\| W\nabla g^* \left(-W[\mathbf{z}_2]^{(i)} - [\mathbf{A}^\top \mathbf{s}]_i\right) - W\nabla g^* \left(-W[\mathbf{z}_1]^{(i)} - [\mathbf{A}^\top \mathbf{s}]_i\right) \right\|_2 \\
&\leq \frac{\lambda_{\max}(W)}{\theta} \sum_{i=1}^m \left\| W([\mathbf{z}_2]^{(i)} - [\mathbf{z}_1]^{(i)}) \right\|_2 \leq \frac{\lambda_{\max}^2(W)}{\theta} \sum_{i=1}^m \left\| [\mathbf{z}_2]^{(i)} - [\mathbf{z}_1]^{(i)} \right\|_2 \\
&\overset{①}{\leq} \frac{\sqrt{m} \lambda_{\max}^2(W)}{\theta} \|\mathbf{z}_2 - \mathbf{z}_1\|_2,
\end{aligned}
$$

where ① holds by convexity of the 2-norm.

We need to bound dual distance for each block of variables, but first we need to claim an useful proposition from functional analysis.

Proposition 1. *Let $p > r \geq 1, x \in \mathbb{R}^d$. It holds*

$$
\|x\|_p \leq \|x\|_r \leq d^{\left(\frac{1}{r} - \frac{1}{p}\right)} \|x\|_p.
$$

Proof. This is a fairly well-known fact with a simple idea of proof. In fact, it is a direct consequence of Hólder's inequality, what means that constant in an inequality are unimprovable.

Now we derive the bound on the norm of the dual solution. The convergence result only relies on the case $p = 1$ ($q = \infty$), but we derive a bound for any $q \geq 1$.

Lemma 4. *Let* $\mathbf{z}^*, \mathbf{s}^*$ *be the solutions of dual problem* (8) *and let* x^* *be a solution of* (1). *It holds*

$$\|\mathbf{z}^*\|_2^2 \le R_{\mathbf{z}}^2 = \frac{2\theta^2 m\|\log x^* + \mathbf{1}_d\|_2^2 + 2\sigma_{\max}^2(\mathcal{A}) \cdot \max\left(1, (mn)^{\left(1-\frac{2}{q}\right)}\right)}{(\lambda_{min}^+(W))^2}.$$

$$\|\mathbf{s}^*\|_2^2 \le R_{\mathbf{s}}^2 = \max\left(1, (mn)^{\left(1-\frac{2}{q}\right)}\right).$$

Proof. Using that the problems (6) and (7) are equal, that means

$$\|\mathbf{s}^*\|_q \le 1. \tag{9}$$

Using Proposition 1 we have

$$\|\mathbf{s}^*\|_2^2 \le \begin{cases} \|\mathbf{s}^*\|_q^2, & q < 2 \\ (mn)^{\left(1-\frac{2}{q}\right)}\|\mathbf{s}^*\|_q^2, & q \ge 2 \end{cases} \tag{10}$$

Combining (9) and (10), we have

$$\|\mathbf{s}^*\|_2^2 \le \begin{cases} 1, & q < 2 \\ (mn)^{\left(1-\frac{2}{q}\right)}, & q \ge 2 \end{cases} \tag{11}$$

With the fact that $(mn)^{\left(1-\frac{2}{q}\right)} < 1$ where $q < 2$ we state the claimed result. Using the fact from proof of Lemma 2 such that

$$-\mathbf{W}\mathbf{z}^* - \mathbf{A}^\top \mathbf{s}^* = \nabla G(\mathbf{x}^*).$$

we have

$$\|\mathbf{W}\mathbf{z}^*\|_2^2 \le 2\|\nabla G(\mathbf{x}^*)\|_2^2 + 2\|\mathbf{A}^\top \mathbf{s}^*\|_2^2 \le 2\theta^2 m\|\log x^* + \mathbf{1}_d\|_2^2 + 2\sigma_{\max}^2(\mathcal{A}) \cdot \|\mathbf{s}^*\|_2^2.$$

As a result

$$\|\mathbf{z}^*\|_2^2 \le \frac{2\theta^2 m\|\log x^* + \mathbf{1}_d\|_2^2 + 2\sigma_{\max}^2(\mathcal{A}) \cdot \|\mathbf{s}^*\|_2^2}{(\lambda_{min}^+(W))^2}. \tag{12}$$

Substituting (11) into (12), we claim the final result.

Proof (Proof of Theorem 2). The proof is based on results in [5]. We have two blocks of variables: $\mathbf{z}$ and $\mathbf{s}$. Firstly, Remark 3 of [5] shows that a block coordinate method is applicable to constrained problems, provided that the constraint set is separable over variable blocks. Secondly, we apply Remark 6 of the same paper with coefficient $\beta = 1/2$. At each step, we randomly choose one of two variable blocks, and factor β rules the probability distribution. In Algorithm 2, the probability of choosing block $\mathbf{z}$ is $\eta = \sqrt{L_{\mathbf{z}}}/(\sqrt{L_{\mathbf{z}}} + \sqrt{L_{\mathbf{s}}})$, and block $\mathbf{s}$ is chosen with probability $(1 - \eta)$. Recall that for accuracy ε in primal problem (1) we need accuracy $m\varepsilon$ in dual problem (6). Combining the two remarks, we obtain that a resulting method makes N iterations to reach ε accuracy with probability at least $1 - \delta$, where

$$N = O\left(\left(\sqrt{L_z} + \sqrt{L_s}\right)\sqrt{\frac{R_z^2 + R_s^2}{m\varepsilon}}\log\left(\frac{1}{\delta}\right)\right).$$

Consequently, the expected number of computations of ∇H_z (that equals the number of communications) is ηN, and the expected number of computations of ∇H_s (that corresponds to the number of local computations) is $(1 - \eta)N$. Substituting the expressions for N and η, we obtain the desired result.

5   Conclusion

In this paper, we considered a particular class of non-smooth decentralized problems. Due to specific problem structure we obtained methods that have a better dependency on problem complexity than general lower bounds. Our approach is based on passing to the dual problem. Moreover, we proposed two accelerated algorithms. The first algorithm is an accelerated primal-dual gradient method that is directly applied to the problem. The second method is a block-coordinate algorithm that allows to split communication and computation complexities.

References

1. Anikin, A., et al.: Modern efficient numerical approaches to regularized regression problems in application to traffic demands matrix calculation from link loads. In: Proceedings of International Conference ITAS-2015. Russia, Sochi (2015)
2. Boyd, S., Parikh, N., Chu, E., Peleato, B., Eckstein, J.: Distributed optimization and statistical learning via the alternating direction method of multipliers. Found. Trends Mach. Learn. **3**(1), 1–122 (2011)
3. Boyd, S., Vandenberghe, L.: Convex Optimization. Cambridge University Press, NY (2004)
4. Dvurechensky, P., Gasnikov, A., Omelchenko, S., Tiurin, A.: Adaptive similar triangles method: a stable alternative to Sinkhorn's algorithm for regularized optimal transport. arXiv preprint arXiv:1706.07622 (2017)
5. Gasnikov, A., Dvurechensky, P., Usmanova, I.: On nontriviality of fast (accelerated) randomized methods. Proc. Moscow Inst. Phys. Technol. **82**(30), 67–100 (2016)
6. Kakade, S., Shalev-Shwartz, S., Tewari, A.: On the duality of strong convexity and strong smoothness: learning applications and matrix regularization (2009). Unpublished Manuscript. http://ttic.uchicago.edu/shai/papers/KakadeShalevTewari09.pdf
7. Kovalev, D., Gasanov, E., Gasnikov, A., Richtarik, P.: Lower bounds and optimal algorithms for smooth and strongly convex decentralized optimization over time-varying networks. Adv. Neural. Inf. Process. Syst. **34**, 22325–22335 (2021)
8. Kovalev, D., Salim, A., Richtárik, P.: Optimal and practical algorithms for smooth and strongly convex decentralized optimization. Adv. Neural. Inf. Process. Syst. **33**, 18342–18352 (2020)

9. Kovalev, D., Shulgin, E., Richtárik, P., Rogozin, A.V., Gasnikov, A.: ADOM: accelerated decentralized optimization method for time-varying networks. In: International Conference on Machine Learning, PMLR, pp. 5784–5793 (2021)
10. Li, G., Chen, Y., Chi, Y., Poor, H.V., Chen, Y.: Fast computation of optimal transport via entropy-regularized extragradient methods. arXiv preprint arXiv:2301.13006 (2023)
11. Li, H., Lin, Z.: Accelerated gradient tracking over time-varying graphs for decentralized optimization. arXiv preprint arXiv:2104.02596 (2021)
12. Nedić, A., Ozdaglar, A.: Distributed subgradient methods for multi-agent optimization. IEEE Trans. Autom. Control 54(1), 48–61 (2009)
13. Scaman, K., Bach, F., Bubeck, S., Lee, Y.T., Massoulié, L.: Optimal algorithms for smooth and strongly convex distributed optimization in networks. In: Proceedings of the 34th International Conference on Machine Learning, vol. 70, pp. 3027–3036 (2017)
14. Scaman, K., Bach, F., Bubeck, S., Massoulié, L., Lee, Y.T.: Optimal algorithms for non-smooth distributed optimization in networks. Adv. Neural. Inf. Process. Syst. 31, 2740–2749 (2018)
15. Zhang, Y., Roughan, M., Lund, C., Donoho, D.L.: Estimating point-to-point and point-to-multipoint traffic matrices: an information-theoretic approach. IEEE/ACM Trans. Netw. 13(5), 947–960 (2005)

Byzantine-Robust Loopless Stochastic Variance-Reduced Gradient

Nikita Fedin[1] and Eduard Gorbunov[2(✉)]

[1] Moscow Institute of Physics and Technology, Dolgoprudny, Russia
`fedin.ng@phystech.edu`
[2] Mohamed bin Zayed University of Artificial Intelligence, Abu Dhabi, UAE
`eduard.gorbunov@mbzuai.ac.ae`

Abstract. Distributed optimization with open collaboration is a popular field since it provides an opportunity for small groups/companies/universities, and individuals to jointly solve huge-scale problems. However, standard optimization algorithms are fragile in such settings due to the possible presence of so-called Byzantine workers – participants that can send (intentionally or not) incorrect information instead of the one prescribed by the protocol (e.g., send anti-gradient instead of stochastic gradients). Thus, the problem of designing distributed methods with provable robustness to Byzantine workers has been receiving a lot of attention recently. In particular, several works consider a very promising way to achieve Byzantine tolerance via exploiting variance reduction and robust aggregation. The existing approaches use SAGA- and SARAH-type variance reduced estimators, while another popular estimator – SVRG – is not studied in the context of Byzantine-robustness. In this work, we close this gap in the literature and propose a new method – Byzantine-Robust Loopless Stochastic Variance Reduced Gradient (BR-LSVRG). We derive non-asymptotic convergence guarantees for the new method in the strongly convex case and compare its performance with existing approaches in numerical experiments.

Keywords: Distributed optimization · Byzantine-robustness · Variance reduction · Stochastic optimization

1 Introduction

In this work, we consider a finite-sum minimization problem

$$\min_{x \in \mathbb{R}^d} \left\{ f(x) = \frac{1}{m} \sum_{j=1}^{m} f_j(x) \right\}. \tag{1}$$

Such problem formulations are very typical for machine learning tasks [9,34], where x represents the model parameters and $f_j(x)$ denotes the loss on the j-th

The research was supported by the Ministry of Science and Higher Education of the Russian Federation (Goszadaniye) 075-00337-20-03, project No. 0714-2020-0005.

element of the dataset. In modern problems of this type, the dataset size m and the dimension of the problem d are typically very large, e.g., several billion [29]. Training of such models on 1 (even very powerful) machine can take years of computations [23]. Therefore, it is inevitable to use distributed (stochastic) approaches to solve such complicated problems, e.g., Parallel Stochastic Gradient Descent (Parallel-SGD) [33, 38].

Distributed optimization is associated with a number of difficulties related to communication efficiency, data privacy, asynchronous updates, and many other aspects that depend on the setup. One of such aspects is the robustness to *Byzantine workers*[1] – the workers that can (intentionally or not) deviate from the prescribed protocol and are assumed to be omniscient (see more details in Sect. 1.1). Byzantine workers can easily destroy the convergence of standard methods based on simple averaging since such workers can send to the server arbitrary vectors. This fact justifies the usage of special methods that are robust to Byzantine attacks.

In particular, one of the existing techniques to achieve Byzantine-robustness is based on the variance reduction mechanism [13]. The key idea behind this approach is based on the fact that variance reduction of stochastic gradients received from regular workers reduces the strengths of Byzantine attacks since it becomes harder to "hide in the noise" for Byzantine workers and easier for the server to reduce the effect of Byzantine attacks. This idea led to the development of such variance-reduced Byzantine-robust methods as Byrd-SAGA [35], which uses celebrated SAGA estimator [6] and geometric median for aggregation, and Byz-VR-MARINA [12], which is based on SARAH estimator [28] and any agnostic robust aggregation [20]. However, there exists no Byzantine-robust version of another popular variance-reduced method called – Stochastic Variance-Reduced Gradient (SVRG) [18]. Moreover, in view of the vulnerability of geometric median to special Byzantine attacks [2, 19, 36], it remains unclear how unbiased variance-reduced estimators (like SVRG/SAGA-estimators) behave in combination with provably robust aggregation rules [19, 20] – the authors [12] focus on biased variance reduction only.

Contributions. We propose a new method called Byzantine-Robust Loopless Stochastic Variance-Reduced Gradient (BR-LSVRG) that uses SVRG-estimator and (provably) robust aggregation rule. We analyze the method for solving smooth strongly convex distributed optimization problems and prove its theoretical convergence. Though our results require the usage of large enough batchsizes, we show that in certain scenarios BR-LSVRG has better convergence guarantees than both Byrd-SAGA and Byz-VR-MARINA. In addition, we study the convergence of BR-LSVRG in several numerical experiments and observe that (i) BR-LSVRG can reach a good accuracy of the solution even with small batchsize and (ii) BR-LSVRG converges better than Byrd-SAGA.

---

[1] This term takes its origin in [22] and has become standard in the literature [26]. By using this term, we do not want to offend any group of people but rather follow standard notation for the community.

1.1 Technical Preliminaries

Notation. We denote the standard Euclidean inner product in $\mathbb{R}^d$ as $\langle x, y \rangle \overset{\text{def}}{=}$ $\sum_{i=1}^d x_i y_i$, where $x = (x_1, \ldots, x_d)^\top, y = (y_1, \ldots, y_d)^\top \in \mathbb{R}^d$ and ℓ_2-norm as $\|x\| \overset{\text{def}}{=} \sqrt{\langle x, x \rangle}$. For any integer $t > 0$ we use $[t]$ to define set $\{1, 2, \ldots, t\}$. Finally, $\mathbb{E}[\cdot]$ denotes full expectation and $\mathbb{E}_k[\cdot]$ denotes the expectation w.r.t. the randomness coming from iteration k.

Byzantine Workers. We assume that the distributed system consists of n workers $[n]$ connected with parameter-server. Each worker can compute gradients of $\nabla f_j(x)$ for any $j \in [m]$ and $x \in \mathbb{R}^d$. Moreover, we assume that workers consist of two groups: $[n] = \mathcal{G} \cup \mathcal{B}$, $\mathcal{G} \cap \mathcal{B} = \varnothing$. Here $\mathcal{G}$ denotes the set of regular workers, and $\mathcal{B}$ is the set of so-called *Byzantine workers*, i.e., the workers that can (intentionally or not) send arbitrary vectors to the server instead of ones prescribed by the algorithm. Moreover, following the classical convention, we assume that Byzantine workers can be omniscient, meaning that they can know exactly what other workers send to the server and what aggregation rule the server uses. Although this assumption is strong, it is quite popular due to the following argument: if the method is robust to the presence of Byzantine workers in these settings, this method is guaranteed to be robust in scenarios when Byzantine workers are less harmful. In addition, one has to assume that $|\mathcal{G}| = G \geq (1 - \delta)n$ or equivalently $|\mathcal{B}| = B \leq \delta n$ for some $\delta < 1/2$ (otherwise Byzantine workers form a majority and provable Byzantine-robustness cannot be achieved in the worst case [20]).

Robust Aggregation. Following [12,20], we use the following definition of the robust aggregator/aggregation.

Definition 1 ((δ, c)-robust aggregator [12,20]). *Let* $x_1, x_2, \ldots, x_n$ *be such that for some subset* $\mathcal{G} \subseteq [n]$ *of size* $|\mathcal{G}| = G \geq (1 - \delta)n$, $\delta < 1/2$ *there exists* $\sigma \geq 0$ *such that* $\frac{1}{G(G-1)} \sum_{i,l \in \mathcal{G}} \mathbb{E}\|x_i - x_l\|^2 \leq \sigma^2$. *The quantity* $\widehat{x}$ *is called* (δ, c)-*robust aggregator* $((\delta, c)$-*RAgg) and denoted as* $\widehat{x} = RAgg(x_1, \ldots, x_n)$ *for some number* $c > 0$ *if the following holds:*

$$\mathbb{E}\|\widehat{x} - \overline{x}\|^2 \leq c\delta\sigma^2, \tag{2}$$

where $\overline{x} = \frac{1}{G} \sum_{i \in \mathcal{G}} x_i$. *In addition, if* $\widehat{x}$ *can be computed without the knowledge of* σ^2, *then* $\widehat{x}$ *is called* (δ, c)-*agnostic robust aggregator* $((\delta, c)$-*ARAgg) and denoted as* $\widehat{x} = ARAgg(x_1, \ldots, x_n)$.

In other words, the aggregator $\widehat{x}$ is called robust if, on average, it is "not far" from $\overline{x}$ – the average of the vectors from regular workers $\mathcal{G}$. Here, the upper bound on how far we allow the robust aggregator to be from the average over regular workers depends on the variance of regular workers and the ratio of Byzantines. It is relatively natural that both characteristics should affect the quality of the aggregation. Moreover, there exists a lower bound stating that for any aggregation rule $\widehat{x}$ there exists a set of vectors satisfying the conditions of the above definition such that $\mathbb{E}\|\widehat{x} - \overline{x}\|^2 = \Omega(\delta\sigma^2)$, which formally establishes the tightness of Definition 1. We provide several examples of robust aggregators in Appendix A.

Assumptions. We make a standard assumption for the analysis of variance-reduced methods [21].

Assumption 1 *Functions* $f_1, \ldots, f_m : \mathbb{R}^d \to \mathbb{R}$ *are convex and L-smooth (L >* 0*), and function* $f : \mathbb{R}^d \to \mathbb{R}$ *is additionally* μ*-strongly convex (*$\mu > 0$*), i.e., for all* $x, y \in \mathbb{R}^d$

$$f_j(y) \geq f_j(x) + \langle \nabla f_j(x), y - x \rangle \quad \forall j \in [m], \tag{3}$$

$$\|\nabla f_j(x) - \nabla f_j(y)\| \leq L\|x - y\| \quad \forall j \in [m], \tag{4}$$

$$f(y) \geq f(x) + \langle \nabla f(x), y - x \rangle + \frac{\mu}{2}\|y - x\|^2. \tag{5}$$

For one of our results, we make the following additional assumption, which is standard for the stochastic optimization literature [1,8].

Assumption 2 *We assume that there exists* $\sigma \geq 0$ *such that for any* $x \in \mathbb{R}^d$ *and* j *being sampled uniformly at random from* $[m]$

$$\mathbb{E}\|\nabla f_j(x) - \nabla f(x)\|^2 \leq \sigma^2. \tag{6}$$

1.2   Related Work

Many existing methods for Byzantine-robust distributed optimization are based on the replacement of averaging with special aggregation rules in Parallel-SGD [3,5,14,30,37]. As it is shown in [2,36], such approaches are not Byzantine-robust and can even perform worse than naïve Parallel-SGD for particular Byzantine-attacks. To circumvent this issue, the authors of [19] introduce a formal definition of robust aggregator (see Definition 1) and propose the first distributed methods with provable Byzantine-robustness. The key ingredient in their method is client heavy-ball-type momentum [31] to make the method non-permutation-invariant that prevents the algorithm from time-coupled Byzantine attacks. In [20], this technique is generalized to heterogeneous problems and robust aggregation agnostic to the noise. An extension to decentralized optimization problems is proposed by [15]. Another approach that ensures Byzantine-robustness both in theory and practice is based on the checks of computations at random moments of time [10]. Finally, there are two approaches based on variance reduction[2] mechanism – Byrd-SAGA [35], which uses well-suited for convex problems SAGA-estimator [7,17,24,28], and Byz-VR-MARINA [12], which employs well-suited for non-convex problems SARAH-estimator. We refer to [10,26] for the extensive summaries of other existing approaches.

---

[2] See [13] for a recent survey on variance-reduced methods.

Algorithm 1. Byzantine-Robust Distributed LSVRG (BR-LSVRG)

1: **Input:** stepsize $\gamma > 0$, batchsize $b \geq 1$, starting point x^0, probability $p \in (0,1]$, (δ, c)-agnostic robust aggregator $\texttt{ARAgg}(x_1, \ldots, x_n)$, number of iterations $K > 0$
2: Set $w_i^0 = x^0$ and compute $\nabla f(w_i^0)$ for all $i \in \mathcal{G}$
3: **for** $k = 0, 1, \ldots, K-1$ **do**
4:      Server sends x^k to all workers
5:      **for all** $i \in \mathcal{G}$ **do**
6:          Choose $j_{i,k}^1, j_{i,k}^2, \ldots, j_{i,k}^b$ from $[m]$ uniformly at random independently from each other and other workers
7:          $g_i^k = \frac{1}{b} \sum\limits_{t=1}^{b} \left(\nabla f_{j_{i,k}^t}(x^k) - \nabla f_{j_{i,k}^t}(w_i^k) \right) + \nabla f(w_i^k)$
8:          $w_i^{k+1} = \begin{cases} x^k, & \text{with probability } p \\ w_i^k, & \text{with probability } 1 - p \end{cases}$
9:          Send g_i^k to the server
10:      **end for**
11:      **for all** $i \in \mathcal{B}$ **do**
12:          Send $g_i^k = *$ (anything) to the server
13:      **end for**
14:      Server receives vectors $g_1^k, \ldots, g_n^k$ from the workers
15:      Server computes $x^{k+1} = x^k - \gamma \cdot \texttt{ARAgg}(g_1^k, \ldots, g_n^k)$
16: **end for**
17: **Output:** x^K

---

2 Main Results

In this section, we introduce the new method called Byzantine-Robust distributed Loopless Stochastic Variance-Reduced Gradient (BR-LSVRG, Algorithm 1). At each iteration of BR-LSVRG, regular workers compute standard SVRG-estimator [18] (line 7) and send it to the server. The algorithm has two noticeable features. First, unlike many existing distributed methods that use averaging or other aggregation rules vulnerable to Byzantine attacks, BR-LSVRG uses a provably robust aggregator (according to Definition 1) on the server. Secondly, following the idea of [16,21], in BR-LSVRG, regular workers update the reference point w_i^{k+1} as x^k with some small probability p. When $w_i^{k+1} = x^k$, worker i has to compute the full gradient during the next step in order to calculate g_i^k; otherwise, only $2b$ gradients of the summands from (1) need to be computed. To make the expected computation cost (number of computed gradients of the summands from (1)) of 1 iteration on each regular worker to be $\mathcal{O}(b)$, probability p is chosen as $p \sim b/m$.

We start the theoretical convergence analysis with the following result.

Theorem 1. *Let Assumptions 1 and 2 hold, batchsize $b \geq 1$, and stepsize $0 < \gamma \leq \min\{1/12L, p/\mu\}$. Then, the iterates produced by* BR-LSVRG *after K iterations satisfy*

$$\mathbb{E}\Psi_K \leq \left(1 - \frac{\gamma\mu}{2}\right)^K \Psi_0 + \gamma\frac{32c\delta\sigma^2}{b\mu} + \frac{32c\delta\sigma^2}{b\mu^2}, \tag{7}$$

where $\Psi_k \overset{def}{=} \|x^k - x^*\|^2 + \frac{8\gamma^2}{p}\sigma_k^2$, $\sigma_k^2 \overset{def}{=} \frac{1}{Gm}\sum_{i \in \mathcal{G}}\sum_{j=1}^{m}\|\nabla f_j(w_i^k) - \nabla f_j(x^*)\|^2$.

Proof. For convenience, we introduce new notation: $\widehat{g}^k = \texttt{ARAgg}(g_1^k, \ldots, g_n^k)$. Then, $x^{k+1} = x^k - \gamma \widehat{g}^k$ and

$$\|x^{k+1} - x^*\|^2 = \|x^k - x^*\|^2 - 2\gamma \langle x^k - x^*, \overline{g}^k \rangle - 2\gamma \langle x^k - x^*, \widehat{g}^k - \overline{g}^k \rangle + \gamma^2 \|\overline{g}^k + (\widehat{g}^k - \overline{g}^k)\|^2.$$

Next, we apply inequalities $\langle a, b \rangle \leq \frac{\alpha}{2}\|a\|^2 + \frac{1}{2\alpha}\|b\|^2$ and $\|a+b\|^2 \leq 2\|a\|^2 + 2\|b\|^2$, which hold for any $a, b \in \mathbb{R}^d$, $\alpha > 0$, to the last two terms and take expectation $\mathbb{E}_k[\cdot]$ from both sides of the above inequality (note that $\mathbb{E}_k[\overline{g}^k] = \nabla f(x^k)$)

$$\mathbb{E}_k \|x^{k+1} - x^*\|^2 \leq \left(1 + \frac{\gamma\mu}{2}\right) \|x^k - x^*\|^2 - 2\gamma \langle x^k - x^*, \nabla f(x^k) \rangle + 2\gamma^2 \mathbb{E}_k \|\overline{g}^k\|^2$$

$$+ 2\gamma \left(\frac{1}{\mu} + \gamma\right) \mathbb{E}_k \|\widehat{g}^k - \overline{g}^k\|^2$$

$$\overset{(5)}{\leq} \left(1 - \frac{\gamma\mu}{2}\right) \|x^k - x^*\|^2 - 2\gamma \left(f(x^k) - f(x^*)\right)$$

$$+ 2\gamma^2 \mathbb{E}_k \|\overline{g}^k\|^2 + 2\gamma \left(\frac{1}{\mu} + \gamma\right) \mathbb{E}_k \|\widehat{g}^k - \overline{g}^k\|^2. \tag{8}$$

To proceed with the derivation, we need to upper bound the last two terms from the above inequality. For the first term, we use Jensen's inequality and the well-known fact that the variance is not larger than the second moment:

$$\mathbb{E}_k \|\overline{g}^k\|^2 \leq \frac{1}{bG} \sum_{i \in \mathcal{G}} \sum_{t=1}^{b} \mathbb{E}_k \|\nabla f_{j_{i,k}^t}(x^k) - \nabla f_{j_{i,k}^t}(w_i^k) + \nabla f(w_i^k)\|^2$$

$$= \frac{1}{G} \sum_{i \in \mathcal{G}} \mathbb{E}_k \|\nabla f_{j_{i,k}^1}(x^k) - \nabla f_{j_{i,k}^1}(w_i^k) + \nabla f(w_i^k)\|^2$$

$$\leq \frac{2}{G} \sum_{i \in \mathcal{G}} \mathbb{E}_k \|\nabla f_{j_{i,k}}(x^k) - \nabla f_{j_{i,k}}(x^*)\|^2$$

$$+ \frac{2}{G} \sum_{i \in \mathcal{G}} \mathbb{E}_k \|\nabla f_{j_{i,k}^1}(w_i^k) - \nabla f_{j_{i,k}^1}(x^*) - \nabla f(w_i^k)\|^2$$

$$\leq \frac{2}{G} \sum_{i \in \mathcal{G}} \mathbb{E}_k \left[\|\nabla f_{j_{i,k}^1}(x^k) - \nabla f_{j_{i,k}^1}(x^*)\|^2 + \|\nabla f_{j_{i,k}^1}(w_i^k) - \nabla f_{j_{i,k}^1}(x^*)\|^2\right]$$

$$= \frac{2}{Gm} \sum_{i \in \mathcal{G}} \sum_{j=1}^{m} \left(\|\nabla f_j(x^k) - \nabla f_j(x^*)\|^2 + \|\nabla f_j(w_i^k) - \nabla f_j(x^*)\|^2\right)$$

$$\overset{(*)}{\leq} \frac{4L}{Gm} \sum_{i \in \mathcal{G}} \sum_{j=1}^{m} \left(f_j(x^k) - f_j(x^*) - \langle \nabla f_j(x^*), x^k - x^* \rangle\right) + 2\sigma_k^2$$

$$= 4L \left(f(x^k) - f(x^*)\right) + 2\sigma_k^2, \tag{9}$$

where in $(*)$ we use that for any convex L-smooth function $h(x)$ and any $x, y \in \mathbb{R}^d$ (e.g., see [27])

$$\|\nabla h(x) - \nabla h(y)\|^2 \leq 2L \left(h(x) - h(y) - \langle \nabla h(y), x - y \rangle\right). \tag{10}$$

To bound the last term from (8), we notice that Assumption 2 gives $\forall i, l \in \mathcal{G}$

$$\mathbb{E}_k \| g_i^k - g_l^k \|^2$$

$$= \mathbb{E}_k \| g_i^k - \nabla f(x^k) \|^2 + \mathbb{E}_k \| g_l^k - \nabla f(x^k) \|^2 + 2\mathbb{E}_k \left[\langle g_i^k - \nabla f(x^k), g_l^k - \nabla f(x^k) \rangle \right]$$

$$= \mathbb{E}_k \left\| \frac{1}{b} \sum_{t=1}^{b} (\nabla f_{j_{i,k}^t}(x^k) - \nabla f(x^k) - (\nabla f_{j_{i,k}^t}(w_i^k) - \nabla f(w_i^k))) \right\|^2$$

$$+ \mathbb{E}_k \left\| \frac{1}{b} \sum_{t=1}^{b} (\nabla f_{j_{l,k}^t}(x^k) - \nabla f(x^k) - (\nabla f_{j_{l,k}^t}(w_l^k) - \nabla f(w_l^k))) \right\|^2$$

$$\leq 2\mathbb{E}_k \left[\left\| \frac{1}{b} \sum_{t=1}^{b} (\nabla f_{j_{i,k}}(x^k) - \nabla f(x^k)) \right\|^2 + \left\| \frac{1}{b} \sum_{t=1}^{b} (\nabla f_{j_{i,k}}(w_i^k) - \nabla f(w_i^k)) \right\|^2 \right]$$

$$+ 2\mathbb{E}_k \left[\left\| \frac{1}{b} \sum_{t=1}^{b} (\nabla f_{j_{l,k}}(x^k) - \nabla f(x^k)) \right\|^2 + \left\| \frac{1}{b} \sum_{t=1}^{b} (\nabla f_{j_{l,k}}(w_l^k) - \nabla f(w_l^k)) \right\|^2 \right]$$

$$\leq \frac{2}{b^2} \sum_{t=1}^{b} \mathbb{E}_k \left[\| \nabla f_{j_{i,k}}(x^k) - \nabla f(x^k) \|^2 + \| \nabla f_{j_{i,k}}(w_i^k) - \nabla f(w_i^k) \|^2 \right]$$

$$+ \frac{2}{b^2} \sum_{t=1}^{b} \mathbb{E}_k \left[\| \nabla f_{j_{l,k}}(x^k) - \nabla f(x^k) \|^2 + \| \nabla f_{j_{l,k}}(w_l^k) - \nabla f(w_l^k) \|^2 \right] \overset{(6)}{\leq} \frac{8\sigma^2}{b}.$$

Then, $\frac{1}{G(G-1)} \sum_{i,l \in \mathcal{G}} \mathbb{E}_k \| g_i^k - g_l^k \|^2 \leq \frac{8\sigma^2}{b}$ and by Definition 1 we have

$$\mathbb{E}_k \| \hat{g}^k - \overline{g}^k \|^2 \overset{(2)}{\leq} \frac{8c\delta\sigma^2}{b}. \tag{11}$$

Putting all together in (8), we arrive at

$$\mathbb{E}_k \Psi_{k+1} \leq \left(1 - \frac{\gamma\mu}{2} \right) \| x^k - x^* \|^2 + 4\gamma^2 \sigma_k^2 - 2\gamma(1 - 4L\gamma) \left(f(x^k) - f(x^*) \right)$$

$$+ 16\gamma \left(\frac{1}{\mu} + \gamma \right) \frac{c\delta\sigma^2}{b} + \frac{8\gamma^2}{p} \mathbb{E}_k [\sigma_{k+1}^2]. \tag{12}$$

Next, we estimate $\mathbb{E}_k [\sigma_{k+1}^2]$:

$$\mathbb{E}_k [\sigma_{k+1}^2] = \frac{1-p}{Gm} \sum_{i \in \mathcal{G}} \sum_{j=1}^{m} \| \nabla f_j(w_i^k) - \nabla f_j(x^*) \|^2 + \frac{p}{m} \sum_{j=1}^{m} \| \nabla f_j(x^k) - \nabla f_j(x^*) \|^2$$

$$\overset{(10)}{\leq} (1-p)\sigma_k^2 + \frac{2Lp}{m} \sum_{j=1}^{m} \left(f_j(x^k) - f_j(x^*) - \langle \nabla f_j(x^*), x^k - x^* \rangle \right)$$

$$= (1-p)\sigma_k^2 + 2Lp \left(f(x^k) - f(x^*) \right). \tag{13}$$

Finally, we combine (12) and (13):

$$
\begin{aligned}
\mathbb{E}_k \Psi_{k+1} &\leq \left(1 - \frac{\gamma\mu}{2}\right) \|x^k - x^*\|^2 + \left(1 - \frac{p}{2}\right) \frac{8\gamma^2}{p} \sigma_k^2 \\
&\quad - 2\gamma(1 - 12L\gamma) \left(f(x^k) - f(x^*)\right) + 16\gamma \left(\frac{1}{\mu} + \gamma\right) \frac{c\delta\sigma^2}{b} \\
&\leq \left(1 - \min\left\{\frac{\gamma\mu}{2}, \frac{p}{2}\right\}\right) \Psi_k - 2\gamma(1 - 12L\gamma)\left(f(x^k) - f(x^*)\right) \\
&\quad + 16\gamma \left(\frac{1}{\mu} + \gamma\right) \frac{c\delta\sigma^2}{b} \\
&\leq \left(1 - \frac{\gamma\mu}{2}\right) \Psi_k + 16\gamma \left(\frac{1}{\mu} + \gamma\right) \frac{c\delta\sigma^2}{b},
\end{aligned}
$$

where in the last step we use $0 < \gamma \leq \min\{1/12L, p/\mu\}$. Taking the full expectation and unrolling the obtained recurrence, we get that for all $K \geq 0$

$$
\begin{aligned}
\mathbb{E}\Psi_K &\leq \left(1 - \frac{\gamma\mu}{2}\right)^K \Psi_0 + 16\gamma \left(\frac{1}{\mu} + \gamma\right) \frac{c\delta\sigma^2}{b} \sum_{k=0}^{K-1} \left(1 - \frac{\gamma\mu}{2}\right)^k \\
&\leq \left(1 - \frac{\gamma\mu}{2}\right)^K \Psi_0 + \gamma \frac{32c\delta\sigma^2}{b\mu} + \frac{32c\delta\sigma^2}{b\mu^2},
\end{aligned}
$$

which concludes the proof.   □

Since $\Psi_K \geq \|x^K - x^*\|^2$, Theorem 1 states that BR-LSVRG converges linearly (in expectation) to the neighborhood of the solution. We notice that the neighborhood's size is proportional to σ^2/b, which is typical for Stochastic Gradient Descent-type methods [11], and also proportional to the ratio of Byzantine workers δ. When $\delta = 0$, BR-LSVRG converges linearly and recovers the rate of LSVRG [21] up to numerical factors. However, in the general case ($\delta > 0$), the last term in (7) can be reduced only via increasing batchsize b. We believe that this is unavoidable for BR-LSVRG in the worst case since robust aggregation creates a bias in the update, and the analysis of LSVRG is very sensitive to the bias in the update direction. Nevertheless, when the size of the neighborhood is small enough, e.g., when δ or σ^2 are small, the method can achieve relatively good accuracy with moderate batchsize – we demonstrate this phenomenon in the experiments.

Next, we present an alternative convergence result that does not rely on the bounded variance assumption.

Theorem 2. *Let Assumption 1 hold, $0 < \gamma \leq \min\{1/144L, p/\mu\}$, and $b \geq \max\{1, c\delta/\gamma\mu\}$. Then, the iterates of* BR-LSVRG *after K iterations satisfy*

$$
\mathbb{E}\Psi_K \leq \left(1 - \frac{\gamma\mu}{2}\right)^K \Psi_0,
$$

where $\Psi_k \overset{def}{=} \|x^k - x^*\|^2 + \frac{72\gamma^2}{p}\sigma_k^2$, $\sigma_k^2 \overset{def}{=} \frac{1}{Gm}\sum_{i\in\mathcal{G}}\sum_{j=1}^m \|\nabla f_j(w_i^k) - \nabla f_j(x^*)\|^2$.

Proof. First, we notice that inequalities (8), (9), and (13) from the proof of Theorem 1 are derived without Assumption 2. We need to derive a version of (11) that does not rely on Assumption 2. Due to the independence of $j_{i,k}^1, j_{i,k}^2, \ldots, j_{i,k}^b$ we have $\forall i \in \mathcal{G}$

$$\mathbb{E}_k \|g_i^k - \nabla f(x^k)\|^2 = \frac{1}{b^2} \sum_{t=1}^{b} \mathbb{E}_k \|\nabla f_{j_{i,k}^t}(x^k) - \nabla f_{j_{i,k}^t}(w_i^k) + \nabla f(w_i^k) - \nabla f(x^k)\|^2$$

$$\leq \frac{1}{b^2} \sum_{t=1}^{b} \mathbb{E}_k \|\nabla f_{j_{i,k}^t}(x^k) - \nabla f_{j_{i,k}^t}(w_i^k)\|^2$$

and

$$\mathbb{E}_k \|g_i^k - \nabla f(x^k)\|^2$$

$$\leq \frac{2}{b^2} \sum_{t=1}^{b} \mathbb{E}_k \left[\|\nabla f_{j_{i,k}^t}(x^k) - \nabla f_{j_{i,k}^t}(x^*)\|^2 + \|\nabla f_{j_{i,k}^t}(w_i^k) - \nabla f_{j_{i,k}^t}(x^*)\|^2 \right]$$

$$= \frac{2}{bm} \sum_{j=1}^{m} \left(\|\nabla f_j(x^k) - \nabla f_j(x^*)\|^2 + \|\nabla f_j(w_i^k) - \nabla f_j(x^*)\|^2 \right)$$

$$\overset{(10)}{\leq} \frac{4L}{b} \left(f(x^k) - f(x^*) \right) + \frac{2}{bm} \sum_{j=1}^{m} \|\nabla f_j(w_i^k) - \nabla f_j(x^*)\|^2. \tag{14}$$

Therefore,

$$\frac{1}{G(G-1)} \sum_{i,l \in \mathcal{G}} \mathbb{E}_k \|g_i^k - g_l^k\|^2 \leq \frac{2}{G(G-1)} \sum_{i,l \in \mathcal{G}, i \neq l} \mathbb{E}_k \|g_i^k - \nabla f(x^k)\|^2$$

$$+ \frac{2}{G(G-1)} \sum_{i,l \in \mathcal{G}, i \neq l} \mathbb{E}_k \|g_l^k - \nabla f(x^k)\|^2$$

$$= \frac{4}{G} \sum_{i \in \mathcal{G}} \mathbb{E}_k \|g_i^k - \nabla f(x^k)\|^2$$

$$\overset{(14)}{\leq} \frac{16L}{b} \left(f(x^k) - f(x^*) \right) + \frac{8}{b} \sigma_k^2,$$

and by definition of (δ, c)-robust aggregator we have $\mathbb{E}_k \|\hat{g}^k - \bar{g}^k\|^2 \overset{(2)}{\leq} \frac{16Lc\delta}{b} \left(f(x^k) - f(x^*) \right) + \frac{8c\delta}{b} \sigma_k^2$. Combining this inequality with (8) and (9), we get

$$\mathbb{E}_k \|x^{k+1} - x^*\|^2 \leq \left(1 - \frac{\gamma\mu}{2}\right) \|x^k - x^*\|^2 - 2\gamma \left(f(x^k) - f(x^*) \right)$$

$$+ 8L\gamma^2 \left(f(x^k) - f(x^*) \right) + 4\gamma^2 \sigma_k^2$$

$$+ \frac{32\gamma Lc\delta}{b} \left(\frac{1}{\mu} + \gamma \right) \left(f(x^k) - f(x^*) \right) + \frac{16\gamma c\delta}{b} \left(\frac{1}{\mu} + \gamma \right) \sigma_k^2$$

$$\leq \left(1 - \frac{\gamma\mu}{2}\right) \|x^k - x^*\|^2 + 36\gamma^2 \sigma_k^2$$

$$- 2\gamma \left(1 - 72\gamma L\right) \left(f(x^k) - f(x^*) \right),$$

where in the last step we use $b \geq \max\{1, c\delta/\gamma\mu\}$. Finally, this inequality and (13) imply that

$$\mathbb{E}_k \Psi_{k+1} \leq \left(1 - \frac{\gamma\mu}{2}\right) \|x^k - x^*\|^2 + 36\gamma^2\sigma_k^2 - 2\gamma\left(1 - 72\gamma L\right)\left(f(x^k) - f(x^*)\right)$$

$$+ (1-p)\frac{72\gamma^2}{p}\sigma_k^2 + 144L\gamma^2\left(f(x^k) - f(x^*)\right)$$

$$= \left(1 - \frac{\gamma\mu}{2}\right)\|x^k - x^*\|^2 + \left(1 - \frac{p}{2}\right)\frac{72\gamma^2}{p}\sigma_k^2$$

$$- 2\gamma\left(1 - 144\gamma L\right)\left(f(x^k) - f(x^*)\right) \leq \left(1 - \frac{\gamma\mu}{2}\right)\Psi_k,$$

where the last step follows from $\gamma \leq \min\{1/144L, p/\mu\}$. Taking the full expectation from both sides and unrolling the recurrence, we get the result.   □

In contrast to Theorem 1, Theorem 2 establishes linear convergence of BR-LSVRG to any accuracy. However, Theorem 2 requires batchsize to satisfy $b \geq \max\{1, c\delta/\gamma\mu\}$, which can be huge in the worst case. If $p = b/m$, $\gamma = \min\{1/144L, b/m\mu\}$, $b = \max\{1, 144c\delta L/\mu, \sqrt{c\delta m}\}$, and $m \geq b$ (for example, these assumptions are satisfied when m is sufficiently large), then, according to Theorem 2, BR-LSVRG finds x^K such that $\mathbb{E}\|x^K - x^*\|^2 \leq \varepsilon\Psi_0$ after

$$\mathcal{O}\left(\left(\frac{L}{\mu} + \frac{m}{b}\right)\log\frac{1}{\varepsilon}\right) \text{ iterations,} \tag{15}$$

$$\mathcal{O}\left(\left(\frac{L}{\mu} + \frac{L^2\sqrt{c\delta}}{\mu^2} + \frac{L\sqrt{c\delta m}}{\mu} + m\right)\log\frac{1}{\varepsilon}\right) \text{ oracle calls.} \tag{16}$$

Under the same assumptions, to achieve the same goal Byrd-SAGA requires [35]

$$\mathcal{O}\left(\frac{m^2L^2}{b^2(1-2\delta)\mu^2}\log\frac{1}{\varepsilon}\right) \text{ iterations,} \tag{17}$$

$$\mathcal{O}\left(\frac{m^2L^2}{b(1-2\delta)\mu^2}\log\frac{1}{\varepsilon}\right) \text{ oracle calls.} \tag{18}$$

Complexity bounds for Byrd-SAGA are inferior to the ones derived for BR-LSVRG as long as our result is applicable. Moreover, when $\delta = 0$ (no Byzantines) our result recovers the known one for LSVRG (up to numerical factors), while the upper bounds (17) and (18) are much larger than the best-known ones for SAGA. This comparison highlights the benefits of our approach compared to the closest one.

Finally, we compare our results against the current state-of-the-art ones obtained for Byz-VR-MARINA in [12]. In particular, under weaker conditions (Polyak-Lojasiewicz condition [25, 32] instead of strong convexity), the authors of [12] prove that to achieve $\mathbb{E}[f(x^K) - f(x^*)] \leq \varepsilon(f(x^0) - f(x^*))$ Byz-VR-MARINA requires

$$\mathcal{O}\left(\left(\frac{L}{\mu} + \frac{L\sqrt{m}}{\mu b\sqrt{n}} + \frac{Lm\sqrt{c\delta}}{\mu\sqrt{b^3}} + \frac{m}{b}\right)\log\frac{1}{\varepsilon}\right) \text{ iterations,} \quad (19)$$

$$\mathcal{O}\left(\left(\frac{bL}{\mu} + \frac{L\sqrt{m}}{\mu\sqrt{n}} + \frac{Lm\sqrt{c\delta}}{\mu\sqrt{b}} + m\right)\log\frac{1}{\varepsilon}\right) \text{ oracle calls.} \quad (20)$$

Complexity bounds for Byz-VR-MARINA are not better than ones derived for BR-LSVRG as long as our result is applicable. Moreover, in the special case, when $m > b\sqrt{n}$ (big data regime), iteration complexity of Byz-VR-MARINA (19) is strictly worse than the one we have for BR-LSVRG (15). When $\delta = 0$, our results are strictly better than the ones for Byz-VR-MARINA. However, it is important to notice that (i) the results for Byz-VR-MARINA are derived under weaker assumptions and (ii) in contrast to the results for Byrd-SAGA and Byz-VR-MARINA, our results require the batchsize to be large enough in general.

3 Numerical Experiments

In our numerical experiments, we consider logistic regression with ℓ_2-regularization – an instance of (1) with $f_j(x) = \ln(1+\exp(-y_j\langle a_j, x\rangle)) + \frac{\ell_2}{2}\|x\|^2$. Here $\{a_j\}_{j\in[m]} \subset \mathbb{R}^d$ are vectors of "features", $\{y_j\}_{j\in[m]} \subset \{-1,1\}^m$ are labels, and $\ell_2 \geq 0$ is a parameter of ℓ_2-regularization. This problem satisfies Assumption 1 (and also Assumption 2 since gradients $\nabla f_j(x)$ are bounded): for each $j \in [m]$ function f_j is ℓ_2-strongly convex and L_j-smooth with $L_j = \ell_2 + \|a_j\|^2/4$ and function f is L-smooth with $L = \ell_2 + \lambda_{\max}(\mathbf{A}^\top\mathbf{A})/4m$, where $\mathbf{A} \in \mathbb{R}^{m\times d}$ is such that the j-th row of $\mathbf{A}$ equals a_j and $\lambda_{\max}(\mathbf{A}^\top\mathbf{A})$ denotes the largest eigenvalue of $\mathbf{A}^\top\mathbf{A}$. We chose $l_2 = L/1000$ in all experiments. We consider 4 datasets from LIBSVM library [4]: a9a ($m = 32561, d = 123$), phishing ($m = 11055, d = 68$), w8a ($m = 49749, d = 300$) and mushrooms ($m = 8124, d = 112$). The total number of workers in our experiments equals $n = 16$ with 3 Byzantine workers among them. Byzantine workers use one of the following baseline attacks: • **Bit Flipping (BF)**: Byzantine workers compute g_i^k following the algorithm and send $-g_i^k$ to the server; • **Label Flipping (LF)**: Byzantine workers compute g_i^k with $y_{j_{i,k}^t}$ replaced by $-y_{j_{i,k}^t}$; • **A Little Is Enough (ALIE)** [2]: Byzantine workers compute empirical mean $\mu_{\mathcal{G}}$ and standard deviation $\sigma_{\mathcal{G}}$ of $\{g_i^k\}_{i\in\mathcal{G}}$ and send vector $\mu_{\mathcal{G}} - z\sigma_{\mathcal{G}}$ to the server, where z controls the strength of the attack ($z = 1.06$ in our experiments); • **Inner Product Manipulation (IPM)** [36]: Byzantine workers send $-\frac{\varepsilon}{G}\sum_{i\in\mathcal{G}} g_i^k$ to the server, where $\varepsilon > 0$ is a parameter ($\varepsilon = 0.1$ in our experiments). Our code is publicly available: https://github.com/Nikosimus/BR-LSVRG.

Experiment 1: BR-LSVRG with Different Batchsizes. In this experiment, we tested BR-LSVRG with two different batchsizes: 1 and 0.01 m. Stepsize was chosen as $\gamma = 1/12L$. In all runs, BR-LSVRG with moderate batchsize $b = 0.01$ m

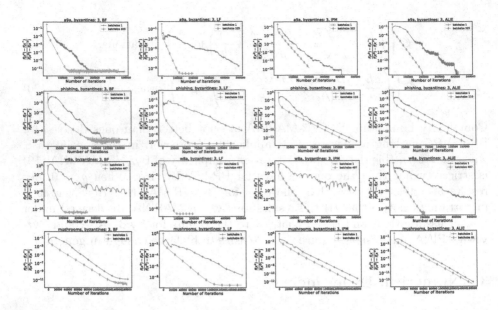

Fig. 1. Trajectories of BR-LSVRG with batchsizes 1 and 0.01 m. Each row corresponds to one of the used datasets with 4 different types of attacks.

achieves a very high accuracy of the solution. In addition, BR-LSVRG with batch-size $b = 1$ always achieves at least 10^{-5} functional suboptimality, which is a relatively good accuracy as well. This experiment illustrates that BR-LSVRG can converge to high accuracy even with small or moderate batchsizes (Fig. 1).

Experiment 2: Comparison with Byz-VR-MARINA **and** Byrd-SAGA. Next, we compare BR-LSVRG with Byz-VR-MARINA and Byrd-SAGA. All the methods were run with stepsize $\gamma = {}^5\!/_{2L}$ and batchsize 0.01 m. In all cases, BR-LSVRG achieves better accuracy than Byrd-SAGA and shows a comparable convergence to Byz-VR-MARINA to the very high accuracy. We also tested Byrd-SAGA with smaller stepsizes, but the method did not achieve better accuracy (Fig. 2).

4   Discussion

In this work, we propose a new Byzantine-robust variance-reduced method based on SVRG-estimator – BR-LSVRG. Our theoretical results show that BR-LSVRG outperforms state-of-the-art methods in certain regimes. Numerical experiments highlight that BR-LSVRG can have a comparable convergence to Byz-VR-MARINA.

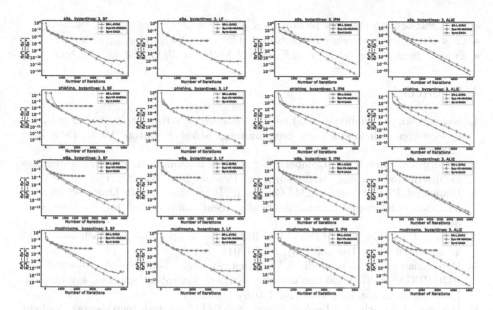

Fig. 2. Trajectories of BR-LSVRG, Byz-VR-MARINA and Byrd-SAGA with batchsize 0.01 m. Each row corresponds to one of the used datasets with 4 different types of attacks.

A   Examples of Robust Aggregators

In [20], the authors propose the procedure called *bucketing* (see Algorithm 2) that robustifies certain aggregation rules such as: • geometric median (GM) $\hat{x} = \arg\min_{x \in \mathbb{R}^d} \sum_{i=1}^{n} \|x - x_i\|$; • coordinate-wise median (CM) $\hat{x} = \arg\min_{x \in \mathbb{R}^d} \sum_{i=1}^{n} \|x - x_i\|_1$; • Krum estimator [3] $\arg\min_{x_i \in \{x_1,\dots,x_n\}} \sum_{j \in S_i} \|x_j - x_i\|^2$, where $S_i \subseteq \{x_1, \dots, x_n\}$ is the subset of $n - |\mathcal{B}| - 2$ closest (w.r.t. ℓ_2-norm) vectors to x_i.

Algorithm 2. Bucketing: Robust Aggregation using bucketing [20]

1: **Input:** $\{x_1, \dots, x_n\}$, $s \in \mathbb{N}$ – bucket size, Aggr – aggregation rule
2: Sample random permutation $\pi = (\pi(1), \dots, \pi(n))$ of $[n]$
3: Compute $y_i = \frac{1}{s} \sum_{k=s(i-1)+1}^{\min\{si,n\}} x_{\pi(k)}$ for $i = 1, \dots, \lceil n/s \rceil$
4: **Return:** $\hat{x} = \text{Aggr}(y_1, \dots, y_{\lceil n/s \rceil})$

The following result establishes the robustness of the aforementioned aggregation rules in combination with Bucketing.

Theorem 3 (Theorem D.1 from [12]). *Assume that $\{x_1, x_2, \dots, x_n\}$ is such that there exists a subset $\mathcal{G} \subseteq [n]$, $|\mathcal{G}| = G \geq (1-\delta)n$ and $\sigma \geq 0$ such that $\frac{1}{G(G-1)} \sum_{i,l \in \mathcal{G}} \mathbb{E}\|x_i - x_l\|^2 \leq \sigma^2$. Assume that $\delta \leq \delta_{\max}$. If Algorithm 2 is run with $s = \lfloor \delta_{\max}/\delta \rfloor$, then*

- *GM* ∘ `Bucketing` *satisfies Definition 1 with* $c = \mathcal{O}(1)$ *and* $\delta_{max} < 1/2$,
- *CM* ∘ `Bucketing` *satisfies Definition 1 with* $c = \mathcal{O}(d)$ *and* $\delta_{max} < 1/2$,
- *Krum* ∘ `Bucketing` *satisfies Definition 1 with* $c = \mathcal{O}(1)$ *and* $\delta_{max} < 1/4$.

References

1. Nemirovski, A., Juditsky, A., Lan, G., Shapiro, A.: Robust stochastic approximation approach to stochastic programming. SIAM J. Optim. **19**(4), 1574–1609 (2009)
2. Baruch, G., Baruch, M., Goldberg, Y.: A little is enough: circumventing defenses for distributed learning. Adv. Neural Inf. Process. Syst. **32** (2019)
3. Blanchard, P., El Mhamdi, E.M., Guerraoui, R., Stainer, J.: Machine learning with adversaries: byzantine tolerant gradient descent. Adv. Neural Inf. Process. Syst. **30** (2017)
4. Chang, C.C., Lin, C.J.: LIBSVM: a library for support vector machines. ACM Trans. Intell. Syst. Technol. **2**(3), 1–27 (2011)
5. Damaskinos, G., El-Mhamdi, E.M., Guerraoui, R., Guirguis, A., Rouault, S.: AGGREGATHOR: byzantine machine learning via robust gradient aggregation. Proc. Mach. Learn. Res. **1**, 81–106 (2019)
6. Defazio, A., Bach, F., Lacoste-Julien, S.: SAGA: a fast incremental gradient method with support for non-strongly convex composite objectives. Adv. Neural Inf. Process. Syst. **27** (2014)
7. Fang, C., Li, C.J., Lin, Z., Zhang, T.: SPIDER: near-optimal non-convex optimization via stochastic path-integrated differential estimator. Adv. Neural Inf. Process. Syst. **31** (2018)
8. Ghadimi, S., Lan, G.: Stochastic first-and zeroth-order methods for nonconvex stochastic programming. SIAM J. Optim. **23**(4), 2341–2368 (2013)
9. Goodfellow, I., Bengio, Y., Courville, A.: Deep learning (2016)
10. Gorbunov, E., Borzunov, A., Diskin, M., Ryabinin, M.: Secure distributed training at scale. In: International Conference on Machine Learning, pp. 7679–7739. PMLR (2022). http://proceedings.mlr.press/v162/gorbunov22a/gorbunov22a.pdf
11. Gorbunov, E., Hanzely, F., Richtárik, P.: A unified theory of SGD: variance reduction, sampling, quantization and coordinate descent. In: International Conference on Artificial Intelligence and Statistics, pp. 680–690. PMLR (2020)
12. Gorbunov, E., Horváth, S., Richtárik, P., Gidel, G.: Variance reduction is an antidote to byzantines: better rates, weaker assumptions and communication compression as a cherry on the top. arXiv preprint arXiv:2206.00529 (2022)
13. Gower, R.M., Schmidt, M., Bach, F., Richtárik, P.: Variance-reduced methods for machine learning. Proc. IEEE **108**(11), 1968–1983 (2020)
14. Guerraoui, R., Rouault, S., et al.: The hidden vulnerability of distributed learning in byzantium. In: International Conference on Machine Learning, pp. 3521–3530. PMLR (2018)
15. He, L., Karimireddy, S.P., Jaggi, M.: Byzantine-robust decentralized learning via self-centered clipping. arXiv preprint arXiv:2202.01545 (2022)
16. Hofmann, T., Lucchi, A., Lacoste-Julien, S., McWilliams, B.: Variance reduced stochastic gradient descent with neighbors. Adv. Neural Inf. Process. Syst. **28** (2015)
17. Horváth, S., Lei, L., Richtárik, P., Jordan, M.I.: Adaptivity of stochastic gradient methods for nonconvex optimization. SIAM J. Math. Data Sci. **4**(2), 634–648 (2022)

18. Johnson, R., Zhang, T.: Accelerating stochastic gradient descent using predictive variance reduction. Adv. Neural Inf. Process. Syst. **26** (2013)
19. Karimireddy, S.P., He, L., Jaggi, M.: Learning from history for byzantine robust optimization. In: International Conference on Machine Learning, pp. 5311–5319. PMLR (2021)
20. Karimireddy, S.P., He, L., Jaggi, M.: Byzantine-robust learning on heterogeneous datasets via bucketing. In: International Conference on Learning Representations (2022). https://arxiv.org/pdf/2006.09365.pdf
21. Kovalev, D., Horváth, S., Richtárik, P.: Don't jump through hoops and remove those loops: SVRG and Katyusha are better without the outer loop. In: Algorithmic Learning Theory, pp. 451–467. PMLR (2020). http://proceedings.mlr.press/v117/kovalev20a/kovalev20a.pdf
22. Lamport, L., Shostak, R., Pease, M.: The byzantine generals problem. ACM Trans. Program. Lang. Syst. **4**(3), 382–401 (1982)
23. Li, C.: Demystifying GPT-3 language model: a technical overview (2020)
24. Li, Z., Bao, H., Zhang, X., Richtárik, P.: PAGE: a simple and optimal probabilistic gradient estimator for nonconvex optimization. In: International Conference on Machine Learning, pp. 6286–6295. PMLR (2021)
25. Lojasiewicz, S.: A topological property of real analytic subsets. Coll. du CNRS, Les équations aux dérivées partielles **117**(87–89), 2 (1963)
26. Lyu, L., et al.: Privacy and robustness in federated learning: attacks and defenses. IEEE Trans. Neural Netw. Learn. Syst. (2022)
27. Nesterov, Y.: Lectures on Convex Optimization. SOIA, vol. 137. Springer, Cham (2018). https://doi.org/10.1007/978-3-319-91578-4
28. Nguyen, L.M., Liu, J., Scheinberg, K., Takáč, M.: SARAH: a novel method for machine learning problems using stochastic recursive gradient. In: International Conference on Machine Learning, pp. 2613–2621. PMLR (2017)
29. Ouyang, L., et al.: Training language models to follow instructions with human feedback. arXiv preprint arXiv:2203.02155 (2022)
30. Pillutla, K., Kakade, S.M., Harchaoui, Z.: Robust aggregation for federated learning. IEEE Trans. Signal Process. **70**, 1142–1154 (2022)
31. Polyak, B.T.: Some methods of speeding up the convergence of iteration methods. USSR Comput. Math. Math. Phys. **4**(5), 1–17 (1964)
32. Polyak, B.T.: Gradient methods for the minimisation of functionals. USSR Comput. Math. Math. Phys. **3**(4), 864–878 (1963)
33. Robbins, H., Monro, S.: A stochastic approximation method. Ann. Math. Stat. 400–407 (1951)
34. Shalev-Shwartz, S., Ben-David, S.: Understanding Machine Learning: From Theory to Algorithms. Cambridge University Press, Cambridge (2014)
35. Wu, Z., Ling, Q., Chen, T., Giannakis, G.B.: Federated variance-reduced stochastic gradient descent with robustness to byzantine attacks. IEEE Trans. Signal Process. **68**, 4583–4596 (2020)
36. Xie, C., Koyejo, O., Gupta, I.: Fall of empires: breaking byzantine-tolerant SGD by inner product manipulation. In: Uncertainty in Artificial Intelligence, pp. 261–270. PMLR (2020)
37. Yin, D., Chen, Y., Kannan, R., Bartlett, P.: Byzantine-robust distributed learning: towards optimal statistical rates. In: International Conference on Machine Learning, pp. 5650–5659. PMLR (2018)
38. Zinkevich, M., Weimer, M., Li, L., Smola, A.: Parallelized stochastic gradient descent. Adv. Neural Inf. Process. Syst. **23** (2010)

Semi-supervised K-Means Clustering via DC Programming Approach

Tatiana V. Gruzdeva$^{(\boxtimes)}$ [iD] and Anton V. Ushakov [iD]

Matrosov Institute for System Dynamics and Control Theory of SB RAS,
134 Lermontov str., 664033 Irkutsk, Russia
{gruzdeva,aushakov}@icc.ru

Abstract. Though clustering is related to unsupervised machine learning and does not require any prior information on data items, in many
real-life settings, there may be some expert knowledge on data labels
or the properties of clusters known in advance. Obviously, such knowledge may be used to guide clustering process and improve the quality of
found partitions. The clustering problems that involve some additional
information on tags are called semi-supervised or constrained clustering
problems. One distinguishes instance-level and cluster-level constraints
usually formalized as the so-called must-link and cannot-link constraints
or minimum/maximum cluster size. In this paper, we consider the constrained minimum sum-of-squares (k-means) clustering (MSSC) problem that incorporates both instance- and cluster-level constraints. As
far as we know, such a semi-supervised MSSC problem has not been
considered in the literature yet. We formulate this clustering problem
and some of its particular cases as DC (difference of convex) optimization problems. Then, we develop a solution approach based on a special
local search method. We carry out computational experiments on test
problem instances demonstrating the efficiency of the proposed solution
approach.

Keywords: semi-supervised clustering · k-means ·
minimum-sum-of-squares clustering · DC programming · local search ·
must-link · cannot-link · constrained clustering

1 Introduction

Clustering, also known as cluster analysis, is the machine learning task aimed
at grouping some set of objects (samples, patterns, data items, observations)
in such a way that each group consists of similar objects, whereas objects from
different groups are dissimilar. The objects are usually represented as points
in a multidimensional space, and the corresponding groups are called clusters.
Though there are many approaches and views on how to determine the similarity
of objects and how to group them into clusters, the most popular approach is to

The research was funded by the Ministry of Education and Science of the Russian
Federation No. 121041300065-9.

consider clustering as a mathematical optimization problem [11]. Thus, it may be solved with effective general optimization tools and algorithms. Among many types of clustering problems known in the literature, the center-based clustering is probably the most popular one. In these problems, a partition into clusters is defined by the so-called cluster centers or representatives. After all representatives are determined, all data items are divided into clusters by assigning to the closest (most similar) cluster center. Note that such clustering problems can be viewed as facility location problems, where one needs to find a set of sites for locating facilities (cluster centers) to minimize the total distance between customers (data items) and their nearest facilities (for a survey, see [34]).

The most popular center-based clustering problem is the minimum-sum-of-squares (MSSC) or k-means clustering problem. Given a set of data items $a^i \in \mathbb{R}^n$, $i \in I$, the problem is to find k cluster representatives $c^j \in R^n$, $j = 1, \ldots, k$ so as the total sum of squared Euclidean distances from data items to cluster centers they assigned to is minimized, i.e.

$$\min_{C \subset \mathbb{R}^n} \left\{ \sum_{j=1}^m \min_{c \in C} \|a^j - c\|^2, \ |C| = k \right\}. \tag{1}$$

All data items assigned to the same center form a cluster. Each data item is supposed to be assigned to only one cluster, hence the clusters are non-overlapped. Although the problem is NP-hard [1] in general dimension even for $k = 2$ and in the plane for arbitrary k [22], it has been very widespread in applications due to the k-means algorithm. The latter is a local search (alternate location-allocation) heuristic for MSSC.

By definition, the cluster analysis problem is related to unsupervised machine learning, i.e. it does not require any prior information on data items and deals with untagged data. On the other hand, in many real-life settings, there is some expert knowledge on data labels or the properties of clusters that must be found in data. Obviously, such knowledge may be used to guide clustering process and improve the quality of found partitions. The clustering problems that involve some additional information on tags are called semi-supervised or constrained clustering problems [12]. In these problems, there is a subset of data items for which the tags are known, and this information is used in a form of constraints on clusters.

In constrained clustering, one usually distinguishes two types of constraints referred to as instance-level and cluster-level constraints [27]. The instance-level constrains are usually formalized as the so-called must-link and cannot-link constraints defined for pairs of data items [35]. A must-link constraint ensures that the corresponding pair of objects is assigned to the same cluster, while a cannot-link constraint enforces the two objects belong to different clusters. In terms of semi-supervised learning, these constraints may be defined using data labels. Note that the exact class labels are not actually necessary, the prior information may include only pairs of data items that must or cannot be in the same cluster. This type of constraints are useful and popular, since they can be used to model many other constraints [4,10]. The most notable examples are the maximum cluster diameter constraint and the minimum cluster separation constraint. The former imposes constraint on the maximal distance between points assigned to the same cluster. Thus, if there are two data items, the distance between which is larger than the required one, they cannot be in the same cluster. The latter imposes constraint on the minimal distance between two clusters. If the distance between two data items is less than the required one, they obviously must be assigned to the same cluster.

The cluster-level constraints incorporate some information on cluster structure, especially minimum or maximum cluster size. Clustering problems with such constraints are also known as capacitated clustering problems closely related to capacitated facility location problems [3, 8, 14, 23, 24].

Though aforementioned constraints may be embedded in many clustering problems, most of research efforts are focused on the semi-supervised MSSC problem. Note that introducing additional instance-level constraints may render a clustering problem infeasible. In general, the feasibility problem under the must-link and cannot=link constraints is NP-complete [10].

There are actually many ways of how to utilize the additional constraints for the cluster analysis problem. For example, must-link and cannot-link constraints may be considered as soft or hard ones. Semi-supervised clustering algorithms may use the information on data labels in initialization step only or ignore it, if it results in worsening the objective value.

As in the case of the unconstrained MSSC problem, the most well-known semi-supervised clustering algorithms are heuristics for the constrained MSSC problem. For example, COP-kmeans, proposed in [35], is a extension of the k-means algorithm, in which cluster assignments are updated, taking into account must-link and cannot link constraints. This is done in a greedy fashion. For example, if for a query point the nearest cluster already includes its cannot-link point, it is assigned to another cluster. As noted in [10, 27], besides inheriting all the drawbacks of k-means, it can fail to find a feasible solution (even if it exists) when the number of both constraints and clusters is large. Though there are several attempts to improve COM-kmeans [28, 33], incorporating such additional constraints into conventional clustering algorithms is quite challenging. In [5], the authors assume the must-link and cannot-link constraints as soft constraints, e.g. they can be violated but this induces penalties.

As to find clustering assignments under the aforementioned constraints is challenging, some authors proposed to formulate this task as an integer program that has to be solved either in each iteration of the k-means-like iterative procedure [7] or in the post-processing step following a conventional unconstrained clustering algorithm [25].

As for MSSC, the drawbacks of k-means-like semi-supervised clustering algorithms attract a lot of attention to devising effective metaheuristics, e.g. dual iterative local search [15], genetic algorithm and column generation combined with path relinking [26], local search [13], etc.

There are also a number of exact algorithms for the constrained MSSC problem, most of which are based on the exact algorithms for the unconstrained counterpart. The first exact algorithm, proposed in [36], is a cutting plane method based on Tuy's concavity cuts. In [2,4] the authors proposed a column generation algorithm extending the sophisticated approach to MSSC proposed in [2]. Recently, a branch-and-cut method based on the semidefinite programming formulation of the problem was developed in [27]. Due to extreme hardness of the constrained MSSP problem, only relatively small problem instances are tractable with exact algorithms.

In this paper, we consider the constrained MSSC problem that incorporates both instance- and cluster-level constraints, i.e. must-link, cannot-link, and cluster capacity constraints. As far as we know, such a semi-supervised MSSC problem has not been considered in the literature yet. We formulate this clustering problem as a DC (difference of convex) optimization problem. Then, we develop a solution approach to this problem based on a special local search method. We carry out computational experiments on test problem instances demonstrating the efficiency of the proposed solution approach.

2 Problem Statement

In this paper, we consider traditional must-link and cannot-link constraints [6] as well as capacity constraints allowing one to avoid local solutions with empty clusters or those with very few points.

The semi-supervised minimum sum-of-squared clustering (SSMSSC) model can be formulated as a mixed integer program. Given a set of data items J that must be divided into clusters. The goal is to find cluster centers such that the overall sum of squared Euclidean distances between data items and their closest centers is minimized subject to the side constraints.

Let us introduce the set $M \subset J$ consisting of pairs (j_1, j_2) of data items $j_1, j_2 \in J$, $j_1 \neq j_2$ that must be clustered together, and the set $L \subset J$ containing pairs (j_1, j_2) of data items that must be assigned to different clusters. To define the cluster-level (capacity) constraints, we introduce the integer parameter $\tau_i > 0$ $\forall i = 1, \ldots, k$ that limits the minimal number of data items in cluster i.

Let us introduce the following binary variables

$$x_{ij} = \begin{cases} 1, & \text{if data item } j \text{ is assigned to cluster } i, \\ 0, & \text{otherwise,} \end{cases} \quad i = 1, \ldots, k, \; j = 1, \ldots, m.$$

which are often referred to as assignment variables.

We also introduce the decision variables $y^i \in \mathbb{R}^n$ that define locations of cluster centers $i = \{1, \ldots, k\}$. With these notations, the SSMSSC can be written in the following statement:

$$\sum_{i=1}^{k} \sum_{j=1}^{m} x_{ij} \| y^i - a^j \|^2 \downarrow \min_{(x,y)}, \tag{2}$$

$$\sum_{i=1}^{k} x_{ij} = 1 \qquad \forall \, j = 1, \ldots, m; \tag{3}$$

$$x_{ij_1} - x_{ij_2} = 0 \qquad\qquad \forall\, i = 1,\ldots,k; \;\; \forall\, (j_1, j_2) \in M; \qquad (4)$$

$$x_{ij_3} + x_{ij_4} \leq 1 \qquad\qquad \forall\, i = 1,\ldots,k; \;\; \forall\, (j_3, j_4) \in L; \qquad (5)$$

$$\sum_{j=1}^{m} x_{ij} \geq \tau_i \qquad\qquad\qquad \forall\, i = 1,\ldots,k; \qquad (6)$$

$$x_{ij} \in \{0,1\} \qquad\qquad \forall\, i = 1,\ldots,k; \;\; \forall\, j = 1,\ldots,m. \qquad (7)$$

The objective function (2) minimizes the sum of squared distances, while constraints (3) guarantee that each data item is assigned to exactly one cluster. Constraints (4) and (5) are the must-link constraints and cannot-link constraints, respectively. Capacity constraints (6) define a lower bound for the number of data items in each cluster.

For the MSSC problem defined by (2), (3), (7), it is convenient to replace the original problem with a natural relaxation where the binary constraints $x_{ij} \in \{0,1\}$ are replaced with continuous box constraints $x_{ij} \in [0,1]$. Indeed, if values of some variables x_{ij} are not integer (data item j is assigned to several equidistant centers), then an equivalent integer x_{ij} can be obtained by fixing one of them to one and the rest to zero. The resultant solution has the same objective value. Thus, any optimal fractional assignments may be represented as equivalent binary ones.

On the other hand, for any fixed assignments x_{ij} of data items, the objective function (2) is convex. Therefore, using the first order optimality conditions

$$\sum_{j=1}^{m} x_{ij}(a_l^j - y_l^i) = 0,$$

we obtain

$$y_l^i = \frac{\sum\limits_{j=1}^{m} x_{ij} a_l^j}{\sum\limits_{j=1}^{m} x_{ij}} \qquad \forall i = 1,\ldots,k; \; l = 1,\ldots,n. \qquad (8)$$

Thus, the optimal centers of clusters are the means (centroids) of the corresponding clusters. Substituting the centroids (8) into the objective function (2), one has the sum of square-error of all clusters.

In the SSMSSC problem, the aforementioned observations in general do not hold. If we replace binary constraints (7) by box constraints $x_{ij} \in [0,1]$, $\forall i = 1,\ldots,k, \forall j = 1,\ldots,m$, and substitute centroids (8) into (2), the obtained problem turns out to be a concave minimization (or convex maximization) problem. Moreover, extreme points of its feasible region are integer iff the problem includes the must-link constraints (4) [36] and the capacity constraints (6) with integer τ_i, $i = 1,\ldots,k$ [9]. But if there exist cannot-link constraints (5) in the problem, the assignment variables may turn out to be non-binary. In other words, the SSMSSC problem and its natural relaxation do not produce equivalent solutions, and the integrality constraints (7) cannot be relaxed.

In order to reduce the mixed integer program (2)–(7) to a non-convex optimization problem, one can replace binary constraints (7) by the following equivalent ones [21]:

$$x_{ij}^2 - x_{ij} \geq 0, \ \ 0 \leq x_{ij} \leq 1, \ \ \forall i = 1,\ldots,k; \ j = 1,\ldots,m. \qquad (9)$$

Summing up the quadratic constraints in (9) by index j, we have

$$0 \leq \sum_{j=1}^{m} \left(x_{ij}^2 - x_{ij}\right) = \sum_{j=1}^{m} \left(x_{ij}^2 - x_{ij} + \frac{1}{4} - \frac{1}{4}\right) = \left\|x^i - \frac{e}{2}\right\|^2 - \frac{m}{4},$$

where x^i denotes the vector of assignment variables for cluster i, $i = 1,\ldots,k$, $e = (1,\ldots,1)^T \in \mathbb{R}^m$. One can see that these inequalities hold if all assignment variables with respect to cluster i take integer values.

Applying these substitutions, our problem is to minimize a nonconvex objective function over a nonconvex feasible set:

$$
\left.
\begin{aligned}
&f(x,y) = \sum_{i=1}^{k} \sum_{j=1}^{m} x_{ij} \|y^i - a^j\|^2 \downarrow \min_{(x,y)}, \ \ y \in \mathbb{R}^{k \times n}, \\
&\left\|x^i - \frac{e}{2}\right\|^2 - \frac{m}{4} \geq 0, \ \ i = 1,\ldots,k, \\
&x \in S \cap S_M \cap S_L \subset \mathbb{R}^{k \times m},
\end{aligned}
\right\} \qquad (10)
$$

where $S = \{x_{ij} \in [0,1] : \sum_{i=1}^{k} x_{ij} = 1, \ \sum_{j=1}^{m} x_{ij} \geq \tau_i, \ i = 1,\ldots,k, \ j = 1,\ldots,m\}$, $S_M = \{x_{ij} : \ x_{ij_1} - x_{ij_2} = 0 \ \forall (j_1,j_2) \in M, \ i = 1,\ldots,k\}$, and $S_L = \{x_{ij} : \ x_{ij_3} + x_{ij_4} \leq 1 \ \forall \ (j_3,j_4) \in L, \ i = 1,\ldots,k\}$ are convex sets.

Thus, the problem (2)–(7) can be reduced to a nonconvex continuous optimization problem (10), where both objective function and feasible set are nonconvex.

3   SSMSSC Problem Without Cannot-Link Constraints

In this section, we consider a particular case of our SSMSSC problem that turns out be much simpler than the original problem. As was noted above, the semi-supervised clustering problem incorporating only the must-link and capacity constraints is much simpler in the sense that it can be reduced to an optimization problem in which only the objective function is non-convex, as the binary constraints on the assignment variables can be replaced with the box constraints. This makes the obtained problem formulation similar to that of the conventional MSSC problem defined by (2), (3), (7). Indeed, the latter can be formulated as a problem of minimizing a DC function over a convex set [16–19]. The objective function of this problem can be decomposed into the difference of two convex functions [17]. Thus, one can apply the special global search scheme based on the global optimality conditions [29,32] and develop an algorithm for finding quality clustering solutions, which demonstrated its effectiveness in comparison with the most widespread k-means algorithms [17].

Adding linear constraints (4) and (6) to the MSSC problem does not greatly affect the structure and difficulty of the resultant problem:

$$f(x, y) = \sum_{i=1}^{k} \sum_{j=1}^{m} x_{ij} \|y^i - a^j\|^2 \downarrow \min_{(x,y)}, \quad y \in \mathbb{R}^{k \times n}, \quad x \in S \cap S_M. \tag{11}$$

We can see that the objective function in problem (11) is the same as in MSSC, therefore we can apply the same DC decomposition as for the MSSC problem [17]. However, recall that it is well-known that a DC decomposition of the problem is not unique and may considerably influence the effectiveness of the solution algorithm.

First of all, in contrast to the MSSC problem, we formulate our SSMSSC without cannot-link constraints as follows. We substitute cluster centers y^i for centroids (8) in the objective function and consider the problem (11) as a problem of maximizing a convex (aka minimizing a concave) objective function (with respect to x) over a convex set, i.e.

$$h(x) \uparrow \max, \quad x \in S \cap S_M, \tag{12}$$

where $h(x) = - \sum_{i=1}^{k} \sum_{j=1}^{m} x_{ij} \left\| \dfrac{\sum_{t=1}^{m} x_{it} a^t}{\sum_{t=1}^{m} x_{it}} - a^j \right\|^2$ is a convex function [36].

To find local solutions of the convex maximization problem (11), one can apply a particular case of the well-known DC Algorithm [20, 29, 30] based on solving a sequence ($s = 0, 1, 2, \ldots$) of the following linearized (at some vector x^s) problems:

$$\langle \nabla h(x^s), x \rangle \uparrow \max_{x}, \quad x \in S \cap S_M, \tag{13}$$

where

$$\nabla_{ij} h(x) = - \left\| \dfrac{\sum_{t=1}^{m} x_{it} a^t}{\sum_{t=1}^{m} x_{it}} - a^j \right\|^2, \quad i = 1, \ldots, k, \; j = 1, \ldots, m. \tag{14}$$

One can see that the linearized problem (13) is a linear programming problem that can easily be solved. Note that making the substitutions for cluster centers render the liearized problem simpler than that used in [17].

We use the following stopping criterion for the local search algorithm [29, 30]

$$\langle \nabla h(x^s), x^{s+1} - x^s \rangle \leq \tau, \tag{15}$$

where $\tau > 0$ is a given accuracy $\tau > 0$. Thus, when the algorithm halts, it provides a local solution $z := x^s$ of the problem (12).

4   SSMSSC Problem

In this Section, we consider the general semi-supervised clustering problem with must-link, cannot-link, and cluster capacity constraints. With the addition of the cannot-link constraints (5), it is no longer possible to guarantee that the optimal solution of the problem

$$h(x) \uparrow \max, \quad x \in S \cap S_M \cap S_L. \tag{16}$$

is integer. Therefore, the semi-supervised clustering problem has to be supplemented by k so-called reverse-convex constraints, see (10). These non-convex constraints guarantee assignment variables to take value from the set $\{0, 1\}$.

The problem (10) is certainly more complicated than the problem (11) or (16), since non-convex is not only the objective function but also the feasible set. Nevertheless, it is a particular case of the general DC optimization problem, hence one can apply the global search search theory [29, 31]. SSMSSC can be written as the following non-convex problem:

$$\left. \begin{array}{l} h(x) \uparrow \max, \quad x \in S \cap S_M \cap S_L, \\ g_i(x) \geq 0, \quad i = 1, \dots, k, \end{array} \right\} \tag{10$'$}$$

where $g_i(x) = \left\| x^i - \dfrac{e}{2} \right\|^2 - \dfrac{m}{4}$, $i = 1, \dots, k$, and $h(x)$ are convex functions.

To find local solutions of the problem (10$'$), we apply a special local search method [30] based on the linearization of both non-convex objective function and constraints, thus making the obtained linearized problem convex. Therefore, the algorithm consists in finding solutions of a sequence ($s = 0, 1, 2, \dots$) of the following linearized (at some vector x^s) problems:

$$\left. \begin{array}{l} \langle \nabla h(x^s), x \rangle \uparrow \max_{x}, \quad x \in S \cap S_M \cap S_L, \\ \langle \nabla g_i(x^s), x - x^s \rangle + g_i(x^s) \geq 0, \quad i = 1, \dots, k, \end{array} \right\} \tag{17}$$

where $\nabla_{ij} g(x) = 2x_{ij} - 1$, $i = 1, \dots, k$, $j = 1, \dots, m$, $\nabla h(x)$ is defined by (14). Note that if $x_{ij}^s \in \{0, 1\}$, then $g_i(x^s) = 0 \; \forall i = 1, \dots, k$.

The problem (17) is obviously convex program and can be solved by any convex optimization method or appropriate solver. To obtain a local solution of the problem (10$'$) with a given accuracy $\tau > 0$, either the stopping criterion (15) or the following criterion must be met [30]:

$$\langle \nabla g(x^s), x^{s+1} - x^s \rangle \leq \tau. \tag{18}$$

Using the presented algorithms and techniques, we propose the following local search approach for the SSMSSC problem that consists of two basic steps:

1. Solving a sequence of the following linearized problems ($s = 0, 1, 2, \dots$)

$$\langle \nabla h(x^s), x \rangle \uparrow \max_{x}, \quad x \in S \cap S_M \cap S_L,$$

find a point z satisfied the stopping criterion (15).

2. If $z_{ij} \notin \{0,1\}$ $\forall i = 1,\ldots,k$, $j = 1,\ldots,m$, search for a feasible solution of the SSMSSC by solving the sequence of the linearized problems (17) starting with $x^0 := z$.

First, it tries to find a feasible solution of the SSMSSC problem with respect to all instance-level and cluster-level constraints (but without integrality constraints) by solving a series of the linearized problems. If the solution found turns out to be non-binary, i.e. there are data items that are assigned to several clusters, it uses the found solutions as a starting one for the local search method based on solving the linearized problems (17).

5   Computational Experiments

In this section, we describe some computational experiments to test the proposed technique for the semi-supervised version of the MSCC with must-link, cannot-link, and capacity constraints. We compare our approach to the k-means clustering algorithm in order to check whether the additional information can allow improving the accuracy of partitions. We implemented all the competing algorithms using C++ (compiled with Visual C++ compiler) and run them on a PC with Intel Core i7-4790K CPU 4.0 GHz. To solve a series of convex (linearized) problems, we used GUROBI 9.1 solver freely available for non-commercial research.

Our test bed consists of several datasets from KEEL-dataset repository[1] which are ordinary used to test clustering algorithms [15]. The authors of paper [15] utilized the true class labels to generate three sets of must-link and cannot-link constraints using the method proposed in [35]. These sets have different number of constraints. The instance name indicates which set of constraints is used. For example, for the instance *zoo*, we test algorithms with three sets of constraints, thus having *zoo*(0.1), *zoo*(0.15), *zoo*(0.2) problem instances. First, we test the effectiveness and efficiency of the special local search method (LSM) and report the computational results in Table 1, where the following denotations are used:

- m is the number of data items;
- k—the number of clusters;
- n—the number of features;
- *Start Obj.Val.*—the value of the objective function at the starting point;
- *Loc Obj.Val.*—the value of the objective function for the solution provided by k-means (MSSC, the first row for each dataset bolded) and by the LSM (for the SSMSSC);
- *Iters* is the number of iterations (linearized problems solved) of the LSM;
- *Time*—run time in seconds.

The starting point was chosen randomly and was the same for the k-means heuristic and LSM.

---

[1] https://sci2s.ugr.es/keel/datasets.php.

Table 1. The testing of LSM

name	m	k	n	Start Obj.Val.	Loc. Obj.Val.	Iters	Time
zoo	101	7	16	1823.00	**149.55**		
zoo(0.10)					149.55	2	0.01
zoo(0.15)					129.39	5	0.03
zoo(0.20)					129.39	5	0.03
iris	150	3	4	3196.29	**142.86**		
iris(0.10)					89.39	2	0.01
iris(0.15)					83.99	3	0.02
iris(0.20)					78.95	3	0.02
glass	214	6	9	1946.45	**375.60**		
glass(0.10)					415.39	4	0.04
glass(0.15)					392.51	3	0.03
glass(0.20)					392.51	3	0.03
ecoli	336	8	7	1792113.00	**296069.86**		
ecoli(0.10)					296069.86	3	0.02
ecoli(0.15)					296069.86	6	0.07
ecoli(0.20)					296069.86	6	0.07
led7digit	500	10	7	1734.00	**269.56**		
led7digit(0.10)					232.28	6	0.09
led7digit(0.15)					271.80	9	0.12
led7digit(0.20)					232.28	10	0.14
hayesroth	160	3	4	1242.00	**375.28**		
hayesroth(0.10)					375.28	2	0.01
hayesroth(0.15)					375.28	2	0.01
hayesroth(0.20)					379.88	3	0.01

As it is shown in Table 1, the LSM finds a solution in very short CPU time. In problems *glass*, *ecoli*, and *hayesroth*, the LSM failed to find assignments better in terms of the objective function value than the *k*-means clustering algorithm. Indeed, a better value of the *k*-means objective does not necessarily correspond to a better partition. In the remaining three problems, new better assignments was found. The most significant improvement was obtained for the *iris* test instance. Here, we can see that the largest number of additional constraints results in the best solution with value 78.95.

Beside the value of the objective function, we also report the values of the following external measures, Pairwise Precision, Recall, and F-measure to assess clustering accuracy (see three last columns in Tables 2 and Table 3):

$$Precision = \frac{TP}{TP + FP}, \ Recall = \frac{TP}{TP + FN},$$

where TP are true positive pairs of data items (correctly clustered), FP—false positive pairs, and FN—false negative pairs. The pairwise F-measure is defined as harmonic mean of pairwise precision and recall:

$$F - measure = \frac{2 \cdot Precision \cdot Recall}{Precision + Recall}.$$

Table 2. Clustering results via external measures: case $S_L = \emptyset$

name	m	k	n	Precision	Recall	F-meas
zoo	101	7	16	**0.75996**	**0.648258**	**0.699679**
zoo(0.10)				0.75996	0.648258	0.699679
zoo(0.15)				0.769313	0.609176	0.679943
zoo(0.20)				0.769313	0.609176	0.679943
iris	150	3	4	**0.548473**	**0.786667**	**0.646322**
iris(0.10)				0.812175	0.849524	0.83043
iris(0.15)				0.798209	0.82449	0.811136
iris(0.20)				0.798209	0.82449	0.811136
glass	214	6	9	**0.417349**	**0.559872**	**0.478217**
glass(0.10)				0.362061	0.42003	0.388898
glass(0.15)				0.415027	0.548556	0.472539
glass(0.20)				0.415027	0.548556	0.472539
ecoli	336	8	7	**0.606874**	**0.339166**	**0.435142**
ecoli(0.10)				0.606874	0.339166	0.435142
ecoli(0.15)				0.606874	0.339166	0.435142
ecoli(0.20)				0.606874	0.339166	0.435142
led7digit	500	10	7	**0.412008**	**0.514263**	**0.457491**
led7digit(0.10)				0.551652	0.551652	0.551652
led7digit(0.15)				0.409698	0.489525	0.446068
led7digit(0.20)				0.551652	0.551652	0.551652
hayesroth	160	3	4	**0.388773**	**0.367463**	**0.377818**
hayesroth(0.10)				0.388773	0.367463	0.377818
hayesroth(0.15)				0.388773	0.367463	0.377818
hayesroth(0.20)				0.371191	0.371191	0.371191

Table 2 demonstrates the clustering results obtained by solving the SSMSSC in the statement (11) (without cannot-link constraints) for all datasets considered, and reports the values of external measures found by the competing algorithms. The *Precision*, *Recall*, and *F−measure* provided by the k-means algorithm are bolded in the first row for each dataset; the next 3 rows correspond to the values of external measures obtained by LSM.

Table 3 shows the clustering results obtained by solving the SSMSSC in the statement (10) (with must-link, cannot-link, and cluster capacity constraints) for only three datasets, since the results for the remaining ones turn out to be the same as in Table 2.

We can see in Table 3 that adding additional constraints results in better partitions for almost all problem instances. However, in many cases, the additional instance-level information does not allow finding better clustering solutions.

Table 3. Clustering results via external measures: case $S \cap S_M \cap S_L \neq \emptyset$

name	m	k	n	Precision	Recall	F-meas
iris	150	3	4	**0.548473**	**0.786667**	**0.646322**
iris(0.10)				0.922805	0.923810	0.923307
iris(0.15)				0.798209	0.824490	0.811136
iris(0.20)				0.824324	0.829932	0.827119
glass	214	6	9	**0.417349**	**0.559872**	**0.478217**
glass(0.10)				0.394282	0.377301	0.385605
glass(0.15)				0.414368	0.537747	0.468063
glass(0.20)				0.415952	0.556663	0.476129
led7digit	500	10	7	**0.412008**	**0.514263**	**0.457491**
led7digit(0.10)				0.410615	0.509347	0.454683
led7digit(0.15)				0.412008	0.514263	0.457491
led7digit(0.20)				0.410615	0.509347	0.454683

6 Conclusion

In this paper, we addressed a variant of the so-called semi-supervised minimum-sum-of-squares clustering problem. It is the constrained MSSC problem that incorporates must-link, cannot-link, and cluster capacity constraints. We formulated this clustering problem as a convex maximization problem which is a particular case of DC (difference of convex) optimization problem. Then, we develop a solution approach to this problem based on a special (DC) local search method. In our computational experiments we demonstrated that the proposed approach is competitive with conventional k-means heuristics and may provide better solutions for problem instances of relatively small size.

Our further research will be focused on improving the proposed methodology to make our algorithm tractable for large-scale problem instances involving thousands of data items.

References

1. Aloise, D., Deshpande, A., Hansen, P., Popat, P.: NP-hardness of Euclidean sum-of-squares clustering. Mach. Learn. **75**, 245–248 (2009). https://doi.org/10.1007/s10994-009-5103-0
2. Aloise, D., Hansen, P., Liberti, L.: An improved column generation algorithm for minimum sum-of-squares clustering. Math. Program. **131**(1–2), 195–220 (2012). https://doi.org/10.1007/s10107-010-0349-7
3. Avella, P., Boccia, M., Sforza, A., Vasilyev, I.: An effective heuristic for large-scale capacitated facility location problems. J. Heuristics **15**(6), 597–615 (2008). https://doi.org/10.1007/s10732-008-9078-y
4. Babaki, B., Guns, T., Nijssen, S.: Constrained clustering using column generation. In: Simonis, H. (ed.) CPAIOR 2014. LNCS, vol. 8451, pp. 438–454. Springer, Cham (2014). https://doi.org/10.1007/978-3-319-07046-9_31

5. Basu, S., Banerjee, A., Mooney, R.J.: Active semi-supervision for pairwise constrained clustering. In: Berry, M.W., Kamath, C., Dayal, U., Skillicorn, D. (eds.) Proceedings of the 2004 SIAM International Conference on Data Mining, pp. 333–344. SIAM (2004). https://doi.org/10.1137/1.9781611972740.31
6. Basu, S., Davidson, I., Wagstaff, K.: Constrained Clustering: Advances in Algorithms, Theory, and Applications. Chapman & Hall, Boca Raton (2008)
7. Baumann, P.: A binary linear programming-based k-means algorithm for clustering with must-link and cannot-link constraints. In: 2020 IEEE International Conference on Industrial Engineering and Engineering Management, pp. 324–328. IEEE, New York (2020). https://doi.org/10.1109/IEEM45057.2020.9309775
8. Boccia, M., Sforza, A., Sterle, C., Vasilyev, I.: A cut and branch approach for the capacitated p-median problem based on fenchel cutting planes. J. Math. Model. Algor. **7**, 43–58 (2008). https://doi.org/10.1007/s10852-007-9074-5
9. Bradley, P.S., Bennett, K.P., Demiriz, A.: Constrained k-means clustering. Microsoft Res. Redmond 1–8 (2000)
10. Davidson, I., Ravi, S.S.: The complexity of non-hierarchical clustering with instance and cluster level constraints. Data Min. Knowl. Disc. **14**, 25–61 (2007). https://doi.org/10.1007/s10618-006-0053-7
11. Gambella, C., Ghaddar, B., Naoum-Sawaya, J.: Optimization problems for machine learning: a survey. Eur. J. Oper. Res. **290**(3), 807–828 (2021). https://doi.org/10.1016/j.ejor.2020.08.045
12. Gançarski, P., Dao, T.-B.-H., Crémilleux, B., Forestier, G., Lampert, T.: Constrained clustering: current and new trends. In: Marquis, P., Papini, O., Prade, H. (eds.) A Guided Tour of Artificial Intelligence Research, pp. 447–484. Springer, Cham (2020). https://doi.org/10.1007/978-3-030-06167-8_14
13. Gao, J., Tao, X., Cai, S.: Towards more efficient local search algorithms for constrained clustering. Inf. Sci. **621**, 287–307 (2023). https://doi.org/10.1016/j.ins.2022.11.107
14. Gnägi, M., Baumann, P.: A matheuristic for large-scale capacitated clustering. Comput. Oper. Res. **132**, 105304 (2021). https://doi.org/10.1016/j.cor.2021.105304
15. González-Almagro, G., Luengo, J., Cano, J.R., García, S.: DILS: constrained clustering through dual iterative local search. Comput. Oper. Res. **121**, 104979 (2020). https://doi.org/10.1016/j.cor.2020.104979
16. Gruzdeva, T.V., Ushakov, A.V.: A computational study of the DC minimization global optimality conditions applied to K-means clustering. In: Olenev, N.N., Evtushenko, Y.G., Jaćimović, M., Khachay, M., Malkova, V. (eds.) OPTIMA 2021. LNCS, vol. 13078, pp. 79–93. Springer, Cham (2021). https://doi.org/10.1007/978-3-030-91059-4_6
17. Gruzdeva, T.V., Ushakov, A.V.: K-means clustering via a nonconvex optimization approach. In: Pardalos, P., Khachay, M., Kazakov, A. (eds.) MOTOR 2021. LNCS, vol. 12755, pp. 462–476. Springer, Cham (2021). https://doi.org/10.1007/978-3-030-77876-7_31
18. Gruzdeva, T.V., Ushakov, A.V.: On a nonconvex distance-based clustering problem. In: Pardalos, P., Khachay, M., Mazalov, V. (eds.) MOTOR 2022. LNCS, vol. 13367, pp. 139–152. Springer, Cham (2022). https://doi.org/10.1007/978-3-031-09607-5_10
19. Hoai An, L.T., Hoai Minh, L., Tao, P.D.: New and efficient DCA based algorithms for minimum sum-of-squares clustering. Pattern Recognit. **47**(1), 388–401 (2014). https://doi.org/10.1016/j.patcog.2013.07.012

20. Hoai An, L.T., Tao, P.D.: The DC (difference of convex functions) programming and DCA revisited with dc models of real world nonconvex optimization problems. Ann. Oper. Res. **133**, 23–46 (2005)
21. Horst, R., Pardalos, P., Thoai, N.: Introduction to Global Optimization. Nonconvex Optimization and Its Applications, Springer, Heidelberg (2001)
22. Mahajan, M., Nimbhorkar, P., Varadarajan, K.: The planar k-means problem is NP-hard. Theor. Comput. Sci. **442**, 13–21 (2012). https://doi.org/10.1016/j.tcs. 2010.05.034.Special Issue on the Workshop on Algorithms and Computation (WALCOM 2009)
23. Mulvey, J.M., Beck, M.P.: Solving capacitated clustering problems. Eur. J. Oper. Res. **18**(3), 339–348 (2003)
24. Negreiros, M., Palhano, A.: The capacitated centred clustering problem. Comput. Oper. Res. **33**(6), 1639–1663 (2006). https://doi.org/10.1016/j.cor.2004.11.011
25. Nghiem, N.-V.-D., Vrain, C., Dao, T.-B.-H., Davidson, I.: Constrained clustering via post-processing. In: Appice, A., Tsoumakas, G., Manolopoulos, Y., Matwin, S. (eds.) DS 2020. LNCS (LNAI), vol. 12323, pp. 53–67. Springer, Cham (2020). https://doi.org/10.1007/978-3-030-61527-7_4
26. de Oliveira, R.M., Chaves, A.A., Lorena, L.A.N.: A comparison of two hybrid methods for constrained clustering problems. Appl. Soft Comput. **54**, 256–266 (2017). https://doi.org/10.1016/j.asoc.2017.01.023
27. Piccialli, V., Russo Russo, A., Sudoso, A.M.: An exact algorithm for semi-supervised minimum sum-of-squares clustering. Comput. Oper. Res. **147**, 105958 (2022). https://doi.org/10.1016/j.cor.2022.105958
28. Rutayisire, T., Yang, Y., Lin, C., Zhang, J.: A modified cop-Kmeans algorithm based on sequenced cannot-link set. In: Yao, J.T., Ramanna, S., Wang, G., Suraj, Z. (eds.) RSKT 2011. LNCS (LNAI), vol. 6954, pp. 217–225. Springer, Heidelberg (2011). https://doi.org/10.1007/978-3-642-24425-4_30
29. Strekalovsky, A.S.: On solving optimization problems with hidden nonconvex structures. In: Rassias, T.M., Floudas, C.A., Butenko, S. (eds.) Optimization in Science and Engineering, pp. 465–502. Springer, New York (2014). https://doi.org/ 10.1007/978-1-4939-0808-0_23
30. Strekalovsky, A.S.: On local search in D.C. optimization problems. Appl. Math. Comput. **255**, 73–83 (2015)
31. Strekalovsky, A.S.: On global optimality conditions for D.C. minimization problems with D.C. constraints. J. Appl. Numer. Optim. **3**, 175–196 (2021)
32. Strekalovsky, A.: On the minimization of the difference of convex functions on a feasible set. Comput. Math. Math. Phys. **43**, 380–390 (2003)
33. Tan, W., Yang, Y., Li, T.: An improved cop-kmeans algorithm for solving constraint violation. In: Ruan, D., Li, T., Chen, G. (eds.) Computational Intelligence, World Scientific Proceedings Series on Computer Engineering and Information Science, vol. 4, pp. 690–696. World Scientific Publishing (2010). https://doi.org/10. 1142/9789814324700_0104
34. Vasilyev, I.L., Ushakov, A.V.: Discrete facility location in machine learning. J. Appl. Ind. Math. **15**(4), 686–710 (2021). https://doi.org/10.1134/ S1990478921040128
35. Wagstaff, K., Cardie, C., Rogers, S., Schrödl, S.: Constrained k-means clustering with background knowledge. In: Brodley, C.E., Pohoreckyj Danyluk, A. (eds.) Proceedings of the Eighteenth International Conference on Machine Learning, pp. 577–584. Morgan Kaufmann Publishers Inc., San Francisco (2001)
36. Xia, Y.: A global optimization method for semi-supervised clustering. Data Min. Knowl. Disc. **18**, 214–256 (2009). https://doi.org/10.1007/s10618-008-0104-3

On the Uniqueness of Identification the Thermal Conductivity and Heat Capacity of Substance

Vladimir Zubov[✉][iD]

Federal Research Center "Computer Science and Control" of the Russian Academy of Sciences, Moscow, Russia
vladimir.zubov@mail.ru

Abstract. The paper describes and investigates an algorithm for simultaneous identification the temperature-dependent volumetric heat capacity and thermal conductivity of a substance that is based on the results of experimental observation of the temperature field dynamics in the object. The algorithm is based on the first boundary value problem for a one-dimensional non-stationary heat equation. The considered inverse coefficient problem is reduced to a variational problem, that is solved by gradient methods. The gradient of the cost functional is calculated with the help of the Fast Automatic Differentiation technique. The results of numerical solution to the considered inverse problem are presented. It is found that some inverse problems have a single solution, the solution of other problems is not the only one. The conditions of uniqueness of the solution to the inverse problem are formulated and justified. A class of temperature fields from the reachable sets is specified, for which the solution to the identification problem under consideration will be non-unique.

Keywords: Thermal conductivity · Inverse coefficient problems · Heat equation · Gradient

1 Introduction

When creating new materials, one has to face a situation where the thermophysical parameters of a substance depend only on temperature, and this dependence is unknown. In this regard, for example, the problem arises of determining the dependence of the volumetric heat capacity and thermal conductivity of a substance on temperature based on the results of experimental observation of the temperature field dynamics. They come to the same problem when they want to describe a complex thermal process by some simplified mathematical model. An example is the mathematical modeling of heat transfer processes in complex porous composite materials, where radiative heat transfer plays a significant role (see [1,2]).

The fact that the problem of determining the thermal conductivity and the volumetric heat capacity of a substance is relevant can be judged, for example,

M. Khachay et al. (Eds.): MOTOR 2023, LNCS 13930, pp. 68–82, 2023.
https://doi.org/10.1007/978-3-031-35305-5_5

from publications [3–5]. It should also be noted that the mathematical formulations of this general problem are different in different papers. The methods used to solve it are also different.

In [6], one of the possible formulations of the inverse coefficient problem is proposed, which is designed to simultaneously identify the temperature-dependent thermal conductivity of a substance and its volumetric heat capacity. The study of this problem was carried out on the basis of the first boundary value problem for a one-dimensional non-stationary heat equation. The inverse coefficient problem was reduced to a variational problem. The standard deviation of the calculated temperature field in the sample from its experimental value was chosen as the cost functional. An algorithm for the numerical solution to the inverse coefficient problem is proposed. This algorithm is based on the modern Fast Automatic Differentiation technique (FAD-technique, see [7,8]), which has made it possible to successfully solve a number of complex optimal control problems for dynamical systems. For example, based on this methodology, the author developed an algorithm for solving the problem of identifying the thermal conductivity of a substance in one-dimensional, two-dimensional and three-dimensional settings (see [9–13]). In all the examples given in [6], the solution to the formulated inverse problem turned out to be nonunique.

In this paper, we give examples for which the solution to the inverse problem is unique, and also study the conditions for the uniqueness of the resulting solution to the formulated inverse problem.

2   Statement of the Problem

A layer of material of width L is considered. The temperature of this layer at the initial time is given. It is also known how the temperature on the boundary of this layer changes in time. The distribution of the temperature field at each instant of time is described by the following initial boundary value (mixed) problem:

$$C(T)\frac{\partial T(x,t)}{\partial t} - \frac{\partial}{\partial x}\left(K(T)\frac{\partial T(x,t)}{\partial x}\right) = 0, \qquad (x,t) \in Q, \qquad (1)$$

$$T(x,0) = w_0(x), \qquad\qquad 0 \le x \le L, \qquad (2)$$

$$T(0,t) = w_1(t), \qquad T(L,t) = w_2(t), \qquad 0 \le t \le \Theta. \qquad (3)$$

Here x is the Cartesian coordinate in the layer; t is time; $T(x,t)$ is the temperature of the material at the point with the coordinates x at time t; $Q = \{(0 < x < L) \times (0 < t < \Theta)\}$; $C(T)$ is the volumetric heat capacity of the material; $K(T)$ is the thermal conductivity of the material; $w_0(x)$ is the given temperature at the initial time $t = 0$; $w_1(t)$ is the given temperature on the left boundary of the layer; $w_2(t)$ is the given temperature on the right boundary of the layer.

If the dependencies of the volumetric heat capacity $C(T)$ of a substance and its thermal conductivity $K(T)$ on the temperature T are known, then we can solve the mixed problem (1)–(3) to find the temperature distribution $T(x,t)$ in $\overline{Q}$. Below, problem (1)–(3) will be called the direct problem.

If the dependencies of the volumetric heat capacity of a substance and its thermal conductivity on the temperature T are not known, it is of great interest to determine these dependencies. A possible statement of this problem (it is classified as an identification problem of the model parameters) is as follows: find the dependencies of $C(T)$ and $K(T)$ on T under which the temperature field $T(x,t)$, obtained by solving the mixed problem (1)–(3), is close to the field $Y(x,t)$ obtained experimentally. The quantity

$$\Phi(C(T),K(T)) = \int_0^\Theta \int_0^L [T(x,t) - Y(x,t)]^2 \cdot \mu(x,t)dx\,dt \qquad (4)$$

can be used as the measure of difference between these functions. Here $\mu(x,t) \geq 0$ is a given weighting function. Thus, the optimal control problem is to find the optimal control $\{C(T),K(T)\}$ and the corresponding solution $T(x,t)$ to the problem (1)–(3) that minimize functional (4).

It should be noted that in the presented formulation, the optimal control problem always has a non-unique solution. Indeed, if $\{C^*(T),K^*(T)\}$ is the solution to the formulated inverse problem, then for any real number λ the control $\{\lambda C^*(T),\lambda K^*(T)\}$ is also a solution to this problem. To single out the unique solution to the formulated identification problem, it is necessary to use an additional condition. Such a condition can be, for example, a condition $K(T_*) = K_*$ with given numbers T_* and K_*. It is also possible to determine the relation $\chi(T) = \frac{C(T)}{K(T)}$ when solving the problem.

3   On the Numerical Solution to the Problem

When solving the optimal control problem numerically the unknown functions $C(T)$ and $K(T)$, $T \in [a,b]$ were approximated by continuous piecewise linear functions. The domain $Q = \{(x,t) : (0 < x < L) \times (0 < t < \Theta)\}$ was decomposed by the grid lines $\{\widetilde{x}_i\}_{i=0}^I$ and $\{\widetilde{t}^j\}_{j=0}^J$ into rectangles with steps $h_i = \widetilde{x}_{i+1} - \widetilde{x}_i$, $(i = \overline{0,I-1})$, $\tau^j = \widetilde{t}^j - \widetilde{t}^{j-1}$, $(j = \overline{1,J})$. At each node $(\widetilde{x}_i, \widetilde{t}^j)$ of Q characterized by the pair of indices (i,j) all the functions are determined by their values at the point $(\widetilde{x}_i, \widetilde{t}^j)$ (e.g., $T(\widetilde{x}_i, \widetilde{t}^j) = T_i^j$). In each rectangle, the thermal balance must be preserved. The result is a finite-difference scheme approximating the mixed problem (1)–(3). The resulting system of nonlinear algebraic equations was solved iteratively using the Gaussian elimination. This approach was used to solve the mixed problem (1)–(3), and the function $T(x,t)$ (more precisely, its approximation T_i^j) was found.

The cost functional (4) was approximated by a function $F(c_0, c_1, \ldots, c_N, k_0, k_1, \ldots, k_N)$ of a finite number of variables as follows:

$$\Phi(C(T),K(T)) \cong F = \sum_{j=1}^J \sum_{i=1}^{I-1} \left((T_i^j - Y_i^j)^2 \cdot \mu_i^j h_i \tau^j \right).$$

For more details on the approximation of the mixed problem and the cost functional, see [9].

To minimize the cost functional, the gradient method was used. It is well known that for gradient methods to work effectively, the exact value of the cost function gradient should be used. Calculate the value of the gradient efficiently and with machine accuracy allows FAD-technique, (see [7,8]). Therefore, in this work, the cost function gradient was calculated using the Adept application software package (see [14]), using the reverse (Reverse, Adjoint) automatic differentiation method. The L-BFGS-B method was used to find the minimum of the cost function (see [15]).

4   Numerical Results

It should be noted that the minimum value of the cost functional (4) obtained by solving the formulated inverse problem depends on the given experimental field $Y(x,t)$. If the experimental field $Y(x,t)$ belongs to the reachable sets of the functions $T(x,t)$ constructed as a result of solving the direct problem (1)–(3) for admissible $C(T)$ and $K(T)$, then the value of the functional is equal to zero. Otherwise, the functional value is greater than zero. When carrying out numerical experiments, the experimental temperature field $Y(x,t)$ was chosen from the reachable sets of solutions to the direct problem.

A large number of test calculations were performed. These calculations can be conditionally divided into two groups. All examples of the first group were based on the fact that the function

$$T_*(x,t) = \left(\frac{n+1}{\beta}\right)^{\frac{1}{n+1}} \cdot [\alpha \cdot (t+\gamma-x)]^{\frac{1}{m-n}}, \qquad \alpha = \frac{m-n}{n+1}, \qquad \beta = (n+1)^{\frac{m-n}{m+1}},$$

is an analytical solution to Eq. (1) for

$$\gamma = const, \qquad C(T) = \beta \cdot T^n, \qquad K(T) = T^m.$$

In accordance with this, the following input data of the problem were used:

$$L = 1, \qquad \Theta = 1, \qquad \gamma = 1.5, \qquad Q = (0,1) \times (0,1),$$

$$w_0(x) = T_*(x,0) = \left(\frac{n+1}{\beta}\right)^{\frac{1}{n+1}} \cdot [\alpha \cdot (\gamma-x)]^{\frac{1}{m-n}}, \qquad (0 \le x \le 1),$$

$$w_1(t) = T_*(0,t) = \left(\frac{n+1}{\beta}\right)^{\frac{1}{n+1}} \cdot [\alpha \cdot (t+\gamma)]^{\frac{1}{m-n}}, \qquad (0 \le t \le 1), \quad (5)$$

$$w_2(t) = T_*(1,t) = \left(\frac{n+1}{\beta}\right)^{\frac{1}{n+1}} \cdot [\alpha \cdot (t+\gamma-1)]^{\frac{1}{m-n}}, \qquad (0 \le t \le 1),$$

$$Y(x,t) = T_*(x,t), \qquad (x,t) \in Q,$$

$$a = \min_{(x,t)\in\overline{Q}} Y(x,t), \qquad b = \max_{(x,t)\in\overline{Q}} Y(x,t).$$

For the solution to the direct problem, we used the uniform grid with the parameters $I = 40$ (the number of intervals along the axis x) and $J = 700$ (the number of intervals along the axis t), which ensures the sufficient accuracy of computed temperatures and pulses. The segment $[a, b]$ was partitioned into 80 intervals ($N = 80$).

Test calculations of the first group were carried out at different values of the parameters m and n, at different initial approximations to the desired functions. Some of the results of the first group of calculations are described in detail and analyzed in [6]. Based on the results obtained by solving the inverse problems of the first group, it was concluded that the use of a function $T_*(x, t)$ as an experimental field leads to non-uniqueness solution to the inverse problem of determining the volumetric heat capacity and thermal conductivity. In all the cases considered, for fixed m and n, the ratio $\chi(T) = C(T)/K(T)$ changed with a change in the initial approximation.

As the first example of the second group, consider the inverse problem (1)–(4) with the following input data:

$$L = 1, \qquad \Theta = 1, \qquad Q = (0, 1) \times (0, 1),$$

$$w_0(x) = x^2, \qquad\qquad (0 \leq x \leq 1),$$

$$w_1(t) = T_*(0, t) = t, \qquad w_2(t) = T_*(1, t) = t+1, \qquad (0 \leq t \leq 1), \quad (6)$$

$$Y(x, t) = T_*(x, t) = x^2 + t, \qquad\qquad (x, t) \in Q,$$

$$a = 0, \qquad b = 2.$$

It is not difficult to verify that the function $T_*(x, t) = x^2 + t$ is an analytical solution to Eq. (1) for $C(T) = 2$ and $K(T) = 1$, so that the pair $\{C(T), K(T)\}$ and the function $T_*(x, t) = x^2 + t$ are the optimal control and the corresponding optimal solution to the inverse problem (1)–(4) with data (6).

The formulated problem was solved numerically for different initial approximations, for example, $C(T) = 2T$, $K(T) = \frac{1}{4}T^2$, $K(2) = 1$. Regardless of the initial approximation, the same solution $C(T) = 2$, $K(T) = 1$ was obtained, with an accuracy 10^{-12} in the norm C. As it turned out, this is due to the specifics of the experimental temperature field used. The uniqueness of the resulting solution can be explained as follows.

Assuming that the function $T(x, t) \in C_{x,t}^{2,1}(Q) \cap C(\overline{Q})$, $C(T) \in C([a, b])$, $K(T) \in C^1([a, b])$, Eq. (1) can be written as

$$C(T) \cdot \frac{\partial T(x, t)}{\partial t} - K'(T) \cdot \left(\frac{\partial T(x, t)}{\partial x} \right)^2 - K(T) \cdot \frac{\partial^2 T(x, t)}{\partial x^2} = 0. \qquad (7)$$

Substituting the function $T_*(x, t) = x^2 + t$ into Eq. (7), we obtain an equality that is valid for all $(x, t) \in Q$

$$C(T_*) \cdot 1 - K'(T_*) \cdot 4x^2 - K(T_*) \cdot 2 = 0.$$

Hence it follows that along an arbitrary isotherm $x^2 + t = \tau$, $(\tau \in (0.0; 2.0))$ of the field $T_*(x, t)$, the relation

$$[C(\tau) - 2 \cdot K(\tau)] = K'(\tau) \cdot 4x^2, \quad \tau = const, \quad x \in \begin{cases} (0.0, \sqrt{\tau}), & 0 < \tau \leq 1, \\ (\sqrt{\tau - 1}, 1.0), & 1 < \tau < 2, \end{cases}$$

must be fulfilled, which is possible only if $K'(\tau) = 0$, $C(\tau) = 2 \cdot K(\tau)$. Taking into account the condition $K(2) = 1$, we obtain the solution to the inverse problem $C(T) = 2$, $K(T) = 1$.

In the second example the inverse problem (1)–(4) was considered with the following input data:

$$L = 1, \qquad \Theta = 1, \qquad\qquad Q = (0, 1) \times (0, 1),$$

$$w_0(x) = 9/(x + 1), \qquad\qquad\qquad (0 \leq x \leq 1),$$
$$w_1(t) = 9/(5t + 1), \qquad\qquad\qquad (0 \leq t \leq 1),$$
$$w_2(t) = 9/(7t + 2), \qquad\qquad\qquad (0 \leq t \leq 1),$$
$$Y(x, t) = T_{**}(x, t), \qquad\qquad\qquad (x, t) \in Q,$$
$$a = 1, \qquad\qquad b = 9, \qquad\qquad K(1) = 1,$$

where the experimental field $Y(x, t) = T_{**}(x, t)$ was constructed numerically as a solution to the following mixed problem (1)–(3):

$$C(T)\frac{\partial T(x, t)}{\partial t} - \frac{\partial}{\partial x}\left(K(T)\frac{\partial T(x, t)}{\partial x}\right) = 0, \qquad (0 < x < 1), \qquad (0 < t < 1),$$

$$T(x, 0) = 9/(x + 1), \qquad\qquad\qquad\qquad (0 \leq x \leq 1),$$
$$T(0, t) = 9/(5t + 1), \qquad\qquad\qquad\qquad (0 \leq t \leq 1),$$
$$T(L, t) = 9/(7t + 2), \qquad\qquad\qquad\qquad (0 \leq t \leq 1),$$

at

$$K(T) = \begin{cases} 0.1 \cdot (T - 3) \cdot (T - 6) \cdot (T - 7) + 3.4, & T \geq 3, \\ 1.2 \cdot (T - 3) + 3.4, & T < 3, \end{cases}$$

$$C(T) = 4 \cdot T^2, \qquad T \in [a, b], \qquad a = 1, \qquad b = 9.$$

Here, as in the first example the inverse problem was solved numerically with different initial approximations. Regardless of the initial approximation, the same solution was obtained with high accuracy. In Fig. 1 and Fig. 2 show the results of solving the problem with the initial approximation $C(T) = 10$, $K(T) = 1$, which confirm that the only solution is determined by the proposed algorithm with high accuracy.

The considered examples allow us to conclude that there are such fields from the reachable sets for which the solution to the inverse problem of determining the volumetric heat capacity and the thermal conductivity is unique.

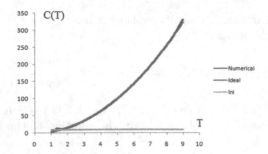

Fig. 1. Distribution of volumetric heat capacity.

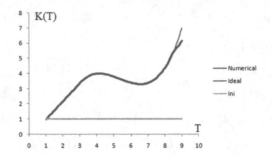

Fig. 2. Distribution of thermal conductivity.

5 Analysis of the Obtained Results

The computational experiments performed have shown that the uniqueness of the solution to the problem of simultaneous identification the volumetric heat capacity and the thermal conductivity in the proposed formulation depends significantly on the experimental field used. These experiments allowed us to draw some conclusions.

We substantiate these conclusions under the following assumptions.

The desired volumetric heat capacity $C(T)$ and thermal conductivity $K(T)$ must belong to the *definition scope*, i.e. satisfy the conditions: $C(T) \in C([a, b])$, $K(T) \in C^1([a, b])$, $C(T) > 0$, $K(T) > 0$, for all $T \in [a, b]$. By the *reachable sets* we mean the set of functions $T(x, t) \in C^{2,1}_{x,t}(Q) \cap C(\overline{Q})$ such that they satisfy conditions (2)–(3) and for each of them in the definition scope there are functions $C(T)$ and $K(T)$, for which $T(x, t)$ is a solution to the mixed problem (1)–(3).

Under these assumptions, Eq. (1) can be represented in the form (7).

Further, by the τ-isoline of the function $T(x, t)$, we will define, as is customary, the set of points (x, t) of the domain Q in which the value of the function $T(x, t)$ has the same value τ. At the same time, we note that an isoline (set) can be a union of several unconnected lines.

The following assertions are true.

Lemma 1. *Let the function $T(x,t)$ belong to the reachable sets. If on some τ-isoline, $T(x,t) = \tau \in [a,b]$, the functions $\frac{\partial T(x,t)}{\partial t}$ and $\left(\frac{\partial T(x,t)}{\partial x}\right)^2$, considered as functions of some parameter along this isoline, are linearly independent, then the ratio $\chi(T) = \frac{C(T)}{K(T)}$ has a single value at $T = \tau$.*

Proof. Indeed, let the inverse problem have two solutions $\{C_1(T), K_1(T)\}$ and $\{C_2(T), K_2(T)\}$ with two different ratios $\chi_1(T) = \frac{C_1(T)}{K_1(T)}$ and $\chi_2(T) = \frac{C_2(T)}{K_2(T)}$. This means that two equalities of type (7) hold for everyone $(x,t) \in Q$:

$$C_1(T)\frac{\partial T(x,t)}{\partial t} - K_1'(T)\left(\frac{\partial T(x,t)}{\partial x}\right)^2 - K_1(T)\frac{\partial^2 T(x,t)}{\partial x^2} = 0, \qquad (8)$$

$$C_2(T)\frac{\partial T(x,t)}{\partial t} - K_2'(T)\left(\frac{\partial T(x,t)}{\partial x}\right)^2 - K_2(T)\frac{\partial^2 T(x,t)}{\partial x^2} = 0. \qquad (9)$$

Let us divide both parts of relation (8) by $K_1(T) > 0$, and both parts of relation (9) by $K_2(T) > 0$. The difference of the relations obtained in this case has the form that is valid for all $(x,t) \in Q$:

$$\left(\frac{C_1(T)}{K_1(T)} - \frac{C_2(T)}{K_2(T)}\right) \cdot \frac{\partial T(x,t)}{\partial t} - \left(\frac{K_1'(T)}{K_1(T)} - \frac{K_2'(T)}{K_2(T)}\right) \cdot \left(\frac{\partial T(x,t)}{\partial x}\right)^2 = 0. \quad (10)$$

At the points $T(x,t) = \tau$ of the τ-isoline the linear independence of the functions $\frac{\partial T(x,t)}{\partial t}$ and $\left(\frac{\partial T(x,t)}{\partial x}\right)^2$ together with condition (10) allows us to conclude that

$$\frac{C_1(\tau)}{K_1(\tau)} = \frac{C_2(\tau)}{K_2(\tau)}, \qquad\qquad \frac{K_1'(\tau)}{K_1(\tau)} = \frac{K_2'(\tau)}{K_2(\tau)}, \qquad (11)$$

that is, the values of the relations $\chi(\tau) = \frac{C(\tau)}{K(\tau)}$ and $\nu(\tau) = \frac{K'(\tau)}{K(\tau)}$ are the same for all possible solutions to the inverse problem.

Theorem 1. *Let the function $T(x,t)$ belong to the reachable sets. If on each τ-isoline of the segment $\tau \in [\alpha, \beta] \subset [a,b]$ the functions $\frac{\partial T(x,t)}{\partial t}$ and $\left(\frac{\partial T(x,t)}{\partial x}\right)^2$, considered as functions of some parameter s along this isoline, are linearly independent, then for any solution $\{C(T), K(T)\}$ to the inverse problem the ratio $\chi(T) = \frac{C(T)}{K(T)}$, $T \in [\alpha, \beta]$, is the same function.*

Proof. To prove Theorem 1, let us assume the existence of two solutions $\{C_1(T), K_1(T)\}$ and $\{C_2(T), K_2(T)\}$ to the inverse problem (1)–(4) with different ratios $\chi_1(T)$ and $\chi_2(T)$. On the basis of Lemma 1, it can be argued that equalities (11) hold for all $\tau \in [\alpha, \beta]$, i.e.

$$\frac{C_1(\tau)}{K_1(\tau)} = \frac{C_2(\tau)}{K_2(\tau)}, \qquad\qquad \frac{K_1'(\tau)}{K_1(\tau)} = \frac{K_2'(\tau)}{K_2(\tau)}, \qquad\qquad \tau \in [\alpha, \beta].$$

The first of the equalities shows that along the entire segment $T \in [\alpha, \beta]$ the ratio $\chi(T) = \frac{C(T)}{K(T)}$ is a function that does not depend on the solution to the inverse problem. Integration of the second equality allows us to conclude that the ratio of different thermal conductivities remains constant on the segment, i.e. $\frac{K_1(T)}{K_2(T)} = const$. Taking into account the remark made at the end of Sect. 2, we can state that, under the conditions formulated in Theorem 1, all solutions to the considered inverse problem are determined up to a multiplier. If a condition $K(T_*) = K_*$ with given numbers $T_* \in [\alpha, \beta]$ and K_* is additionally specified, then the solution to the inverse problem on the segment $T \in [\alpha, \beta]$ will be unique, although outside this segment $T \in [a, b] \setminus [\alpha, \beta]$ the solutions $\{C_1(T), K_1(T)\}$ and $\{C_2(T), K_2(T)\}$ may differ.

Remark 1. An example illustrating Theorem 1 is the first example of the second group (see (6)). In it, the function $T(x, t) = x^2 + t$, derivatives $\frac{\partial T(x,t)}{\partial t} = 1$, $\frac{\partial T(x,t)}{\partial x} = 2x$, and functions 1 and $4x^2$ are linearly independent along any isoline $x^2 + t = \tau$, ($\tau \in (0.0; 2.0)$) of the field $T(x, t)$. As shown in Sect. 4, under an additional condition, the solution to the inverse problem is unique.

Theorem 2. *Let the function $T(x, t)$ belongs to the reachable sets. If on each $\tau-$isoline of the segment $\tau \in [\alpha, \beta] \subset [a, b]$ the functions $\frac{\partial T(x,t)}{\partial t}$ and $\left(\frac{\partial T(x,t)}{\partial x} \right)^2$, considered as functions of some parameter s along this isoline, are linearly dependent and the function $\frac{\partial T(x,t)}{\partial x}$ is not identically equal to zero, then the solution $\{C(T), K(T)\}$ to the inverse problem is not unique (there are solutions with different ratio $\chi(T) = \frac{C(T)}{K(T)}$).*

Proof. To prove Theorem 2, we first pay attention to the linear dependence of the functions $\frac{\partial T(x,t)}{\partial t}$ and $\left(\frac{\partial T(x,t)}{\partial x} \right)^2$ along each $\tau-$isoline of the segment $\tau \in [\alpha, \beta] \subset [a, b]$. This means that there are functions $A(T)$ and $B(T)$, $T \in [\alpha, \beta]$, such that $A(T)\frac{\partial T(x,t)}{\partial t} + B(T) \left(\frac{\partial T(x,t)}{\partial x} \right)^2 = 0$, and $A^2(T) + B^2(T) > 0$. Moreover, $A(T) \neq 0$, for everyone $T \in [\alpha, \beta]$. Indeed, assuming that $A(\tau) = 0$ for some value $\tau \in [\alpha, \beta]$, we arrive at a contradiction with the condition that the derivative $\frac{\partial T(x,t)}{\partial x}$ differs from the identical zero. Thus, the following equality is true

$$\frac{\partial T(x,t)}{\partial t} = \tilde{B}(T) \cdot \left(\frac{\partial T(x,t)}{\partial x} \right)^2, \tag{12}$$

where $\tilde{B}(T) = -\frac{B(T)}{A(T)} \in C([\alpha, \beta])$.

Since the function $T(x, t)$ belongs to the reachable sets, there are also functions $C_*(T)$ and $K_*(T)$ from the definition scope for which the function $T(x, t)$ is a solution to the mixed problem (1)–(3). Let us show that in the definition scope, in addition to them, there are other functions $C(T)$ and $K(T)$ for which this function $T(x, t)$ is a solution to the mixed problem (1)–(3).

Let $\Psi(T) \in C([\alpha, \beta])$ be an arbitrary positive function and $\Psi(\alpha) = \Psi(\beta) = 0$. We define functions $f(T) \in C([\alpha, \beta])$ and $\sigma(T) \in C^1([\alpha, \beta])$ by relations

$$f(T) = \frac{K'_*(T)}{K_*(T)} + \tilde{B}(T) \cdot \Psi(T), \qquad \sigma(T) = \exp \left[\int_\alpha^T f(\xi) d\xi \right].$$

Let us construct new volumetric heat capacity $C(T)$ and thermal conductivity $K(T)$ as follows

$$K(T) = \begin{cases} K_*(T), & a \leq T < \alpha, \\ K_*(\alpha) \cdot \sigma(T), & \alpha \leq T \leq \beta, \\ K_*(T) \cdot K_*(\alpha) \cdot \sigma(\beta)/K_*(\beta), & \beta < T \leq b, \end{cases} \tag{13}$$

$$C(T) = \begin{cases} C_*(T), & a \leq T < \alpha, \\ K(T) \cdot (C_*(T)/K_*(T) + \Psi(T)), & \alpha \leq T \leq \beta, \\ C_*(T) \cdot K(T)/K_*(T), & \beta < T \leq b, \end{cases}$$

and show that they are also solutions to the inverse problem for the considered function-field $T(x,t)$ from the reachable sets.

First, it is not difficult to verify by direct calculation that the constructed functions $K(T) \in C^1([a,b])$, $C(T) \in C([a,b])$ and that the relation

$$\frac{K'(T)}{K(T)} - \tilde{B}(T) \cdot \frac{C(T)}{K(T)} = \frac{K'_*(T)}{K_*(T)} - \tilde{B}(T) \cdot \frac{C_*(T)}{K_*(T)}, \tag{14}$$

is valid for all $T \in [\alpha, \beta]$, and for all $T \in [a, \alpha]$ and $T \in [\beta, b]$ the relations

$$\frac{K'(T)}{K(T)} = \frac{K'_*(T)}{K_*(T)}, \qquad \frac{C(T)}{K(T)} = \frac{C_*(T)}{K_*(T)}. \tag{15}$$

are valid.

Second, using these relations, we show that for the function $T(x,t)$ given in the condition of THEOREM 2, Eq. (7) is satisfied not only for given $(C_*(T), K_*(T))$, but also for constructed $(C(T), K(T))$. Indeed, at those points of the domain Q where $T \in [a, \alpha]$ or $T \in [\beta, b]$, we have

$$\Psi(T) \equiv C(T) \cdot \frac{\partial T(x,t)}{\partial t} - K'(T) \cdot \left(\frac{\partial T(x,t)}{\partial x} \right)^2 - K(T) \cdot \frac{\partial^2 T(x,t)}{\partial x^2}$$

$$= K(T) \cdot \left[\frac{C(T)}{K(T)} \cdot \frac{\partial T(x,t)}{\partial t} - \frac{K'(T)}{K(T)} \cdot \left(\frac{\partial T(x,t)}{\partial x} \right)^2 - \frac{\partial^2 T(x,t)}{\partial x^2} \right]$$

$$= K(T) \cdot \left[\frac{C_*(T)}{K_*(T)} \cdot \frac{\partial T(x,t)}{\partial t} - \frac{K'_*(T)}{K_*(T)} \cdot \left(\frac{\partial T(x,t)}{\partial x} \right)^2 - \frac{\partial^2 T(x,t)}{\partial x^2} \right]$$

$$= \frac{K(T)}{K_*(T)} \cdot \left[C_*(T) \cdot \frac{\partial T(x,t)}{\partial t} - K'_*(T) \cdot \left(\frac{\partial T(x,t)}{\partial x} \right)^2 - K_*(T) \cdot \frac{\partial^2 T(x,t)}{\partial x^2} \right] = 0.$$

Relations (15) were used here.

At the points of the domain Q where $T \in [\alpha, \beta]$, we have

$$\Psi(T) \equiv C(T) \cdot \frac{\partial T(x,t)}{\partial t} - K'(T) \cdot \left(\frac{\partial T(x,t)}{\partial x} \right)^2 - K(T) \cdot \frac{\partial^2 T(x,t)}{\partial x^2}$$

$$= K(T) \cdot \left[\frac{C(T)}{K(T)} \cdot \frac{\partial T(x,t)}{\partial t} - \frac{K'(T)}{K(T)} \cdot \left(\frac{\partial T(x,t)}{\partial x} \right)^2 - \frac{\partial^2 T(x,t)}{\partial x^2} \right]$$

(we use condition (14))

$$= K(T) \cdot \left\{ \frac{C(T)}{K(T)} \cdot \frac{\partial T(x,t)}{\partial t} - \left[\frac{K'_*(T)}{K_*(T)} - \tilde{B}(T) \cdot \left(\frac{C_*(T)}{K_*(T)} - \frac{C(T)}{K(T)} \right) \right] \cdot \right.$$
$$\left. \cdot \left(\frac{\partial T(x,t)}{\partial x} \right)^2 - \frac{\partial^2 T(x,t)}{\partial x^2} \right\} =$$

$$= K(T) \cdot \left\{ \frac{C(T)}{K(T)} \cdot \left[\frac{\partial T(x,t)}{\partial t} - \tilde{B}(T) \cdot \left(\frac{\partial T(x,t)}{\partial x} \right)^2 \right] - \left[\frac{K'_*(T)}{K_*(T)} - \tilde{B}(T) \cdot \frac{C_*(T)}{K_*(T)} \right] \cdot \right.$$
$$\left. \cdot \left(\frac{\partial T(x,t)}{\partial x} \right)^2 - \frac{\partial^2 T(x,t)}{\partial x^2} \right\} =$$

(we use condition (12) in the first square bracket)

$$= K(T) \cdot \left\{ \frac{C_*(T)}{K_*(T)} \cdot \tilde{B}(T) \cdot \left(\frac{\partial T(x,t)}{\partial x} \right)^2 - \frac{K'_*(T)}{K_*(T)} \cdot \left(\frac{\partial T(x,t)}{\partial x} \right)^2 - \frac{\partial^2 T(x,t)}{\partial x^2} \right\} =$$

(we use condition (12) in the first term)

$$= K(T) \cdot \left\{ \frac{C_*(T)}{K_*(T)} \cdot \frac{\partial T(x,t)}{\partial t} - \frac{K'_*(T)}{K_*(T)} \cdot \left(\frac{\partial T(x,t)}{\partial x} \right)^2 - \frac{\partial^2 T(x,t)}{\partial x^2} \right\}$$

$$= \frac{K(T)}{K_*(T)} \cdot \left[C_*(T) \cdot \frac{\partial T(x,t)}{\partial t} - K'_*(T) \cdot \left(\frac{\partial T(x,t)}{\partial x} \right)^2 - K_*(T) \cdot \frac{\partial^2 T(x,t)}{\partial x^2} \right] = 0.$$

Thus, in addition to the functions $C_*(T)$ and $K_*(T)$, the inverse problem has another constructed solution $C(T)$ and $K(T)$, and $\frac{C(T)}{K(T)} \neq \frac{C_*(T)}{K_*(T)}$.

Remark 2. Let the given function $T(x,t)$ belongs to the reachable sets. Let on each τ−isoline of a segment $\tau \in [a, b]$ the functions $\frac{\partial T(x,t)}{\partial t}$ and $\left(\frac{\partial T(x,t)}{\partial x} \right)^2$, considered as functions of some parameter s along this isoline, be linearly dependent and related by relation (12). Then for all $T \in [a, b]$ the function $\frac{\partial^2 T(x,t)}{\partial x^2}$ must have the form

$$\frac{\partial^2 T(x,t)}{\partial x^2} = R(T) \cdot \left(\frac{\partial T(x,t)}{\partial x} \right)^2, \tag{16}$$

where $R(T) \in C([a, b])$ is uniquely determined by the field $T(x,t)$.

Indeed, using expression (12) in Eq. (7), one can obtain

$$\frac{\partial^2 T(x,t)}{\partial x^2} = \left(\frac{C(T)}{K(T)} \cdot \tilde{B}(T) - \frac{K'(T)}{K(T)} \right) \cdot \left(\frac{\partial T(x,t)}{\partial x} \right)^2 = R(T) \cdot \left(\frac{\partial T(x,t)}{\partial x} \right)^2.$$

Since the functions $\widetilde{B}(T)$ and $R(T)$ are uniquely determined by a given field $T(x,t)$ from the reachable sets, then $C(T)$ and $K(T)$ must be related by the following relation

$$K'(T) + R(T) \cdot K(T) - C(T) \cdot \widetilde{B}(T) = 0. \tag{17}$$

Consequently, the functions $C(T)$ and $K(T)$ can be any that satisfy Eq. (17). Given arbitrarily $K(T) \in C^1([a,b])$, from (17) we can uniquely determine $C(T) \in C([a,b])$. And vice versa, by setting $C(T) \in C([a,b])$ and the value $K(T)$ at some point of the segment $[a,b]$, it can be unambiguously determined $K(T) \in C^1([a,b])$ as a solution to the Cauchy problem. It was precisely this uncertainty that had to be encountered during identification $C(T)$ and $K(T)$ in the first group of calculations.

Let us now ask the opposite question: what temperature fields $T(x,t)$ from the reachable sets satisfy the conditions of Theorem 2? As noted in REMARK 2, if the functions $\frac{\partial T(x,t)}{\partial t}$ and $\left(\frac{\partial T(x,t)}{\partial x}\right)^2$ are linearly dependent along each τ−isoline of the segment $\tau \in [a,b]$, then equality (16) is valid. Using the function $V(T) = \int\limits_a^T R(\xi)d\xi$, equality (16) can be transformed as follows:

$$\frac{\partial^2 T(x,t)}{\partial x^2} = R(T) \cdot \frac{\partial T(x,t)}{\partial x} \cdot \frac{\partial T(x,t)}{\partial x} = \frac{\partial T(x,t)}{\partial x} \cdot \frac{\partial V(T(x,t))}{\partial x}, \qquad (x,t) \in Q,$$

$$\frac{\partial}{\partial x}\left(\ln\left|\frac{\partial T(x,t)}{\partial x}\right| - V(T(x,t))\right) = 0,$$

where from

$$\frac{\partial T(x,t)}{\partial x} = \rho_1(t) \cdot \exp\left(V(T(x,t))\right).$$

If we introduce a new function $G(T) = \int\limits_a^T \exp(-V(\xi))d\xi$, then the last equality can be written as

$$\exp\left(-V(T(x,t))\right) \cdot \frac{\partial T(x,t)}{\partial x} = \frac{\partial G(T(x,t))}{\partial x} = \rho_1(t),$$

integrating which we obtain $G(T(x,t)) = q(x,t) = \rho_1(t) \cdot x + \rho_2(t)$, where $\rho_1(t)$ and $\rho_2(t)$ are arbitrary smooth functions. Since $G(T) \in C^2([a,b])$ and is monotonic, there exists a function $S(T) = G^{-1}(T) \in C^2([a,b])$ inverse to $G(T)$. That's why

$$T(x,t) = S(q) = S(\rho_1(t) \cdot x + \rho_2(t)). \tag{18}$$

By the conditions of Theorem 2, the functions $\frac{\partial T(x,t)}{\partial t}$ and $\left(\frac{\partial T(x,t)}{\partial x}\right)^2$ are linearly dependent on each τ−isotherm. Therefore, they are related by relation (12). Calculating the derivatives of the function $T(x,t)$, using equality (18) and substituting into relation (12), we obtain

$$S'(q) \cdot (\rho_1'(t) \cdot x + \rho_2'(t)) = \widetilde{B}(T) \cdot (S'(q))^2 \cdot \rho_1^2(t), \qquad \text{or}$$

$$\rho_1'(t) \cdot x + \rho_2'(t) = \widetilde{B}(S(q)) \cdot (S'(q)) \cdot \rho_1^2(t). \tag{19}$$

Since the left side of equality (19) is a linear function of the variable x, the function $\widetilde{B}(S(q)) \cdot S'(q)$ must be linear in the variable q, i.e. have the form $\widetilde{B}(S(q)) \cdot S'(q) = \omega_1 q + \omega_2$, where ω_1 and ω_2 are arbitrary constants. Using the obtained representation of the function $\widetilde{B}(S(q)) \cdot S'(q)$ into equality (19), we find

$$\rho_1'(t) \cdot x + \rho_2'(t) = [\omega_1 \cdot (\rho_1(t) \cdot x + \rho_2(t)) + \omega_2] \cdot \rho_1^2(t). \tag{20}$$

Equality (20) is valid for all $(x,t) \in Q$ if the unknown functions satisfy the system of equations

$$\begin{cases} \rho_1'(t) = \omega_1 \cdot \rho_1^3(t), \\ \rho_2'(t) = \rho_1^2(t) \cdot (\omega_1 \cdot \rho_2(t) + \omega_2). \end{cases} \tag{21}$$

At $\omega_1 = 0$, the system of Eqs. (21) has the following solution

$$\begin{cases} \rho_1(t) = \lambda_1, \\ \rho_2(t) = \omega_2 \lambda_1^2 t + \lambda_2, \end{cases} \tag{22}$$

and at $\omega_1 \neq 0$

$$\begin{cases} \rho_1(t) = \frac{1}{\sqrt{\sigma_1 - 2\omega_1 t}}, \\ \rho_2(t) = \frac{\sigma_2}{\sqrt{\sigma_1 - 2\omega_1 t}} - \frac{\omega_2}{\omega_1}, \end{cases} \qquad \sigma_1 - 2\omega_1 t > 0, \tag{23}$$

where $\lambda_1, \lambda_2, \sigma_1, \sigma_2$ are arbitrary constants.

Thus, the function $q(x,t) = \rho_1(t) \cdot x + \rho_2(t)$ defined above can have only such a form $q(x,t) = \varepsilon_1 x + \varepsilon_2 t + \varepsilon_3$, or such a form $q(x,t) = \frac{x + \varepsilon_3}{\sqrt{\varepsilon_1 - \varepsilon_2 t}} + \varepsilon_4$, where $\varepsilon_1 - \varepsilon_2 t > 0$, $\varepsilon_1, \varepsilon_2, \varepsilon_3, \varepsilon_4$ are arbitrary constants, and a field $T(x,t)$ from the reachable sets for which the solution to the inverse problem is not unique can have the form $T(x,t) = S(q(x,t))$. By direct calculation, it is easy to verify that this statement is true for any constants $\varepsilon_1, \varepsilon_2, \varepsilon_3, \varepsilon_4$, provided that the condition $\varepsilon_1 - \varepsilon_2 t > 0$ is satisfied in the domain $(x,t) \in Q$.

Remark 3. All examples of the first group of calculations illustrate the statement obtained for $q(x,t) = \varepsilon_1 x + \varepsilon_2 t + \varepsilon_3$. Indeed, the experimental fields used there had the form $T(x,t) = S(q(x,t)) = S(\varepsilon_1 x + \varepsilon_2 t + \varepsilon_3)$ and a non-unique solutions were obtained. As for the case $q(x,t) = \frac{x + \varepsilon_3}{\sqrt{\varepsilon_1 - \varepsilon_2 t}} + \varepsilon_4$, then, as the calculations have confirmed, with an experimental field $T(x,t) = \frac{x^2}{t} = S(q(x,t)) = q^2(x,t) = \left(\frac{x}{\sqrt{t}}\right)^2$ the inverse problem has a non-unique solution.

6   Conclusion

The paper considers one of the possible formulations of the inverse coefficient problem, which allows, based on the results of monitoring the dynamics of the temperature field, to simultaneously identify the temperature-dependent volumetric heat capacity of a substance and its thermal conductivity. It is noted that in such a statement the desired coefficients can be found with an accuracy of one and the same factor. A numerical algorithm for solving this problem is proposed. The result of solving the inverse problem is both the desired parameters of the substance and the temperature field, in the general case, not coinciding with the experimentally given field. The found field belongs to the reachable sets and is the closest to the experimental field in the space norm L_2. The properties of just this field from the reachable sets determine whether the solution to the inverse problem is unique. A class of temperature fields from the reachable sets is found for which the inverse problem has a non-unique solution.

References

1. Zverev, V.G., Gol'din, V.D., Nazarenko, V.A.: Radiation-conduction heat transfer in fibrous heat-resistant insulation under thermal effect. High Temp. **46**, 108–114 (2008). https://doi.org/10.1134/s10740-008-1015-0
2. Alifanov, O.M., Cherepanov, V.V.: Mathematical simulation of high-porosity fibrous materials and determination of their physical properties. High Temp. **47**, 438–447 (2009). https://doi.org/10.1134/S0018151X09030183
3. Huang, C.H., Yan, J.Y.: An inverse problem in simultaneously measuring temperature-dependent thermal conductivity and heat capacity. Int. J. Heat Mass Transf. **38**, 3433–3441 (1995). https://doi.org/10.1016/0017-9310(95)00059-I
4. Imani, A., Ranjbar, A.A., Esmkhani, M.: Simultaneous estimation of temperature-dependent thermal conductivity and heat capacity based on modified genetic algorithm. Inverse Probl. Sci. Eng. **14**(7), 767–783 (2006). https://doi.org/10.1080/17415970600844242
5. Cui, M., Yang, K., Xiao-liang, X., Wang, S.: Xiao-weiGao: a modified Levenberg-Marquardt algorithm for simultaneous estimation of multi-parameters of boundary heat flux by solving transient nonlinear inverse heat conduction problems. Int. J. Heat Mass Transf. **97**, 908–916 (2016). https://doi.org/10.1016/j.ijheatmasstransfer.2016.02.085
6. Zubov, V.I., Gorchakov, A.Y., Albu, A.F.: On the simultaneous identification of the volumetric heat capacity and the thermal conductivity of a substance. In: Olenev, N., Evtushenko, Y., Jacimovic, M., Khachay, M., Malkova, V., Pospelov, I. (eds.) Optimization and Applications. OPTIMA 2022. LNCS, vol. 13781, pp. 207–220. Springer, Cham (2022). https://doi.org/10.1007/978-3-031-22543-7_15
7. Evtushenko, Y.G.: Optimization and Fast Automatic Differentiation, Vychisl. Tsentr Ross. Akad. Nauk, Moscow (2013). [in Russian]
8. Evtushenko, Y.G., Zubov, V.I.: Generalized fast automatic differentiation technique. Comp. Math. Math. Phys. **56**(11), 1819–1833 (2016). https://doi.org/10.1134/S0965542516110075
9. Zubov, V.I.: Application of fast automatic differentiation for solving the inverse coeficient problem for the heat equation. Comp. Math. Math. Phys. **56**(10), 1743–1757 (2016). https://doi.org/10.7868/S0044466916100148
10. Albu, A.F., Evtushenko, Yu.G., Zubov, V.I.: Application of the fast automatic differentiation technique for solving inverse coefficient problems. Comp. Math. Math. Phys. **60**(1), 15–25 (2020). https://doi.org/10.1134/S0965542520010042

11. Albu, A.F., Zubov, V.I.: Identification of the thermal conductivity coefficient in the three-dimensional case by solving a corresponding optimization problem. Comp. Math. Math. Phys. **61**(9), 1416–1431 (2021). https://doi.org/10.1134/S0965542521090037
12. Albu, A.F., Zubov, V.I.: Determination of the thermal conductivity from the heat flux on the surface of a three-dimensional body. Comp. Math. Math. Phys. **61**(10), 1567–1581 (2021). https://doi.org/10.1134/S096554252110002X
13. Albu, A., Zubov, V.: Application of second-order optimization methods to solving the inverse coefficient problems. In: Olenev, N.N., Evtushenko, Y.G., Jaćimović, M., Khachay, M., Malkova, V. (eds.) OPTIMA 2021. LNCS, vol. 13078, pp. 351–364. Springer, Cham (2021). https://doi.org/10.1007/978-3-030-91059-4_25
14. Hogan, R.J.: Fast reverse-mode automatic differentiation using expression templates in C++. ACM Trans. Math. Softw. (TOMS) **40**(4), 26–42 (2014)
15. Qiu, Y.: L-BFGS++ (2021). https://github.com/yixuan/LBFGSpp/

Discrete and Combinatorial Optimization

Constant-Factor Approximation Algorithms for Some Maximin Multi-clustering Problems

Vladimir Khandeev[1]($\boxtimes$) and Sergey Neshchadim[2]

[1] Sobolev Institute of Mathematics, 4 Koptyug Ave., 630090 Novosibirsk, Russia
khandeev@math.nsc.ru
[2] Novosibirsk State University, 2 Pirogova St., 630090 Novosibirsk, Russia
s.neshchadim@g.nsu.ru

Abstract. We consider several problems of searching for a family of non-intersecting subsets in a finite set of points of Euclidean space. In these problems, it is required to maximize the minimum cluster's cardinality under constraint on each cluster's scatter. The scatter is the sum of the distances from the cluster elements to the center, which is defined differently in each of the problems. We show that these problems are NP-hard in the case of an arbitrary fixed number of clusters and propose polynomial constant-factor approximation algorithms for this case.

Keywords: Euclidean space · Clustering · Max-min problem · NP-hardness · Bounded scatter · Approximation algorithm

1 Introduction

The paper considers the problem of finding a family of disjoint subsets in a finite set of points in Euclidean space. In each of the problems, the minimum of the cardinalities of the unknown subsets should be maximal under the constraint of the upper bound on the scatter of each set. In this formulation, these problems were previously considered only in the case when the number of clusters is equal to two. This paper is devoted to the case when the number of clusters is at least two (when it is an arbitrary fixed number). The main results of the work are the proof of the NP-hardness of three considered problems, as well as polynomial approximation algorithms with guaranteed accuracy.

The considered optimization criterion (maximization of the cardinalities of the subsets) is fundamentally different from such well-known clustering optimization problems as *k-means* (also known as *MSSC*) [3], *k-medians* [13], *k-centers* [7], *k-centers with outliers* [5], etc. In such clustering problems, the function that determines the quality of the partition is usually optimized.

A less common approach to clustering is to find a subset (or subsets) that satisfies some bound on such a function, but contains as many elements as possible. Examples of optimization problems corresponding to this approach can

M. Khachay et al. (Eds.): MOTOR 2023, LNCS 13930, pp. 85–100, 2023.
https://doi.org/10.1007/978-3-031-35305-5_6

be found, for example, in [1,6,8,9]. This approach is applicable in those practical situations when the number of required similar elements in the initial set is unknown, but some "quality" of the cluster of these similar elements is known. One of the ways to eliminate outliers present in the input data is to use optimization problems that maximize the cardinalities of the desired subsets. This is an actual applied problem (see, for example, [2,4] and references therein).

In this paper we will use the following definitions.

Definition 1. $\mathcal{P}(\mathbb{R}^d)$ *is the set of all finite subsets of* $\mathbb{R}^d$.

Definition 2. $F(\mathcal{C}, z)$: $\mathcal{P}(\mathbb{R}^d) \times \mathbb{R}^d \to \mathbb{R}_+$ *is a function such that*

$$F(\mathcal{C}, z) := \sum_{y \in \mathcal{C}} \|y - z\|_2.$$

Definition 3. *Centroid of a set* $\mathcal{C} \in \mathcal{P}(\mathbb{R}^d)$ *is a point* $\bar{y}(\mathcal{C})$ *from* $\mathbb{R}^d$ *such that*

$$\bar{y}(\mathcal{C}) = \frac{1}{|\mathcal{C}|} \sum_{y \in \mathcal{C}} y.$$

Considered problems are formulated below.

Problem 1M. *Given* a set $\mathcal{Y} = \{y_1, \dots, y_N\} \subset \mathbb{R}^d$, *points* $z_1, z_2, \dots, z_K \in \mathbb{R}^d$, *and a real number* $A \in \mathbb{R}_+$. *Find* non-empty non-intersecting subsets $\mathcal{C}_1, \mathcal{C}_2, \dots, \mathcal{C}_K \subset \mathcal{Y}$ *such that*

$$\min(|\mathcal{C}_1|, |\mathcal{C}_2|, \dots, |\mathcal{C}_K|) \to \max \tag{1}$$

holds under the constraints

$$F_1(\mathcal{C}_i) = F(\mathcal{C}_i, z_i) \le A, \ i = 1, \dots, K. \tag{2}$$

Problem 2M. *Given* a set $\mathcal{Y} = \{y_1, \dots, y_N\} \subset \mathbb{R}^d$ *and a real number* $A \in \mathbb{R}_+$. *Find* non-empty non-intersecting subsets $\mathcal{C}_1, \mathcal{C}_2, \dots, \mathcal{C}_K \subset \mathcal{Y}$ *and points (also called medoids)* $u_1, u_2, \dots, u_K \in \mathcal{Y}$ *such that* (1) holds under the constraints

$$F_2(\mathcal{C}_i, u_i) = F(\mathcal{C}_i, u_i) \le A, \ i = 1, \dots, K. \tag{3}$$

Problem 3M. *Given* a set $\mathcal{Y} = \{y_1, \dots, y_N\} \subset \mathbb{R}^d$ *and a real number* $A \in \mathbb{R}_+$. *Find* non-empty non-intersecting subsets $\mathcal{C}_1, \mathcal{C}_2, \dots, \mathcal{C}_K \subset \mathcal{Y}$ *such that* (1) holds under the constraints

$$F_3(\mathcal{C}_i) = F(\mathcal{C}_i, \bar{y}(\mathcal{C}_i)) \le A, \ i = 1, \dots, K. \tag{4}$$

The interpretation of these considered problems, as well as the interpretation of related problems with other functions in the constraints (2)–(4), can be found, for example, in [10,12].

Further in this paper, by problems $1M(K)$–$3M(K)$ we denote modifications of problems $1M$–$3M$ in which the number K of clusters is fixed and is not a part of the input of the problem.

It was shown [12] that the problems under consideration are NP-hard in the case when the number of clusters equals two and the dimension d of the space equals one. In other words, the following theorem holds.

Theorem 1 ([12]). *Problems* $1M(2)$, $2M(2)$, $3M(2)$ *are NP-hard even in the one-dimensional case.*

An obvious consequence of Theorem 1 is that problems $1M$–$3M$ (when the number of clusters is a part of the input) are NP-hard even in the one-dimensional case.

In addition, for problems $1M(2)$, $2M(2)$, and for the one-dimensional version of problem $3M(2)$, polynomial $1/2$-approximation algorithms were constructed in [11].

In this paper, we generalize the complexity and approximability results of problems $1M(2)$–$3M(2)$ to the case of an arbitrary fixed number of clusters. In Sect. 2, we prove that in the case of an arbitrary fixed number K of clusters, problems $1M(K)$–$3M(K)$ are NP-hard. For each problem, this is proved by polynomial reduction of the decision version of the problem with K clusters to the problem with $K+1$ clusters.

In the next section, we consider a clustering problem (for K clusters) that generalizes problems $1M(K)$–$3M(K)$—problem **CLUST-K**. In addition, we consider modifications of this problem with fixed cluster sizes (**FCLUST-K**) and with one cluster of fixed size (**CLUST1**). After that, we show how, having an exact polynomial algorithm for solving **CLUST1** problem, we can construct an exact algorithm for solving **FCLUST-K** problem for a certain value of the cluster size and an approximate algorithm for **CLUST-K** problem.

Finally, in Sect. 4, we show that the approach described in the previous section allows to build polynomial algorithms for problems $1M(K)$–$3M(K)$ with guaranteed accuracy.

2 NP-Hardness of Multiclustering Problems

Let us prove the NP-hardness of the problems under consideration for an arbitrary fixed number K of clusters.

In this section, we will assume that $\mathcal{Y}$, $\mathcal{C}_1, \ldots, \mathcal{C}_K$ are multisets, which means that it is allowed to include the same element multiple times.

The NP-completeness will be proved by constructing a polynomial reduction of the already known NP-complete problems to the decision version of the considered problems.

Theorem 2. *For any fixed parameter $K \geq 2$, problem $1M(K)$ is NP-hard even in the one-dimensional case.*

Proof. Let us formulate the decision version of problem $1M(K)$ and show its NP-completeness.

Problem 1MA(K). *Given a set* $\mathcal{Y} = \{y_1, \ldots, y_N\} \subset \mathbb{R}^d$, *a positive integer* $M \in \mathbb{N}$, *points* $z_1, z_2, \ldots, z_K \in \mathbb{R}^d$, *and a real number* $A \in \mathbb{R}_+$.

Question: are there non-empty non-intersecting subsets $\mathcal{C}_1, \mathcal{C}_2, \ldots, \mathcal{C}_K \subset \mathcal{Y}$ such that

$$\min \left(|\mathcal{C}_1|, |\mathcal{C}_2|, \ldots, |\mathcal{C}_K| \right) \geq M \tag{5}$$

holds under the constraints (2).

Let us prove the NP-completeness of problem $1MA(K)$ by induction on the number K of clusters. The case $K = 2$ is proved in Theorem 1.

Let us construct a polynomial reduction of problem $1MA(K)$ to problem $1MA(K + 1)$. Given the instance of problem $1MA(K)$, construct the following instance of problem $1MA(K + 1)$. In problem $1MA(K + 1)$, we put

$$\tilde{\mathcal{Y}} = \mathcal{Y} \cup \mathcal{G}, \ \tilde{N} = N + M, \ \tilde{A} = A, \ \tilde{M} = M, \qquad (6)$$

$$\mathcal{G} = \{g_1, \dots, g_M\}, \ g_i = L, \ i = 1, \dots, M, \qquad (7)$$

$$\tilde{z}_i = z_i, i = 1, \dots, K, \ \tilde{z}_{K+1} = L,$$

where $L = \max\limits_{i-1, \dots, K} z_i + A + 1$.

Assume the required in problem $1MA(K)$ multisubsets $\mathcal{C}_i$, $i = 1, \dots, K$, exist. Then, if we put $\tilde{\mathcal{C}}_i = \mathcal{C}_i$, $i = 1, \dots, K$, and $\tilde{\mathcal{C}}_{K+1} = \mathcal{G}$, we obtain a solution to problem $1MA(K + 1)$.

Assume now the required multisubsets $\tilde{\mathcal{C}}_1, \dots, \tilde{\mathcal{C}}_{K+1}$ exist in problem $1MA(K +1)$. Clearly, g_i cannot be in any multisubset $\tilde{\mathcal{C}}_1, \dots, \tilde{\mathcal{C}}_K$. Therefore, if we put $\mathcal{C}_i = \tilde{\mathcal{C}}_i$, $i = 1, \dots, K$, then we obtain the required multisubsets for problem $1MA(K)$. Thus, the induction step is proved. ∎

Proofs of the following two theorems are similar to the proof of Theorem 2, but with different polynomial reductions.

Theorem 3. *For any fixed parameter $K \geq 2$, problem $2M(K)$ is NP-hard even in the one-dimensional case.*

Proof. Let us formulate the decision version of problem $2M(K)$ and show its NP-completeness.

Problem 2MA(K). *Given a set $\mathcal{Y} = \{y_1, \dots, y_N\} \subset \mathbb{R}^d$, a positive integer $M \in \mathbb{N}$, and a real number $A \in \mathbb{R}_+$. Question:* are there non-empty non-intersecting subsets $\mathcal{C}_1, \mathcal{C}_2, \dots, \mathcal{C}_K \subset \mathcal{Y}$ and points $u_1, u_2, \dots, u_K \in \mathcal{Y}$ such that (5) holds under the constraints (3).

By analogy with Theorem 2, the case $K = 2$ is proved.

Let us construct a polynomial reduction of problem $2MA(K)$ to problem $2MA(K+1)$. Given an instance of problem $2MA(K)$, we construct the instance of problem $2MA(K+1)$ defined by equalities (6) and (7), in which $L = \max\limits_{y \in \mathcal{Y}} y + A + 1$.

Assume that the required multisubsets $\mathcal{C}_i$, $i = 1, \dots, K$, exist in problem $2MA(K)$. Then, if we put $\tilde{\mathcal{C}}_i = \mathcal{C}_i$, $\tilde{u}_i = u_i$, $i = 1, \dots, K$, and $\tilde{\mathcal{C}}_{K+1} = \mathcal{G}$, $\tilde{u}_{K+1} = L$, we will get the solution to problem $2MA(K + 1)$.

Suppose the required multisubsets $\tilde{\mathcal{C}}_1, \dots, \tilde{\mathcal{C}}_{K+1}$ exist in problem $2MA(K + 1)$. Let us show that the elements g_i cannot be in a multisubset that contains at least one element different from L. Assume the opposite: let there be a multisubset $\tilde{\mathcal{C}}_v$ such that some elements g_w and $\hat{y} \in \mathcal{Y}$ are in it. Consider the possible values of $\tilde{u}_v$.

If $\tilde{u}_v = L$, we obtain (by the definition of L)

$$F_2(\tilde{\mathcal{C}}_v, \tilde{u}_v) = \sum_{y \in \tilde{\mathcal{C}}_v} |y - L| \geq |\hat{y} - L| > \tilde{A}.$$

If $\tilde{u}_v \in \mathcal{Y}$, then we get

$$F_2(\tilde{\mathcal{C}}_v, \tilde{u}_v) \geq |g_w - \tilde{u}_v| > \tilde{A}.$$

Thus, for any value of $\tilde{u}_v$, we obtain inequality $F_2(\tilde{\mathcal{C}}_v, \tilde{u}_v) > \tilde{A}$, which contradicts the condition $F_2(\tilde{\mathcal{C}}_v, \tilde{u}_v) \leq \tilde{A}$.

Therefore, all elements g_i can be contained only in one of the clusters (they cannot be in several clusters due to the restrictions on the size of the cluster and the number of such elements). We can assume that the clusters $\tilde{\mathcal{C}}_1, \ldots, \tilde{\mathcal{C}}_K$ do not contain elements g_i.

Then, if we put $\mathcal{C}_i = \tilde{\mathcal{C}}_i$, $u_i = \tilde{u}_i$, $i = 1, \ldots, K$, we obtain the desired multisubsets for problem $2MA(K)$. Thus, the theorem is proved. ∎

Theorem 4. *For any fixed parameter $K \geq 2$, problem $3M(K)$ is NP-hard even in the one-dimensional case.*

Proof. Let us formulate the decision version of problem $3M(K)$ and show its NP-completeness.

Problem 3MA(K). *Given a set $\mathcal{Y} = \{y_1, \ldots, y_N\} \subset \mathbb{R}^d$, a positive integer $M \in \mathbb{N}$, and a real number $A \in \mathbb{R}_+$. Question: are there non-empty non-intersecting subsets $\mathcal{C}_1, \mathcal{C}_2, \ldots, \mathcal{C}_K \subset \mathcal{Y}$ such that (5) holds under the constraints (4).*

The case $K = 2$ follows from Theorem 1.

Let us construct a polynomial reduction of problem $3MA(K)$ to problem $3MA(K+1)$. Given an instance of problem $3MA(K)$, we construct the instance of problem $3MA(K+1)$ defined by equalities (6) and (7), in which $L = \max\limits_{y \in \mathcal{Y}} y + \tilde{N}\tilde{A} + 1$.

Assume that the required multisubsets $\mathcal{C}_i$, $i = 1, \ldots, K$, exist in problem $3MA(K)$. Then, if we put $\tilde{\mathcal{C}}_i = \mathcal{C}_i$, $i = 1, \ldots, K$, and $\tilde{\mathcal{C}}_{K+1} = \mathcal{G}$, we will obtain a solution to problem $3MA(K+1)$.

Suppose the required multisubsets $\tilde{\mathcal{C}}_1, \ldots, \tilde{\mathcal{C}}_{K+1}$ exist in problem $3MA(K+1)$. By analogy with Theorem 3, we will show that the elements g_i cannot be in a multisubset that contains at least one element different from L. Assume the contrary, let there be a multisubset $\tilde{\mathcal{C}}_v$ such that some elements g_w and $\hat{y} \in \mathcal{Y}$ are in it. Then

$$F_3(\tilde{\mathcal{C}}_v) = \sum_{y \in \tilde{\mathcal{C}}_v} |y - \bar{y}(\tilde{\mathcal{C}}_v)| \geq g_w - \bar{y}(\tilde{\mathcal{C}}_v) \geq L - \frac{\hat{y} + (|\tilde{\mathcal{C}}_v| - 1)L}{|\tilde{\mathcal{C}}_v|}$$

$$\geq \frac{1}{|\tilde{\mathcal{C}}_v|}(L - \max_{y \in \mathcal{Y}} y) \geq \frac{1}{\tilde{N}}(L - \max_{y \in \mathcal{Y}} y) > \tilde{A},$$

which contradicts the condition $F_3(\mathcal{C}_v) \leq \tilde{A}$.

Therefore, all elements g_i can only be in one of the clusters. We can assume that the clusters $\tilde{C}_1, \ldots, \tilde{C}_K$ do not contain elements g_i.

Then, if we set $C_i = \tilde{C}_i$, we obtain the required multisubsets for problem $3MA(K)$. Thus, the theorem is proved. ∎

Thus, the NP-hardness in the one-dimensional case of problems $1M(K)$–$3M(K)$ is proved. Let us note that the following corollary follow from this.

Corollary 1. *Problems $1M(K)$–$3M(K)$ are NP-hard for any fixed dimension d of Euclidean space.*

3   Generalized Problem

Definition 4. *A d-dimensional scatter function is as an arbitrary function $\mathcal{F}:$ $\mathcal{P}(\mathbb{R}^d) \rightarrow \mathbb{R}_+$, such that for any $\mathcal{C}' \subset \mathcal{C}'' \subset \mathbb{R}^d$ inequality $F(\mathcal{C}') \leq F(\mathcal{C}'')$ holds.*

Consider a problem that generalizes problems $1M(K)$–$3M(K)$.

Problem CLUST-K($\mathcal{Y}, \mathcal{F}_1, \ldots, \mathcal{F}_K, A$). *Given a set $\mathcal{Y} = \{y_1, \ldots, y_N\} \subset \mathbb{R}^d$, d-dimensional scatter functions $\mathcal{F}_1, \ldots, \mathcal{F}_K$, and a non-negative number $A \in \mathbb{R}^+$. Find non-empty disjoint subsets $\mathcal{C}_1, \ldots, \mathcal{C}_K \subset \mathcal{Y}$ such that the minimum subset size is maximal:*

$$\min(|\mathcal{C}_1|, \ldots, |\mathcal{C}_K|) \rightarrow \max,$$

where

$$\mathcal{F}_i(\mathcal{C}_i) \leq A, \ i = 1, \ldots, K.$$

We also consider a modification of this problem in which the cardinalities of all sets must be equal to a given number (in other words, the next problem can be considered as problem **CLUST-K** with fixed cluster sizes).

Problem FCLUST-K($\mathcal{Y}, \mathcal{F}_1, \ldots, \mathcal{F}_K, A, M$). *Given a set $\mathcal{Y} = \{y_1, \ldots, y_N\} \subset \mathbb{R}^d$, d-dimensional scatter functions $\mathcal{F}_1, \ldots, \mathcal{F}_K$, a positive integer $M \in \mathbb{N}$, and a non-negative number $A \in \mathbb{R}^+$. Find non-empty disjoint subsets $\mathcal{C}_1, \ldots, \mathcal{C}_K \subset \mathcal{Y}$ with cardinalities equal to M such that*

$$\mathcal{F}_i(\mathcal{C}_i) \leq A, \ i = 1, \ldots, K,$$

or prove that they do not exist.

Finally, let us consider the one-cluster version of the previous problem.

Problem CLUST1($\mathcal{X}, \mathcal{F}, M$). *Given a set $\mathcal{X} = \{y_1, \ldots, y_k\} \subset \mathbb{R}^d$, a d-dimensional scatter function $\mathcal{F}$ and a positive integer $M \in \mathbb{N}$. Find M-element subset $\mathcal{C}$ of $\mathcal{X}$ with minimal scatter $\mathcal{F}(\mathcal{C})$.*

Let us consider an arbitrary instance of problem **CLUST-K**($\mathcal{Y}, \mathcal{F}_1, \ldots, \mathcal{F}_K, A$), where $\mathcal{Y} \subset \mathbb{R}^d$; $\mathcal{F}_1, \ldots, \mathcal{F}_K$ are d-dimensional scatter functions; $A \in \mathbb{R}_+$. Let $\mathcal{C}_1^*, \ldots, \mathcal{C}_K^*$ be an arbitrary feasible solution to this problem. Denote the minimal cardinality of $\mathcal{C}_1^*, \ldots, \mathcal{C}_K^*$ by M^*.

If σ is some permutation of natural numbers from 1 to N, then by $\sigma[i]$ we mean the natural number into which σ maps i.

Proposition 1. *Let $M^* = KS^* + d$, where $1 \leq d \leq K-1$ and $S^* \geq K-1$; let σ be an arbitrary permutation of the numbers $1, \ldots, K$, and let the sets $C_1, \ldots, C_K$ be such that $C_{\sigma[k]}$ is the optimal solution to problem **CLUST1**$(\mathcal{Y} \setminus \cup_{j=1}^{k-1} C_{\sigma[j]}, \mathcal{F}_{\sigma[k]}, S^* + 1)$, $i = 1, \ldots, K$. Then $\max\{\mathcal{F}_{\sigma[1]}(C_{\sigma[1]}), \ldots, \mathcal{F}_{\sigma[K-1]}(C_{\sigma[K-1]})\} \leq A$. Moreover, if $\mathcal{F}_{\sigma[K]}(C_{\sigma[K]}) > A$, then at least one of the sets $C_{\sigma[1]}, \ldots, C_{\sigma[K-1]}$ is contained in $C^*_{\sigma[K]}$.*

Proof. Suppose $\max\{\mathcal{F}_{\sigma[1]}(C_{\sigma[1]}), \ldots, \mathcal{F}_{\sigma[K]}(C_{\sigma[K]})\} > A$ (if this is not the case, then the proposition holds). Then for a permutation $\sigma = (i_1, \ldots, i_K)$ of $1, \ldots, K$ there exists an index i_σ such that for $j = 1, \ldots, i_\sigma - 1$ the scatter $\mathcal{F}_{\sigma[j]}$ of the set $C_{\sigma[j]}$ does not exceed A, while the scatter of the set $C_{\sigma[i_\sigma]}$ does. In this case, problem **CLUST1**$(\mathcal{Y} \setminus \cup_{j=1}^{i_\sigma-1} C_{\sigma[j]}, \mathcal{F}_{\sigma[i_\sigma]}, S^* + 1)$ has no feasible solutions with a scatter less than or equal to A.

Let us estimate i_σ from below. Since the scatter of the set $C_{\sigma[i_\sigma]}$ is greater than A, then

$$|C^*_{\sigma[i_\sigma]} \setminus \cup_{j=1}^{i_\sigma-1} C_{\sigma[j]}| < S^* + 1. \qquad (8)$$

Indeed, the monotonicity of the scatter function $\mathcal{F}_{\sigma[i_\sigma]}$ implies that the scatter of the set $C^*_{\sigma[i_\sigma]} \setminus \cup_{j=1}^{i_\sigma-1} C_{\sigma[j]} \subset C^*_{\sigma[i_\sigma]}$ does not exceed A, and hence any of its $(S^* + 1)$-element subsets (if such subset exists) would have a scatter of at most A, contradicting the optimality of $C_{\sigma[i_\sigma]}$.

Note that the following inequality holds.

$$|C^*_{\sigma[i_\sigma]} \setminus \cup_{j=1}^{i_\sigma-1} C_{\sigma[j]}| \geq |C^*_{\sigma[i_\sigma]}| - \sum_{j=1}^{i_\sigma-1} |C_{\sigma[j]}|$$

$$= M^* - \sum_{j=1}^{i_\sigma-1} (S^* + 1) = M^* - (i_\sigma - 1)(S^* + 1). \qquad (9)$$

From inequalities (8) and (9) it follows that

$$M^* - (i_\sigma - 1)(S^* + 1) < S^* + 1.$$

Transforming the previous inequality, we get

$$\frac{M^*}{S^* + 1} < i_\sigma. \qquad (10)$$

The left-hand side of (10) can be estimated from below as follows:

$$\frac{M^*}{S^* + 1} = \frac{KS^* + d}{S^* + 1} = K + \frac{d - K}{S^* + 1} \geq K - \frac{K}{S^* + 1} \geq K - \frac{K}{K} = K - 1. \qquad (11)$$

Combining (10) and (11), we get $i_\sigma > K - 1$. Since i_σ is an integer, then $i_\sigma \geq K$, and hence the scatters $\mathcal{F}_{\sigma[1]}(C_{\sigma[1]}), \ldots, \mathcal{F}_{\sigma[K-1]}(C_{\sigma[K-1]})$ do not exceed A.

Now suppose $\mathcal{F}_{\sigma[K]}(C_{\sigma[K]}) > A$ (i.e., $i_\sigma = K$). In this case, (8) means that

$$|C^*_{\sigma[K]} \setminus \cup_{j=1}^{K-1} C_{\sigma[j]}| < S^* + 1. \qquad (12)$$

The left-hand side of (12) can be written as follows:

$$|\mathcal{C}^*_{\sigma[K]} \setminus \cup_{j=1}^{K-1} \mathcal{C}_{\sigma[j]}| = |\mathcal{C}^*_{\sigma[K]}| - \sum_{j=1}^{K-1} |\mathcal{C}^*_{\sigma[K]} \cap \mathcal{C}_{\sigma[j]}|. \tag{13}$$

Combining (12) and (13), we get the inequality

$$|\mathcal{C}^*_{\sigma[K]}| - \sum_{j=1}^{K-1} |\mathcal{C}^*_{\sigma[K]} \cap \mathcal{C}_{\sigma[j]}| < S^* + 1.$$

Using $|\mathcal{C}^*_{\sigma[K]}| = KS^* + d$, we can transform the previous inequality as follows

$$KS^* + d - S^* - 1 < \sum_{j=1}^{K-1} |\mathcal{C}^*_{\sigma[K]} \cap \mathcal{C}_{\sigma[j]}|. \tag{14}$$

Since $d \geq 1$, (14) implies

$$(K-1)S^* < \sum_{j=1}^{K-1} |\mathcal{C}^*_{\sigma[K]} \cap \mathcal{C}_{\sigma[j]}|. \tag{15}$$

Finally, if we assume that none of the sets $\mathcal{C}_{\sigma[1]}$, ..., $\mathcal{C}_{\sigma[K-1]}$ is contained in $\mathcal{C}^*_{\sigma[K]}$ (i.e., if $|\mathcal{C}^*_{\sigma[K]} \cap \mathcal{C}_{\sigma[j]}| \leq S^*, j = 1, \ldots, K-1$), then (15) gives a contradiction. Thus, at least one of the sets $\mathcal{C}_{\sigma[1]}$, ..., $\mathcal{C}_{\sigma[K-1]}$ is contained in $\mathcal{C}^*_{\sigma[K]}$.   ∎

Thus, if we construct sets $\mathcal{C}_i$ in some order defined by σ, then only the scatter of the last set can exceed the boundary A.

Let us consider K arbitrary scatter functions $\mathcal{F}_1$, ..., $\mathcal{F}_K$ and let us assume that there is an algorithm that allows us to find optimal solutions to problems **CLUST1**$(\mathcal{X}, \mathcal{F}_i, M), i = 1, \ldots, K$. Then we can propose the following algorithm for solving problem **FCLUST-K**$(\mathcal{Y}, \mathcal{F}_1, \ldots, \mathcal{F}_K, A, M)$.

Algorithm 1. $\mathcal{A}^{FCL-K}_{appr}(\mathcal{Y}, A, M)$

Input: $\mathcal{Y} \subset \mathbb{R}^d$, $A \in \mathbb{R}_+$, $M \in \mathbb{N}$.

1: If $KM > N$, terminate the algorithm (no solution is constructed).
2: Construct a family $\mathcal{C}'_1$, ..., $\mathcal{C}'_K$ of sets where $\mathcal{C}'_k$ is the optimal solution to problem **CLUST1**$(\mathcal{Y} \setminus \cup_{j=1}^{k-1} \mathcal{C}'_j, \mathcal{F}_k, M), i = 1, \ldots, K$.
   If $\max\{\mathcal{F}_1(\mathcal{C}'_1), \ldots, \mathcal{F}_K(\mathcal{C}'_K)\} \leq A$, then go to Step 5.
3: If $K = 2$, then construct set $\mathcal{C}_2$ as an optimal solution to **CLUST1**$(\mathcal{Y}, \mathcal{F}_2, M)$, and set $\mathcal{C}_1$ — optimal solution to **CLUST1**$(\mathcal{Y} \setminus \mathcal{C}_1, \mathcal{F}_1, M)$.
   If $\max\{\mathcal{F}_1(\mathcal{C}_1), \mathcal{F}_2(\mathcal{C}_2)\} \leq A$, then go to Step 5.
   Otherwise, for each $m = 1, \ldots, K - 1$:
      Let $\sigma = (m, K, 1, \ldots, m-1, m+1, \ldots, K-1)$.
      Construct a family $\mathcal{C}_1$, ..., $\mathcal{C}_K$ of sets where $\mathcal{C}_{\sigma[1]} = \mathcal{C}'_m$ and $\mathcal{C}_{\sigma[k]}, k = 2, \ldots, K$, is the optimal solution to the problem **CLUST1**$(\mathcal{Y} \setminus \cup_{j=1}^{k-1} \mathcal{C}_{\sigma[j]}, \mathcal{F}_{\sigma[k]}, M)$.
      If $\max\{\mathcal{F}_1(\mathcal{C}_1), \ldots, \mathcal{F}_K(\mathcal{C}_K)\} \leq A$, then go to Step 5.
4: Terminate the algorithm (no solution has been constructed).
5: Return the last constructed family of sets as the result of the algorithm.

The following two statements establish the quality of the solution constructed by Algorithm $\mathcal{A}_{appr}^{FCL-K}$ in the case when $\lfloor \frac{M^*}{K} \rfloor \geq K - 1$.

Proposition 2. *If $M^* = KS^* + d$, where $1 \leq d \leq K - 1$ and $S^* \geq K - 1$, then the algorithm $\mathcal{A}_{appr}^{FCL-K}(\mathcal{Y}, A, S^* + 1)$ applied to problem FCLUST-K($\mathcal{Y}$, $\mathcal{F}_1$, ..., $\mathcal{F}_K$, A, $S^* + 1$), will construct a feasible solution.*

Proof. Assume that a feasible solution has not been constructed. In particular, this means that the sets $\mathcal{C}_1'$, ..., $\mathcal{C}_K'$ constructed at Step 2 are not feasible solutions for the problem under consideration. Then by Proposition 1 we get that there exists $K^* \in \{1, \dots, K - 1\}$ such that $\mathcal{C}_{K^*}' \subset \mathcal{C}_{K^*}^*$.

Let us start with a case $K = 2$. Then, at step Step 2, two sets $\mathcal{C}_1'$, $\mathcal{C}_2'$ were constructed, and $\mathcal{C}_1' \subset \mathcal{C}_2'$. Next, at Step 3, $\mathcal{C}_1$, $\mathcal{C}_2$ will be built, and since no solutions were found, from Proposition 1 it follows that $\mathcal{C}_2 \subset \mathcal{C}_1^*$. But, since $\mathcal{C}_1^* \subset \mathcal{Y} \setminus \mathcal{C}_2^* \subset \mathcal{Y} \setminus \mathcal{C}_1'$, the following inclusions holds $\mathcal{C}_2 \subset \mathcal{Y} \setminus \mathcal{C}_1'$ and $\mathcal{F}_2(\mathcal{C}_2) \leq A < \mathcal{F}_2(\mathcal{C}_2')$, i.e. $\mathcal{C}_2'$ is not an optimal solution to CLUST1($\mathcal{Y} \setminus \mathcal{C}_1$, $\mathcal{F}_1$, M), which is a contradicts our assumptions.

Consider Step 3 when $m = K^*$ and the permutation σ constructed at this step. Let $\mathcal{C}_1, \dots, \mathcal{C}_K$ be the sets constructed at this iteration. Then $\mathcal{C}_{\sigma[1]} = \mathcal{C}_m = \mathcal{C}_{K^*}'$, and $\mathcal{C}_{\sigma[k]}$, $k = 2, \dots, K$, are optimal solution to problem CLUST1($\mathcal{Y} \setminus \cup_{j=1}^{k-1} \mathcal{C}_{\sigma[j]}$, $\mathcal{F}_{\sigma[k]}$, $S^* + 1$).

Let us estimate the cardinality $|\mathcal{C}_{\sigma[k]}^* \setminus \cup_{j=1}^{k-1} \mathcal{C}_{\sigma[j]}|$, $k = 2, \dots, K$, from below. Note that the following inequality holds.

$$|\mathcal{C}_{\sigma[k]}^* \setminus \cup_{j=1}^{k-1} \mathcal{C}_{\sigma[j]}| = |\mathcal{C}_{\sigma[k]}^*| - \sum_{j=1}^{k-1} |\mathcal{C}_{\sigma[k]}^* \cap \mathcal{C}_{\sigma[j]}|. \tag{16}$$

If $k \leq K - 1$, then (16) can be continued as follows:

$$|\mathcal{C}_{\sigma[k]}^* \setminus \cup_{j=1}^{k-1} \mathcal{C}_{\sigma[j]}| \geq |\mathcal{C}_{\sigma[k]}^*| - \sum_{j=1}^{k-1} |\mathcal{C}_{\sigma[k]}^* \cap \mathcal{C}_{\sigma[j]}|$$

$$\geq KS^* + d - (k-1)(S^* + 1) \geq KS^* + d - (K-2)(S^* + 1). \tag{17}$$

Let us separately consider the case when $k = K$. Note that $|\mathcal{C}_{\sigma[K]}^* \cap \mathcal{C}_{\sigma[1]}| = 0$ since $\mathcal{C}_{\sigma[1]} = \mathcal{C}_m = \mathcal{C}_{K^*}' \subset \mathcal{C}_K^*$, $\sigma[K] \neq K$, and sets $\mathcal{C}_1^*$, ..., $\mathcal{C}_K^*$ are pairwise disjoint. Using this, for $k = K$ we obtain the following inequality.

$$|\mathcal{C}_{\sigma[k]}^* \setminus \cup_{j=1}^{k-1} \mathcal{C}_{\sigma[j]}| = |\mathcal{C}_{\sigma[K]}^*| - |\mathcal{C}_{\sigma[K]}^* \cap \mathcal{C}_{\sigma[1]}| - \sum_{j=2}^{K-1} |\mathcal{C}_{\sigma[K]}^* \cap \mathcal{C}_{\sigma[j]}|$$

$$= |\mathcal{C}_{\sigma[K]}^*| - \sum_{j=2}^{K-1} |\mathcal{C}_{\sigma[K]}^* \cap \mathcal{C}_{\sigma[j]}| \geq KS^* + d - (K-2)(S^* + 1). \tag{18}$$

Thus, combining (17) and (18), we get that for any $k \in \{2, \dots, K\}$, the following is true:

$$|\mathcal{C}_{\sigma[k]}^* \setminus \cup_{j=1}^{k-1} \mathcal{C}_{\sigma[j]}| \geq KS^* + d - (K-2)(S^* + 1). \tag{19}$$

Transforming the right-hand side of (19) using the assumption $S^* \geq K - 1$, we get

$$|\mathcal{C}^*_{\sigma[k]} \setminus \cup_{j=1}^{k-1} \mathcal{C}_{\sigma[j]}| \geq 2S^* + d + 2 - K \geq S^* + d + 1 > S^* + 1.$$

Thus, for any $k \in \{2, \ldots, K\}$, any $(S^* + 1)$-element subset of $\mathcal{C}^*_{\sigma[k]} \setminus \cup_{j=1}^{k-1} \mathcal{C}_{\sigma[j]}$ is a valid solution to problem **CLUST1**($\mathcal{Y} \setminus \cup_{j=1}^{k-1} \mathcal{C}_{\sigma[j]}$, $\mathcal{F}_{\sigma[k]}$, $S^* + 1$), the scatter of which, by the monotonicity of the scatter function $\mathcal{F}_{\sigma[k]}$, does not exceed A. Thus, the scatter of the sets $\mathcal{C}_{\sigma[k]}$, $k = 2, \ldots, K$, does not exceed A, since these sets are optimal solutions of this one-cluster problems.

For $k = 1$ this is also true. Indeed, since $\mathcal{C}_{\sigma[1]} = \mathcal{C}'_m = \mathcal{C}'_{K^*}$, from Proposition 1 we get inequality $\mathcal{F}_{\sigma[1]}(\mathcal{C}_{\sigma[1]}) \leq A$.

Thus, for $m = K^*$, all the sets $\mathcal{C}_1, \ldots, \mathcal{C}_K$ have a scatter that does not exceed A, which contradicts the assumption that the solution has not been constructed. ∎

Proposition 3. *If $M^* = KS^*$ and $S^* \geq K - 1$, then the algorithm $\mathcal{A}^{FCL-K}_{appr}$ applied to problem **FCLUST-K**($\mathcal{Y}$, $\mathcal{F}_1$, ..., $\mathcal{F}_K$, A, S^*) will construct a feasible solution.*

Proof. The proof is similar to the proof of Proposition 2: it can be shown that a feasible solution will be constructed at Step 2, since the inequality

$$|\mathcal{C}^*_k \setminus \cup_{j=1}^{k-1} \mathcal{C}_j| \geq S^*$$

holds for all $k = 1, \ldots, K$. ∎

Propositions 2 and 3 can be combined as follows.

Proposition 4. *If $\lfloor \frac{M^*}{K} \rfloor \geq K - 1$, then the algorithm $\mathcal{A}^{FCL-K}_{appr}$ applied to problem **FCLUST-K**($\mathcal{Y}$, $\mathcal{F}_1$, ..., $\mathcal{F}_K$, A, $\lceil \frac{M^*}{K} \rceil$) will construct a feasible solution.*

The complexity of the $\mathcal{A}^{FCL-K}_{appr}$ algorithm is established by the following proposition.

Proposition 5. *Consider $\mathcal{Y} \subset \mathbb{R}^d$. If problems **CLUST1**($\mathcal{X}$, $\mathcal{F}_k$, M), $\mathcal{X} \subset \mathcal{Y}$, $k = 1, \ldots, K$, $M = 1, \ldots, |\mathcal{Y}|$, are solvable in time $O(T_k(|\mathcal{Y}|))$, then the running time of the $\mathcal{A}^{FCL-K}_{appr}$ algorithm can be estimated as $O(T_1(|\mathcal{Y}|) + \ldots + T_K(|\mathcal{Y}|))$.*

Proof. It follows from the fact that the solution of each of problems **CLUST1**(*, $\mathcal{F}_k$, *) is constructed at most K times (once at Step 2, and $K - 1$ times at Step 3). But since K is a fixed parameter, the total complexity is $O(T_1(|\mathcal{Y}|) + \ldots + T_K(|\mathcal{Y}|))$. ∎

Let us now formulate an algorithm that will allow us to find the exact solution of the **FCLUST-K** problem in the case of $\lfloor \frac{M^*}{K} \rfloor < K - 1$ by iterating over all feasible subsets.

Algorithm 2. $\mathcal{A}_{ex}^{FCL-K}(\mathcal{Y}, A, M)$

Input: $\mathcal{Y} \subset \mathbb{R}^d$, $A \in \mathbb{R}_+$, $M \in \mathbb{N}$.

1: If $KM > N$, terminate the algorithm (no solution is constructed).
2: For each (KM)-element ordered set of distinct indices $i_1^1, \ldots, i_M^1, \ldots, i_1^K, \ldots, i_M^K$:
   Construct a family $\mathcal{C}_k = \{y_{i_j^k}, j = 1, \ldots, M\}$ of sets.
   If $\max\{\mathcal{F}_1(\mathcal{C}_i), \ldots, \mathcal{F}_K(\mathcal{C}_K)\} \leq A$, then go to Step 4.
3: Terminate the algorithm (no solution has been constructed).
4: Return the last constructed family of sets as the result of the algorithm.

The complexity and quality of the $\mathcal{A}_{ex}^{FCL-K}$ algorithm is established by the following proposition.

Proposition 6. *Let $\mathcal{Y} \subset \mathbb{R}^s$, $|\mathcal{Y}| = N$ and $\lfloor \frac{M^*}{K} \rfloor \leq K-1$, $A \in \mathbb{R}_+$. Let also $\mathcal{F}_1$, ..., $\mathcal{F}_K$ be d-dimensional scatter functions such that $\mathcal{F}_i(\hat{\mathcal{Y}})$, $i = 1, \ldots, K$, can be computed in $O(T_i^{\mathcal{F}}(|\hat{\mathcal{Y}}|))$ time for every arbitrary subset $\hat{\mathcal{Y}} \subset \mathcal{Y}$. Then feasible solution to problem* **FCLUST-K***$(\mathcal{Y}, \mathcal{F}_1, \ldots, \mathcal{F}_K, A, \lceil \frac{M^*}{K} \rceil)$ is constructed by the algorithm $\mathcal{A}_{ex}^{FCL-K}(\mathcal{Y}, A, \lceil \frac{M^*}{K} \rceil)$ in time $O((T_1^{\mathcal{F}}(N) + \ldots + T_K^{\mathcal{F}}(N))N^{K^2})$.*

Proof. Put $S^* = \lceil \frac{M^*}{K} \rceil$. Consider $\hat{\mathcal{C}}_k \subset \mathcal{C}_k^*$, $k = 1, \ldots, K$, which are arbitrary S^*-element subsets of sets $\mathcal{C}_k^*$. We define the indices $\hat{i}_s^k$ so that $\hat{\mathcal{C}}_k = \{y_{i_s^k} \mid s = 1, \ldots, S^*\}$, $k = 1, \ldots, K^*$.

Consider the iteration of Step 2 with indices $i_s^k = \hat{i}_s^k$, $s = 1, \ldots, S^*$, $k = 1, \ldots, K$. Then, a feasible solution will be obtained for problem **FCLUST-K**$(\mathcal{Y}, \mathcal{F}_1, \ldots, \mathcal{F}_K, A, S^*)$, since at this step the constructed family of sets $\mathcal{C}_1, \ldots, \mathcal{C}_K$ will be equal to $\hat{\mathcal{C}}_1, \ldots, \hat{\mathcal{C}}_K$, which are a feasible solution to the considered problem.

All permutations on Step 2 can be iterated over in $N^{S^*K} \leq N^{K^2}$. To calculate the scatter of $\mathcal{F}_i$ no more than $O(T_i(N))$ time is needed, according to the condition of the proposition. Thus, in $O((T_1(N) + \ldots + T_K(N))N^{K^2})$ time we can iterate over all possible solutions to problem **FCLUST-K**$(\mathcal{Y}, \mathcal{F}_1, \ldots, \mathcal{F}_K, A, S^*)$ and calculate their scatters, i.e., find a feasible solution. ∎

Let us combine algorithms $\mathcal{A}_{ex}^{FCL-K}$, $\mathcal{A}_{appr}^{FCL-K}$ into one algorithm and establish its quality.

Algorithm 3. $\mathcal{A}_{full}^{FCL-K}(\mathcal{Y}, A, M)$

Input: $\mathcal{Y} \subset \mathbb{R}^d$, $A \in \mathbb{R}_+$, $M \in \mathbb{N}$

1: If $M \leq K - 1$, then return the result of algorithm $\mathcal{A}_{ex}^{FCL-K}(\mathcal{Y}, A, M)$.
2: Otherwise, return $\mathcal{A}_{appr}^{FCL-K}(\mathcal{Y}, A, M)$ as a result.

Proposition 7. *If the requirements of Proposition 5 are satisfied with the polynomial functions $T_i^{\mathcal{F}}$, $i = 1, \ldots, K$, and the scatter functions $\mathcal{F}_i$, $i = 1, \ldots, K$*

*can be computed in polynomial time, then algorithm $\mathcal{A}_{full}^{FCL-K}(\mathcal{Y}, A, M)$ applied to problem **FCLUST-K***($\mathcal{Y}$, $\mathcal{F}_1$, ..., $\mathcal{F}_K$, A, $\lceil \frac{M^*}{K} \rceil$), *will construct a feasible solution in polynomial time.*

Proof. Follows from Propositions 4 and 6. ∎

Finally, similarly to [11], using algorithm $\mathcal{A}_{full}^{FCL-K}$, we formulate an algorithm that allows us to find a $\frac{1}{K}$-approximate (Proposition 8) solution to problem **CLUST-K** in polynomial time (Proposition 9).

Algorithm 4. $\mathcal{A}^{CL-K}(\mathcal{Y}, A)$

Input: $\mathcal{Y} \subset \mathbb{R}^d$, $A \in \mathbb{R}_+$
1: Let $M_f = 1$, $M_t = \lceil \frac{N}{K} \rceil + 1$. Construct a solution $\mathcal{C}_1, ..., \mathcal{C}_K$ to problem **FCLUST-K**($\mathcal{Y}$, $\mathcal{F}_1$, ..., $\mathcal{F}_K$, A, M) for $M = 1$. If there is no solution, then terminate the algorithm (problem **CLUST-K** has no solutions).
2: If $M_f + 1 = M_t$, then go to Step 4, otherwise go to Step 3.
3: Let $M = \lceil \frac{M_f + M_t}{2} \rceil$. Construct a solution to problem **FCLUST-K**($\mathcal{Y}$, $\mathcal{F}_1$, ..., $\mathcal{F}_K$, A, M). If the solution has been constructed, then put $M_f = M$ and store the constructed solution in $\mathcal{C}_1, ..., \mathcal{C}_K$ (an approximate solution of size M has been constructed); otherwise put $M_t = M$. Go to Step 2.
4: Return the sets $\mathcal{C}_1, ..., \mathcal{C}_K$ as the result of the algorithm.

Proposition 8. *Algorithm* $\mathcal{A}^{CL-K}(\mathcal{Y}, A)$ *constructs a $\frac{1}{K}$-approximate solution to problem **CLUST-K**.*

Proof. Let $\mathcal{C}_1^*, ..., \mathcal{C}_K^* \subset \mathcal{Y}$ be the optimal solution to problem **CLUST-K** and $M^* = \min(|\mathcal{C}_1^*|, ..., |\mathcal{C}_K^*|)$.

Note that during the execution of Steps 2–4 of the algorithm, the variable M_f always contains the cardinality for which algorithm $\mathcal{A}_{full}^{FCL-K}$ constructs a solution and the variable M_t always contains the cardinality for which the algorithm finds no solution. Indeed, since the problem we are considering has at least one feasible solution (the sets $\mathcal{C}_1^*, ..., \mathcal{C}_K^*$), then the monotonicity of the scatter function implies that there also exists a feasible solution with one-element clusters (arbitrary one-element subsets of sets $\mathcal{C}_1^*, ..., \mathcal{C}_K^*$). Then it follows from Proposition 7 that at the first step, algorithm $\mathcal{A}_{full}^{FCL-K}$ will construct a feasible solution of cardinality 1, i.e., M_f will contain the cardinality for which the algorithm has constructed a solution. The initial property for M_t is also true because at Step 1, the value stored in this variable is more than $\frac{1}{K}$ of the cardinality of the original set. The fact that these properties are preserved at subsequent steps follows from the structure of Step 3.

After $\mathcal{O}(\log N)$ iterations of Step 3, the algorithm will reach the state where $M_f + 1 = M_t$. Since there are feasible solutions with cardinalities from 1 to M^*, it follows from Proposition 7 that algorithm $\mathcal{A}_{full}^{FCL-K}$ is able to construct feasible solutions for $M = 1, ..., \lceil \frac{M^*}{K} \rceil$. Therefore, at the end of the algorithm, $M_t >$

$\lceil \frac{M^*}{K} \rceil$, which implies that $M_t = M_f - 1 \geq \lceil \frac{M^*}{K} \rceil$. Thus, the final minimum cardinality of the constructed sets, that are feasible solutions, is at least $\lceil \frac{M^*}{K} \rceil$, and the result of the algorithm is a $\frac{1}{K}$-approximate solution. ∎

Proposition 9. *If* $\mathcal{A}_{full}^{FCL-K}$ *constructs solution to problem* **FCLUST-K** *in* $O(T(N))$ *time, then the total complexity of* $\mathcal{A}^{CL-K}$ *is* $O(T(N) \log N)$.

Proof. The number of repetitions of Steps 2 and 3 of algorithm $\mathcal{A}^{CL-K}$ is determined by the complexity of the binary search on the interval $[1, \lceil \frac{N}{K} \rceil + 1]$ and can be estimated by $\mathcal{O}(\log N)$. By the assumption of the proposition, at each repetition the number of operations is equal to $\mathcal{O}(T(N))$, so the total complexity of the $\mathcal{A}^{CL-K}$ algorithm is $\mathcal{O}(T(N) \log N)$. ∎

Propositions 8 and 9 imply the following theorem.

Theorem 5. *Let* $\mathcal{Y} \subset \mathbb{R}^s$, $A \in \mathbb{R}_+$, *let* $\mathcal{F}_1$, ..., $\mathcal{F}_K$ *be d-dimensional scatter functions computable in polynomial time, and let the requirements of the statement 5 with polynomial functions* $T_i^{\mathcal{F}}$ *hold. Then algorithm* $\mathcal{A}^{CL-K}(\mathcal{Y}, A)$ *constructs a* $\frac{1}{K}$-*approximate solution to problem* **CLUST-K**$(\mathcal{Y}, \mathcal{F}_1, \ldots, \mathcal{F}_K, A)$ *in polynomial time.*

4   Approximate Algorithms for Multicluster Case

In this section, we show how the approach described in Sect. 3 can be applied to the considered problems $1M(K)$–$3M(K)$.

Problems $1M(K)$–$3M(K)$ are special cases of the considered in Sect. 3 generalized problem **CLUST-K**. Indeed, problem $1M(K)$ is the same as **CLUST-K**$(\mathcal{Y}, \mathcal{F}(\mathcal{C}, z_1), \ldots, \mathcal{F}(\mathcal{C}, z_k), A)$. Problem $2M(K)$ can be reduced (see [11]) to **CLUST-K**$(\mathcal{Y}, \mathcal{F}_2^M(\mathcal{C}), \ldots, \mathcal{F}_2^M(\mathcal{C}), A)$, where

$$\mathcal{F}_2^M(\mathcal{C}) = \min_{u \in \mathcal{Y}} \mathcal{F}(\mathcal{C}, u).$$

Problem $3M(K)$ is the same as **CLUST-K**$(\mathcal{Y}, \mathcal{F}(\mathcal{C}, \overline{y}(\mathcal{C})), \ldots, \mathcal{F}(\mathcal{C}, \overline{y}(\mathcal{C})), A)$.

In [11], it was proved that **CLUST1** problems with the scatter functions considered in this paper could be solved in polynomial time.

Theorem 6 ([11]). *The following* **CLUST1** *problems can be solved in polynomial time:*

1. *Problem* **CLUST1**$(\mathcal{X}, \mathcal{F}(\mathcal{C}, z), M)$ *can be solved in* $\mathcal{O}(N \log N)$.
2. *Problem* **CLUST1**$(\mathcal{X}, \mathcal{F}_2^M(\mathcal{C}), M)$ *can be solved in* $\mathcal{O}(N^2 \log N)$.
3. *Problem* **CLUST1**$(\mathcal{X}, \mathcal{F}(\mathcal{C}, \overline{y}(\mathcal{C})), M)$ *can be solved in* $\mathcal{O}(N \log N)$ *in the one-dimensional* $(\mathcal{Y} \subset \mathbb{R})$ *case.*

Theorems 5 and 6 imply the following theorem, which is the main result of the paper.

Theorem 7. *For problems* $1M(K)$, $2M(K)$, *and one-dimensional case of problem* $3M(K)$, *there are* $\frac{1}{K}$*-approximation algorithms with polynomial complexity.*

A similar approach can be applied to other scatter functions as well. For example, consider the analogues of problems $1M(K)$ and $2M(K)$—problems $1M^2(K)$ and $2M^2(K)$, in which the squared distances to the center are summed to calculate the scatter, i.e., the following scatter functions are used:

$$F^{(2)}\left(\mathcal{C}, u\right) := \sum_{y \in \mathcal{C}} \|y - u\|_2^2,$$

where u is either some fixed point in Euclidean space or an unknown element of the input set. Let us denote these scatter functions as $F_1^{(2)}$ and $F_2^{(2)}$, respectively. These problems [10] are known to be NP-hard for any fixed number of clusters $K \geq 2$.

It is not difficult to show that **CLUST1** problems generated by $1M^2(K)$ and $2M^2(K)$ can be solved in polynomial time. Thus, using the approach described in this paper, $\frac{1}{K}$-approximation polynomial algorithms for these problems can be obtained, where K is a fixed number of clusters. Also, by analogy with problem $3M(K)$, it is possible to construct a $\frac{1}{K}$-approximation algorithm for the one-dimensional case of problem $3M^2(K)$ with the scatter function $F_3^{(2)}(\mathcal{C}) = F^{(2)}\left(\mathcal{C}, \bar{y}(\mathcal{C})\right)$.

Also, in the **CLUST1** problem, it is not necessary to consider scatter functions of the same type. For example, let us consider problems **CLUST-K** $(\mathcal{Y}, \mathcal{F}_1, \dots, \mathcal{F}_K, A)$, where

$$\mathcal{F}_i \in \{F_1, F_2, F_1^{(2)}, F_2^{(2)}\}. \tag{20}$$

For these problems, there will also exist $\frac{1}{K}$-approximation polynomial algorithms. The set of considered (20) scatter functions can also be extended with F_3 and $F_3^{(2)}$ if the one-dimensional (1D) case is considered. In total, the following table summarizes this paper's (**bold text**) and known results for the considered problems.

Problem	Type of scatter	$K = 2$		$K \geq 3$	
		NP-hardness	Polynomial algorithms	NP-hardness	Polynomial algorithms
$1M(K)$ $2M(K)$	$\|...\|$	in 1D [12]	$\frac{1}{2}$-appr. [11]	**in 1D**	$\frac{1}{K}$**-appr.**
$3M(K)$			$\frac{1}{2}$-appr. (in 1D) [11]		$\frac{1}{K}$**-appr. (in 1D)**
$1M(K)$ $2M(K)$	$\|...\|^2$	in 1D [10]	$\frac{1}{2}$**-appr.**	in 1D [10]	$\frac{1}{K}$**-appr.**
$3M(K)$		unknown		unknown	$\frac{1}{K}$**-appr. (in 1D)**

5   Conclusion

In this article, we considered three problems of finding several clusters of the possible largest size in a finite set of points in Euclidean space. In all considered problems, the desired sets must satisfy the scatter constraint, that is, the constraint on the sum of the distances from the elements of the set to its center (the center in each problem is determined in its own way). We have proved that the considered problems are NP-hard even in one-dimensional space and for an arbitrary fixed number of clusters. In addition, we proposed polynomial algorithms with guaranteed accuracy for the case of a fixed number of clusters. Since the complexity of algorithms depends exponentially on the number of clusters, it is of interest to construct algorithms with guaranteed accuracy that are polynomial with respect to the number of clusters.

Acknowledgments. The study presented was supported by the Russian Academy of Science (the Program of basic research), project FWNF-2022-0015.

References

1. Ageeva, A.A., Kel'manov, A.V., Pyatkin, A.V., Khamidullin, S.A., Shenmaier, V.V.: Approximation polynomial algorithm for the data editing and data cleaning problem. Pattern Recogn. Image Anal. **27**(3), 365–370 (2017). https://doi.org/10.1134/S1054661817030038
2. Aggarwal, C.C.: Outlier Analysis. Springer, Cham (2017)
3. Aloise, D., Deshpande, A., Hansen, P., Popat, P.: NP-hardness of Euclidean sum-of-squares clustering. Mach. Learn. **75**(2), 245–248 (2009)
4. Chandola, V., Banerjee, A., Kumar, V.: Anomaly detection: a survey. ACM Comput. Surv. **41**(3), 15:1–15:58 (2009)
5. Charikar, M., Khuller, S., Mount, D.M., Narasimhan, G.: Algorithms for facility location problems with outliers. In: Proceedings of the Twelfth Annual ACM-SIAM Symposium on Discrete Algorithms (SODA 2001), USA, pp. 642–651. Society for Industrial and Applied Mathematics (2001)
6. Eremeev, A.V., Kelmanov, A.V., Pyatkin, A.V., Ziegler, I.A.: On finding maximum cardinality subset of vectors with a constraint on normalized squared length of vectors sum. In: van der Aalst, W.M.P., et al. (eds.) AIST 2017. LNCS, vol. 10716, pp. 142–151. Springer, Cham (2018). https://doi.org/10.1007/978-3-319-73013-4_13
7. Gonzalez, T.: Clustering to minimize the maximum inter cluster distance. Theor. Comput. Sci. **38**, 293–306 (1985)
8. Kel'manov, A.V., Panasenko, A.V., Khandeev, V.I.: Exact algorithms of search for a cluster of the largest size in two integer 2-clustering problems. Numer. Anal. Appl. **12**, 105–115 (2019)
9. Kel'manov, A., Pyatkin, A., Khamidullin, S., Khandeev, V., Shamardin, Y.V., Shenmaier, V.: An approximation polynomial algorithm for a problem of searching for the longest subsequence in a finite sequence of points in Euclidean space. In: Eremeev, A., Khachay, M., Kochetov, Y., Pardalos, P. (eds.) OPTA 2018. CCIS, vol. 871, pp. 120–130. Springer, Cham (2018). https://doi.org/10.1007/978-3-319-93800-4_10

10. Kel'manov, A.V., Pyatkin, A.V., Khandeev, V.I.: On the complexity of some max-min clustering problems. Proc. Steklov Inst. Math. **309**(Suppl 1), S65–S73 (2020)
11. Khandeev, V., Neshchadim, S.: Approximate algorithms for some maximin clustering problems. In: Kochetov, Y., Eremeev, A., Khamisov, O., Rettieva, A. (eds.) MOTOR 2022. CCIS, vol. 1661, pp. 89–103. Springer, Cham (2022). https://doi.org/10.1007/978-3-031-16224-4_6
12. Khandeev, V., Neshchadim, S.: Max-min problems of searching for two disjoint subsets. In: Olenev, N.N., Evtushenko, Y.G., Jaćimović, M., Khachay, M., Malkova, V. (eds.) OPTIMA 2021. LNCS, vol. 13078, pp. 231–245. Springer, Cham (2021). https://doi.org/10.1007/978-3-030-91059-4_17
13. Lin, J.-H., Vitter, J.S.: Approximation algorithms for geometric median problems. Inform. Proc. Lett. **44**, 245–249 (1992)

Aggregation Tree Construction Using Hierarchical Structures

Adil Erzin[1,2,3]($\boxtimes$) , Roman Plotnikov[1,3] , and Ilya Ladygin[2]

[1] Sobolev Institute of Mathematics, SB RAS, Novosibirsk 630090, Russia
adilerzin@math.nsc.ru
[2] Novosibirsk State University, Novosibirsk 630090, Russia
[3] St. Petersburg State University, St. Petersburg 199034, Russia

Abstract. In the problem under consideration it is necessary to find a schedule of conflict-free data aggregation of a minimum length, i.e. to determine the transmission moments for each vertex of the communication graph in such a way that there are no conflicts, and all data gets into the base station within the minimum time using the arcs of unknown spanning aggregation tree (AT). Although the problem remains NP-hard even with a known AT, finding an AT is an important step in conflict-free data aggregation. This paper proposes a new algorithm for constructing the ATs in the special hierarchical structures, k-HS, which initially may contain up to $k \geq 1$, copies of the same vertex located at k neighbouring levels. Then for different k we build a spanning tree k-HST and propose a heuristic to leave in k-HST a single copy of each node. The result is an aggregation tree k-AT. Using k-ATs constructed for different k, we find a conflict-free schedule applying the best-known algorithms. The outcome of our method is the best schedule found on k-ATs with different values of k. To assess the quality of the proposed approach, we carried out a numerical experiment on the randomly constructed unit-disk graphs, and can conclude that the construction of k-ATs speeds up the aggregation.

Keywords: Hierarchical structure · Aggregation tree · Conflict-free aggregation · Convergecast

1 Introduction

In some applications, such as security, environmental protection, fire alarm systems, area or targets monitoring, and others, the nodes of a distributed wireless network (DWN) send data to a designated node called a *sink* or *base station* (BS) [1]. Often it is highly desirable to minimize the delay between the moment of detection of some event and the moment of receiving the information about it by the sink.

The research was supported by the Russian Science Foundation (grant No. 22-71-10063 "Development of intelligent tools for optimization multimodal flow assignment systems in congested networks of heterogeneous products").

The all-to-one communication pattern used in DWNs is also known as a *convergecast*. If the nodes use radio transmitters with limited transmission range, hop-by-hop communications are used to deliver the information from the nodes to the BS. DWN is commonly represented as a communication graph. The convergecasting, in this case, is modelled by building a logical tree with the sink located at the root, assuming that packets are routing along the tree's arcs.

Since energy consumption is the most critical issue in DWNs, energy efficiency becomes one of the primary design goals for a convergecasting. Obviously, the convergecast of all raw data will cause a burst in traffic load. To reduce it, a data aggregation can be applied. Here aggregation refers to the process when the relay nodes merge the received data with their own data [17]. In this case, each node has to transmit only once and the transmission links form an *aggregation tree* (AT).

When a node sends the data to the receiver, a *collision* or *interference* can occur at the receiver if the transmission interferes with signals concurrently sent by other nodes, and thus the data should be retransmitted. Since the retransmissions cause both extra energy consumption and an increase of convergence time, protocols able to eliminate their collisions are in great demand. A common approach is to assign a sending time slot to each node in such a way that all data can be aggregated without any collision on the way to the sink node, a technique known as *time division multiple access* (TDMA)-based scheduling. Most of the scheduling algorithms adopt the *protocol interference model* [14], which enables the use of simple graph-based scheduling schemes [11].

In terms of common objectives of TDMA-based scheduling algorithms, the following two are most fundamental with respect to data aggregation in DWNs: minimizing schedule length or latency and minimizing energy consumption. In terms of design assumptions, the algorithms differ mainly in the following categories: use of communication and interference models, centralized or distributed implementation, topology assumption, and types of data collection [16].

In this paper, we consider the problem of minimization of aggregation latency assuming the conflict-free transmission. This problem is known as the Minimum-Latency Aggregation Scheduling (MLAS) [28] or Aggregation Convergecast Scheduling [19]. The MLAS problem is NP-hard [3], so most of the existing results in the literature are heuristic algorithms that are usually comprised of two independent phases: construction of an AT and link transmission scheduling. It is worth mentioning that both of these problems are very hard to solve. On the one hand, there is no result describing the structure of an optimal AT for a given graph; on the other hand, even on a given AT, the optimal time slot allocation is still NP-hard [5], except for some special cases, in particular, on lattices [6–8].

1.1 Related Work

In [3], Chen et al. consider a generalized version of the MLAS, Minimum Data Aggregation Time (MDAT) problem, where only a subset of nodes generates data. They proved that MDAT is NP-hard and designed an approximation algorithm with guaranteed performance ratio $\Delta - 1$, where Δ is the graph degree. In [29], the authors propose an approximation algorithm with guaranteed performance ratio $7\Delta/\log_2|S| + c$, where S is the set of nodes containing data, and c is a constant.

Most algorithms solve the MLAS problem in two consecutive steps: AT construction and conflict-free scheduling. The Shortest Path Tree (SPT) and Connected Dominating Set (CDS) are the usual patterns for the AT. In SPT based algorithms, data goes along a min-length path, which reduces aggregation delay [3,16,17], but they do not take into consideration the potential collisions. As for CDS based algorithms, due to the topological properties of CDS, it is often possible to find the upper bound of data aggregation delay, which depends on the network radius R and degree Δ. Thus Huang et al. [15], proposed an aggregation scheduling method using CDS with the latency bound $23R + \Delta - 18$. Based on the deeper study of the properties of neighbouring dominators in CDS, Nguyen et al. [18] provided proof of an upper bound $12R + \Delta - 12$ for their algorithm. Despite the ability to have a delay upper bound for CDS based algorithms, a dominating node is frequently a node of big degree, which may have a negative effect on aggregation delay. The SPT has similar problems for the sink node. It was shown in [27] that an optimal solution could be neither SPT nor CDS based. De Souza et al. [25] constructed an AT by combining an SPT and a minimum interference tree built by Edmond's algorithm [4]. In [19], the authors proposed a Minimum Lower bound Spanning Tree (MLST) algorithm for AT construction. To achieve a small delay lower bound, they use the sum of the receiver's depth and child number as the cost of the transmission link. However, the problem of finding the optimal AT for the MLAS problem remains unsolved.

One of the most efficient heuristics for AT construction is Round Heuristic (RH) from [2] developed for solving the telephone gossiping problem. In this algorithm, the edges are weighted in a special way. On each iteration, having the partially constructed tree, a maximum weighted matching is found on a bipartite subgraph induced by the vertices of a tree and their neighbouring vertices that are not included in the tree. Although This method was developed for the telephone gossiping problem, it appeared to be very efficient for the MLAS problem in practice [23]. There were also a lot of scheduling heuristics developed for the MLAS problem. One of the most efficient is the Neighbour Degree Ranking (NDR) [19]. In this algorithm, for a time slot assignment, non-conflicting transmissions from the subset of leaves are chosen, with priorities dependent on the vertices' degrees in the communication graph. Also, due to the additional subroutine Supplementary Scheduling, some leaf nodes are allowed to send the data to non-parent nodes, which changes the aggregation tree.

The genetic algorithm (GA) is one of the most common metaheuristics used for the approximate solution of NP-hard discrete optimization problems including those in the DWN domain [24]. In [21] a GA based algorithm was proposed to solve the convergecast scheduling problem with an unbounded number of channels, where only conflicts between the transmitters to the same addressee were taken into account. In [9] two channels are used during convergecasting.

A variable neighbourhood search (VNS) is a metaheuristic approach proposed by Hansen and Mladenovic [12,13] and was further developed in several papers. The basic idea of VNS is to find better solutions based on a dynamic neighbourhood model. VNS-based heuristics are applied to different combinatorial optimization problems including those in the DWN domain. In [10], the authors jointly solve the point coverage problem, sink location problem and data routing problem on heterogeneous sensor networks. Su et al. [26] proposed the VNS heuristic to minimize transmission tardiness in data aggregation scheduling. Plotnikov et al. [20] investigated the problem of finding an optimal communication subgraph in a given edge-weighted graph, which reduces to the problem of a minimization of transmission energy consumption in a DWN. In [22,23], the heuristic algorithms based on genetic local search and variable neighbourhood search metaheuristics were presented. An extensive simulation demonstrates the superiority of these algorithms compared with the best of the previous approaches.

1.2 Our Contribution

In this paper, we propose a new $O(n^2)$-time method for constructing several ATs, where n is the number of nodes in the communication graph. To build an AT for some k, we first construct the special hierarchical structure k-HS from the given communication graph. In k-HS, each vertex can initially be on k adjacent levels $l, l + 1, \ldots, l + k - 1$, where l is the minimum number of edges in the path from the sink to this vertex. Then, using a proposed heuristic algorithm, we form a spanning tree k-HST in k-HS. After that, all copies of each vertex are removed, except one that is located at a certain level of the hierarchical structure. As a result, for each vertex one parent node is defined, located at the *neighbouring* level. Applying the described procedure for different k, we obtain different aggregation trees k-ATs. Note that the k-ATs will not always be constructed for all $k \geq 2$. But for $k = 1$, 1-AT will always be built, because 1-HS contains a shortest paths tree which always exists. Various k-ATs can be used in the algorithms for constructing a conflict-free data aggregation schedule. The quality of the k-AT can be assessed after scheduling. For this purpose, we use the NDR algorithm from [19], as well as VNS from [23], which have proven themselves well. Our numerical experiment shows that using the k-ATs constructed in the hierarchical structures as initial trees speeds up the process of conflict-free data aggregation.

The rest of this paper is organized as follows. Section 2 provides assumptions and the formulation of the problem. In Sect. 3 an algorithm to build k-HS is proposed, and in Sect. 4, we show how to construct an ATs. Section 5 contains the results and analysis of an experimental study, and Sect. 6 concludes the paper.

2   Problem Formulation

We consider a DWN consisting of homogeneous nodes with one sink. We use the protocol interference model [14], which is a graph theoretic approach that assumes correct reception of a message if and only if there are no simultaneous transmissions within proximity of the receiver. The DWN can be represented as a graph $G = (V, E)$, where V denotes the set of nodes and BS is represented as the vertex $s \in V$. An edge (i, j) belongs to E if and only if the distance between the nodes i and j is within the transmission range. We also assume that time is divided into equal-length slots under the assumption that each slot is long enough to send or receive one packet along any edge. The problem considered in this paper is defined as follows. Given a connected undirected graph $G = (V, E)$, $|V| = n$ and a sink node $s \in V$, find the min-length schedule of data aggregation from all the vertices of $V \setminus \{s\}$ to s (i.e., assign a sending time slot and a recipient to each vertex) under the following conditions: (i) each vertex sends a message only once during the aggregation session (except the sink which always only receives messages); (ii) once a vertex sends a message, it can no longer be a destination of any transmission; (iii) if any vertex sends a message, then during the same time slot none of the other vertices within a receiver's interference range can send a message; (iv) a vertex cannot receive and transmit at the same time slot.

As follows from this formulation, the data aggregation has to be performed along the arcs of a spanning tree – aggregation tree (AT) rooted in s.

3 k-Hierarchical Structures

As mentioned above, the problem of minimizing the time of conflict-free data aggregation is divided into two problems: building an AT rooted in the sink and finding a schedule for conflict-free data aggregation. This section provides a procedure to build ATs using hierarchical structures (HSs).

k-HS is constructed by the following steps: (i) place the vertex s to the level 0; (ii) for each vertex $v \in V \setminus \{s\}$ find the length l_v of the shortest (by the number of edges) path from s to v; (iii) for each vertex $v \in V \setminus \{s\}$ put its copies on the levels $l_v, l_v + 1, \ldots, l_v + k - 1$ if there exists a path from s to v, whose length matches the level number; (iv) add the edges between the vertices of *neighbouring* levels if they exist in E.

Since only the vertices of neighbouring levels are connected, some edges of the set E may not be included in the 1-HS, so it makes sense to consider the k-HS with $k > 1$. An example of 1-HS is shown in Fig. 1b (the red edges are not included in the 1-HS). An example of 3-HS is shown in Fig. 1c. Here, s has index 0, the copies of vertices 3, 4, 5, and 6 are located at the levels 2, 3, and 4, the copies of the vertices 7 and 8 are at the levels 3, 4, and 5, and the copies of the vertex 9 are at the levels 4, 5, and 6. It is necessary to notice that some edges are duplicated. For example, two copies exist of the edge $(3, 4)$: one connects the copies of vertices at levels 2 and 3, and another one—at levels 3 and 4.

Algorithm 1. Construction of k-HS

1: *Input*: Graph $G = (V, E)$, $V = \{0, \ldots, n\}$, integer $k > 0$;
2: *Output*: k-HS;
3: Create an empty array of arrays $nodesByLevels$;
4: Construct a shortest-path tree;
5: **for all** $v \in V$ **do**
6:     Create u — a copy of v: $u.index = v$, $u.copyIndex = 1$;
7:     Push u into $nodesByLevels[level(v)]$;
8: **end for**
9: **for** $i = 2, \ldots, k$ **do**
10:    **for** $l = k - 1, \ldots, nodesByLevels.size - 1$ **do**
11:       **for all** $u \in nodesByLevels[l]$: $u.copyIndex = i - 1$ **do**
12:          **for all** $w \in nodesByLevels[l]$: $w.index \neq u.index$ **do**
13:             **if** $(w.index, u.index) \in E$ **then**
14:                **if** $l = nodesByLevels.size - 1$ **then**
15:                   Add a new empty array to the end of $nodesByLevels$ and increment $nodesByLevels.size$;
16:                **end if**
17:                Create u' — a copy of $u.index$: $u'.index = u.index$, $u'.copyIndex = i$;
18:                Push u' into $nodesByLevels[l + 1]$;
19:             **end if**
20:          **end for**
21:       **end for**
22:    **end for**
23: **end for**
24: Push 0 into the 0-th level of k-HS;
25: **for** $l = 1, \ldots, nodesByLevels.size - 1$ **do**
26:    **for all** $u \in nodesByLevels[l]$ **do**
27:       Push u into the l-th level k-HS;
28:       **for all** $w \in nodesByLevels[l - 1]$ **do**
29:          **if** $(w.index, u.index) \in E$ **then**
30:             Push (w, u) into k-HS;
31:          **end if**
32:       **end for**
33:    **end for**
34: **end for**

We assume that the vertices in V are indexed, and s has index 0. The pseudocode of the k-HS construction algorithm is presented in Algorithm 1. In this algorithm, at first, a shortest-path tree is constructed in order to define for each $v \in V$ the number of edges in the shortest path to the root (it is denoted as $level(v)$). After that, (lines 5–23), the copies of vertices are placed into the array $nodesByLevels$. Each element of this array is a set of nodes that belong to the corresponding level of HS. Then, in lines 24–33, all the nodes and edges between the nodes of neighbouring levels are added to the k-HS. For convenience, each node u stores the index of the corresponding vertex in V $u.index$, and the index

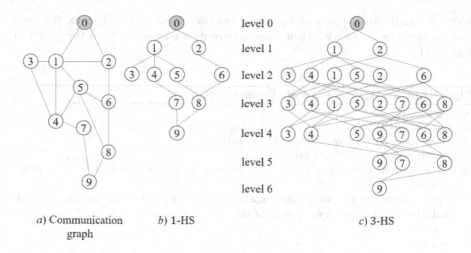

a) Communication   b) 1-HS   c) 3-HS
graph

Fig. 1. Construction of hierarchical structures

of a copy $u.copyIndex$. The index of a copy at the lowest level equals 1, at the next level (if it exists) it equals 2, and so on.

The most time-consuming procedure of the Algorithm 1 is construction of the shortest paths tree. It can be done with time complexity equals $O(n^2)$.

If $k > 1$ then in order to build an AT from k-HS it is required to remove extra copies of vertices and edges. The corresponding heuristic is given in the next section.

4   Aggregation Tree Construction

Construction of AT from k-HS consists of two stages: (i) choosing the parent-nodes and (ii) removing the extra copies of the vertices. At the first stage, for each copy v of a vertex in k-HS that lies at some level $l > 0$, we choose its parent – a copy of some vertex u at the previous level $l - 1$ that is connected with v by an edge. All other edges ending in v are removed. As a result, a spanning tree on all of the vertices copies of k-HS is constructed, we call it k-HST. At the second stage, for each vertex in V, all its copies except one are removed and we get an AT, which we denote as k-AT.

The choice of the parent-node is implemented as follows. (i) If a node has only one potential parent, then this parent is chosen; (ii) if a node has several potential parents, then the node with the minimum degree is chosen from a set of potential parents if the path from it to 0 does not contain any copy of v. In Algorithm 2, the pseudo-code of the parent choosing procedure is presented. At first (line 5), a set of parent candidates is found for each vertex copy u. This set contains the nodes of the previous level in the k-HS that are adjacent to the considered vertex copy. After that, in line 6, the only parent node is selected from the candidates set according to the mentioned rules. All edges except ones that connect the nodes with their chosen parents are removed from the graph in lines 9–11.

Algorithm 2. Algorithm of k-HST construction

1: *Input*: k-HS;
2: *Output*: k-HST;
3: Create an empty array *parents* of size $|k$-HS$|$;
4: **for all** node $u \in k$-HS $\{0\}$ **do**
5:    Find *parent_candidates* — a set of nodes in level $Level(u)$ that are connected with u;
6:    Select from the set *parent_candidates* a node p with the smallest degree, such that $Path(p)$ does not have a copy of the same vertex as u;
7:    $parents[u] = p$;
8: **end for**
9: **for all** $u \in k$-HS **do**
10:    Remove the edges from k-HS that connect u with any node of the previous level except $parent[u]$;
11: **end for**

To obtain an AT, it is necessary to leave one copy of each vertex, while maintaining the connectivity of the AT under construction. Let us introduce a definition:

Definition 1. *A copy of the vertex v we call* critical *if there is a vertex u, such that a subtree in k-HST rooted in v contains all copies of u.*

Obviously, a critical node cannot be removed, otherwise the AT will not be built. We say that a node in k-HS is *fixed* if it is the only copy of this vertex remaining in k-HST, the other its copies are removed.

The pseudocode of Algorithm 3 describes the removing of copies of vertices in k-HST. If a copy of a vertex is critical, then it is fixed, and the remaining copies of this vertex are deleted. For the remaining nodes, we calculate the penalty by the *CalcPenalty* procedure. Penalty of a node is equal to the number of conflicts in which this node participates. A copy of the vertex with maximum penalty is removed. Removing a copy of a vertex may cause fixation of another copy of the vertex if it becomes critical. For this, we call a special procedure *FixationAndRemoval* that fixes critical nodes and removes their copies with the subtrees. It also checks if the copies being deleted are critical. In this case, the procedure aborts the entire algorithm of construction an AT. *FixationAndRemoval* is recursively repeated after a node was removed or fixed. The process is continued while the number of not fixed nodes is less than n, which means that there exist vertices with more than one copies.

Algorithm 3. Algorithm of k-AT construction

1: *Input*: $G = (V, A)$ – the communication graph, $|V| = n$, k-HST;
2: *Output*: k-AT;
3: $fixed\_nodes = 0$;
4: $FixationAndRemoval(k\text{-HST})$;
5: Calculate $fixed\_nodes$;
6: **for all** $u \in k$-HST **do**
7:    $penalties[u] = CalcPenalty(u)$;
8: **end for**
9: **while** $fixed\_nodes < n$ **do**
10:   Select copy a node $u \in k$-HST with maximum value of $penalty[.]$;
11:   **if** u is not critical **then**
12:     Remove from k-HST a node u with a subtree;
13:   **end if**
14:   $FixationAndRemoval(k\text{-HST})$;
15:   Calculate $fixed\_nodes$;
16: **end while**
17: **if** k-HST is not a spanning tree **then**
18:   Emergency terminate the procedure;
19: **end if**
20: Return k-HST.

Time complexity of Algorithm 3 equals $O(n^2)$ since $k = const$, and after fixation/deletion of any node it is necessary to check the connectivity of the HS, what is done in linear time by one scan of the vertices.

As shown by numerical experiments, this heuristic does not guarantee obtaining an AT for any $k > 1$.

In Fig. 2, the described algorithms are applied to the 3-HS. Figure 2a represents 3-HS for the network in Fig. 1a. Here, green edges are incident to the copies of vertices that are adjacent to only one potential parent, blue edges connect the copies of vertices with parents having a minimum degree, and red edges are deleted. Note that the copies of vertices 1 and 2 at level 3 have been removed together with incident edges, because the path from the only potential parent to 0 contains their copies. Similarly, a copy of 3 at level 4 was removed – the path from its parent contains a copy of 3.

Figure 2b represents a 3-HST. The critical vertex at the beginning of the algorithm is highlighted in green (its copies with subtrees have been removed). Each vertex's penalty is indicated by a number to the right of the vertex. Figure 2c shows a 3-AT built from 3-HST using Algorithm 3.

Having an AT, one can start building a conflict-free schedule. We used two algorithms for this: NDR [19] and VNS [23].

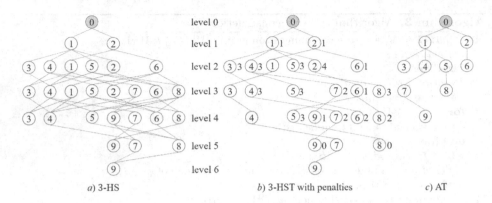

Fig. 2. Building AT: *a*) choosing the parent-nodes; *b*) deleting the copies; *c*) 3-AT

5   Simulation

The proposed algorithms have been implemented in the programming language C++ using the Visual Studio 2019 IDE. The experiment was carried out on the Intel Core i3-9100 CPU 3.60 GHz 8 Gb RAM, Windows 10 × 64.

The test instances were generated as follows. The nodes are uniformly and randomly spread over the square using a fixed transmission distance and a unit-disk graph (UDG) is constructed to generate a network topology. In each instance, we construct k-AT for $k = 1, \ldots, 5$ and use the algorithm NDR for scheduling as one of the most efficient fast heuristics. For the comparison, we also use the RH from [2], and MLST introduced in [19]. We also compare performance of the VNS based algorithm from [23], as the best-known metaheuristics that solves this problem, with different aggregation trees taken as an initial solution: VSN_RH, where the RH is used for the initial aggregation tree construction, and VNS_HS, where the best of considered k-AT is used as the initial tree.

We tested 4 combinations (n, d) of number of nodes n and transmission distance d: (1) $n = 50$, $d = 0.4$; (2) $n = 100$, $d = 0.3$; (3) $n = 250, d = 0.2$; and (4) $n = 1000$, $d = 0.15$, in 5 different instances for each (n, d). The results are presented in Tables 1 and 2. For those cases where k-AT was not constructed, the values are replaced with dashes. Since VNS is a randomized metaheuristic, we launch it 10 times on each instance and calculate the mean values of objective function and CPU time. The results for 1-AT and 2-AT are not included in the tables because they never outperform 3-AT, 4-AT, and 5-AT. We marked the minimal schedule lengths in bold in Table 1. Note that we do not compare VNS based methods with other (*fast*) heuristic algorithms because these approaches are completely different: metaheuristics, such as the presented VNS based ones, can be used only when calculation speed does not pay a significant role.

It can be seen in cases when $n < 250$, HS-based heuristics steadily give better results than algorithms that use other tree construction approaches, but the situation is reversed for greater n. The k-AT based fast heuristics consume

Table 1. Schedule lengths comparison. Fast heuristics (AT is constructed by different algorithms: 3-AT, 4-AT, 5-AT, MLST, and RH; scheduling is performed by NDR algorithm) and average values for 10 runs of VNS based algorithms with different initial solutions: VNS_HS and VNS_RH.

n	d	3-AT	4-AT	5-AT	MLST	RH	VNS_HS	VNS_RH
50	0.4	15	15	**14**	19	17	**14**	14.1
		17	–	16	16	14	**12.2**	12.6
		19	14	15	18	16	**13.3**	13.4
		16	14	**14**	18	15	**13**	13.5
		14	13	**12**	17	14	**11.8**	12.6
100	0.3	29	20	–	21	19	**17.8**	18.7
		25	–	**19**	20	21	**17.4**	17.8
		24	23	**20**	23	22	18.3	**17.7**
		26	19	**18**	20	19	**17**	17.2
		22	21	**20**	21	22	**18**	18.3
250	0.2	29	30	**25**	31	26	**23.1**	23.5
		29	28	–	33	24	**23**	23.1
		28	28	–	28	23	22.8	**21.8**
		31	29	30	33	27	22.8	**22.6**
		30	29	–	30	25	23.8	**22.9**
1000	0.15	69	72	–	**63**	48	46	**45**
		63	71	–	64	48	47	**46.1**
		66	66	–	66	**51**	47	**46.1**
		63	64	–	61	47	45.2	**44.8**
		63	65	–	64	50	46.1	**46**

Table 2. CPU time comparison (in seconds). Fast heuristics (AT is constructed by different algorithms: 3-AT, 4-AT, 5-AT, MLST, and RH; scheduling is performed by NDR algorithm) and average values for 10 runs of VNS based algorithms with different initial solutions: VNS_HS and VNS_RH.

n	d	3-AT	4-AT	5-AT	MLST	RH	VNS_HS	VNS_RH
50	0.4	0.01	0.01	0.01	0.00	0.00	0.63	1.28
		0.01	–	0.01	0.00	0.00	0.85	0.76
		0.01	0.01	0.01	0.00	0.00	1.06	1.28
		0.01	0.01	0.01	0.00	0.00	1.03	0.87
		0.01	0.01	0.01	0.00	0.00	0.62	0.77
100	0.3	0.01	0.01	–	0.01	0.00	4.12	3.44
		0.01	–	0.02	0.01	0.00	3.75	2.65
		0.01	0.02	0.02	0.00	0.00	3.09	2.65
		0.01	0.01	0.02	0.00	0.00	2.40	2.64
		0.01	0.02	0.02	0.01	0.00	2.97	2.94
250	0.2	0.03	0.04	0.06	0.02	0.01	12.28	8.99
		0.03	0.04	–	0.06	0.01	12.15	10.01
		0.03	0.04	–	0.02	0.01	9.48	8.01
		0.03	0.04	0.06	0.02	0.01	13.90	8.25
		0.03	0.04	–	0.03	0.01	12.08	9.03
1000	0.15	0.24	0.37	–	0.80	0.07	186.87	132.91
		0.26	0.37	–	0.93	0.07	164.20	155.57
		0.26	0.38	–	0.94	0.07	183.42	147.28
		0.25	0.38	–	1.16	0.07	160.06	165.79
		0.26	0.38	–	0.94	0.07	186.38	132.62

at most 0.1 s as well as other tested algorithms for $n \leq 250$, and their running time does not exceed 0.4 s for $n = 1000$, which is faster than the MLST based approach that takes about 0.8–1.2 s but slower than the RH based one that takes about 0.07 s. VNS_HS gives better average results than VNS_RH for $n < 250$ and takes about the same amount of time on average. We may conclude that usage of the HS based aggregation tree construction method is justified for small and moderate size instances both for fast aggregation heuristics and for generating the initial solution in VNS metaheuristic.

6   Conclusion

In this paper, we considered an NP-hard problem of conflict-free min-length aggregation scheduling in a distributed wireless network. One of the most common approaches to solve this problem approximately consists of two stages: constructing an appropriate AT at the first stage and finding a near-optimal schedule on this tree at the second stage. We noticed that such an approach has the following disadvantage; although the corresponding algorithms are often rather fast, the obtained solution could be bad, because of the chosen AT. We propose a new heuristic for AT construction using the special hierarchical structures, k-HS, when initially up to k copies of each node can be located at different neighbouring levels. When $k = 1$, we always construct an aggregation tree 1-AT. If $k \geq 2$, then we built, at most, one aggregation tree k-AT for each value of k. We propose new heuristics to construct the trees in k-HS and finally to find the aggregation trees k-ATs for different k. The outcome of our approach is the best schedule found on k-ATs, $k = 1, \ldots, 5$. To assess the quality of the proposed approach, we carried out a numerical experiment on the randomly constructed unit-disk graphs, and can conclude that the construction of k-ATs in the hierarchical structures speeds up the aggregation.

In the future, we plan to prepare an extended version of the paper, which will include a new algorithm for building a multilevel AT, in which vertices of the same level can transmit messages simultaneously without conflicts. Then the number of levels in AT will be equal to the length of the schedule. To assess the quality of this algorithm and previously considered approaches, a numerical experiment will be carried out not only on the unit-disk graphs, but also on the lattices.

References

1. Bagaa, M., Challal, Y., Ksentini, A., Derhab, A., Badache, N.: Data aggregation scheduling algorithms in wireless sensor networks: solutions and challenges. IEEE Commun. Surv. Tutor. **16**, 1339–1367 (2014)
2. Beier, R., Sibeyn, J.F.: A powerful heuristic for telephone gossiping. In: Proceedings of 17th International Colloquium on Structural Information & Communication Complexity (SIROCCO 2000), pp. 17–36 (2000)

3. Chen, X., Hu, X., Zhu, J.: Minimum data aggregation time problem in wireless sensor networks. In: Jia, X., Wu, J., He, Y. (eds.) MSN 2005. LNCS, vol. 3794, pp. 133–142. Springer, Heidelberg (2005). https://doi.org/10.1007/11599463_14
4. Edmonds, J.: Optimum branchings. J. Res. Natl. Bureau Stand. B **71**, 233–240 (1967)
5. Erzin, A., Pyatkin, A.: Convergecast scheduling problem in case of given aggregation tree: the complexity status and some special cases. In: Proceedings of 10th International Symposium on Communication Systems Networks and Digital Signal Processing (CSNDSP), pp. 1–6. IEEE (2016). https://doi.org/10.1109/CSNDSP. 2016.7574007
6. Erzin, A.: Solution of the convergecast scheduling problem on a square unit grid when the transmission range is 2. In: Battiti, R., Kvasov, D.E., Sergeyev, Y.D. (eds.) LION 2017. LNCS, vol. 10556, pp. 50–63. Springer, Cham (2017). https://doi.org/10.1007/978-3-319-69404-7_4
7. Erzin, A., Plotnikov, R.: Conflict-free data aggregation on a square grid when transmission distance is not less than 3. In: Fernández Anta, A., Jurdzinski, T., Mosteiro, M.A., Zhang, Y. (eds.) ALGOSENSORS 2017. LNCS, vol. 10718, pp. 141–154. Springer, Cham (2017). https://doi.org/10.1007/978-3-319-72751-6_11
8. Erzin, A., Plotnikov, R.: The accuracy of one polynomial algorithm for the convergecast scheduling problem on a square grid with rectangular obstacles. In: Battiti, R., Brunato, M., Kotsireas, I., Pardalos, P.M. (eds.) LION 12 2018. LNCS, vol. 11353, pp. 131–140. Springer, Cham (2019). https://doi.org/10.1007/978-3-030-05348-2_11
9. Erzin, A., Plotnikov, R.: Two-channel conflict-free square grid aggregation. In: Kotsireas, I.S., Pardalos, P.M. (eds.) LION 2020. LNCS, vol. 12096, pp. 168–183. Springer, Cham (2020). https://doi.org/10.1007/978-3-030-53552-0_18
10. Guney, E., Altinel, I.K., Aras, N., Ersoy, C.: A variable neighbourhood search heuristic for point coverage, sink location and data routing in wireless sensor networks. In: Proceedings of 2nd International Conference on Communication Theory, Reliability, and Quality of Service (CTRQ 2009), pp. 81–86 (2009)
11. Gupta, P., Kumar, P.R.: The capacity of wireless networks. IEEE Trans. Inf. Theory **46**, 388–404 (2000)
12. Hansen, P., Mladenovic, N.: Variable neighborhood search: principles and applications. Eur. J. Oper. Res. **130**, 449–467 (2001)
13. Hansen, P., Mladenovic, N., Perez-Britos, D.: Variable neighborhood decomposition search. J. Heuristics **7**(4), 335–350 (2001). https://doi.org/10.1023/A: 1011336210885
14. Hromkovic, J., Klasing, R., Monien, B., Peine, R.: Dissemination of information in interconnection networks (broadcasting & gossiping). In: Du, D.Z., Hsu, D.F. (eds.) Combinatorial Network Theory. APOP, vol. 1, pp. 125–212. Springer, Boston (1996). https://doi.org/10.1007/978-1-4757-2491-2_5
15. Huang, S.C.-H., Wan, P.J., Vu, C.T., Li, Y., Yao, F.: Nearly constant approximation for data aggregation scheduling in wireless sensor networks. In: IEEE Conference on Computer Communications (INFOCOM 2007), pp. 366–472 (2007)
16. Incel, O.D., Ghosh, A., Krishnamachari, B., Chintalapudi, K.: Fast data collection in tree-based wireless sensor networks. IEEE Trans. Mob. Comput. **11**, 86–99 (2011)
17. Malhotra, B., Nikolaidis, I., Nascimento, M.A.: Aggregation convergecast scheduling in wireless sensor networks. Wirel. Netw. **17**, 319–335 (2011)

18. Nguyen, T.D., Zalyubovskiy, V., Choo, H.: Efficient time latency of data aggregation based on neighboring dominators in WSNs. In: IEEE Globecom 6133827 (2011)
19. Pan, C., Zhang, H.: A time efficient aggregation convergecast scheduling algorithm for wireless sensor networks. Wirel. Netw. **22**(7), 2469–2483 (2016). https://doi.org/10.1007/s11276-016-1337-5
20. Plotnikov, R., Erzin, A., Mladenovic, N.: Variable neighborhood search-based heuristics for min-power symmetric connectivity problem in wireless networks. In: Kochetov, Y., Khachay, M., Beresnev, V., Nurminski, E., Pardalos, P. (eds.) DOOR 2016. LNCS, vol. 9869, pp. 220–232. Springer, Cham (2016). https://doi.org/10.1007/978-3-319-44914-2_18
21. Plotnikov, R., Erzin, A., Zalyubovsky, V.: Convergecast with unbounded number of channels. In: MATEC Web of Conferences, vol. 125, p. 03001 (2017). https://doi.org/10.1051/matecconf/201712503001
22. Plotnikov, R., Erzin, A., Zalyubovskiy, V.: Genetic local search for conflict-free minimum-latency aggregation scheduling in wireless sensor networks. In: Evtushenko, Y., Jaćimović, M., Khachay, M., Kochetov, Y., Malkova, V., Posypkin, M. (eds.) OPTIMA 2018. CCIS, vol. 974, pp. 216–231. Springer, Cham (2019). https://doi.org/10.1007/978-3-030-10934-9_16
23. Plotnikov, R., Erzin, A., Zalyubovskiy, V.: GLS and VNS based heuristics for conflict-free minimum-latency aggregation scheduling in WSN. Optim. Methods Softw. **36**(4), 697–719 (2021)
24. Sivanandam, S., Deepa, S.: Introduction to Genetic Algorithms. Springer, Heidelberg (2008). https://doi.org/10.1007/978-3-540-73190-0
25. de Souza, E., Nikolaidis, I.: An exploration of aggregation convergecast scheduling. Ad Hoc Netw. **11**, 2391–2407 (2013)
26. Su, S., Yu, H.: Minimizing tardiness in data aggregation scheduling with due date consideration for single-hop wireless sensor networks. Wirel. Netw. **21**, 1259–1273 (2015)
27. Tian, C., Jiang, H., Wang, C., Wu, Z., Chen, J., Liu, W.: Neither shortest path nor dominating set: aggregation scheduling by greedy growing tree in multihop wireless sensor networks. IEEE Trans. Veh. Technol. **60**, 3462–3472 (2011)
28. Xu, X., Li, X.Y., Mao, X., Tang, S., Wang, S.: A delay-efficient algorithm for data aggregation in multihop wireless sensor networks. IEEE Trans. Parallel Distrib. Syst. **22**, 163–175 (2011)
29. Zhu, J., Hu, X.: Improved algorithm for minimum data aggregation time problem in wireless sensor networks. J. Syst. Sci. Complex. **21**, 626–636 (2018)

Enumeration and Unimodular Equivalence of Empty Delta-Modular Simplices

D. V. Gribanov[1,2]([☒])(iD)

[1] Lobachevsky State University of Nizhny Novgorod,
23 Gagarina Avenue, Nizhny Novgorod 603950, Russian Federation
dimitry.gribanov@gmail.com
[2] National Research University Higher School of Economics,
25/12 Bolshaja Pecherskaja Ulitsa, Nizhny Novgorod 603155, Russian Federation

Abstract. Consider a class of simplices defined by systems $Ax \leq b$ of linear inequalities with Δ-*modular* matrices. A matrix is called Δ-*modular*, if all its rank-order sub-determinants are bounded by Δ in an absolute value. In our work we call a simplex Δ-*modular*, if it can be defined by a system $Ax \leq b$ with a Δ-modular matrix A. And we call a simplex *empty*, if it contains no points with integer coordinates. In literature, a simplex is called *lattice*, if all its vertices have integer coordinates. And a lattice-simplex is called *empty*, if it contains no points with integer coordinates excluding its vertices. Recently, assuming that Δ is fixed, it was shown in [37] that the number of Δ-modular empty simplices modulo the unimodular equivalence relation is bounded by a polynomial on dimension. We show that an analogous fact holds for the class of Δ-modular empty lattice-simplices. As the main result, assuming again that the value of the parameter Δ is fixed, we show that the all unimodular equivalence classes of simplices of both types can be enumerated by a polynomial-time algorithm. As the secondary result, we show the existence of a polynomial-time algorithm for the problem to check the unimodular equivalence relation for a given pair of Δ-modular (not necessarily empty) simplices.

Keywords: Lattice Simplex · Empty Simplex · Delta-Modular Matrix · Bounded Sub-determinants · Unimodular Equivalence · Enumeration Algorithm

1 Introduction

On the one hand, simplices are quite simple objects, since they form a class of solid polytopes with the minimum possible number of vertices. At the same time, the simplices are substantial and fundamental geometric and combinatorial

The article was prepared under financial support of Russian Science Foundation grant No 21-11-00194.

M. Khachay et al. (Eds.): MOTOR 2023, LNCS 13930, pp. 115–132, 2023.
https://doi.org/10.1007/978-3-031-35305-5_8

objects. For example, the unbounded knapsack problem can be modeled as the integer linear optimization problem on a simplex; Simplices are fundamental building blocks of any triangulation of a given polytope; Simplices can be used as universal and quite simple bounding regions; And etc.

In our work we consider only the simplices $\mathcal{S}$ defined by $\mathcal{S} = \{x \in \mathbb{R}^n : Ax \leq b\}$, where $A \in \mathbb{Z}^{(n+1) \times n}$, $\operatorname{rank}(A) = n$ and $b \in \mathbb{Z}^{n+1}$. Our main interest consists of various problems associated with the set $\mathcal{S} \cap \mathbb{Z}^n$. For example, in the *integer feasibility problem* we need to decide, ether $\mathcal{S} \cap \mathbb{Z}^n \neq \emptyset$ or not. This problem is naturally *NP*-complete, because the classical *NP*-complete unbounded subset-sum problem can be represented as an integer feasibility problem in a simplex. In the more general *NP*-hard *integer linear optimization problem* we need to find a point $x^* \in \mathcal{S} \cap \mathbb{Z}^n$ such that $c^\top x^* = \max\{c^\top x : x \in \mathcal{S} \cap \mathbb{Z}^n\}$, where $c \in \mathbb{Z}^n$ is an arbitrary vector, or decide that $\mathcal{S} \cap \mathbb{Z}^n = \emptyset$. However, assuming that dimension is fixed, due to seminal work of Lenstra [48] (for more up-to-date and universal algorithms see, for example, [2,22,24,25,30,31,61]), the integer feasibility and integer linear optimization problems, for any polytope defined by a linear inequalities system, can be solved by a polynomial-time algorithm. In the *integer points counting problem* we need to calculate the value of $|\mathcal{S} \cap \mathbb{Z}^n|$, which is, by the same reasons, a #*P*-hard problem. Due to the seminal work of Barvinok [12] (see also the works [11,13,29,45] and the books [10,16,26,47]), assuming that dimension is fixed, the last problem can be solved by a polynomial-time algorithm for general polytopes.

Important classes of simplices (and even more general polytopes) are the *lattice-simplices* and *empty lattice-simplices*. A simplex $\mathcal{S}$ (or a general polytope) is called *lattice*, if $\operatorname{vert}(\mathcal{S}) \subseteq \mathbb{Z}^n$. A lattice-simplex $\mathcal{S}$ (or a general lattice-polytope) is called *empty*, if $\mathcal{S} \cap \mathbb{Z}^n = \operatorname{vert}(\mathcal{S})$. In our work a simplex $\mathcal{S}$, which is not a lattice-simplex, is called *empty*, if $\mathcal{S} \cap \mathbb{Z}^n = \emptyset$.

Empty lattice-polytopes appear in many important works concerning the theory of integer programming. They are one of the central objects in the theory of cutting planes [3,6,15,20,27], and they can be used to construct optimality certificates in convex integer optimization [1,5,14,17,18,50,53,63]. From the perspective of algebraic geometry, empty simplices are almost in bijection with the terminal quotient singularities, see, for example, [19,41,51,52]. The problem to classify all 4-dimensional empty lattice-simplices was completely solved in the paper [41] due to Iglesias-Valino & Santos. Earlier results in this direction could be found, for example, in the works [39,51,57,59]. The 3-dimensional classification problem was solved by White [62] (see also [52,57]). Important structural properties of general 3-dimensional empty lattice-polytopes are presented in the paper [60] due to Veselov & Chirkov.

An important parameter that is strongly connected with the emptiness property of a simplex, lattice-simplex, or a general lattice-free convex body is the *lattice-width* or, simply, the *width*:

Definition 1. *For a convex body* $\mathcal{P} \subseteq \mathbb{R}^n$ *and a vector* $c \in \mathbb{Z}^n \setminus \{0\}$, *define*

$$\text{width}_c(\mathcal{P}) = \max_{x \in \mathcal{P}} c^\top x - \min_{x \in \mathcal{P}} c^\top x, \quad and$$
$$\text{width}(\mathcal{P}) = \min_{c \in \mathbb{Z}^n \setminus \{0\}} \{\text{width}_c(\mathcal{P})\}.$$

The value of $\text{width}(\mathcal{P})$ *is called the* lattice-width *(or simply the* width*) of* $\mathcal{P}$.

This connection can be derived from the famous work of A. Khinchin, and it is known by the name *Flatness Theorem*.

Theorem 1 (Flatness Theorem, A. Khinchin [44]). *Let* $\mathcal{P} \subseteq \mathbb{R}^n$ *be a convex body with* $\mathcal{P} \cap \mathbb{Z}^n = \emptyset$.

$$Then, \quad \text{width}(\mathcal{P}) \leq \omega(n),$$

where $\omega(n)$ *is some function that depends only on* n.

Many proofs of the Flatness Theorem have been presented in [4,7–9,42,46], each giving different asymptotic estimates on $\omega(n)$. For example, combining the works of Banaszczyk, Litvak, Pajor & Szarek [9] and Rudelson [55], it can be seen that $\omega(n) = n^{4/3} \cdot \log^{O(1)}(n)$. Very recently the last bound was improved to $\omega(n) = O(n \cdot \log^8(n))$ in the breakthrough work [54] due to Reis & Rothvoss. With restriction to simplices, it was shown [9] that $\omega(n) = O(n \log(n))$. The same bound on $\omega(n)$ with restriction to symmetric convex bodies, was shown by Banaszczyk [8]. There is a sequence of works [23,28,43,49,57] devoted to establish a lower bound for $\omega(n)$. The state of the art result of this kind belongs to the work [49] of Mayrhofer, Schade & Weltge, where the construction of an empty lattice-simplex of the width $2n - o(n)$ is presented. It is conjectured in [9] that $\omega(n) = \Theta(n)$. To the best of our knowledge, the only one non-trivial case, when the conjecture was verified, is the case of ellipsoids, due to [8].

Due to Sebő [57], it is a *NP*-complete problem to decide that $\text{width}(\mathcal{S}) \leq 1$ for a given lattice-simplex $\mathcal{S}$. However, as it was noted in [57], assuming that dimension is fixed, the problem again can be solved by a polynomial-time algorithm.

1.1 The Class of Δ-Modular Simplices

It turns out that it is possible to define an interesting and substantial parameter of the system $Ax \leq b$, which significantly affects complexity of the considered problems.

Definition 2. *For a matrix* $A \in \mathbb{Z}^{m \times n}$, *by*

$$\Delta_k(A) = \max \Big\{ |\det(A_{\mathcal{I}\mathcal{J}})| : \mathcal{I} \subseteq \{1,\dots,m\}, \ \mathcal{J} \subseteq \{1,\dots,n\}, \ |\mathcal{I}| = |\mathcal{J}| = k \Big\},$$

we denote the maximum absolute value of determinants of all the $k \times k$ *sub-matrices of* A. *Here, the symbol* $A_{\mathcal{I}\mathcal{J}}$ *denotes the sub-matrix of* A, *which is* *generated by all the rows with indices in $\mathcal{I}$ and all the columns with indices in $\mathcal{J}$.*

Note that $\Delta_1(A) = \|A\|_{\max} = \max_{ij} |A_{ij}|$. Additionally, let $\Delta(A) = \Delta_{\text{rank}(A)}(A)$. The matrix A with $\Delta(A) \leq \Delta$, for some $\Delta > 0$, is called Δ-modular. The polytope $\mathcal{P}$, which is defined by a system $Ax \leq b$ with a Δ-modular matrix A, is called Δ-modular.

Surprisingly, all the mentioned problems restricted on Δ-modular simplices with $\Delta = \text{poly}(\phi)$, where ϕ is the input length, can be solved by a polynomial time algorithms. Such algorithms are also known as the FPT-algorithms. Definitely, restricting the problems on simplices, due to Gribanov, Shumilov, Malyshev, Pardalos [37, Theorem 14] the integer feasibility problem can be solved with $O(n + \min\{n, \Delta\} \cdot \Delta \cdot \log(\Delta))$ operations. Due to [37, Corollary 9] (see also [35]), the integer linear optimization problem can be solved with $O(n + \Delta^2 \cdot \log(\Delta) \cdot \log(\Delta_{\gcd}))$ operations. Due to Gribanov & Malyshev [34] and a modification from Gribanov, Malyshev & Zolotykh [32], the integer points counting problem can be solved with $O(n^4 \cdot \Delta^3)$ operations. Due to Gribanov, Malyshev, Pardalos & Veselov [35] (see also [33]), the lattice-width can be computed with $\text{poly}(n, \Delta, \Delta_{ext})$-operations, where $\Delta_{ext} = \Delta(A\,b)$ and $(A\,b)$ is the system's extended matrix. Additionally, for empty simplices, the complexity dependence on Δ_{ext} can be avoided, which gives the bound $\text{poly}(n, \Delta)$. The analogous result for lattice-simplices defined by convex hulls of their vertices is presented in Gribanov, Malyshev & Veselov [36].

Additionally, it was shown by Gribanov & Veselov [38] that it is possible to prove a variant of the Flatness Theorem for simplices such that the function ω will depend only on the sub-determinants spectrum of A instead of n. More precisely, the following statement is true: for any empty simplex $\mathcal{S}$, the inequality $\text{width}(\mathcal{S}) < \Delta_{\min}(A)$ holds, where $\Delta_{\min}(A)$ is the minimum between all rank-order nonzero sub-determinants of A taken by an absolute value. Additionally, if the inequality is not satisfied, then some integer point inside $\mathcal{S}$ can be found by a polynomial-time algorithm. The last inequality for empty simplices was improved to $\text{width}(\mathcal{S}) < \lfloor \Delta_{\min}/2 \rfloor$ in the work [40] of Henk, Kuhlmann & Weismantel. For a general Δ-modular empty polytope $\mathcal{P}$, the inequality $\text{width}(\mathcal{S}) \leq (\Delta_{\text{lcm}}(A) - 1) \cdot \frac{\Delta(A)}{\Delta_{\gcd}(A)} \cdot (n+1)$ was shown in [38], where $\Delta_{\gcd}(A)$ and $\Delta_{\text{lcm}}(A)$ are the gcd and lcm functions of rank-order sub-determinants of A taken by an absolute value. Additionally, if the inequality is not satisfied, then $\mathcal{P}$ contains a lattice-simplex, and some vertex of this simplex can be found by a polynomial-time algorithm. Recently, in the work [21] of Celaya, Kuhlmann, Paat & Weismantel, the last inequality for empty polytopes was improved to $\text{width}(\mathcal{S}) < \frac{4n+2}{9} \cdot \Delta(A)$.

1.2   The Main Motivation of the Work and Results

To formulate the main problems and results, we need first to make some additional definitions. For a unimodular matrix $U \in \mathbb{Z}^{n \times n}$ and an integer vector $x_0 \in \mathbb{Z}^n$, the affine map $\mathcal{U}(x) = Ux + x_0$ is called *unimodular*. Simplices $\mathcal{S}_1$ and $\mathcal{S}_2$ are called *unimodular equivalent*, if there exists an unimodular affine map

$\mathcal{U}$ such that $\mathcal{U}(\mathcal{S}_1) = \mathcal{S}_2$. It is easy to see that this relation partitions all the simplices into equivalence classes.

It was shown in [37, Theorem 16] that the number of empty Δ-modular simplices modulo the unimodular equivalence relation is bounded by

$$\binom{n + \Delta - 1}{\Delta - 1} \cdot \Delta^{\log_2(\Delta)+2}, \tag{1}$$

which is a polynomial, for fixed Δ. This result motivates the following

Problem 1 (Empty Δ-modular Simplices Enumeration). For a given value of the parameter Δ, enumerate all the unimodular equivalence classes of Δ-modular empty simplices and empty lattice-simplices.

To solve the last problem, it is important to have an algorithm that can check the unimodular equivalence relation.

Problem 2 (Unimodular Equivalence Checking). For given simplices $\mathcal{S}$ and $\mathcal{T}$, determine whether the simplices $\mathcal{S}$ and $\mathcal{T}$ are unimodular equivalent.

As the main results we show that the both problems can be solved by polynomial-time algorithms, assuming that the value of Δ is fixed. Unfortunately, the new algorithms do not belong to the FPT-class. The formal definitions of our results are presented in the following theorems.

Theorem 2. *Let $\mathcal{S}$ and $\mathcal{T}$ be Δ-modular simplices. The unimodular equivalence relation for $\mathcal{S}$ and $\mathcal{T}$ can be checked by an algorithm with the complexity*

$$n^{\log_2(\Delta)} \cdot \Delta! \cdot \text{poly}(n, \Delta),$$

which is polynomial, for fixed values of Δ.

Theorem 3. *We can enumerate all representatives of the unimodular equivalence classes of the Δ-modular empty simplices and empty lattice-simplices by an algorithm with the complexity bound*

$$O(n + \Delta)^{\Delta-1} \cdot n^{\log_2(\Delta)+O(1)},$$

which is polynomial, for fixed values of Δ.

The proof is given in the next Sect. 2.

2   Proof of Theorems 2 and 3

2.1   The Hermite Normal Form

Let $A \in \mathbb{Z}^{m \times n}$ be an integer matrix of rank n, assuming that the first n rows of A are linearly independent. It is a known fact (see, for example, [56,58]) that there exists a unimodular matrix $Q \in \mathbb{Z}^{n \times n}$, such that $A = \binom{H}{B}Q$, where $B \in$

$\mathbb{Z}^{(m-n)\times n}$ and $H \in \mathbb{Z}_{\geq 0}^{n\times n}$ is a lower-triangular matrix, such that $0 \leq H_{ij} < H_{ii}$, for any $i \in \{1,\ldots,n\}$ and $j \in \{1,\ldots,i-1\}$. The matrix $\binom{H}{B}$ is called the *Hermite Normal Form* (or, shortly, the HNF) of the matrix A. Additionally, it was shown in [35] that $\|B\|_{\max} \leq \Delta(A)$ and, consequently, $\left\|\binom{H}{B}\right\|_{\max} \leq \Delta(A)$. Near-optimal polynomial-time algorithm to construct the HNF of A is given in the work [58] of Storjohann & Labahn.

2.2 Systems in the Normalized Form and Their Enumeration

Definition 3. *Assume that a simplex S is defined by a system $Ax \leq b$ with $A \in \mathbb{Z}^{(n+1)\times n}$ and $b \in \mathbb{Z}^{n+1}$. The system $Ax \leq b$ is called* normalized, *if it has the following form:*

$$\binom{H}{c^\top} x \leq \binom{h}{c_0}, \quad where$$

1. *The matrix $A = \binom{H}{c^\top}$ is the HNF of some integer $(n+1) \times n$ matrix and $|\det(H)| = \Delta(A)$;*
2. *The matrix H has the form $H = \begin{pmatrix} I_s & \mathbf{0}_{s\times k} \\ B & T \end{pmatrix}$, where $k+s = n$, $k \leq \log_2(\Delta)$, the columns of B are lexicographically sorted, T has a lower triangular form and $T_{ii} \geq 2$, for any $i \in \{1,\ldots,k\}$;*
3. *For $i \in \{1,\ldots,n\}$, $0 \leq h_i < H_{ii}$. Consequently, $h_i = 0$, for $i \in \{1,\ldots,s\}$;*
4. *For any inequality $a^\top x \leq a_0$ of the system $Ax \leq b$, $\gcd(a, a_0) = 1$;*
5. *$c \in \mathrm{Par}(-H^\top)$, where $\mathrm{Par}(M) = \{Mt: t \in (0,1]^n\}$, for arbitrary $M \in \mathbb{Z}^{n\times n}$;*
6. *$\|A\|_{\max} \leq \Delta$.*

Lemma 1. *The following properties of normalized systems hold:*

1. *The 5-th and 6-th conditions of Definition 3 are redundant, i.e. they follow from the remaining conditions and properties of S;*
2. *Let S be the simplex used in Definition 3. Assume that S is an empty simplex or an empty lattice-simplex, then $|c_0 - c^\top v| \leq \Delta$, where $v = H^{-1}h$ is the opposite vertex of S with respect to the facet induced by the inequality $c^\top x \leq c_0$.*

Proof. Let us show that the condition $c \in \mathrm{Par}(-H^\top)$ is redundant. Definitely, since S is a simplex, it follows that $c = -H^\top t$, for some $t \in \mathbb{R}_{>0}^n$. Assume that there exists $i \in \{1,\ldots,n\}$ such that $t_i > 1$. Let M be the matrix formed by rows of H with indices in the set $\{1,\ldots,n\} \setminus \{i\}$ and a row $c^\top$. Then, $|\det(M)| = t_i \cdot |\det(H)| > \Delta$, which contradicts to the Δ-modularity property of S. Hence, $0 < t_i \leq 1$, for all $i \in \{1,\ldots,n\}$, and $c \in \mathrm{Par}(-H^\top)$.

As it was mentioned in Subsect. 2.1, for each Δ-modular matrix A that is reduced to the HNF, the inequality $\|A\|_{\max} \leq \Delta$ holds. So the 6-th condition just follows from the 1-st condition and the Δ-modularity property of A.

Let us prove the 2-nd property claimed in the lemma. In the case when S is empty, the inequality $|c_0 - c^\top v| < \Delta$ was proven in [38, Theorem 7] (see also [37, Corollary 10]). Let us prove the inequality $|c^\top v - c_0| \leq \Delta$ in the case when

S is empty lattice-simplex. Consider a point $p = v - u$, where u is the first column of the adjugate matrix $H^* = \det(H) \cdot H^{-1}$. Note that p is an integer feasible solution for the sub-system $Hx \leq h$. Since $\|c^\top H^*\|_\infty \leq \Delta$, it holds that $|c^\top p - c^\top v| \leq \Delta$. Then, the inequality $|c^\top v - c_0| > \Delta$ implies that $p \in S \setminus \text{vert}(S)$, which is the contradiction.

The following lemma is one of the elementary building blocks for an algorithm to construct normalized systems.

Lemma 2. *Let $H \in \mathbb{Z}^{n \times n}$ be a non-degenerate matrix reduced to the HNF and $b \in \mathbb{Z}^n$. Then, there exists a polynomial-time algorithm that computes the unique translation $x \to x + x_0$ that maps the system $Hx \leq b$ to the system $Hx \leq h$ with the property $0 \leq h_i < H_{ii}$, for any $i \in \{1, \dots, n\}$.*

Proof. Denote the i-th standard basis vector of $\mathbb{R}^n$ by e_i. Consider the following algorithm: set $b^{(0)} := b$ and, for the index i increasing from 1 to n, we sequentially apply the integer translation $x \to x + \lfloor b^{(i-1)}/H_{ii} \rfloor \cdot e_i$ to the system $Ax \leq b$, where $b^{(i)}$ denotes the r.h.s. of $Ax \leq b$ after the i-th iteration. We have $b^{(i+1)} := b^{(i)} - \lfloor b^{(i)}/H_{ii} \rfloor H_i$, where H_i is the i-th column of H. If we denote the resulting system r.h.s. by h, then $h = b^{(n)}$ and $h_i = b_i \bmod H_{ii}$, for $i \in \{1, \dots, n\}$. Clearly, the algorithm that returns h and the cumulative integer translation vector x_0 is polynomial-time.

Note that the vectors b and h are additionally connected by the formula $h = b - Hx_0$. Due to the HNF triangle structure, and since $h_1 = b_1 \bmod H_{11}$, it follows that $(x_0)_1 = \lfloor b_0/H_{11} \rfloor$, so the first component of x_0 is unique. Now, using the induction principle, assume that the first k components of x_0 are uniquely determined by H and b, and let us show that the same fact holds for $(x_0)_{k+1}$. Clearly,

$$h = b - H_{\{1,\dots,k\}}(x_0)_{\{1,\dots,k\}} - H_{\{k+1,\dots,n\}}(x_0)_{\{k+1,\dots,n\}}.$$

Denoting $\hat{h} = h + H_{\{1,\dots,k\}}(x_0)_{\{1,\dots,k\}}$, we have

$$\hat{h} = b - H_{\{k+1,\dots,n\}}(x_0)_{\{k+1,\dots,n\}}.$$

Note that $\hat{h}_{k+1} = h_{k+1} = b_{k+1} \bmod H_{(k+1)(k+1)}$. Again, due to the HNF triangle structure, it holds that $(x_0)_{k+1} = \lfloor b_{k+1}/H_{(k+1)(k+1)} \rfloor$, which finishes the proof.

The following lemma proposes an algorithm to construct normalized systems. Additionally, it proves that all equivalent normalized systems can be constructed by this way.

Lemma 3 (The Normalization Algorithm). *Let S be a Δ-modular simplex defined by a system $Ax \leq b$. The following propositions hold:*

1. *For a given base $\mathcal{B}$ of A with $|\det(A_{\mathcal{B}})| = \Delta$, there exists a polynomial-time algorithm that returns a unimodular equivalent normalized system $\binom{H}{c^\top} x \leq \binom{h}{c_0}$ and a permutation matrix $P \in \mathbb{Z}^{n \times n}$ such that H is the HNF of $PA_{\mathcal{B}}$. Let this algorithm be named as the Normalization Algorithm;*

2. *Any normalized system $\left(\begin{smallmatrix} H \\ c^\top \end{smallmatrix}\right)x \le \left(\begin{smallmatrix} h \\ c_0 \end{smallmatrix}\right)$ that is unimodular equivalent to $\mathcal{S}$ can be obtained as an output of the normalization algorithm with the input base $\{1, \ldots, n\}$ of a permuted system $PAx \le Pb$, for some permutation matrix $P \in \mathbb{Z}^{(n+1) \times (n+1)}$.*

Proof. Let us describe the normalization algorithm. First of all, we represent $A_{\mathcal{B}} = HQ$, where $H \in \mathbb{Z}^{n \times n}$ is the HNF of $A_{\mathcal{B}}$ and $Q \in \mathbb{Z}^{n \times n}$ is a unimodular matrix. After some straightforward additional permutations of rows and columns, we can assume that the original system is equivalent to the system

$$\begin{pmatrix} H \\ c^\top \end{pmatrix} x \le \begin{pmatrix} h \\ c_0 \end{pmatrix}, \quad \text{where } H = \begin{pmatrix} I_s & \mathbf{0}_{s \times k} \\ B & T \end{pmatrix},$$

$s + k = n$, $k \le \log_2(n)$ and T has a lower triangular form with $T_{ii} \ge 2$, for $i \in \{1, \ldots, k\}$.

To fulfil all the conditions of Definition 3, we need some additional work with the matrix B and vector h. We can fulfill the 2-nd condition in Definition 3 for B just using any polynomial-time sorting algorithm. The new system will be unimodular equivalent to the original one, because column-permutations of the first s columns can be undone by row-permutations of the first s rows. Next, the 3-rd condition for h can be satisfied just using the algorithm of Lemma 2. Finally, the 4-th condition can be easily satisfied just by dividing each line of the system on the corresponding value of the gcd function. Since all the steps can be done using polynomial-time algorithms, the whole algorithm is polynomial-time, which proves the first proposition of the lemma.

Let us prove the second proposition. Assume that some unimodular equivalent normalized system $A'x \le b'$ is given. Let $\mathcal{S}'$ be the simplex defined by $A'x \le b'$, and let $\mathcal{U}(x) = Ux' + x_0$ be the unique unimodular affine map that maps $\mathcal{S}$ into $\mathcal{S}'$. It follows that there exists a bijection between the facets of $\mathcal{S}$ and $\mathcal{S}'$ induced by $\mathcal{U}$. Since any inequality of the system $Ax \le b$ defines some facet of $\mathcal{S}$, it directly follows that there exists a permutation matrix $P \in \mathbb{Z}^{(n+1) \times (n+1)}$ such that the systems $PAx \le Pb$ and $A'x \le b'$ are equivalent with one to one correspondence of each inequality of both systems. Since two arbitrary inequalities are equivalent modulo some positive homogeneity multiplier, and due to the condition 4 of Definition 3, it follows that there exists a diagonal matrix $D \in \mathbb{Z}^{(n+1) \times (n+1)}$ with strictly positive integer diagonal elements such that

$$PAU = DA' \quad \text{and} \quad P(b - Ax_0) = Db'. \tag{2}$$

Let us use the normalization algorithm with the input system $PAx \le Pb$ and base $\mathcal{B} = \{1, \ldots, n\}$. As is was described, the algorithm consists of the three parts: the first part constructs the resulting matrix, the second part constructs the resulting r.h.s., the third part removes homogeneity multipliers from rows of the resulting system. Let us consider how the input system $PAx \le Pb$ will look after the first part of the algorithm. Since DA' is the HNF of PA, the system will be $DA'x \le Pb$. Now, we apply the second part of the algorithm using Lemma 2 to the system $DA'x \le Pb$. Due to the equalities (2),

$$Db' = Pb - PAx_0 = Pb - PAUt = Pb - DA't, \quad \text{for } t = U^{-1}x_0 \in \mathbb{Z}^n.$$

Due to Lemma 2, it holds that t is the unique integer translation vector that transforms the system $DA'x \leq Pb$ to the system $DA'x \leq Db'$. Consequently, the second part of the algorithm will produce the system $DA'x \leq Db'$, which will be transformed to the resulting normalized system $A'x \leq b'$ after the third part of the algorithm. The proof is finished.

The following lemma helps to enumerate all the equivalent normalized systems.

Lemma 4. *For a given parameter Δ, there exists an algorithm that constructs two families of simplices $\mathscr{S}$ and $\mathscr{L}$ with the following properties:*

1. *Each simplex of the families $\mathscr{S}$ and $\mathscr{L}$ is Δ-modular, and is defined by a normalized system;*
2. *The simplices of $\mathscr{S}$ are empty, and all the unimodular equivalence classes of Δ-modular empty simplices are represented as simplices from $\mathscr{S}$;*
3. *The simplices of $\mathscr{L}$ are lattice and empty, and all the unimodular equivalence classes of Δ-modular empty lattice-simplices are represented as simplices from $\mathscr{L}$.*

The computational complexity of the algorithm is

$$\binom{n + \Delta - 1}{\Delta - 1} \cdot \Delta^{\log_2(\Delta)} \cdot \mathrm{poly}(n, \Delta) =$$

$$= O\left(\frac{n + \Delta}{\Delta}\right)^{\Delta - 1} \cdot n^{O(1)},$$

which is polynomial, for any fixed Δ.

Proof. Due to Lemma 3, to construct simplices of the sets $\mathscr{S}$ and $\mathscr{L}$, it is sufficient to enumerate all the normalized systems

$$\binom{H}{c^\top} x \leq \binom{h}{c_0}, \quad \text{where } H = \begin{pmatrix} I_s & \mathbf{0}_{s \times k} \\ B & T \end{pmatrix}.$$

Let $\mathcal{D}$ be the set of all nontrivial divisors of Δ. Clearly, $\mathcal{D}$ can be constructed in $\mathrm{poly}(\Delta)$-time. To enumerate all the normalized systems, the following simple scheme can be used:

1. Enumerate all the possible tuples $(d_1, d_2, \ldots, d_k)$ of $\mathcal{D}$ such that $d_1 \cdot \ldots \cdot d_k = \Delta$. Note that tuples allow duplicate elements;
2. For any fixed tuple $(d_1, d_2, \ldots, d_k)$, enumerate the matrices $H = \begin{pmatrix} I_s & \mathbf{0}_{s \times k} \\ B & T \end{pmatrix}$ and the vectors h, where the tuple $(d_1, d_2, \ldots, d_k)$ forms the diagonal of T;
3. For any fixed H, enumerate the vectors $c \in \mathrm{Par}(-H^\top)$;
4. For any fixed triplet (H, h, c), enumerate all the values c_0 such that the corresponding simplex stays empty.

The family of all tuples $(d_1, d_2, \ldots, d_k)$ can be enumerated by the following way: assuming that values $d_1, d_2, \ldots, d_j$ are already chosen from $\mathscr{D}$ and $d_1 \cdot \ldots \cdot d_j < \Delta$, we chose d_{j+1} such that the product $d_1 \cdot \ldots \cdot d_j \cdot d_{j+1}$ divides Δ. Since this procedure consists of at most $\log_2(\Delta)$ steps, the enumeration complexity can be roughly bounded by

$$\frac{\Delta}{2^0} \cdot \frac{\Delta}{2^1} \cdot \ldots \cdot \frac{\Delta}{2^{\lfloor \log_2(\Delta) \rfloor - 1}} \leq \Delta^{\log_2(\Delta)} / 2^{\frac{(\log_2(\Delta) - 2)(\log_2(\Delta) - 1)}{2}} =$$

$$= \frac{1}{2} \cdot \Delta^{3/2} \cdot \Delta^{\log_2(\Delta)/2}.$$

Here the inequalities $x - 1 \leq \lfloor x \rfloor \leq x$ has been used.

For a chosen tuple $(d_1, d_2, \ldots, d_k)$, the lower-triangular matrices $T \subset \mathbb{Z}^{k \times k}$ can be enumerated by the following way. We put $(d_1, d_2, \ldots, d_k)$ into the diagonal of T. When the diagonal of T is defined, for any $j \in \{1, \ldots, k\}$ and $i \in \{j + 1, \ldots, k\}$, we need to choose $T_{ij} \in \{0, \ldots, T_{ii} - 1\}$. Note that, for any $j \in \{1, \ldots, k\}$, it holds $\prod_{i=j+1}^{k} T_{ij} \leq \Delta/2^j$. Consequently, for a given diagonal, the matrices T can be enumerated with $\Delta^{\log_2(\Delta)/2}$ operations. Together with the enumeration of diagonals, the arithmetic cost is $\Delta^{3/2} \cdot \Delta^{\log_2(\Delta)}$.

To finish enumeration of the matrices H, we need to enumerate the matrices $B \in \mathbb{Z}^{k \times s}$. Let $\mathscr{B} = \{x \in \mathbb{Z}^k : 0 \leq x < T_{ii}, \text{for } i \in \{1, \ldots, k\}\}$. Then, since the columns of B are lexicographically sorted, they are in one to one correspondence with the multi-subsets of $\mathscr{B}$ of cardinality s. Since $|\mathscr{B}| = \Delta$, the total number of matrices B is

$$\binom{s + \Delta - 1}{\Delta - 1} \leq \binom{n + \Delta - 1}{\Delta - 1}.$$

The enumeration of such objects is straightforward.

Consequently, the arithmetic complexity to enumerate the matrices $H = \begin{pmatrix} I_s & \mathbf{0}_{s \times k} \\ B & T \end{pmatrix}$ is bounded by $\binom{n + \Delta - 1}{\Delta - 1} \cdot \Delta^{\log_2(\Delta) + 3/2}$. Since $h \in \mathscr{B}$, the vectors h can be enumerated just with $O(\Delta)$ operations. Therefore, the total number of operations to enumerate subsystems $Hx \leq h$ is bounded by

$$\binom{n + \Delta - 1}{\Delta - 1} \cdot \Delta^{\log_2(\Delta) + 5/2}. \tag{3}$$

Due to [35, Lemma 6] (see also [45, Lemma 9], [57] or [37, Lemma 1]), the vectors $c \in \operatorname{Par}(-H^\top)$ can be enumerate with $O(n \cdot \Delta \cdot \min\{n, \log_2(\Delta)\})$ operations. Now, let us discuss how to enumerate the values of c_0. Let $v = H^{-1}h$ and denote

$$\mathcal{S}(c_0) = \left\{x \in \mathbb{R}^n : \begin{pmatrix} H \\ c^\top \end{pmatrix} x \leq \begin{pmatrix} h \\ c_0 \end{pmatrix}\right\}.$$

Note that if $\mathcal{S}(c_0)$ is full-dimensional, then it forms a simplex, and v is the opposite vertex of $\mathcal{S}(c_0)$ with respect to the facet induced by the inequality $c^\top x \leq c_0$. Consider the following two cases: $v \notin \mathbb{Z}^n$ and $v \in \mathbb{Z}^n$.

The Case $v \notin \mathbb{Z}^n$. Consider the problem

$$c^\top x \to \min$$
$$\begin{cases} Hx \leq h \\ x \in \mathbb{Z}^n, \end{cases} \tag{4}$$

and let f^* be the optimal value of the objective function. Due to [37, Theorem 12], the problem (4) can be solved with $O(n + \min\{n, \Delta\} \cdot \Delta \cdot \log(\Delta))$ operations. Denote

$$l^* = \begin{cases} \lceil c^\top v \rceil, & \text{if } c^\top v \notin \mathbb{Z} \\ c^\top v + 1, & \text{if } c^\top v \notin \mathbb{Z}. \end{cases}$$

Since $v \notin \mathbb{Z}^n$, and, due to Lemma 3, we have $0 < f^* - l^* \leq \Delta$. By the construction, for any value of $c_0 \in \{l^*, \ldots, f^* - 1\}$, the set $\mathcal{S}(c_0)$ forms an empty simplex. Due to the definition of f^*, for the values $c_0 \geq f^*$, the simplex $\mathcal{S}(c_0)$ is not empty. Whenever, for $c_0 \leq l^* - 1$, we have $\mathcal{S}(c_0) = \emptyset$. Therefore, for any $c_0 \in \{l^*, \ldots, f^* - 1\}$, we put the corresponding simplex $\mathcal{S}(c_0)$ into the family $\mathcal{S}$.

The Case $v \in \mathbb{Z}^n$. Consider the following problem

$$c^\top x \to \min$$
$$\begin{cases} Hx \leq h \\ x \in \mathbb{Z}^n \setminus \{0\}, \end{cases} \tag{5}$$

and let again f^* be the optimal value of the objective function. To find f^* we can solve the following set of problems: for any $j \in \{1, \ldots, n\}$, the problem is

$$c^\top x \to \min$$
$$\begin{cases} Hx \leq h - e_j \\ x \in \mathbb{Z}^n. \end{cases} \tag{6}$$

Clearly, the value of f^* can be identified as the minimum of optimum objectives of the defined set of problems. As it was already mentioned, each of these n problems can be solved with $O(n + \min\{n, \Delta\} \cdot \Delta \cdot \log(\Delta))$ operations. By the construction, for $c_0 < f^*$, the set $\mathcal{S}(c_0)$ does not form a lattice-simplex or just empty. For $c_0 > f^*$, even if $\mathcal{S}(c_0)$ forms a lattice-simplex, it is not empty. Consequently, only the simplex $\mathcal{S}(f^*)$ is interesting. Next, if $\text{vert}(\mathcal{S}(f^*)) \subseteq \mathbb{Z}^n$, we put the simplex $\mathcal{S}(c_0)$ into the family $\mathcal{L}$. In the opposite case, we just skip the current value of the vector c, and move to the next value (if it exists).

Due to Lemma 1, for a fixed vector c, there are at most Δ possibilities to chose c_0. Consequently, with the proposed algorithms the complexity to enumerate the values of c and c_0 is bounded by $\text{poly}(n, \Delta)$. Therefore, recalling the complexity bound (3) to enumerate the systems $Hx \leq h$, the total enumeration complexity is the same as in the lemma definition, which completes the proof.

The following lemma helps to check the unimodular equivalence relation for two simplices.

Lemma 5. *Let S be an arbitrary simplex defined by a system $Ax \leq b$ with a Δ-modular matrix A. Then, there exists an algorithm, which finds the set of all simplices of the family $\mathscr{S}$ (or $\mathscr{L}$, respectively) that are unimodular equivalent to S. The algorithm complexity is*

$$n^{\log_2(\Delta)} \cdot \Delta! \cdot \mathrm{poly}(\phi),$$

where ϕ is the bit-encoding length of $Ax \leq b$.

Proof. Denote the set of resulting simplices by $\mathscr{R}$. Due to the second proposition of Lemma 3, we can use the following simple scheme to generate the family $\mathscr{R}$:

1. Enumerate all the bases $\mathcal{B}$ of A such that $|\det(A_{\mathcal{B}})| = \Delta$;
2. For a fixed base $\mathcal{B}$, use the normalization algorithm of Lemma 3 to the system $\binom{A_{\mathcal{B}}}{A_l}x \leq \binom{b_{\mathcal{B}}}{b_l}$, where l is the row-index of the line of $Ax \leq b$, which is not contained in $\mathcal{B}$. Let

$$\binom{H}{c^\top}x \leq \binom{h}{c_0}, \quad \text{where } H = \begin{pmatrix} I_s & \mathbf{0}_{s \times k} \\ B & T \end{pmatrix}.$$

   be the resulting normalized system;
3. Enumerate all the $n \times n$ permutation matrices P;
4. For a fixed P, use normalization algorithm of Lemma 3 to the system

$$\binom{PH}{c^\top}x \leq \binom{Ph}{c_0}.$$

The last step will produce a new normalized system that is unimodular equivalent to S, put it into $\mathscr{R}$.

This approach will generate the family $\mathscr{R}$, but the enumeration of all permutations is expensive with respect to n. We will show next that it is not necessary to enumerate all the permutation matrices, and it is enough to enumerate only a relatively small part of them.

Let us fix a base $\mathcal{B}$ of A, which can be done in $n + 1$ ways, and consider the normalized system $\binom{H}{c^\top}x \leq \binom{h}{c_0}$ that was produced by the second step of the scheme. Note that the rows of H can be partitioned into two parts: each row of the first part is a row of $n \times n$ identity matrix I_n, each row of the second part is a row of the matrix $(B \ T)$. Let $\mathcal{I}$ be the set of indices of the first part, and $\mathcal{J}$ be the set of indices of the second. Any permutation π of the rows of $A_{\mathcal{B}}$ can be generated as follows:

1. Chose $|\mathcal{J}|$ positions from n positions where the rows with indexes $\mathcal{J}$ will be located, which can be done in $\binom{n}{k}$ ways;
2. Permute rows indexed by $\mathcal{J}$, which can be done in $k!$ ways;
3. Permute the rows with indexes in $\mathcal{I}$.

Let us take a closer look at permutations of the third type. We call a permutation π *redundant*, if $P_\pi H = HQ$, for some unimodular matrix $Q \in \mathbb{Z}^{n \times n}$.

Clearly, all the redundant permutations π of the rows with indices in $\mathcal{I}$ can be skipped, because H and $P_\pi H$ have the same HNFs. There is a natural one to one correspondence between the rows with indices in $\mathcal{I}$ and columns of B: any row-permutation of this rows can be undone by a corresponding column-permutation of B. Since, for any $i \in \{1, \ldots, k\}$ and $j \in \{1, \ldots, s\}$, $B_{ij} \leq T_{ii}$ and $\prod_{i=1}^k T_{ii} = \Delta$, the matrix B has at most Δ different columns. It is easy to see now that any row-permutation π of row indices that correspond to equal columns of B are redundant. Therefore, it is sufficient to consider only the row-permutations of $\mathcal{I}$ that correspond to different columns of B. Since B consists of at most Δ different columns, there are at most $\Delta!$ of such permutations. It is relatively easy to compute these permutations: we just need to locate the classes of equal rows in B, locate any representative index of each class, and enumerate the permutations of these representatives by a standard way.

Combining permutations of all three types and the complexity to construct the HNF, which is $\mathrm{poly}(\phi)$, the complexity of the procedure for a fixed base $\mathcal{B}$ is

$$\binom{n}{k} \cdot k! \cdot \Delta! \cdot \mathrm{poly}(\phi) \;\lesssim\; n^{\log_2(\Delta)} \cdot \Delta! \cdot \mathrm{poly}(\phi).$$

Since there are at most $n + 1$ ways to chose $\mathcal{B}$, the total algorithm complexity is the same as in the lemma definition, which finishes the proof.

2.3   Finishing the Proof of Theorems 2 and 3

The following proposition is straightforward, but it is convenient to emphasise it for a further use.

Proposition 1. *The simplices $\mathcal{S}_1$ and $\mathcal{S}_2$ are equivalent if and only if, then there exists a simplex $\mathcal{S}'$ defined by a normalized system, such that $\mathcal{S}_1$ and $\mathcal{S}_2$ are both unimodular equivalent to $\mathcal{S}'$.*

The Proof of Theorem 2.

Proof. First of all, we construct a unimodular equivalent normalized system for $\mathcal{T}$. Due to Lemma 3, it can be done by a polynomial-time algorithm. Next, we use Lemma 5 to construct the family $\mathcal{R}$ of all the normalized systems that are equivalent to $\mathcal{S}$. Due to Proposition 1, if the simplices $\mathcal{S}$ and $\mathcal{T}$ are unimodular equivalent, then the family $\mathcal{R}$ must contain a normalized system of $\mathcal{T}$. To store the set $\mathcal{R}$ effectively, we can use any well-balanced search-tree data structure with logarithmic-cost search, insert and delete operations. Clearly, we can compare two normalized systems just by representing them as two vectors of length $(n + 1)^2$ and using the lexicographic order on these vectors. Due to complexity bound of Lemma 5, and since $\log_2(|\mathcal{R}|) = \mathrm{poly}(n, \Delta)$, the full algorithm complexity is the same as in the theorem definition, which completes the proof.

The Proof of Theorem 3.

Proof. First of all, using Lemma 4, we generate the families $\mathscr{S}$ and $\mathscr{L}$ of Δ-modular empty simplicies and empty lattice-simplices respectively. As it was shown by Lemma 4, all the unimodular equivalence classes are already represented by $\mathscr{S}$ and $\mathscr{L}$, but some different normalized systems in $\mathscr{S}$ or $\mathscr{L}$ can represent the same equivalence class. Let us show how to remove such duplicates from $\mathscr{S}$, the same procedure works for $\mathscr{L}$.

To store the set $\mathscr{S}$ effectively, we can use any well-balanced search-tree data structure with logarithmic-cost search, insert and delete operations. We can compare two normalized systems just by representing them as two vectors of length $(n+1)^2$ and using the lexicographic order on these vectors. Due to Lemma 4, we have $|\mathscr{S}| = O\left(\frac{n+\Delta}{\Delta}\right)^{\Delta-1} \cdot n^{O(1)}$. So, all operations with the search-tree can be performed with an $\operatorname{poly}(n, \Delta)$-operations cost.

We enumerate all the normalized systems $\mathcal{S} \in \mathscr{S}$, and using algorithm of Lemma 5, for each $\mathcal{S}$, enumerate all the normalized systems in $\mathscr{S}$ that are unimodular equivalent to $\mathcal{S}$. For each system that was generated by this way, we remove the corresponding entry from the search-tree that represents $\mathscr{S}$. Due to Proposition 1, after this procedure all the duplicates will be removed from the search-tree, and only the normalized systems representing the unique equivalence classes will remain. Clearly, the total algorithm complexity equals to the product of the algorithm complexities of Lemmas 4 and 5:

$$O\left(\frac{n+\Delta}{\Delta}\right)^{\Delta-1} \cdot n^{\log_2(\Delta)+O(1)} \cdot \Delta! =$$

$$= O(n+\Delta)^{\Delta-1} \cdot n^{\log_2(\Delta)+O(1)},$$

which completes the proof.

Conclusion

The paper considers two problems:

- The problem to enumerate all empty Δ-modular simplices and empty Δ-modular lattice-simplices modulo the unimodular equivalence relation;
- The problem to check the unimodular equivalence of two Δ-modular simplices that are not necessarily empty.

It was shown, assuming that the value of Δ is fixed, that the both problems can be solved by polynomial-time algorithms.

These results can be used to construct a data base containing all unimodular equivalence classes of empty Δ-modular simplices and empty Δ-modular lattice-simplices, for small values of Δ and moderate values of n. Due to [35], the lattice-width of empty simplex or empty lattice-simplex can be computed by an FPT-algorithm with respect to the parameter Δ. Hence, the lattice-width is also can be precomputed for each simplex of the base.

The author hopes that such a base and results of the paper will be helpful for studying the properties of empty simplices of both types, and general empty lattice-polytopes. Construction of the base and improvement of existing algorithms to deal with simplices is an interesting direction for further research.

Acknowledgments. The results was prepared under financial support of Russian Science Foundation grant No 21-11-00194.

References

1. The b-hull of an integer program. Discret. Appl. Math. **3**(3), 193–201 (1981)
2. Aardal, K., Eisenbrand, F.: Integer programming, lattices, and results in fixed dimension. Handb. Oper. Res. Manag. Sci. **12**, 171–243 (2005)
3. Andersen, K., Louveaux, Q., Weismantel, R., Wolsey, L.A.: Inequalities from two rows of a simplex tableau. In: Fischetti, M., Williamson, D.P. (eds.) IPCO 2007. LNCS, vol. 4513, pp. 1–15. Springer, Heidelberg (2007). https://doi.org/10.1007/978-3-540-72792-7_1
4. Babai, L.: On lovász'lattice reduction and the nearest lattice point problem. Combinatorica **6**, 1–13 (1986)
5. Baes, M., Oertel, T., Weismantel, R.: Duality for mixed-integer convex minimization. Math. Program. **158**, 547–564 (2016)
6. Balas, E.: Intersection cuts-a new type of cutting planes for integer programming. Oper. Res. **19**(1), 19–39 (1971)
7. Banaszczyk, W.: New bounds in some transference theorems in the geometry of numbers. Math. Ann. **296**, 625–635 (1993)
8. Banaszczyk, W.: Inequalities for convex bodies and polar reciprocal lattices in r^n. Discret. Comput. Geom. **13**, 217–231 (1995)
9. Banaszczyk, W., Litvak, A.E., Pajor, A., Szarek, S.J.: The flatness theorem for nonsymmetric convex bodies via the local theory of banach spaces. Math. Oper. Res. **24**(3), 728–750 (1999)
10. Barvinok, A.: Integer Points in Polyhedra. European Mathematical Society, ETH-Zentrum, Zürich, Switzerland (2008)
11. Barvinok, A., Pommersheim, J.: An algorithmic theory of lattice points in polyhedra. New Perspect. Algebraic Combin. **38** (1999)
12. Barvinok, A.: A polynomial time algorithm for counting integral points in polyhedra when the dimension is fixed. In: Proceedings of 1993 IEEE 34th Annual Foundations of Computer Science, pp. 566–572 (1993). https://doi.org/10.1109/SFCS.1993.366830
13. Barvinok, A., Woods, K.: Short rational generating functions for lattice point problems. J. Am. Math. Soc. **16**(4), 957–979 (2003). http://www.jstor.org/stable/30041461
14. Basu, A., Conforti, M., Cornuéjols, G., Weismantel, R., Weltge, S.: Optimality certificates for convex minimization and helly numbers. Oper. Res. Lett. **45**(6), 671–674 (2017)
15. Basu, A., Conforti, M., Di Summa, M.: A geometric approach to cut-generating functions. Math. Program. **151**(1), 153–189 (2015). https://doi.org/10.1007/s10107-015-0890-5
16. Beck, M., Robins, S.: Computing the Continuous Discretely. Springer, New York (2015). https://doi.org/10.1007/978-1-4939-2969-6

17. Blair, C.E., Jeroslow, R.G.: Constructive characterizations of the value-function of a mixed-integer program i. Discret. Appl. Math. **9**(3), 217–233 (1984)
18. Blair, C.E., Jeroslow, R.G.: Constructive characterizations of the value function of a mixed-integer program ii. Discret. Appl. Math. **10**(3), 227–240 (1985)
19. Borisov, A.: Quotient singularities, integer ratios of factorials, and the Riemann hypothesis. Int. Math. Res. Notices **2008**(9), rnn052–rnn052 (2008)
20. Borozan, V., Cornuéjols, G.: Minimal valid inequalities for integer constraints. Math. Oper. Res. **34**(3), 538–546 (2009)
21. Celaya, M., Kuhlmann, S., Paat, J., Weismantel, R.: Proximity and flatness bounds for linear integer optimization (2022)
22. Chirkov, A.Y., Gribanov, D.V., Malyshev, D.S., Pardalos, P.M., Veselov, S.I., Zolotykh, N.Y.: On the complexity of quasiconvex integer minimization problem. J. Glob. Optim. **73**(4), 761–788 (2018). https://doi.org/10.1007/s10898-018-0729-8
23. Codenotti, G., Santos, F.: Hollow polytopes of large width. Proc. Am. Math. Soc. **148**(2), 835–850 (2020)
24. Dadush, D.: Integer programming, lattice algorithms, and deterministic volume estimation. Georgia Institute of Technology, ProQuest Dissertations Publishing, Ann Arbor (2012)
25. Dadush, D., Peikert, C., Vempala, S.: Enumerative lattice algorithms in any norm via m-ellipsoid coverings. In: 2011 IEEE 52nd Annual Symposium on Foundations of Computer Science, pp. 580–589 (2011). https://doi.org/10.1109/FOCS.2011.31
26. De Loera, Jesús, A., Hemmecke, R., Köppe, M.: Algebraic and geometric ideas in the theory of discrete optimization. Society for Industrial and Applied Mathematics, Philadelphia, USA (2013)
27. Del Pia, A., Weismantel, R.: Relaxations of mixed integer sets from lattice-free polyhedra. 4OR **10**, 221–244 (2012)
28. Doolittle, J., Katthän, L., Nill, B., Santos, F.: Empty simplices of large width (2021)
29. Dyer, M., Kannan, R.: On Barvinok's algorithm for counting lattice points in fixed dimension. Math. Oper. Res. **22**(3), 545–549 (1997). https://doi.org/10.1287/moor.22.3.545
30. Eisenbrand, F.: Fast integer programming in fixed dimension. In: Di Battista, G., Zwick, U. (eds.) ESA 2003. LNCS, vol. 2832, pp. 196–207. Springer, Heidelberg (2003). https://doi.org/10.1007/978-3-540-39658-1_20
31. Gribanov, D.V., Malyshev, D.S.: Integer conic function minimization based on the comparison oracle. In: Khachay, M., Kochetov, Y., Pardalos, P. (eds.) MOTOR 2019. LNCS, vol. 11548, pp. 218–231. Springer, Cham (2019). https://doi.org/10.1007/978-3-030-22629-9_16
32. Gribanov, D., Malyshev, D., Zolotykh, N.: Faster algorithms for sparse ilp and hypergraph multi-packing/multi-cover problems. arXiv preprint arXiv:2201.08988v2 [cs.CC] (2022)
33. Gribanov, D.V., Chirkov, A.Y.: The width and integer optimization on simplices with bounded minors of the constraint matrices. Optim. Lett. **10**(6), 1179–1189 (2016). https://doi.org/10.1007/s11590-016-1048-y
34. Gribanov, D., V., Malyshev, D., S.: A faster algorithm for counting the integer points number in δ-modular polyhedra. Siberian Electron. Math. Rep. (2022). https://doi.org/10.33048/semi.2022.19.051
35. Gribanov, D.V., Malyshev, D.S., Pardalos, P.M., Veselov, S.I.: FPT-algorithms for some problems related to integer programming. J. Comb. Optim. **35**(4), 1128–1146 (2018). https://doi.org/10.1007/s10878-018-0264-z

36. Veselov, S.I., Gribanov, D.V., Malyshev, D.S.: FPT-algorithm for computing the width of a simplex given by a convex hull. Mosc. Univ. Comput. Math. Cybern. **43**(1), 1–11 (2019). https://doi.org/10.3103/S0278641919010084
37. Gribanov, D., Shumilov, I., Malyshev, D., Pardalos, P.: On δ-modular integer linear problems in the canonical form and equivalent problems. J. Glob. Optim. (2022). https://doi.org/10.1007/s10898-022-01165-9
38. Gribanov, D.V., Veselov, S.I.: On integer programming with bounded determinants. Optim. Lett. **10**(6), 1169–1177 (2015). https://doi.org/10.1007/s11590-015-0943-y
39. Haase, C., Ziegler, G.M.: On the maximal width of empty lattice simplices. Eur. J. Comb. **21**(1), 111–119 (2000)
40. Henk, M., Kuhlmann, S., Weismantel, R.: On lattice width of lattice-free polyhedra and height of hilbert bases. SIAM J. Discret. Math. **36**(3), 1918–1942 (2022)
41. Iglesias-Valiño, O., Santos, F.: The complete classification of empty lattice 4-simplices. Revista matemática iberoamericana **37**(6), 2399–2432 (2021)
42. Kannan, R., Lovász, L.: Covering minima and lattice-point-free convex bodies. Ann. Math. 577–602 (1988)
43. Kantor, J.M.: On the width of lattice-free simplices. Compos. Math. **118**, 235–241 (1999)
44. Khinchine, A.: A quantitative formulation of kronecker's theory of approximation. Izvestiya Akademii Nauk SSR Seriya Matematika **12**(2), 113–122 (1948). [in russian]
45. Köppe, M., Verdoolaege, S.: Computing parametric rational generating functions with a primal barvinok algorithm. Electron. J. Comb. **15** (2008). https://doi.org/10.37236/740
46. Lagarias, J.C., Lenstra, H.W., Schnorr, C.P.: Korkin-zolotarev bases and successive minima of a lattice and its reciprocal lattice. Combinatorica **10**, 333–348 (1990)
47. Lasserre, J.B.: Linear and Integer Programming vs Linear Integration and Counting: A Duality Viewpoint. Springer Science & Business Media, New York (2009). https://doi.org/10.1007/978-0-387-09414-4
48. Lenstra, H., W.: Integer programming with a fixed number of variables. Math. Oper. Res. **8**(4), 538–548 (1983). https://doi.org/10.1287/moor.8.4.538
49. Mayrhofer, L., Schade, J., Weltge, S.: Lattice-free simplices with lattice width $2d - o(d)$. In: Integer Programming and Combinatorial Optimization: 23rd International Conference, IPCO 2022, Eindhoven, The Netherlands, 27–29 June 2022, Proceedings, pp. 375–386. Springer (2022)
50. Morán R, D.A., Dey, S.S., Vielma, J.P.: A strong dual for conic mixed-integer programs. SIAM J. Optim. **22**(3), 1136–1150 (2012)
51. Mori, S., Morrison, D.R., Morrison, I.: On four-dimensional terminal quotient singularities. Math. Comput. **51**(184), 769–786 (1988)
52. Morrison, D.R., Stevens, G.: Terminal quotient singularities in dimensions three and four. Proc. Am. Math. Soc. **90**(1), 15–20 (1984)
53. Paat, J., Schlöter, M., Speakman, E.: Constructing lattice-free gradient polyhedra in dimension two. Math. Program. **192**(1–2), 293–317 (2022)
54. Reis, V., Rothvoss, T.: The subspace flatness conjecture and faster integer programming (2023)
55. Rudelson, M.: Distances between non-symmetric convex bodies and the mm^*-estimate. Positivity **2**(4), 161–178 (2000)
56. Schrijver, A.: Theory of Linear and Integer Programming. John Wiley & Sons, Chichester, Hoboken (1998)

57. Sebő, A.: An introduction to empty lattice simplices. In: Cornuéjols, G., Burkard, R.E., Woeginger, G.J. (eds.) IPCO 1999. LNCS, vol. 1610, pp. 400–414. Springer, Heidelberg (1999). https://doi.org/10.1007/3-540-48777-8_30

58. Storjohann, A., Labahn, G.: Asymptotically fast computation of Hermite normal forms of integer matrices. In: Proceedings of the 1996 International Symposium on Symbolic and Algebraic Computation, pp. 259–266. ISSAC 1996, Association for Computing Machinery, New York, NY, USA (1996). https://doi.org/10.1145/236869.237083

59. Valiño, Ó.I., Santos, F.: Classification of empty lattice 4-simplices of width larger than 2. Electron. Notes Discret. Math. **61**, 647–653 (2017)

60. Veselov, S., Chirkov, Y.: The structure of simple sets in $\mathbb{Z}^3$. Autom. Remote Control **65**(3), 396–400 (2004). http://www.uic.unn.ru/~vesi/SimpleSet.pdf

61. Veselov, S., Gribanov, D., Zolotykh, N., Chirkov, A.: A polynomial algorithm for minimizing discrete convic functions in fixed dimension. Discrct. Appl. Math. **283**, 11–19 (2020). https://doi.org/10.1016/j.dam.2019.10.006, https://www.sciencedirect.com/science/article/pii/S0166218X19304561

62. White, G.K.: Lattice tetrahedra. Can. J. Math. **16**, 389–396 (1964)

63. Wolsey, L.A.: Integer programming duality: price functions and sensitivity analysis. Math. Program. **20**, 173–195 (1981)

PTAS for p-Means q-Medoids r-Given Clustering Problem

Artem V. Pyatkin[✉][iD]

Sobolev Institute of Mathematics, Koptyug Ave., 4, Novosibirsk 630090, Russia
artem@math.nsc.ru

Abstract. We consider the problem of clustering (partitioning) a given set of d-dimensional Euclidean vectors into a fixed number of clusters with different center types so that the total dispersion would be minimal, where by dispersion we mean the sum of squared distances from the elements of the cluster to its center. We consider the case where p centers are means (or centroids) of the cluster elements, q centers are medoids (some vectors from the initial sets) and r centers are given (fixed points of the space). We assume that the sizes of the clusters are also given in advance. It is known that this problem is NP-hard even for $p \geq 1$ and $p + q + r \geq 2$. In this paper a PTAS of complexity $\mathcal{O}((d + n)n^{q+2+p/\varepsilon})$ for this problem is suggested.

Keywords: Clustering · k-means · centroid · medoid · PTAS

1 Introduction

The object of study in this paper is clustering, i. e. partitioning a set of Euclidean vectors into several non-empty clusters according to some similarity criteria. Such kind of problems is quite actual in data analysis, data mining, computational geometry, mathematical statistics and discrete optimization [3,4,11]. It is indeed a wide class of problems differing by the number of clusters, similarity criteria, cluster cardinalities constraints, etc.

One of the most popular similarity criteria is the minimum dispersion that is the sum of the squared distances from the elements of the cluster to some point called a *center* of the cluster. Namely, for a cluster $\mathcal{C}$ and a center x define

$$f(\mathcal{C}, x) = \sum_{y \in \mathcal{C}} \|x - y\|^2.$$

Usually the following types of the center are considered:

1. An arbitrary point (no restrictions)
2. A point from the given set
3. A given (fixed) point of the space

The study was carried out within the framework of the state contract of the Sobolev Institute of Mathematics (project FWNF-2022-0019).

M. Khachay et al. (Eds.): MOTOR 2023, LNCS 13930, pp. 133–141, 2023.
https://doi.org/10.1007/978-3-031-35305-5_9

This choice of centers types can be motivated as follows. Asssume that several sensors should be placed for monitoring the situation in towns in some geographical area. Some sensors are autonomous and could be placed anywhere; others need regular service and so they should be put in towns. There could be also sensors that had been put earlier and cannot be moved (fixed). Since the energy consumption is proportional to the square of the distance, the clustering problem with different centers types can be interpreted as a location problem for the sensors of two types taking into account existing sensors and minimizing the total energy consumption.

If all points of a cluster C are known and the center can be chosen arbitrarily then it is easy to prove by taking partial derivations that the optimal center coincides with the so-called *centroid* defined as

$$\overline{y}(C) = \frac{\sum_{y \in C} y}{|C|}.$$

If this constraint must hold for all cluster centers then one gets a classical problem MSSC (minimum sum-squared clustering) [8,19] also known as k-*means* where k is the number of clusters. So, we call such centers *means*. If the center of the cluster must coincide with one of the points from the initial set then we call it *medoid*. Note that the term medoid was introduced in [14] as a representative object of a cluster within a data set whose average dissimilarity to all the objects in the cluster is minimal. Although the term medoid is usually associated with the sum of distances, applying the same term for the sum of squared distances looks quite natural and does not contradict the initial definition. Finally, the third type of centers (a fixed point) is called *given*.

The main object of study in this paper is the following

Problem 1 (*p*-means *q*-medoids *r*-given problem). Given a finite set $\mathcal{Y} = \{y_1, \ldots, y_n\} \subset \mathbb{R}^d$ of Euclidean vectors, set of points $\{z_1, \ldots, z_r\} \subset \mathbb{R}^d$ and positive inegers $m_1, \ldots, m_k$, such that $m_1 + \ldots + m_k = n$ and $p + q + r = k$, partition $\mathcal{Y}$ into k clusters $\mathcal{C}_1, \ldots, \mathcal{C}_k$ with $|\mathcal{C}_i| = m_i$ for all $i = 1, \ldots, k$ so that

$$F(\mathcal{C}_1, \ldots, \mathcal{C}_k; X) = \sum_{i=1}^{k} f(\mathcal{C}_i, x_i)$$

would be minimal, where $X = \{x_1, \ldots, x_k\}$ is the set of clusters' centers satisfying the following constraints:

- $x_i = \overline{y}(\mathcal{C}_i)$ if $1 \le i \le p$;
- $x_i \in \mathcal{Y}$ if $p + 1 \le i \le p + q$;
- $x_i = z_{i-p-q}$ if $p + q + 1 \le i \le k$.

In other words, first p cluster centers must be means, next q cluster centers must be medoids, and the last r clusters must have the given centers.

This problem is known to be NP-hard even in case of $p \ge 1$ and $k = 2$. Namely, if $p = 2$ and $q = r = 0$, we get a 2-means problem whose NP-hardness was proved in [1] even in the case of non-fixed clusters cardinalities (clearly, if a version of a problem with fixed cardinalities is polynomially solvable, then version of a problem with non-fixed cardinalities is also polynomially solvable; so, the NP-hardness of the version with non-fixed cardinalities implies the NP-hardness of the version with fixed ones). The first PTAS for 2-means with non-fixed cardinalities was suggested in [12]; it finds $(1 + \varepsilon)$-approximate solution in time $\mathcal{O}(n(1/\varepsilon)^d)$. In [18] a PTAS for k-means with non-fixed cardinalities is provided, that finds $(1 + \varepsilon)$-approximate solution in time $\mathcal{O}(dn2^{(k/\varepsilon)^{\mathcal{O}(1)}})$. Both these PTAS's are randomized.

The case $p = r = 1$, $q = 0$ was considered in [2,10,15,16] where its NP-hardness was proved first for fixed [2,10] and then for non-fixed [15,16] cardinalities of the clusters. For both variants 2-approximation algorithms of complexity $\mathcal{O}(n^2 d)$ are known [5,13]. A PTAS for this problem was suggested in [6]. It has complexity $\mathcal{O}(dn^{1+2/\varepsilon}(9/\varepsilon)^{3/\varepsilon})$. Finally, NP-completeness of Problem 1 in the case of $p = q = 1$, $r = 0$ was proved in [17] for fixed cardinalities and in [20] for non-fixed cardinalities. The paper [20] also contains 2-approximation algorithm of complexity $\mathcal{O}(n^2 d + n^3)$ for the latter problem. However, in case of $p = 0$ Problem 1 is polynomially solvable for every k, as it is shown in the next section.

In the current paper we provide a PTAS for the general case of Problem 1. For the best of our knowledge, no such schemes are known for $p > 0$ and $q > 0$.

Note that in [21] a PTAS for the following close vector subset choice problem was presented.

Problem 2 (vector subset choice problem). Given a set $\mathcal{Y} = \{y_1, \ldots, y_n\} \subset \mathbb{R}^d$ of Euclidean vectors and a positive integer $m \leq n$, find a subset $\mathcal{C} \subseteq \mathcal{Y}$ of cardinality m so that $f(\mathcal{C}, \overline{y}(\mathcal{C}))$ would be minimal.

The PTAS in [21] finds a $(1 + \varepsilon)$-approximate solution of Problem 2 in time $\mathcal{O}(dn^{1+2/\varepsilon}(9/\varepsilon)^{3/\varepsilon})$. The idea of the method is as follows. First, it is proved that the linear hull of some subset of small cardinality contains a point that is not too far from the center of the optimal cluster. Then a grid is considered at each of such hulls and the bound is proved on the number of points of the grid to be considered in order to find a good approximation for rhe center of the optimal cluster. However, this bound only works for the optimal cluster and cannot be applied for an arbitrary cluster of a given size that can appear in Problem 1. This fact does not allow to apply directly the PTAS from [21] to get a PTAS for Problem 1, that can be shown by the following simple example. Let $d = 1$ and $\mathcal{Y}$ contain four numbers $y_1 = -2$, $y_2 = y_3 = 0$, and $y_4 = 2$. Let also $p = q = 1$, $r = 0$ and $m_1 = m_2 = 2$. Clearly, for some small enough $\varepsilon > 0$ the PTAS finds a solution $\mathcal{C} = \{y_2, y_3\}$ (that is indeed the optimal solution of Problem 2 for this $\mathcal{Y}$ and $m = 2$). Then $F(\mathcal{C}, \mathcal{Y} \backslash \mathcal{C}; \{0, 0\}) = 8$ (take y_2 as a center of the second cluster), while the optimal solution is $\mathcal{C}_1 = \{y_1, y_2\}$ and $\mathcal{C}_2 = \{y_3, y_4\}$ with $F(\mathcal{C}_1, \mathcal{C}_2; \{-1, 0\}) = 6$, i. e. for arbitrary small $\varepsilon > 0$ the relative error is still $1/3$. However, using some geometrical properties proved in [21] we may apply another technique for constructing a PTAS for Problem 1.

The paper is organized as follows. In the next section some preliminary results are presented. The main result of the paper (namely, the PTAS for Problem 1) can be found in Sect. 3. The last section concludes the paper.

2   Preliminary Results and Known Facts

In this section we provide some preliminary results that are necessary for PTAS specification. Note that some of these results are either known or folklore, but we need to formulate them here implicitly for the sake of completeness.

First of all we need to show that r-given problem is polynomially solvable for any fixed r. It is a folklore result, but we provide its proof here since it is very short.

Proposition 1. *In case of $p = q = 0$, Problem 1 is solvable in time $\mathcal{O}(n^2 d + n^3)$.*

Proof. Let $z_1, \ldots, z_r$ and $m_1, \ldots, m_r$ be, respectively, the centers and cardinalities of the sought clusters. Note that $m_1 + \ldots + m_r = n$. For each $i = 1, \ldots, r$ consider a vector $a_i \in \mathbb{R}^n$ such that $a_i(j) = \|z_i - y_j\|^2$ for all $j = 1, \ldots, n$. Take m_i copies of the vector a_i for each $i = 1, \ldots, r$ and consider them as the columns of the square matrix A of order n. Then, clearly, $F(\mathcal{Y})$ is equal to the minimum cost assignment for the matrix A. Since the assignment problem can be solved by the classical Hungarian Method [7] in time $\mathcal{O}(n^3)$ and constructing the matrix A requires $\mathcal{O}(n^2 d)$ time, the initial problem is solvable in time $\mathcal{O}(n^2 d + n^3)$. $\square$

Note that in the case when all elements of the matrix A are positive integers at most C, then a faster algorithm [9] of time complexity $\mathcal{O}(M\sqrt{n}\log(nC))$ can be applied for solving an assignment problem (here C is a constant and M is the number of non-zero entries of the matrix A). However, in our case this algorithm may work worse since $M = n^2 - n$ and even in the case of integral coordinates of all vecrors the constant C could be larger than 2^n.

Note also that Proposition 1 implies the polynomial solvability of Problem 1 in the case of $p = 0$. Indeed, q-medoid r-given problem can be reduced to at most n^q instances of $(q + r)$-given problems (just apply the brute force for searching the medoids).

We require the following

Lemma 1. *Let $\mathcal{C} \subset \mathcal{Y}$ be an arbitrary cluster of cardinality m. Then for an arbitrary positive integer $t \leq m$ there exists a subset $\mathcal{T}^* \subset \mathcal{C}$ of cardinality t such that*

$$\|\overline{y}(\mathcal{C}) - \overline{y}(\mathcal{T}^*)\|^2 \leq \frac{f(\mathcal{C}, \overline{y}(\mathcal{C}))}{mt}.$$

Proof. Introduce the notation $g(\mathcal{T}) = \|\overline{y}(\mathcal{C}) - \overline{y}(\mathcal{T})\|^2$. Consider an arbitrary subset $\mathcal{T} = \{y_1, \ldots, y_t\}$. Then

$$
\begin{aligned}
g(\mathcal{T}) &= \|\overline{y}(\mathcal{C}) - \tfrac{\sum_{i=1}^t y_i}{t}\|^2 = \frac{\|t\overline{y}(\mathcal{C}) - \sum_{i=1}^t y_i\|^2}{t^2} = \frac{\|\sum_{i=1}^t (\overline{y}(\mathcal{C}) - y_i)\|^2}{t^2} \\
&= \frac{\sum_{i=1}^t \|\overline{y}(\mathcal{C}) - y_i\|^2 + 2\sum_{i=1}^{t-1}\sum_{j=i+1}^t \langle \overline{y}(\mathcal{C}) - y_i, \overline{y}(\mathcal{C}) - y_j \rangle}{t^2} \\
&= \frac{\sum_{y \in \mathcal{T}} \|\overline{y}(\mathcal{C}) - y\|^2 + \sum_{y \in \mathcal{T}}\sum_{y' \in \mathcal{T}, y' \neq y} \langle \overline{y}(\mathcal{C}) - y, \overline{y}(\mathcal{C}) - y' \rangle}{t^2}.
\end{aligned}
$$

Then the average value of $g(\mathcal{T})$ among all subsets $\mathcal{T} \subset \mathcal{C}$ of cardinality t is

$$\frac{\sum_{\mathcal{T} \subset \mathcal{C}, |\mathcal{T}|=t} (\sum_{y \in \mathcal{T}} \|\overline{y}(\mathcal{C}) - y\|^2 + \sum_{y \in \mathcal{T}} \sum_{y' \in \mathcal{T}, y' \neq y} \langle \overline{y}(\mathcal{C}) - y, \overline{y}(\mathcal{C}) - y' \rangle)}{\binom{m}{t} t^2}.$$

Since each $y \in \mathcal{C}$ belongs to $\binom{m-1}{t-1}$ subsets $\mathcal{T} \subset \mathcal{C}$ of size t, we have

$$\sum_{\mathcal{T} \subset \mathcal{C}, |\mathcal{T}|=t} \sum_{y \in \mathcal{T}} \|\overline{y}(\mathcal{C}) - y\|^2 = \binom{m-1}{t-1} \sum_{y \in \mathcal{C}} \|\overline{y}(\mathcal{C}) - y\|^2$$

$$= \binom{m-1}{t-1} f(\mathcal{C}, \overline{y}(\mathcal{C})). \tag{1}$$

Clearly, for every $z \in \mathbb{R}^d$ the equality

$$\sum_{y \in \mathcal{C}} \langle \overline{y}(\mathcal{C}) - y, z \rangle = \langle m\overline{y}(\mathcal{C}) - \sum_{y \in \mathcal{C}} y, z \rangle = 0$$

holds. Hence,

$$0 = \sum_{y \in \mathcal{C}} \sum_{y' \in \mathcal{C}} \langle \overline{y}(\mathcal{C}) - y, \overline{y}(\mathcal{C}) - y' \rangle$$

$$= \sum_{y \in \mathcal{C}} \|\overline{y}(\mathcal{C}) - y\|^2 + \sum_{y \in \mathcal{C}} \sum_{y' \in \mathcal{C}, y' \neq y} \langle \overline{y}(\mathcal{C}) - y, \overline{y}(\mathcal{C}) - y' \rangle.$$

Note that each ordered pair of $y, y' \in \mathcal{C}$ such that $y \neq y'$ appears in $\binom{m-2}{t-2}$ subsets $\mathcal{T} \subset \mathcal{C}$ of cardinality t. Therefore

$$\sum_{\mathcal{T} \subset \mathcal{C}, |\mathcal{T}|=t} \sum_{y \in \mathcal{T}} \sum_{y' \in \mathcal{T}, y' \neq y} \langle \overline{y}(\mathcal{C}) - y, \overline{y}(\mathcal{C}) - y' \rangle$$

$$= \binom{m-2}{t-2} \sum_{y \in \mathcal{C}} \sum_{y' \in \mathcal{C}, y' \neq y} \langle \overline{y}(\mathcal{C}) - y, \overline{y}(\mathcal{C}) - y' \rangle = -\binom{m-2}{t-2} \sum_{y \in \mathcal{C}} \|\overline{y}(\mathcal{C}) - y\|^2$$

$$= -\binom{m-2}{t-2} f(\mathcal{C}, \overline{y}(\mathcal{C})). \tag{2}$$

Due to (1) and (2), the average value of $g(\mathcal{T})$ among all subsets of cardinality t equals to

$$\frac{(\binom{m-1}{t-1} - \binom{m-2}{t-2}) f(\mathcal{C}, \overline{y}(\mathcal{C}))}{\binom{m}{t} t^2} = \left(\frac{1}{mt} - \frac{t-1}{(m-1)mt} \right) f(\mathcal{C}, \overline{y}(\mathcal{C})) = \frac{(m-t) f(\mathcal{C}, \overline{y}(\mathcal{C}))}{(m-1)mt}.$$

Since the minimum value of the function $g(\mathcal{T})$ is at most its average value, there exists a subset $\mathcal{T}^* \subset \mathcal{C}$ of cardinality t such that

$$g(\mathcal{T}^*) \leq \frac{(m-t) f(\mathcal{C}, \overline{y}(\mathcal{C}))}{(m-1)mt} \leq \frac{f(\mathcal{C}, \overline{y}(\mathcal{C}))}{mt}.$$

$\square$

The following statement was proved in [21]:

Proposition 2 [21]. *For every cluster $C \subset \mathcal{Y}$ of cardinality m and any point $z \in \mathbb{R}^d$ the equality*

$$f(C, z) = f(C, \overline{y}(C)) + m\|\overline{y}(C) - z\|^2$$

holds.

Combining Proposition 2 with Lemma 1 we obtain the main tool on which our PTAS is based.

Lemma 2. *Let $\mathcal{Z}$ be the set of centroids of all clusters of cardinality at most t in $\mathcal{Y}$. Then for every cluster $C \subset \mathcal{Y}$ there exists a point $z \in \mathcal{Z}$ such that $f(C, z) \leq (1 + 1/t)f(C, \overline{y}(C))$.*

Proof. Consider an arbitrary cluster $C \subset \mathcal{Y}$ and let $m = |C|$. Clearly, if $m \leq t$ then the set $\mathcal{Z}$ contains the centroid of C and thus the point $z = \overline{y}(C)$ satisfies the statement. Otherwise, choose for a cluster C a subset T^* of cardinality t from Lemma 1. Then the point $z = \overline{y}(T^*)$ belongs to $\mathcal{Z}$ and by Proposition 2 and Lemma 1,

$$f(C, z) = f(C, \overline{y}(C)) + m\|\overline{y}(C) - z\|^2 \leq (1 + \frac{1}{t})f(C, \overline{y}(C)),$$

as required.   □

So, the set $\mathcal{Z}$ constructed in Lemma 2 contains at most $2n^t$ candidates for the approximate centers of the first p clusters in Problem 1; moreover, for each such cluster C at least one of these centers gives $(1 + \varepsilon)$-approximation for $f(C, \overline{y}(C))$ where $\varepsilon = 1/t$.

3   The Main Result

Informally, the suggested PTAS for Problem 1 does the following. After choosing the parameter t we construct the set $\mathcal{Z}$ of centroids of all subsets of cardinality at most t in $\mathcal{Y}$. Consider this $\mathcal{Z}$ as the set of possible centers for the first p clusters (that have means as centers) in Problem 1. Consider $\mathcal{Y}$ as the set of possible centers for the next q clusters (medoids) and $\{z_j\}$ as a set of possible centers for each corresponding cluster among the last r clusters (given). Then consider all possible combinations of these centers and solve for them the k-given problem using Proposition 1. Then the best of these solutions provides a $(1 + \varepsilon)$-approximation for Problem 1.

A formal description of the PTAS is as follows.

Algorithm 1. *Input*: A set $\mathcal{Y} = \{y_1, \ldots, y_n\} \subset \mathbb{R}^d$ of Euclidean vectors, set of points $\{z_1, \ldots, z_r\} \subset \mathbb{R}^d$, positive integers $p, q, r, m_1, \ldots, m_k$, such that $m_1 + \ldots + m_k = n$ and $p + q + r = k$, and a positive integer parameter t.

Step 0. Put $\mathcal{Z} = \emptyset$. Consider all subsets $\mathcal{T} \subset \mathcal{Y}$ of cardinality at most t and add to $\mathcal{Z}$ all points $\overline{y}(\mathcal{T})$. Put $\mathcal{Z}_i = \mathcal{Z}$ for all $i = 1, \dots, p$, and $\mathcal{Z}_i = \mathcal{Y}$ for all $i = p+1 \dots, p+q$, and $\mathcal{Z}_i = \{z_{i-p-q}\}$ for all $i = p+q+1, \dots, k$.

Step 1. Consider all possible choices $x_i \in \mathcal{Z}_i$ and solve for each of them the k-given problem for set $\mathcal{Y}$ using Proposition 1. Denote by $\mathcal{C}_1, \dots, \mathcal{C}_k$ and $X = \{x_1, \dots, x_k\}$ the sets of clusters and their centers in the best solution found.

Output: the sets $\mathcal{C}_1, \dots, \mathcal{C}_k$ and $X = \{x_1, \dots, x_k\}$.

The following theorem shows that the provided algorithm is indeed a PTAS.

Theorem 2. *Algorithm 1 finds* $(1 + \varepsilon)$-*approximate solution of Problem 1 in time* $\mathcal{O}((d+n)n^{q+2+p/\varepsilon})$ *where* $\varepsilon = 1/t$.

Proof. Let us first show the correctness of Algorithm 1. Let $\mathcal{C}_1, \dots, \mathcal{C}_k$ and $X = \{x_1, \dots, x_k\}$ be the sets of clusters and their centers found by Algorithm 1. Denote by $\mathcal{C}_1^*, \dots, \mathcal{C}_k^*$ and $X^* = \{x_1^*, \dots, x_k^*\}$ the sets of optimal clusters and their centers in Problem 1. By Lemma 2, for every $i = 1, \dots, p$ there is $x_i' \in \mathcal{Z}_i$ such that $f(\mathcal{C}_i^*, x_i') \leq (1 + \varepsilon)f(\mathcal{C}_i^*, x_i^*)$. Put $x_i' = x_i^*$ for all $i = p+1, \dots, k$ and let $X' = \{x_1', \dots, x_k'\}$. Then, clearly,

$$F(\mathcal{C}_1^*, \dots, \mathcal{C}_k^*; X') \leq (1 + \varepsilon)F(\mathcal{C}_1^*, \dots, \mathcal{C}_k^*; X^*). \tag{3}$$

Note that the set of centers X' was considered at Step 2 of Algorithm 1. Denote by $\mathcal{C}_1', \dots, \mathcal{C}_k'$ the partition found for X' at that step. Since it is the optimal solution of k-given problem for X' and $\mathcal{Y}$, and partition $\mathcal{C}_1, \dots, \mathcal{C}_k$ with the set of centers X is the best solution found by Algorithm 1, we have

$$F(\mathcal{C}_1, \dots, \mathcal{C}_k; X) \leq F(\mathcal{C}_1', \dots, \mathcal{C}_k'; X') \leq F(\mathcal{C}_1^*, \dots, \mathcal{C}_k^*; X'). \tag{4}$$

Combining (3) and (4) we get the desired approximation ratio.

Clearly, Step 0 of Algorithm 1 requires $\mathcal{O}(dn^t)$ time, while Step 1 is more time-consuming since we consider at most $(2n^t)^p n^q$ possible choices of the centers and solve for each of them a k-given problem in time $\mathcal{O}(n^2 d + n^3)$ by Proposition 1. So, the total complexity is $\mathcal{O}((d+n)n^{pt+q+2})$. Note that since p and q are fixed parameters (constants, i. e. not the part of the input) we obtained a PTAS. □

4 Conclusion

In this paper we suggested a PTAS of complexity $\mathcal{O}((d+n)n^{q+2+p/\varepsilon})$ for the quite general clustering problem where the centers of the first p clusters are means, of the next q clusters are medoids, and of the last r clusters are given. The cardinalities of all clusters are fixed. The key idea of the method is in constructing a not too large set containing good approximations of all clusters' centroids. The PTAS is based on choosing all possible combinations of centers and solving for them the k-given problem via the known algorithm for the assignment problem.

References

1. Aloise, D., Deshpande, A., Hansen, P., Popat, P.: NP-hardness of Euclidean sum-of-squares clustering. Mach. Learn. **75**(2), 245–248 (2009)
2. Baburin, A.E., Gimadi, E.K., Glebov, N.I., Pyatkin, A.V.: The problem of finding a subset of vectors with the maximum total weight. J. Appl. Ind. Math. **2**(1), 32–38 (2008). https://doi.org/10.1134/S1990478908010043
3. Berkhin, P.: A survey of clustering data mining techniques. In: Kogan, J., Nicholas, C., Teboulle, M. (eds.) Grouping Multidimensional Data, pp. 25–71. Springer, Heidelberg (2006). https://doi.org/10.1007/3-540-28349-8_2
4. Dubes, R.C., Jain, A.K.: Algorithms for Clustering Data. Prentice Hall, Hoboken (1988)
5. Dolgushev, A.V., Kelmanov, A.V.: An approximation algorithm for solving a problem of cluster analysis. J. Appl. Ind. Math. **5**(4), 551–558 (2011)
6. Dolgushev, A.V., Kel'manov, A.V., Shenmaier, V.V.: Polynomial-time Approximation Scheme for a Problem of Partitioning a Fnite Set into Two Clusters. Trudy Instituta Matematiki i Mekhaniki UrO RAN. **21** (3), 100–109 (2015). (in Russian)
7. Edmonds, J., Karp, R.M.: Theoretical improvements in algorithmic efficiency for network flow problems. J. ACM **19**(2), 248–264 (1972)
8. Fisher, W.D.: On grouping for maximum homogeneity. J. Amer. Statist. Assoc. **53**(284), 789–798 (1958)
9. Gabow, H.N., Tarjan, R.E.: Faster scaling algorithms for network problems. SIAM J. Comput. **18**(5), 1013–1036 (1989)
10. Gimadi, E.K., Kelmanov, A.V., Kelmanova, M.A., Khamidullin, S.A.: A posteriori detection of a quasi periodic fragment in numerical sequences with given number of recurrences. Sib. Zh. Ind. Mat. **9**(1), 55–74 (2006). (in Russian)
11. Ghoreyshi, S., Hosseinkhani, J.: Developing a Clustering Model based on K-Means Algorithm in order to Creating Different Policies for Policyholders. Intern. J. of Adv. Comp. Sci. Inf. Tech. **4**(2), 46–53 (2015)
12. Inaba, M. Katoh, N., Imai H.: Applications of weighted Voronoi diagrams and randomization to variance-based kclustering: (extended abstract). In Proceedings of the Tenth Annual Symposium on Computational Geometry, pp. 332–339. ACM Press (1994)
13. Kel'manov, A.V., Khandeev, V.I.: A 2-approximation polynomial algorithm for a clustering problem. J. Appl. Ind. Math. **7**(4), 515–521 (2013). https://doi.org/10.1134/S1990478913040066
14. Kaufman, L., Rousseeuw, P.J.: Clustering by means of medoids. In: Dodge, Y. (ed.) Statistical Data Analysis based on the L_1 Norm, pp. 405–416. North-Holland, Amsterdam (1987)
15. Kelmanov, A.V., Pyatkin, A.V.: On the complexity of a search for a subset of similar vectors. Dokl. Math. **78**(1), 574–575 (2008)
16. Kelmanov, A.V., Pyatkin, A.V.: On a version of the problem of choosing a vector subset. J. Appl. Ind. Math. **3**(4), 447–455 (2009)
17. Kelmanov, A.V., Pyatkin, A.V., Khandeev, V.I.: NP-hardness of quadratic Euclidean 1-mean and 1-median 2-clustering problem with constraints on the cluster sizes. Dokl. Math. **100**(3), 545–548 (2019)
18. Kumar, A., Sabharwal, Y., Sen, S.: A simple linear time $(1 + \varepsilon)$-approximation algorithm for geometric k-means clustering in any dimensions. In: Proceedings of Annual Symposium on Foundations of Computer Science, pp. 454–462 (2004)

19. MacQueen, J.: Some methods for classification and analysis of multivariate observations. In: ProcEEDINGS 5-th Berkeley Symposium. on Mathematics, Statistics and Probability, vol. 1, pp. 281–297 (1967)
20. Pyatkin, A.V.: 1-mean and 1-medoid 2-clustering problem with arbitrary cluster sizes: complexity and approximation. Yugoslav J. Oper. Res. **33**, 59–69 (2022). https://doi.org/10.2298/YJOR211018008P
21. Shenmaier, V.V.: An approximation scheme for a problem of search for a vector subset. J. Appl. Ind. Math. **6**(3), 381–386 (2012)

Nested (2,3)-Instances of the Cutting Stock Problem

Artem V. Ripatti$^{(\boxtimes)}$ (iD)

Ufa University of Science and Technology, Zaki Validy str. 32, 450076 Ufa, Russia
ripatti@inbox.ru

Abstract. We consider the well-known cutting stock problem (CSP). A $(2,3)$-instance of CSP is an instance for which each of its items can only be packed into the container either two or three times. The gap of a CSP instance is the difference between its optimal function value and the optimal value of its continuous relaxation. For most instances of CSP the gap is less than 1, and the maximal gap currently known is $77/64 = 1.203125$. For the $(2,3)$- case the best known construction gives the gap greater than $73/72 - \varepsilon \approx 1.013888$ for any $\varepsilon > 0$. In this paper we give a construction of $(2,3)$-instances having recursive (nested) structure and the gap greater than $7/6 - \varepsilon \approx 1.166666$ for any $\varepsilon > 0$. We also give a construction of CSP instances with the gap greater than $59/48 - \varepsilon \approx 1.229166$ for any $\varepsilon > 0$.

Keywords: Cutting stock problem · Integer round up property · Integrality gap

1 Introduction

In the classical formulation, the cutting stock problem (CSP) is stated as follows: there are infinite pieces of stock material of fixed length L. We have to produce $m \in \mathbb{N}$ groups of pieces of different lengths $l_1, \cdots, l_m$ and demanded quantities $b_1, \cdots, b_m$ by cutting initial pieces of stock material in such a way that the number of used initial pieces is minimized. It can also be formulated in terms of packing problems: the set of m types of items with lengths $l_1, \cdots, l_m$ and availabilities $b_1, \cdots, b_m$ should be packed into the minimal number of identical containers having capacity L each.

The cutting stock problem is one of the earliest problems that have been studied through methods of operational research [8]. This problem has many real-world applications, especially in industries where high-value material is being cut [3] (steel industry, paper industry). No exact algorithm is known that solves practical problem instances optimally, so there are lots of heuristic approaches. The number of publications about this problem increases each year, so we refer the reader to bibliography [23] and the most recent survey [2].

Supported by RFBR, project 19-07-00895.

Throughout this paper we abbreviate an instance of CSP as $E := (L, l, b)$. The total number of items is $n = \sum_{i=1}^{m} b_i$. W.l.o.g., we assume that all numbers in the input data are positive integers and $L \geq l_1 > \cdots > l_m > 0$.

The classical approach for solving CSP is based on the formulation by Gilmore and Gomory [6]. Any subset of items (called a *pattern*) is formalized as a vector $a = (a_1, \cdots, a_m)^\top \in \mathbb{Z}_+^m$ where $a_i \in \mathbb{Z}_+$ denotes the number of items i in the pattern a. A pattern a of E is *feasible* if $a^\top l \leq L$. So, we can define the set of all feasible patterns $P^f(E) = \{a \in \mathbb{Z}_+^m \mid a^\top l \leq L\}$. For a given set of patterns $P = \{a^1, \cdots, a^r\}$, let $A(P)$ be the $(n \times r)$-matrix whose columns are given by the patterns a^i. Then the CSP can be formulated as follows:

$$z(E) := \sum_{i=1}^{r} x_i \rightarrow \min \text{ subject to } A(P^f(E))x = b, x \in \mathbb{Z}_+^r.$$

The common approximate solution approach involves considering *the continuous relaxation* of CSP

$$z_C(E) := \sum_{i=1}^{r} x_i^C \rightarrow \min \text{ subject to } A(P^f(E))x^C = b, x^C \in \mathbb{R}_+^r.$$

Here $z(E)$ and $z_C(E)$ are called *the optimal function values* for the instance E. The difference $\Delta(E) = z(E) - z_C(E)$ is called *the gap* of instance E. Practical experience and numerous computations have shown that for most instances the gap is very small. An instance E has *the integer round up property* (IRUP) if $\Delta(E) < 1$. Otherwise, E is called a non-IRUP instance. This notation was introduced by Baum and Trotter [1].

Subsequently, the largest known gap was increased. In 1986 Marcotte constructed the first known non-IRUP instance with the gap of exactly 1 [11]. Fieldhouse found an instance with gap 31/30 in 1990 [4]. In 1991 Scheithauer and Terno slightly improved this result to 137/132 [20]. An instance with gap 16/15 was found by Gau [5][1]. Rietz, Scheithauer and Terno subsequently constructed non-IRUP instances with gaps 10/9 and 7/6 in 1998 and 2000 respectively [15, 16] (both papers were published in 2002). Rietz constructed an instance with gap 6/5 and published it in his PhD thesis in 2003 [13] and a slightly smaller instance with the same gap together with Dempe in 2008 [14]. Recently an instance with gap 77/64 was constructed by Ripatti and Kartak [19].

Table 1 summarizes the progress of searching for instances with large gaps.

For the $(2, 3)$-instances of CSP that will be defined below the maximal known gap (as we were able to find) is $73/72 - \varepsilon \approx 1.013888$ constructed by Rietz, Scheithauer and Terno [16].

The MIRUP (modified IRUP) conjecture [21] asserts that $\Delta(E) < 2$ for all CSP instances E, but it is still open. It is proved for several classes of CSP instances [12, 16, 22]. The MIRUP conjecture implies existence of fast algorithms

---

[1] Looks like this paper is lost in times, but we mention it here for the sake of completeness. An instance $(10000, (5000, 3750, 3250, 3001, 2000)^\top, (1, 1, 1, 1, 2)^\top)$ from this paper is mentioned in [15].

Table 1. The progress of searching for instances with large gaps.

year	the gap		found by
1986	1	1.000000	Marcotte [11]
1990	31/30	1.033333	Fieldhouse [4]
1991	137/132	1.037879	Scheithauer, Terno [20]
1994	16/15	1.066666	Gau [5]
1998	10/9	1.111111	Rietz, Scheithauer, Terno [15]
2000	7/6	1.166666	Rietz, Scheithauer, Terno [16]
2003	6/5	1.200000	Rietz [13]
2020	77/64	1.203125	Ripatti, Kartak [19]

for solving the CSP [7]. Further research of non-IRUP instances can be found in [9,10,17].

In this paper we construct new CSP instances with a gap larger than previously known. The paper has the following structure. In Sect. 2, we introduce the nesting operation and various **gadgets** that allow to construct $(2,3)$-instances with large gaps. In Sect. 3, we expand our construction for $(2,3)$-instances to the common case and obtain instances with much larger gaps. We draw a conclusion in Sect. 4.

Because the text of the paper is rather technical, we suggest to skip proof of optimality on first reading, namely Theorems 4, 6, 8, 12, and related Lemmas. Without them our construction provides a lower bound for the gap. Subsection 2.1 describes in detail a core element of our construction. Subsections 2.2 and 2.3 provide extended versions of the core element, and structure of these Subsections is similar to Subsect. 2.1. Structure of proofs in Sect. 3 is also similar to one in Subsect. 2.1.

2   Nested (2, 3)-Instances

We start with several definitions and theorems which we will use in our constructions.

Definition 1. *A CSP instance $E = (L, l, b)$ is called a $(2,3)$-**instance** if each of its items can only be packed into the container either two or three times, i.e. $L/4 < l_i \leq L/2$ for all $1 \leq i \leq m$.*

Definition 2 (The composition of instances [14,16]). *Let $E_1 = (L^1, l^1, b^1)$ and $E_2 = (L^2, l^2, b^2)$ denote two instances of CSP having n^1 and n^2 items respectively and with $L^1 = L^2$. The composed instance $E := E_1 \oplus E_2$ of CSP consists of the task of packing all the $n^1 + n^2$ items of lengths from vectors l^1 and l^2 and with demands from vectors b^1 and b^2. In case when L^1 and L^2 are different, they can be multiplied by a common multiplier (together with item lengths)* to adjust the capacity of containers of both instances. For example, instances $(2, (1)^\top, (1)^\top)$ and $(5, (2)^\top, (2)^\top)$ can be composed into the new instance

$$(2, (1)^\top, (1)^\top) \oplus (5, (2)^\top, (2)^\top) = (10, (5, 4)^\top, (1, 2)^\top).$$

Definition 3 (The nesting operation). *Consider a CSP instance* $E_1 = (L^1, l^1, b^1)$ *and a* $(2,3)$*-instance* $E_2 = (L^2, l^2, b^2)$. *Assuming that item and container sizes are fractional, we scale the size of the container* L^2 *of* E_2 *to* L^1 *and then scale item sizes of* E^2 *to be in range* $[L^1/3 - 1, L^1/3 + 1]$. *For example, each size* l_i^2 *can be changed to*

$$l_i^{2*} = L^1/3 + (l_i^2 - L^2/3)/(L^2/6),$$

and we obtain a $(2,3)$*-instance* E_2^* *which is equivalent to* E_2. *After that we compose* E_1 *and* E_2^* *into instance* $E = E_1[E_2] = E_1 \oplus E_2^*$. *Indeed, the resulting instance* E *can be equivalently transformed into an instance with integer sizes. For example, for* $E_1 = (30, (12, 8)^\top, (1, 1)^\top)$ *we have*

$$\begin{aligned} E_1[E_1] &= (30, (12, 8)^\top, (1, 1)^\top) \oplus (30, (11, 9)^\top, (1, 1)^\top) \\ &= (30, (12, 11, 9, 8)^\top, (1, 1, 1, 1)^\top). \end{aligned}$$

If E_1 is a $(2, 3)$-instance too, then the nesting operation can be applied repeatedly. We write it as $E_1^t[E_2]$, where t is the number of times the nesting operation is applied. So, $E_1^0[E_2] = E_2$, and $E_1^t[E_2] = E_1[E_1^{t-1}[E_2]]$ for all integer $t > 0$.

Definition 4. *Let* $\alpha(E, x^C)$ *be the coefficient of a feasible continuous solution* x^C *of* E *that corresponds to pattern* $(2, 0, \cdots, 0)^\top$. *Let* $z'_C(E, x^C)$ *be a function value of* E *that corresponds to the feasible continuous solution* x^C. *Let* $\Delta'(E, x^C)$ *be* $z(E) - z'_C(E, x^C)$, *and we have* $\Delta(E) \geq \Delta'(E, x^C)$. *Finally, let the* (α, Δ)-**value** *of a* $(2, 3)$*-instance* E *be* $\alpha(E, x^C) + 3\Delta'(E, x^C)$ *for* x^C.

Theorem 1. *If* E *is a* $(2, 3)$*-instance with* $3k + p$ *items and* $z(E) = k + s$, *where* k, p *and* s *are non-negative integers, then* (α, Δ)*-value of* E *does not exceed* $3s - p$ *for any feasible solution* x^C.

Proof. Consider a feasible continuous solution $x^C = (x_1^C, \cdots, x_{r-1}^C, \alpha)^\top$. Let X be $\sum_{i=1}^{r-1} x_i^C$, then $z'_C(E, x^C) = X + \alpha(E, x^C)$, and $\Delta'(E, x^C) = k + s - X - \alpha(E, x^C)$. Because E is a $(2, 3)$-instance, each of its patterns contains no more than 3 items, it implies $3k + p \leq 3X + 2\alpha(E, x^C)$, or after rearranging $3(k - X) \leq 2\alpha(E, x^C) - p$. Finally, $\alpha(E, x^C) + 3\Delta'(E, x^C) = \alpha(E, x^C) + 3(k - X) + 3s - 3\alpha(E, x^C) \leq 3s - p$. $\square$

Corollary 1. *If* E *is a* $(2, 3)$*-instance with* $3k + 1$ *items and* $z(E) = k + 1$ *then* $\Delta(E) \leq 2/3$.

Definition 5. *A* $(2, 3)$*-instance* E *holds the* $(2, 3)$-**packing property** *if it has* $3k + 1$ *items,* $z(E) = k + 1$ *and any* $3k$ *of its items cannot be packed into* k *containers.*

As a starting point we can consider a $(2,3)$-instance

$$E_I = (3p, (p+2, p, p-1)^\top, (2, 1, 1)^\top), p > 4,$$

We have $p > 4$ here because for $p \leq 4$ E_I is not a $(2,3)$-instance by definition (the items are not in range $(\frac{3}{4}p, \frac{3}{2}p]$).

The instance E_I has the following feasible continuous solution (which is optimal):

$$\frac{1}{3}(0,3,0)^\top + \frac{1}{2}(1,0,2)^\top + \frac{3}{4}(2,0,0)^\top = (2,1,1)^\top.$$

Throughout the paper we will write the solution in the following form to illustrate its nested structure:

$$
\begin{array}{c}
b = (\begin{array}{ccc} 2 & 1 & 1 \end{array})^\top \quad \text{solution} \\
\begin{array}{|cc|}
\hline
\multicolumn{2}{|c|}{3} \\
1 & 2 \\
2 & \\
\hline
\end{array}
\quad
\begin{array}{c}
1/3 \\
1/2 \\
\alpha \ 3/4
\end{array}
\end{array}
$$

$z_C(E_I) = 19/12$, $z(E_I) = 2$, $\Delta(E_I) = 5/12$. E_I holds the $(2,3)$-packing property and its (α, Δ)-value is 2 for the presented feasible continuous solution.

Now we will transform E_I into another $(2,3)$-instances E with $3k+1$ items using various **gadgets** that decrease $\alpha(E, x^C)$ and increase $\Delta'(E, x^C)$. For the constructed instances we will have $\Delta(E) \geq \Delta'(E, x^C)$.

By term "gadget" we mean a small construction unit which preserves certain properties. Gadgets are typically used in computational complexity theory to construct reductions from one computational problem to another, as part of proofs of NP-completeness or other types of computational hardness. In this paper a gadget is an instance of CSP which can be combined with another CSP instances by composition or nesting operation to preserve the (α, Δ)-value and the $(2,3)$-packing propery, increase the gap, or make the instance to be so called "sensitive".

2.1   A Gadget with 6 Items

Consider a $(2,3)$-instance

$$G_{1/4} = (3p, (p+8, p+6, p-4, p-12)^\top, (2, 1, 2, 1)^\top),$$

where $p > 48$ (otherwise $G_{1/4}$ is not a $(2,3)$-instance). We call this instance $G_{1/4}$-**gadget**. The $G_{1/4}$-gadget and all gadgets in the following subsections were found computationally using the method described in [19] and [18].

Theorem 2. *If E holds the $(2,3)$-packing property, then $G_{1/4}[E]$ holds the $(2,3)$-packing property too.*

Proof. Let E be a $(2,3)$-instance with $3k+1$ items of lengths in range $[p-1, p+1]$ that holds the $(2,3)$-packing property. We have to show that $G_{1/4}[E]$ with $3(k+2)+1$ items holds the $(2,3)$-packing property too.

$z(G_{1/4}[E]) \geq k + 3$ because $G_{1/4}[E]$ is $(2,3)$-instance with $3k + 7$ items. All items of $G_{1/4}[E]$ can be easily packed into $k + 3$ containers in the following way. First, pack $3k+1$ items of E into $k+1$ containers. There is empty space of size at least $p - 2$ because lengths of nested items are in range $[p - 1, p + 1]$. Insert item $(p - 12)$ here. The remaining items of $G_{1/4}$ pack as $(p + 6) + (p - 4) + (p - 4) = 3p - 2 < 3p$ and $(p + 8) + (p + 8) = 2p + 16 < 3p$. So, $z(G_{1/4}[E]) = k + 3$.

Now we have to show that any $3(k + 2)$ items of $G_{1/4}[E]$ cannot be packed into $k + 2$ containers. W.l.o.g. we drop item $(p + 8)$. Now each container should contain exactly 3 items. We say that items $(p + 8)$ and $(p + 6)$ are big and items $(p - 4)$ and $(p - 12)$ are small.

Two big items cannot be in a single container because $(p + 8) + (p + 6) + (p - 12) = 3p + 2 > 3p$. None of big items can be packed together with 2 items from E because $(p + 6) + 2(p - 1) = 3p + 4 > 3p$. So, any big item should be packed together with at least one small item. At least one of items $(p - 4)$ should be packed together with a big item and we cannot insert here an item from E because $(p + 6) + (p - 4) + (p - 1) = 3p + 1 > 3p$. So, we have to pack a big item together with two small ones. It means that all 5 items of $G_{1/4}$ must be packed into 2 containers. Even if we insert here an item from E, we cannot pack the remaining $3k$ items of E into k containers because E holds the $(2,3)$-packing property. □

Let E be a $(2,3)$-instance with $1/4 \leq \alpha(E) \leq 9/4$. Setting $\alpha' = \alpha(E, x^C)$, where x^C is a feasible solution of E, we have the following feasible continuous solution x_*^C of $G_{1/4}[E]$:

$$
b = \begin{pmatrix} 2 & 1 & \cdots & 2 & 1 \end{pmatrix}^{\mathsf{T}} \quad
\begin{array}{l} \text{solution} \\ \hline \end{array}
$$

$$
\begin{bmatrix}
 & 2 & \cdots & & 1 & \\
 & 2 & & & 1 & \\
1 & & & & 2 & \\
1 & 1 & & & 1 & \\
2 & & & & &
\end{bmatrix}
\quad
\begin{array}{ll}
\beta & \alpha' - 1/4 \\
 & 1/2 \\
\gamma & 9/8 - \alpha'/2 \\
C & 1/2 \\
\alpha & 3/16 + \alpha'/4
\end{array}
$$

The feasible continuous solution of the $G_{1/4}$-gadget and all other G_ψ-gadgets presented in this paper have a similar structure. There is a set of patterns with constant solution values that denote the value C. This value corresponds to "1" under "2" from solution of the nested instance. Solution values for this set of patterns can be found by solving a system of linear equations. Value β replaces $\alpha' = \alpha(E, x^C)$ of the nested solution. There are also parameters γ and $\alpha = \alpha(G_{1/4}[E], x_*^C)$.

The feasible solution of a G_ψ-gadget with $3k$ items satisfies the following properties:

1. Sum of all constant values is $k - 1$.
2. $C = 2\psi$.
3. $\beta = \alpha' - C/2$.
4. $\gamma = 1 + C/4 - \alpha'/2$.
5. $\alpha = 3C/8 + \alpha'/4$.

For any G_ψ-gadget we have $\alpha + \beta + \gamma = 3\alpha'/4 + 1 + C/8$. Let $z'_C(G_\psi[E], x_*^C)$ be a function value of the continuous relaxation that corresponds to the feasible continuous solution:

$$z'_C(G_\psi[E], x_*^C) = z'_C(E, x^C) - \alpha' + (k-1) + \frac{3\alpha'}{4} + 1 + \frac{C}{8} = z'_C(E, x^C) + k - \frac{\alpha'}{4} + \frac{C}{8}.$$

If E holds the $(2,3)$-packing property and G_ψ preserves it, then $z(G_\psi[E]) = z(E) + k$. We have

$$\Delta'(G_\psi[E], x_*^C) = z(E) + k - z'_C(G_\psi[E], x_*^C) = \Delta'(E, x^C) + \frac{\alpha'}{4} - \frac{C}{8}$$

and, using Property 5, we get

$$\begin{aligned}
\alpha(G_\psi[E], x_*^C) + 3\Delta'(G_\psi[E], x_*^C) &= \frac{3C}{8} + \frac{\alpha'}{4} + 3\Delta'(E, x^C) + \frac{3\alpha'}{4} - \frac{3C}{8} \\
&= \alpha(E, x^C) + 3\Delta'(E, x^C).
\end{aligned}$$

So, any G_ψ-gadget preserves the (α, Δ)-value for the constructed feasible continuous solution.

The $G_{1/4}$-gadget can be applied to an instance E multiple times. For example, there is a feasible continuous solution of instance $G^2_{1/4}[E_I]$:

$b =$ (2 1 2 1 2 1 1 2 1 2 1)$^\top$	solution	
3	1/3	0.333(3)
1 2	1/2	0.5
2 1	1/2	0.5
2 1	1/2	0.5
1 2	3/4	0.75
1 1 1	1/2	0.5
2 1	1/8	0.125
2 1	1/2	0.5
1 2	15/16	0.9375
1 1 1	1/2	0.5
2	9/32	0.28125

We remark, that if $1/4 \le \alpha(E, x^C) \le 9/4$, then $\alpha(G^t_{1/4}[E], x^C_{(t)})$ for the constructed feasible continuous solution $x^C_{(t)}$ of $G^t_{1/4}[E]$ is also in this range for all $t > 0$, and all values α, β and γ in the feasible continuous solution are non-negative. Applying $G_{1/4}$ to an instance E with $\alpha(E, x^C)$ beyond this range leads to a feasible solution of a different structure.

Theorem 3. *Let E be a $(2,3)$-instance, for which the $(2,3)$-packing property holds. Suppose, its (α, Δ)-value is 2 and $\alpha(E, x^C) \in [\psi, \psi + 2]$ for the feasible continuous solution x^C. Then*

$$\lim_{t \to \infty} \alpha\left(G^t_\psi[E], x^C_{(t)}\right) = \psi,$$

where $x^C_{(t)}$ is the feasible continuous solution of $G^t_\psi[E]$ constructed by the G_ψ-gadget.

Proof. If $\alpha(E, x^C) = \psi + x$, $x \in [0, 2]$, then by Property 5 $\alpha(G_\psi[E], x_*^C) = (3\psi + \alpha(E, x^C))/4 = \psi + x/4$, and $\alpha(G_\psi^t[E], x_{(t)}^C) = \psi + x/4^t$, where t is integer > 0. $\alpha(G_\psi^t[E], x_{(t)}^C)$ is still in range $[\psi, \psi + 2]$ for all $t > 0$, so values α, β, γ are non-negative. $\quad\square$

Because G_ψ preserves the (α, Δ)-value, $\lim_{t\to\infty} \Delta'(G_\psi^t[E_I], x_{(t)}^C) = (2 - \psi)/3$. Namely, for $G_{1/4}$ we have $\lim_{t\to\infty} \Delta'(G_{1/4}^t[E_I], x_{(t)}^C) = 7/12$.

The values $\alpha\left(G_\psi^t[E_I], x_{(t)}^C\right)$ and $\Delta'\left(G_\psi^t[E_I], x_{(t)}^C\right)$ can also be written in closed form. For E_I we have $\alpha(E_I, x^C) = 3/4 = \psi + x$ and $x = 3/4 - \psi$. Then

$$\alpha\left(G_\psi^t[E_I], x_{(t)}^C\right) = \psi + x/4^t = \psi + (3/4 - \psi)/4^t,$$

and

$$\Delta'\left(G_\psi^t[E_I], x_{(t)}^C\right) = \left(2 - \alpha\left(G_\psi^t[E_I], x_{(t)}^C\right)\right)/3 = \frac{2}{3} - \frac{\psi}{3} - \frac{3/4 - \psi}{3 \cdot 4^t}. \quad (1)$$

Now we will prove that our constructed feasible continuous solution $x_{(t)}^C$ for $G_{1/4}^t[E_I]$ is optimal. To this end, we state the following

Definition 6 (Domination relation [10, 15]). *A pattern a is dominated by b ($a \preceq b$) if $\sum_{i=1}^j a_i \leq \sum_{i=1}^j b_i$ for all $1 \leq j \leq m$. For example,*

$$(0, 0, 1, 0, 0) \preceq (0, 0, 1, 2, 0) \preceq (1, 1, 0, 1, 0) \preceq (2, 0, 1, 0, 0) \preceq (3, 0, 0, 0, 0).$$

Lemma 1. *If $a \preceq b$ then $a^\top l \leq b^\top l$ for all $l \in \mathbb{R}_+^m$ such that $l_1 \geq \cdots \geq l_m > 0$.*

Proof. Setting $k_i = l_i - l_{i+1}$ for $1 \leq i < m$ and $k_m = l_m$ we have $k_i \geq 0$ and

$$a^\top l = \sum_{i=1}^m \left(k_i \sum_{j=1}^i a_i\right) \leq \sum_{i=1}^m \left(k_i \sum_{j=1}^i b_i\right) = b^\top l.$$

$\quad\square$

We also need the following technical lemma:

Lemma 2. *Let be $f^0(\omega) = 5/6 + 3\omega/2$ and $f^i(\omega) = D + H\omega + f^{i-1}((1 + \omega)/4)$ for all integer $i > 0$. Then*

$$f^i(\omega) = (D + H/3)i + 4/3 + (\omega - 1/3)\left(\frac{4H}{3} + \frac{9 - 8H}{6 \cdot 4^i}\right).$$

Proof. Let ω_0 be ω and $\omega_i = (1 + \omega_{i-1})/4$ for all integer $i > 0$. Then

$$\omega_i = 1/4^i + 1/4^{i-1} + \cdots + 1/4 + \omega/4^i = (1 - 1/4^i)/3 + \omega/4^i = 1/3 + (\omega - 1/3)/4^i,$$

and

$$f^i(\omega) = Di + H(\omega_0 + \omega_1 + \cdots + \omega_{t-1}) + \frac{5}{6} + \frac{3\omega_i}{2}$$

$$= Di + H\left(\frac{i}{3} + \frac{4(\omega - 1/3)(1 - 1/4^i)}{3}\right) + \frac{5}{6} + \frac{1}{2} + \frac{3(\omega - 1/3)}{2 \cdot 4^i}$$

$$= (D + H/3)i + (\omega - 1/3)\left(\frac{4H(1 - 1/4^i)}{3} + \frac{3}{2 \cdot 4^i}\right) + \frac{4}{3}$$

$$= (D + H/3)i + 4/3 + (\omega - 1/3)\left(\frac{4H}{3} + \frac{9 - 8H}{6 \cdot 4^i}\right).$$

$\square$

Finally, we state the following

Theorem 4. *The constructed feasible continuous solution $x_{(t)}^C$ for $G_{1/4}^t[E_I]$ is optimal.*

Proof. The function value of the feasible continuous solution $x_{(t)}^C$ of $G_{1/4}^t[E_I]$ can be written in closed form using (1):

$$z_C'(G_{1/4}^t[E_I], x_{(t)}^C) = z(G_{1/4}^t[E_I]) - \Delta'(G_{1/4}^t[E_I], x_{(t)}^C)$$

$$= (2t + 2) - \left(\frac{2}{3} - \frac{1}{12} - \frac{1}{6 \cdot 4^t}\right) = 2t + \frac{17}{12} + \frac{1}{6 \cdot 4^t}.$$

To prove that $z_C'(G_{1/4}^t[E_I], x_{(t)}^C)$ indeed is optimal, we formulate the dual problem for the continuous relaxation of CSP:

$$y_C(E) := b^\top u^C \to \max \text{ subject to } a^\top u^C \leq 1 \text{ for all } a \in P^f(E), u^C \in \mathbb{R}_+^m.$$

Let B be the base matrix for the feasible solution $x_{(t)}^C$, which consists vectors from A corresponding to positive elements (basis) of $x_{(t)}^C$. Now we solve the system of linear equations $B^\top u = e$ from outside to inside to obtain a feasible dual solution u.

Suppose $G_{1/4}$ is applied to an instance E. Then $u = (u_1, u_2, u_1', \ldots, u_{m'}', u_3, u_3)^\top$, where m' is the number of different item lengths of E. The system of equations is the following:

$$
\begin{aligned}
1 &= 2u_1' + u_3 \\
1 &= 2u_2 + u_4 \\
1 &= u_1 + 2u_3 \\
1 &= u_1 + u_1' + u_4 \\
1 &= 2u_1
\end{aligned}
\tag{2}
$$

If $G_{1/4}$ is the nested instance, u_1 depends on the outer solution (like u_1' depends on u_1 as we show below), so w.l.o.g. we omit the Eq. (2) and set $u_1 = \omega$.

Now the solution u depends on ω and we write it as $u(\omega)$. The function value of the constructed feasible solution is $y'_C(G_{1/4}[E], \omega)$. On the final step we will set $\omega = 1/2$.

Solving the system of equations, we get $u_1 = \omega$, $u_3 = (1 - \omega)/2$, $u'_1 = (1 + \omega)/4$, $u_4 = (3 - 5\omega)/4$, and $u_2 = (1 + 5\omega)/8$. For $\omega > 1/3$ we have

$$\omega > (1 + 5\omega)/8 > (1 + \omega)/4 > 1/3 > (1 - \omega)/2 > (3 - 5\omega)/4,$$

that yields $u_1 > u_2 > u'_1 > 1/3 > u_3 > u_4$.

$u'_1 = (1 + \omega)/4$ is the new ω' for the nested instances E. If $E = G_{1/4}[E']$ then $u'_{m'} = (3 - 5(1 + \omega)/4)/4 = (7 - 5\omega)/16$. For $\omega > 1/3$ we have $(7 - 5\omega)/16 > (1 - \omega)/2$ and $u'_{m'} > u_3$. If $E = E_I$ then the nested solution is $u'(\omega') = (u'_1, u'_2, u'_3)^\top = (\omega', 1/3, (1 - \omega')/2)^\top$, and for $\omega' > 1/3$ we have $\omega' > 1/3 > (1 - \omega')/2$. In terms of ω we have $u'_3 = (1 - (1 + \omega)/4)/2 = (3 - \omega)/8$ which is greater than $u_3 = (1 - \omega)/2$ for $\omega > 1/3$.

The inequalities above show that the constructed dual solution $u^{(t)}(\omega) = (u_1, u_2, \ldots, u_s)$ for the instance $G^t_{1/4}[E_I]$ has the property $u_1 > u_2 > \cdots > u_s$ for all integer $t \geq 0$. Also for all patterns $a \in B$ we have $a^\top u^{(t)}(\omega) = 1$.

Each pattern from $P^f(E)$ is dominated by some pattern from B. Indeed, for $G_{1/4}[E]$ all patterns having 2 items and less are dominated by pattern $(2, 0, \cdots, 0, 0)^\top$; all patterns without items from E are dominated either by $(1, 0, \cdots, 2, 0)^\top$ or by $(0, 2, \cdots, 0, 1)^\top$; patterns having one or two items from E are dominated by $(1, 0, 1, \cdots, 0, 1)^\top$; patterns with all items from E can be processed recursively. Because elements of $u^{(t)}(\omega)$ are in decreasing order, $a_*^\top u^{(t)}(\omega) \leq 1$ for all patterns $a_* \in P^f(E)$ by Lemma 1. So, $u^{(t)}(\omega)$ is indeed a feasible dual solution.

Now we are going to evaluate $y'_C(G^t_{1/4}[E_I], \omega)$. We have

$$y'_C(G_{1/4}[E], \omega) = 2\omega + \frac{1 + 5\omega}{8} + 2\frac{1 - \omega}{2} + \frac{3 - 5\omega}{4} + y'_C(E, (1 + \omega)/4)$$

$$= \frac{15}{8} + \frac{3}{8}\omega + y'_C(E, (1 + \omega)/4)$$

and

$$y'_C(E_I, \omega) = 5/6 + 3\omega/2.$$

Using Lemma 2 and setting $\omega = 1/2$ we get

$$y'_C(G^t_{1/4}[E_I], 1/2) = \left(\frac{15}{8} + \frac{3}{8 \cdot 3} \right)t + \frac{4}{3} + \left(\frac{1}{2} - \frac{1}{3} \right)\left(\frac{4 \cdot 3}{3 \cdot 8} + \frac{9 - 8 \cdot 3/8}{6 \cdot 4^i} \right)$$

$$= 2t + \frac{4}{3} + \frac{1}{6}\left(\frac{1}{2} + \frac{1}{4^t} \right) = 2t + \frac{17}{12} + \frac{1}{6 \cdot 4^t}$$

$$= z'_C(G^t_{1/4}[E_I], x^C_{(t)}).$$

We have $y'_C(G^t_{1/4}[E_I], 1/2) \leq z_C(G^t_{1/4}[E_I]) \leq z'_C(G^t_{1/4}[E_I], x^C_{(t)})$ by the duality theorem, so $x^C_{(t)}$ is optimal. $\square$

2.2   Gadgets with 9 Items

Let $e = (1, 1, \ldots, 1)^\top$ be a vector of dimension m. Now we present several G_ψ gadgets having 9 items together with their feasible continuous solutions:

$$G_{1/6} = (3p, (8, 7, 6, -4, -13, -14)^\top + pe, (2, 2, 1, 2, 1, 1)^\top), p > 56.$$

$$
b = \begin{pmatrix} 2 & 2 & 1 & \cdots & 2 & 1 & 1 \end{pmatrix}^\top
\quad
\begin{array}{ccccccc}
 & & \lfloor 2 & \cdots & \rfloor 1 & & \\
 & 1 & 1 & & & 1 & \\
 & 2 & & & & & 1 \\
 1 & & & & 2 & & \\
 1 & & 1 & & & 1 & \\
 1 & & 1 & & & & 1 \\
 2 & & & & & & \\
\end{array}
\quad
\begin{array}{ll}
\text{solution} \\ \hline
\beta & \alpha' - 1/6 \\
 & 2/3 \\
 & 2/3 \\
\gamma & 13/12 - \alpha'/2 \\
C & 1/3 \\
 & 1/3 \\
\alpha & 3/24 + \alpha'/4 \\
\end{array}
$$

$$G_{1/8} = (3p, (8, 7, 6, -4, -12, -14)^\top + pe, (2, 1, 2, 2, 1, 1)^\top), p > 56.$$

$$
b = \begin{pmatrix} 2 & 1 & 2 & \cdots & 2 & 1 & 1 \end{pmatrix}^\top
\quad
\begin{array}{ccccccc}
 & & \lfloor 2 & \cdots & \rfloor 1 & & \\
 & 2 & & & & 1 & \\
 & 2 & & & & & 1 \\
 1 & & & & 2 & & \\
 1 & & 1 & & & 1 & \\
 1 & & 1 & & & & 1 \\
 2 & & & & & & \\
\end{array}
\quad
\begin{array}{ll}
\text{solution} \\ \hline
\beta & \alpha' - 1/8 \\
 & 3/4 \\
 & 1/2 \\
\gamma & 17/16 - \alpha'/2 \\
C & 1/4 \\
 & 1/2 \\
\alpha & 3/32 + \alpha'/4 \\
\end{array}
$$

$$G_{1/12} = (3p, (12, 10, 9, 8, -6, -18, -20)^\top + pe, (2, 1, 1, 1, 2, 1, 1)^\top), p > 80.$$

$$
b = \begin{pmatrix} 2 & 1 & 1 & 1 & \cdots & 2 & 1 & 1 \end{pmatrix}^\top
\quad
\begin{array}{cccccccc}
 & & & \lfloor 2 & \cdots & \rfloor 1 & & \\
 & & 2 & & & 1 & & \\
 & 1 & 1 & & & 1 & & \\
 & 2 & & & & & 1 & \\
 1 & & & & 2 & & & \\
 1 & & 1 & & & 1 & & \\
 1 & & 1 & & & & 1 & \\
 2 & & & & & & & \\
\end{array}
\quad
\begin{array}{ll}
\text{solution} \\ \hline
\beta & \alpha' - 1/12 \\
 & 1/2 \\
 & 1/3 \\
 & 1/3 \\
\gamma & 25/24 - \alpha'/2 \\
C & 1/6 \\
 & 2/3 \\
\alpha & 3/48 + \alpha'/4 \\
\end{array}
$$

Theorem 5. $G_{1/6}$, $G_{1/8}$ and $G_{1/12}$ preserve the $(2, 3)$-packing property.

Proof. Let E be a $(2, 3)$-instance with $3k + 1$ items of lengths in range $[p - 1, p + 1]$ that holds the $(2, 3)$-packing property. We wish to show that $G_a[E]$ holds the $(2, 3)$-packing property too for all $a \in \{1/6, 1/8, 1/12\}$.

First, we prove that $z(G_a[E]) = k + 4$ for all a. $z(G_a[E]) \geq k + 4$ because $G_a[E]$ is $(2,3)$-instance with $3(k + 3) + 1$ items for all possible a. The packing scheme of all items of $G_a[E]$ into $k+4$ containers is the following. First, we pack $3k + 1$ items of E into $k + 1$ containers. In this packing there is an empty space of size at least $p - 2$. Next, we insert the minimal item of G_a into this space. The remaining items of G_a are packed in the following way:

$G_{1/6}$: $(p + 8) + 2(p - 4)$, $(p + 7) + (p + 6) + (p - 13)$, $(p + 8) + (p + 7)$;
$G_{1/8}$: $(p + 8) + 2(p - 4)$, $2(p + 6) + (p - 12)$, $(p + 8) + (p + 7)$;
$G_{1/12}$: $(p + 12) + 2(p - 6)$, $(p + 10) + (p + 8) + (p - 18)$, $(p + 12) + (p + 9)$.

Now we have to show that any $3(k+3)$ items of $G_a[E]$ cannot be packed into $k + 3$ containers. W.l.o.g. we drop the maximal item. Now any container should contain exactly 3 items. We say that items greater than $p + 1$ are big, and items less than $p - 1$ are small.

Four big items cannot be packed into 2 containers because together with two smallest items they have total length $6p + 1$ for all G_a. So, all big items should be packed either into 3 or 4 containers.

No big item can be placed together with two items from E because we need a container of capacity at least $3p + 4$ for this. It means that any big item should be packed together with a small item.

The maximal of small items x ($(p-4)$ or $(p-6)$ for corresponding G_a) cannot be packed together with a big item and an item from E because their total length is at least $3p + 1$. Then x must either be packed together with another small item and a big item, or placed into a container without a big item. The latter case is impossible because we have 2 items x. If both of them are not packed with big items, we have to pack all big items into 2 containers which is impossible. If only one of x is not packed with a big item, another x should be packed with small and big items, and again we have to pack all big items into 2 containers. Actually, the only way is to pack both items x together with a big item. It implies that all big items must be packed into 3 containers together with all small items.

So, all 8 items of G_a should be packed into 3 containers, and there is a place only for at most one item from E. But items of E without any of them cannot be packed into k containers because E holds the $(2,3)$-packing property.   □

Theorem 6. *The constructed feasible continuous solutions* $x_{(t)}^C$ *for* $G_{1/6}^t[E_I]$, $G_{1/8}^t[E_I]$ *and* $G_{1/12}^t[E_I]$ *are optimal.*

Sketch of Proof. The proof if similar to one in Theorem 4. First, using (1), we write down the function values for the constructed feasible continuous solutions:

$$z_C'(G_{1/6}^t[E_I], x_{(t)}^C) = z(G_{1/6}^t[E_I]) - \Delta'(G_{1/6}^t[E_I], x_{(t)}^C) = 3t + \frac{25}{18} + \frac{7}{36 \cdot 4^t},$$

$$z_C'(G_{1/8}^t[E_I], x_{(t)}^C) = 3t + \frac{11}{8} + \frac{5}{24 \cdot 4^t} \text{ and } z_C'(G_{1/12}^t[E_I], x_{(t)}^C) = 3t + \frac{49}{36} + \frac{2}{9 \cdot 4^t}.$$

Next, we consider the dual problem for the continuous relaxation of CSP. We denote B to be the base matrix for the feasible solution $x_{(t)}^C$ and solve the system of linear equations $B^\top u = e$. The solution has the following form:

$$u = \left(\omega, \frac{1+9\omega}{12}, \frac{1+3\omega}{6}, \frac{1+\omega}{4}, \cdots, \frac{1-\omega}{2}, \frac{3-5\omega}{4}, \frac{5-9\omega}{6}\right) \text{ for } G_{1/6}[E],$$

$$u = \left(\omega, \frac{1+13\omega}{16}, \frac{1+5\omega}{8}, \frac{1+\omega}{4}, \cdots, \frac{1-\omega}{2}, \frac{3-5\omega}{4}, \frac{7-13\omega}{8}\right) \text{ for } G_{1/8}[E],$$

$$u = \left(\omega, \frac{1+9\omega}{12}, \frac{1+5\omega}{8}, \frac{1+3\omega}{6}, \frac{1+\omega}{4}, \cdots, \frac{1-\omega}{2}, \frac{3-5\omega}{4}, \frac{5-9\omega}{6}\right) \text{ for } G_{1/12}[E].$$

For $\omega > 1/3$ all presented elements of u go in descending order. Also for $\omega > 1/3$ we have $\dfrac{7 - 13(1+\omega)/4}{8} = \dfrac{15 - 13\omega}{32} > \dfrac{1-\omega}{2}$ if $E = G_{1/8}[E']$, $\dfrac{5 - 9(1+\omega)/4}{6} = \dfrac{11 - 9\omega}{24} > \dfrac{1-\omega}{2}$ if $E = G_{1/6}[E']$ or $E = G_{1/12}[E']$, and $\dfrac{3-\omega}{8} > \dfrac{1-\omega}{2}$ if $E = E_I$. So, for $\omega > 1/3$ all elements of $u^{(t)}(\omega)$ go in descending order for $G_{1/6}^t[E_I]$, $G_{1/8}^t[E_I]$ and $G_{1/12}^t[E_I]$.

Each pattern from $P^f(E)$ is dominated by some pattern from B (it can be verified considering several cases like in proof of Theorem 4). Because elements of $u^{(t)}(\omega)$ are in descending order by Lemma 1 $a_*^\top u^{(t)}(\omega) \le 1$ for all $a_* \in P^f(E)$. So, $u^{(t)}(\omega)$ is indeed a feasible dual solution.

The function values for the feasible dual solutions are the following

$$y_C'(G_{1/6}[E], \omega) = \frac{35}{12} + \frac{1}{4}\omega + y_C'(E, (1+\omega)/4),$$

$$y_C'(G_{1/8}[E], \omega) = \frac{47}{16} + \frac{3}{16}\omega + y_C'(E, (1+\omega)/4),$$

$$y_C'(G_{1/12}[E], \omega) = \frac{71}{24} + \frac{1}{8}\omega + y_C'(E, (1+\omega)/4),$$

and

$$y_C'(E_I, \omega) = \frac{5}{6} + \frac{3}{2}\omega.$$

Using Lemma 2 we evaluate the function values in closed form for $\omega = 1/2$

$$y_C'(G_{1/6}^t[E_I], 1/2) = 3t + \frac{4}{3} + \frac{1}{6}\left(\frac{1}{3} + \frac{9-2}{6 \cdot 4^t}\right) = 3t + \frac{25}{18} + \frac{7}{36 \cdot 4^t},$$

$$y_C'(G_{1/8}^t[E_I], 1/2) = 3t + \frac{4}{3} + \frac{1}{6}\left(\frac{1}{4} + \frac{9-3/2}{6 \cdot 4^t}\right) = 3t + \frac{11}{8} + \frac{5}{24 \cdot 4^t},$$

$$y_C'(G_{1/12}^t[E_I], 1, 2) = 3t + \frac{4}{3} + \frac{1}{6}\left(\frac{1}{6} + \frac{9-1}{6 \cdot 4^t}\right) = 3t + \frac{49}{36} + \frac{2}{9 \cdot 4^t}.$$

For all $\psi \in \{1/6, 1/8, 1/12\}$ $y_C'(G_\psi^t[E_I], 1/2) = z_C'(G_\psi^t[E_I], x_{(t)}^C)$, so $x_{(t)}^C$ is optimal. $\square$

2.3 Gadgets with 12 Items

First, we present three G_ψ-gadgets with $\psi = \frac{1}{10}, \frac{1}{14}, \frac{1}{16}$ together with their feasible continuous solutions:

$$G_{1/10} = (3p, (10, 9, 8, 7, -5, -15, -17, -18)^\top + pe, (2, 2, 2, 1, 2, 1, 1, 1)^\top), p > 72.$$

$b = (\ 2 \ \ 2 \ \ 2 \ \ 1 \quad \cdots \quad 2 \ \ 1 \ \ 1 \ \ 1 \ )^\top$

matrix								solution	
			$\lfloor 2 \ \cdots \ \rfloor 1$					β	$\alpha' - 1/10$
	1 1				1				4/5
	1 1					1			4/5
	2						1		3/5
1				2				γ	$21/20 - \alpha'/2$
1		1			1			C	1/5
1	1					1			1/5
1	1						1		2/5
2								α	$3/40 + \alpha'/4$

$$G_{1/14} = (3p, (14, 12, 11, 10, -7, -21, -22, -24)^\top + pe, (2, 1, 2, 2, 2, 1, 1, 1)^\top), p > 96.$$

$b = (\ 2 \ \ 1 \ \ 2 \ \ 2 \quad \cdots \quad 2 \ \ 1 \ \ 1 \ \ 1 \ )^\top$

matrix								solution	
			$\lfloor 2 \ \cdots \ \rfloor 1$					β	$\alpha' - 1/14$
	1 1				1				6/7
	2					1			4/7
1		1				1			3/7
2							1		2/7
1				2				γ	$29/28 - \alpha'/2$
1		1			1			C	1/7
1	1						1		5/7
2								α	$3/56 + \alpha'/4$

$$G_{1/16} = (3p, (14, 13, 12, 10, -7, -21, -24, -26)^\top + pe, (2, 1, 2, 2, 2, 1, 1, 1)^\top), p > 104.$$

$b = (\ 2 \ \ 1 \ \ 2 \ \ 2 \quad \cdots \quad 2 \ \ 1 \ \ 1 \ \ 1 \ )^\top$

matrix								solution	
			$\lfloor 2 \ \cdots \ \rfloor 1$					β	$\alpha' - 1/16$
		2				1			7/8
		2					1		3/4
	2						1		1/2
1				2				γ	$33/32 - \alpha'/2$
1			1		1			C	1/8
1	1					1			1/4
1	1						1		1/2
2								α	$3/64 + \alpha'/4$

Finally, we present yet another G_ψ gadget with 12 items. Surprisingly, there exists a G_ψ-gadget with $\psi = 0$, and its structure is the following:

$$G_0 = (3p, (16, 15, 13, 12, 10, -8, -24, -26, -30)^\top + pe, (2, 1, 2, 1, 1, 2, 1, 1, 1)^\top), p > 120.$$

$$
\begin{array}{c}
b = (\ 2\ 1\ 2\ 1\ 1 \quad \cdots \quad 2\ 1\ 1\ 1\ )^\top \\
\begin{array}{|ccccccccc|}
\hline
 & & & 2 & \cdots & & 1 & & \\
 & 2 & & & & 1 & & & \\
 & & 1 & 1 & & 1 & & & \\
 & & 2 & & & & 1 & & \\
 & 2 & & & & & & 1 & \\
1 & & & & & 2 & & & \\
1 & & & 1 & & & 1 & & \\
1 & & 1 & & & & & 1 & \\
1 & 1 & & & & & & & 1 \\
2 & & & & & & & & \\
\hline
\end{array}
\end{array}
\qquad
\begin{array}{ll}
\text{solution} \\
\beta & \alpha' \\
 & 1/2 \\
 & 1/2 \\
 & 1/2 \\
 & 1/2 \\
\gamma & 1 - \alpha'/2 \\
C & 0 \\
 & 1/2 \\
 & 1/2 \\
\alpha & \alpha'/4 \\
\end{array}
$$

Theorem 7. $G_{1/10}$, $G_{1/14}$, $G_{1/16}$ and G_0 preserve the $(2,3)$-packing property.

Proof. Let E be a $(2,3)$-instance with $3k + 1$ items of lengths in range $[p - 1, p + 1]$ that holds the $(2,3)$-packing property. We wish to show that $G_a[E]$ holds the $(2,3)$-packing property too for all $a \in \{1/10, 1/14, 1/16, 0\}$.

First, we prove that $z(G_a[E]) = k + 5$ for all a:

- $z(G_a[E]) \geq k + 5$ because $G_a[E]$ is $(2,3)$-instance with $3(k + 4) + 1$ items for all possible a.
- All items of $G_a[E]$ can be packed into $k + 5$ containers. We pack the minimal item of $G_a[E]$ and all items of E into $k + 1$ containers. Two maximal items of $G_a[E]$ are packed into the $(k + 2)$-th container. The remaining items are packed in the following way:
  $G_{1/10}$: $(p+9)+2(p-5)$, $(p+9)+(p+8)+(p-17)$, $(p+8)+(p+7)+(p-15)$;
  $G_{1/14}$: $(p+12)+2(p-7)$, $2(p+11)+(p-22)$, $2(p+10)+(p-21)$;
  $G_{1/16}$: $(p+13)+2(p-7)$, $2(p+12)+(p-24)$, $2(p+10)+(p-21)$;
  G_0: $(p+15)+2(p-8)$, $2(p+13)+(p-26)$, $(p+12)+(p+10)+(p-24)$.

Now we say that items greater than $p + 1$ are big, and those less than $p - 1$ are small. To show that any $3(k+4)$ items of $G_a[E]$ cannot be packed into $k + 4$ containers we do the following steps:

- W.l.o.g. we drop the maximal item. Any container should contain exactly 3 items.
- Six big items cannot be packed into 1 or 2 containers because each of them is greater than p.
- Six big items cannot be packed into 3 containers together with three smallest items:
  - For $G_{1/10}$ and $G_{1/14}$ they have total length $9p + 1$.
  - For $G_{1/16}$ the total length is $9p$, so all containers should be fully filled. By parity argument items $(p + 13)$ and $(p - 21)$ should be in a single container, but there is no item $(p + 8)$.

- For G_0 the total length is $9p - 1$. The only way to pack item $(p - 30)$ is $(p + 16) + (p + 13) + (p - 30) = 3p - 1$. The remaining items should fully fill 2 containers. By parity argument items $(p + 15)$ and $(p + 13)$ should by in a single container, but there is no item $(p - 28)$.
- Six big items cannot be packed into 3 containers together with any other three items because then their total length is greater than $9p$. So, six big items should be packed into 4, 5 or 6 containers.
- Any big item cannot be packed together with two items from E. It means that any big item should be packed together with a small item. Because we have only 5 small items, six big items cannot be packed into 6 containers.
- The maximal small item x $((p - 5)$ for $G_{1/10}$, $(p - 7)$ for $G_{1/14}$ and $G_{1/16}$, and $(p - 8)$ for G_0) cannot be packed together with a big item and an item from E. Then x must either be packed together with big and small items, or placed into a container without a big item. The latter case is impossible because we have 2 items x:
  - If both items x are in container/containers without a big item, then all big items should be packed into 3 containers which is impossible.
  - If only one item x is in a container without a big item, then other item x should be packed together with another small item. And again we have to pack all big items into 3 containers.
  So, actually, the only way is to pack both items x together with a big item. And the only way to pack all big items is to pack them into 4 containers together with all small items.
- All 11 items of G_a should be packed into 4 containers and there is a place for at most one item from E. But any remaining $3k$ items of E cannot be packed into k containers because E holds the $(2, 3)$-packing property. □

By Theorem 3 we have $\lim_{t \to \infty} \alpha(G_0^t[E_I], x_{(t)}^C) = 0$. Since G_0 preserves the (α, Δ)-value, $\lim_{t \to \infty} \Delta'(G_0^t[E_I], x_{(t)}^C) = 2/3$, which gives a $(2, 3)$-instance near to upper bound given by Corollary 1.

Theorem 8. *The constructed feasible continuous solutions $x_{(t)}^C$ for $G_{1/10}^t[E_I]$, $G_{1/14}^t[E_I]$, $G_{1/16}^t[E_I]$ and $G_0^t[E_I]$ are optimal.*

Sketch of Proof. The proof is analogous to ones in Theorem 4 and 6. Using (1) we write down the function values for the constructed feasible continuous solution:

$$z_C'(G_{1/10}^t[E_I], x_{(t)}^C) = 4t + \frac{41}{30} + \frac{13}{60 \cdot 4^t}, \quad z_C'(G_{1/14}^t[E_I], x_{(t)}^C) = 4t + \frac{19}{14} + \frac{19}{84 \cdot 4^t},$$

$$z_C'(G_{1/16}^t[E_I], x_{(t)}^C) = 4t + \frac{65}{48} + \frac{11}{48 \cdot 4^t} \quad \text{and} \quad z_C'(G_0^t[E_I], x_{(t)}^C) = 4t + \frac{4}{3} + \frac{3}{12 \cdot 4^t}.$$

Solving the system of linear equations $B^\top u = e$, where B is the base matrix for the feasible continuous solution $x^C_{(t)}$, we get

$$u = \left(\omega, \tfrac{1+17\omega}{20}, \tfrac{1+7\omega}{10}, \tfrac{3+11\omega}{20}, \tfrac{1+\omega}{4}, \ldots, \tfrac{1-\omega}{2}, \tfrac{3-5\omega}{4}, \tfrac{17-31\omega}{20}, \tfrac{9-17\omega}{10}\right) \text{ for } G_{1/10}[E],$$

$$u = \left(\omega, \tfrac{1+11\omega}{14}, \tfrac{3+19\omega}{28}, \tfrac{1+4\omega}{7}, \tfrac{1+\omega}{4}, \ldots, \tfrac{1-\omega}{2}, \tfrac{3-5\omega}{4}, \tfrac{11-19\omega}{14}, \tfrac{6-11\omega}{7}\right) \text{ for } G_{1/14}[E],$$

$$u = \left(\omega, \tfrac{1+29\omega}{32}, \tfrac{1+13\omega}{16}, \tfrac{1+5\omega}{8}, \tfrac{1+\omega}{4}, \ldots, \tfrac{1-\omega}{2}, \tfrac{3-5\omega}{4}, \tfrac{7-13\omega}{8}, \tfrac{15-29\omega}{16}\right) \text{ for } G_{1/16}[E],$$

$$u = \left(\omega, \tfrac{1+21\omega}{24}, \tfrac{1+9\omega}{12}, \tfrac{1+5\omega}{8}, \tfrac{1+3\omega}{6}, \tfrac{1+\omega}{4}, \ldots, \tfrac{1-\omega}{2}, \tfrac{3-5\omega}{4}, \tfrac{5-9\omega}{6}, \tfrac{11-21\omega}{12}\right) \text{ for } G_0[E].$$

It is easy to verify that elements of $u^{(t)}(\omega)$ go in decreasing order for $\omega > 1/3$. It can also be verified that all patterns from $P^f(E)$ are dominated by patterns from B. It implies that $u^{(t)}(\omega)$ is indeed a feasible solution of the dual problem.

Writing down the function values for the feasible dual solution we get

$$y'_C(G_\psi[E], \omega) = 4 - \psi/2 + 3\psi\omega/2 + y'_C(E, (1+\omega)/4) \text{ for all } \psi \in \{1/10, 1/14, 1/16, 0\}.$$

Finally, using Lemma 2, we verify that $y'_C(G^t_\psi[E_I], 1/2) = z'_C(G^t_\psi[E_I], x^C_{(t)})$ for all $\psi \in \{1/10, 1/14, 1/16, 0\}$.

We leave all omitted technical details of the proof to the reader.   □

2.4 (2, 3)-Instances with Large Gap

We introduce an I-gadget here. It just adds one item of maximal length. In other words, it increases b_1 by one (e_i means a vector of dimension m in which "1" is set on position i and all other elements are "0"):

$$I \circ (L, l, b) = (L, l, b + e_1).$$

Theorem 9. *If a* (2, 3)-*instance E holds the* (2, 3)-*packing property, then* $z(I \circ E) = z(E) + 1$.

Proof. Let E be a (2, 3)-instance with $3k + 1$ items. Because E holds the (2, 3)-packing property, $z(E) = k + 1$. So, $z(I \circ E) \geq k + 1$. Packing all items of $I \circ E$ into $k + 2$ containers is trivial.

Now we have to show that all $3k + 2$ items of $I \circ E$ cannot be packed into $k + 1$ containers. Suppose we can. Then all containers in this packing store exactly 3 items except a container x which stores two items. We swap these items with two maximal items and drop container x. Now $3k$ items of E are packed into k containers, but this is impossible because E holds the (2, 3)-packing property. □

Setting $\alpha(I \circ E, x^C_*) = \alpha(E, x^C) + 1/2$ we obtain a feasible continuous solution with the function value $z'_C(I \circ E, x^C) = z'_C(E, x^C) + 1/2$. It gives $\Delta'(I \circ E, x^C_*) = \Delta'(E, x^C) + 1/2$. It also increases $y'_C(I \circ E, 1/2)$ by $1/2$ without changing the feasible dual solution. So, actually x^C_* is optimal.

Now we can construct several families of (2, 3)-instances with large gaps in the following way: $I \circ G^t_\psi[E_I]$, for $t = 0, 1, \cdots$ and $\psi \in \{1/4, 1/6, 1/8, 1/10, 1/12, 1/14, 1/16, 0\}$. We have

$$\lim_{t \to \infty} \Delta(I \circ G^t_\psi[E_I]) = (2 - \psi)/3 + 1/2 = 7/6 - \psi/3.$$

For all values of ψ the limits for Δ can be found in Table 2.

Table 2. The limits for $\Delta(I \circ G_\psi^t[E_I])$.

ψ	$7/6 - \psi/3$		ψ	$7/6 - \psi/3$	
1/4	13/12	1.0833333	1/12	41/36	1.1388889
1/6	10/9	1.1111111	1/14	8/7	1.1428571
1/8	9/8	1.1250000	1/16	55/48	1.1458333
1/10	34/30	1.1333333	0	7/6	1.1666667

3 Instances with Large Gap in Common Case

The construction principles of Rietz and Dempe [14] are based on the instance

$$E_0(p,q) = (33 + 3p + q, (21 + q, 19 + q, 15 + q, 10, 9, 7, 6, 4)^\top + pe, e + e_6),$$

where p and q are positive integers, and the following theorem:

Theorem 10 (Rietz and Dempe). *Consider an instance $E = (L, l, b)$ of CSP with the following properties: $l_1 > l_2 > \cdots > l_{m-1} > 2l_m$ and $l_m \leq L/4$. Moreover, assume that this instance is sensitive, i.e. its optimal function value increases if b_m is increased by 1. Then, there are integers p and q such that instance $E' = E \oplus E_0(p,q)$ has gap $\Delta(E') = 1 + \Delta(E)$.*

Here we introduce a RD-gadget that makes a composition of instance E and $E_0(p,q)$ with appropriate values p and q to increase the gap by 1:

$$RD \circ E = E \oplus E_0(p,q).$$

There is also a S-gadget that transforms a $(2,3)$-instance into a sensitive one:

$$S = (7q + 2, (3q + 2, 3q + 1, q)^\top, (1, 1, 4)^\top).$$

Items of the nested $(2,3)$-instance E in $S[E]$ are in range $[(7q + 2)/3 - 1, (7q + 2)/3 + 1]$ by definition. But actually they may be scaled to be in range $[2q + 2, \frac{5}{2}q + 1]$. Item lengths of $S[E]$ satisfy required properties in Theorem 10.

Theorem 11. *If a $(2,3)$-instance E holds the $(2,3)$-packing property, then $z(S[E]) = k + 2$ and $S[E]$ is sensitive.*

Proof. First, we prove that $z(S[E]) = k + 2$.

- Consider a packing of E into $k + 1$ containers. Because maximal item length of E is no more than $\frac{5}{2}q + 1$, there are 2 empty slots (in different containers because E holds the $(2,3)$-packing property) of size at least $2q$. Insert here all the smallest items. Items $(3q + 2)$ and $(3q + 1)$ are packed into the last container. So, all items of $S[E]$ can be packed into $k + 2$ containers.
- There are $3k + 3$ items of length at least $2q + 2$, and to pack them into $k + 1$ container we have to store them 3 items per container. And there is no free space for smallest items because $(7q + 2) - 3(2q + 2) = q - 4 < q$. So, all items of $S[E]$ cannot be packed into $k + 1$ containers.

Next, let $S'[E]$ be instance $S[E]$ where b_m is increased by 1. We have to prove that $z(S'[E]) = k + 3$.

- Packing of $S'[E]$ into $k + 3$ containers exists: place one item q into a single container, the remaining items can be packed into $k + 2$ containers as shown above.
- Let us try to pack all items of $S'[E]$ into $k + 2$ containers.
  - If items $(3q + 2)$ and $(3q + 1)$ are in a single container then this container has no more place for any other item. $3k + 1$ items of E can be packed into $k + 1$ containers and there are 2 empty slots (in different containers because E holds the $(2, 3)$-packing property) of length no more than $(7q + 2) - 2(2q + 2) = 3q - 2$ each. We cannot insert 5 items of length q here.
  - If items $(3q + 2)$ and $(3q + 1)$ are in different containers, then there is a place for no more than one item from E for each of these containers because $(3q + 1) + 2(2q + 2) = 7q + 5 > 7q + 2$. In k remaining containers we can pack only $3k - 1$ items of E because of the $(2, 3)$-packing property, so 2 remaining items of E we must pack with items $(3q + 2)$ and $(3q + 1)$. There are 3 empty slots of the following maximal lengths: $(7q + 2) - (3q + 2) - (2q + 2) = 2q - 2$, $(7q + 2) - (3q + 1) - (2q + 2) = 2q - 1$, and $(7q + 2) - 2(2q + 2) = 3q - 2$. We can insert only 4 items q here, not 5.

Thus, all items of $S'[E]$ cannot be packed into $k + 2$ containers.   □

Let E be a $(2, 3)$-instance with $0 \le \alpha' \le 7/4$, where $\alpha' = \alpha(E, x^C)$. Then the feasible continuous solution x_*^C of $S[E]$ is the following:

$$
b = \begin{pmatrix} 1 & 1 & \cdots & 4 \end{pmatrix}^\top \quad \text{solution}
$$

				solution
	$\lfloor 2$	$\cdots$	$2 \rfloor$	α'
	2		1	$1/2$
1			4	$7/8 - \alpha'/2$
2				$1/16 + \alpha'/4$

We have

$$
z'_C(S[E], x_*^C) = z'_C(E, x^C) + \frac{23}{16} - \frac{\alpha'}{4}.
$$

If E holds the $(2, 3)$-packing property, we can use Theorem 11. And we obtain

$$
\Delta'(S[E], x_*^C) = z(E) + 1 - z'_C(E, x^C) - \frac{23}{16} + \frac{\alpha'}{4} = \Delta'(E, x^C) - \frac{7}{16} + \frac{\alpha'}{4}.
$$

If for E the (α, Δ)-value is 2, then

$$
\Delta'(S[E], x_*^C) = \left(\frac{3\Delta'(E, x^C)}{4} + \frac{\alpha'}{4} \right) + \frac{\Delta'(E, x^C)}{4} - \frac{7}{16} = \frac{\Delta'(E, x^C)}{4} + \frac{1}{16}. \quad (3)
$$

Theorem 12. *The constructed feasible continuous solution $x_{(t)}^C$ for $S[G_\psi^t[E_I]]$ is optimal, where $\psi \in \{1/4, 1/6, 1/8, 1/10, 1/12, 1/14, 1/16, 0\}$.*

Proof. Let E^t_ψ be $S[G^t_\psi[E_I]]$. Using (1) and (3), we get

$$z'_C(E^t_\psi, x^C_{(t)}) = z(E^t_\psi) - \Delta'(E^t_\psi, x^C_{(t)}) = dt + 3 - \frac{1}{16} - \frac{1}{6} + \frac{\psi}{12} + \frac{3/4 - \psi}{12 \cdot 4^t}$$

$$= dt + \frac{133}{48} + \frac{\psi}{12} + \frac{3 - 4\psi}{48 \cdot 4^t},$$

where $d = 2$ for $\psi = 1/4$, $d = 3$ for $\psi \in \{1/6, 1/8, 1/12\}$ and $d = 4$ for $\psi \in \{1/10, 1/14, 1/16, 0\}$.

Consider the dual problem for the continuous relaxation of CSP:

$$y_C(E) := b^\top u^C \to \max \text{ subject to } a^\top u^C \le 1 \text{ for all } a \in P^f(E), u^C \in \mathbb{R}^m_+.$$

For the instance $S[E]$ a feasible dual solution is the following:

$$u = (u_1, u_2, u'_1, \cdots, u'_{m'}, u_3)^\top = (1/2, 7/16, 3/8, \cdots, u'_{m'}, 1/8)^\top,$$

where m' is the number of different inem lengths of E. From proofs of Theorems 4, 6 and 8 we have $u'_1 > u'_2 > \cdots u'_{m'}$ for $E = G^t_\psi[E_I]$. All patterns of $S[E]$ with at least one item from S have no more than 2 items from E and for such items the coefficients in u are no more than $3/8$ for $E = G^t_\psi[E_I]$. Considering all such patterns a_* we verify that $a^\top_* u \le 1$ for all of them. Patterns from E can be processed recursively. So, u is indeed a feasible solution for the dual problem.

Now we evaluate the function value for the dual problem:

$$y'_C(S[G^t_\psi[E_I]]) = 23/16 + y'_C(G^t_\psi[E_I], 3/8).$$

From proofs of Theorems 4, 6 and 8 we have

$$y'_C(G_\psi[E], \omega) = d - \psi/2 + 3\psi\omega/2 + y'_C(E, (1 + \omega)/4).$$

Finally, using Lemma 2 we obtain

$$y'_C(S[G^t_\psi[E_I]]) = \frac{23}{16} + \left(d - \frac{\psi}{2} + \frac{3\psi}{2 \cdot 3}\right)t + \frac{4}{3} + \left(\frac{3}{8} - \frac{1}{3}\right)\left(\frac{4 \cdot 3\psi}{2 \cdot 3} + \frac{9 - 12\psi}{6 \cdot 4^t}\right)$$

$$= dt + \frac{23}{16} + \frac{4}{3} + \frac{1}{24}\left(2\psi + \frac{3 - 4\psi}{2 \cdot 4^t}\right) = dt + \frac{133}{48} + \frac{\psi}{12} + \frac{3 - 4\psi}{48 \cdot 4^t}.$$

So, $y'_C(S[G^t_\psi[E_I]]) = z'_C(E^t_\psi, x^C_{(t)})$, and $x^C_{(t)}$ is optimal. □

The Eq. (3) can also be rewritten as

$$\Delta'(S[E], x^C_*) = \left(\Delta'(E, x^C) + \frac{\alpha'}{3}\right) - \frac{7}{16} - \frac{\alpha'}{12} = \frac{11}{48} - \frac{\alpha'}{12}.$$

Finally, applying Theorems 3, 10 and 12 for the last equation, for all $\psi \in \{1/4, 1/6, 1/8, 1/10, 1/12, 1/14, 1/16, 0\}$ we obtain

$$\lim_{t \to \infty} \Delta(RD \circ S[G^t_\psi[E_I]]) = 59/48 - \psi/12.$$

For all values of ψ the limits for Δ can be found in Table 3.

Table 3. The limits for $\Delta(RD \circ S[G_\psi^t[E_I]])$.

ψ	$59/48 - \psi/12$		ψ	$59/48 - \psi/12$	
1/4	29/24	1.2083333	1/12	11/9	1.2222222
1/6	175/144	1.2152778	1/14	137/112	1.2232143
1/8	39/32	1.2187500	1/16	235/192	1.2239583
1/10	293/240	1.2208333	0	59/48	1.2291667

We also present several instances $RD \circ S[G_0^t[E_I]]$ for integer $t \in [0, 5]$:

$E_1 = RD \circ S[E_I] = (1110, l_1, b_1)$,
$l_1 = (818, 816, 812, 510, 480, 390, 370, 360, 150, 149, 147, 146, 144)^\top$,
$b_1(1, 1, 1, 1, 1, 2, 1, 1, 5, 1, 2, 1, 1)^\top$,
$z(E_1) = 7, \Delta(E_1) = 7/6 \approx 1.1666667$.

$E_2 = RD \circ S[G_0[E_I]] = (13500, l_2, b_2)$,
$l_2 = (9668, 9666, 9662, 5820, 5790, 4820, 4800, 4760, 4740, 4700, 4520, 4500,$
$4490, 4340, 4020, 3980, 3900, 1920, 1919, 1917, 1916, 1914)^\top$,
$b_2 = (1, 1, 1, 1, 1, 2, 1, 2, 1, 1, 2, 1, 1, 2, 1, 1, 1, 5, 1, 2, 1, 1)^\top$,
$z(E_2) = 11, \Delta(E_2) = 233/192 \approx 1.2135417$.

$E_3 = RD \circ S[G_0^2[E_I]] = (402840, l_3, b_3)$,
$l_3 = (287768, 287766, 287762, 172680, 172650, 143880, 143280, 142080, 141480,$
$140280, 134600, 134580, 134540, 134520, 134480, 134300, 134280, 134270, 134120,$
$133800, 133760, 133680, 129480, 119880, 118680, 116280, 57540, 57539, 57537,$
$57536, 57534)^\top$,
$b_3 = (1, 1, 1, 1, 1, 2, 1, 2, 1, 1, 2, 1, 2, 1, 1, 2, 1, 1, 2, 1, 1, 1, 2, 1, 1, 1, 5, 1, 2, 1, 1)^\top$,
$z(E_3) = 15, \Delta(E_3) = 941/768 \approx 1.2252604$.

$E_4 = RD \circ S[G_0^3[E_I]] = (12095640, l_4, b_4)$,
$l_4 = (8639768, 8639766, 8639762, 5183880, 5183850, 4319880, 4301880, 4265880,$
$4247880, 4211880, 4041480, 4040880, 4039680, 4039080, 4037880, 4032200, 4032180,$
$4032140, 4032120, 4032080, 4031900, 4031880, 4031870, 4031720, 4031400, 4031360,$
$4031280, 4027080, 4017480, 4016280, 4013880, 3887880, 3599880, 3563880, 3491880,$
$1727940, 1727939, 1727937, 1727936, 1727934)^\top$,
$b_4 = (1, 1, 1, 1, 1, 2, 1, 2, 1, 1, 2, 1, 2, 1, 1, 2, 1, 2, 1, 1, 2, 1, 1, 2, 1, 1, 1, 2, 1, 1, 1, 2,$
$1, 1, 1, 5, 1, 2, 1, 1)^\top$,
$z(E_4) = 19, \Delta(E_4) = 3773/3072 \approx 1.2281901$.

$E_5 = RD \circ S[G_0^4[E_I]] = (362879640, l_5, b_5)$,
$l_5 = (259199768, 259199766, 259199762, 155519880, 155519850, 129599880,$
$129059880, 127979880, 127439880, 126359880, 121247880, 121229880, 121193880,$
$121175880, 121139880, 120969480, 120968880, 120967680, 120967080, 120965880,$
$120960200, 120960180, 120960140, 120960120, 120960080, 120959900, 120959880,$
$120959870, 120959720, 120959400, 120959360, 120959280, 120955080, 120945480,$

120944280, 120941880, 120815880, 120527880, 120491880, 120419880, 116639880,
107999880, 106919880, 104759880, 51839940, 51839939, 51839937, 51839936,
$51839934)^{\top}$,

$b_5 = (1,1,1,1,1,2,1,2,1,1,2,1,2,1,1,2,1,2,1,1,2,1,2,1,1,2,1,1,2,1,1,1,$
$2,1,1,1,2,1,1,1,2,1,1,1,5,1,2,1,1)^{\top}$,

$z(E_5) = 23, \Delta(E_5) = 15101/12288 \approx 1.2289225$.

$E_6 = RD \circ S[G_0^5[E_I]] = (10886399640, l_6, b_6)$,

$l_6 = (7775999768, 7775999766, 7775999762, 4665599880, 4665599850,$
3887999880, 3871799880, 3839399880, 3823199880, 3790799880, 3637439880,
3636899880, 3635819880, 3635279880, 3634199880, 3629087880, 3629069880,
3629033880, 3629015880, 3628979880, 3628809480, 3628808880, 3628807680,
3628807080, 3628805880, 3628800200, 3628800180, 3628800140, 3628800120,
3628800080, 3628799900, 3628799880, 3628799870, 3628799720, 3628799400,
3628799360, 3628799280, 3628795080, 3628785480, 3628784280, 3628781880,
3628655880, 3628367880, 3628331880, 3628259880, 3624479880, 3615839880,
3614759880, 3612599880, 3499199880, 3239999880, 3207599880, 3142799880,
$1555199940, 1555199939, 1555199937, 1555199936, 1555199934)^{\top}$,

$b_6 = (1,1,1,1,1,2,1,2,1,1,2,1,2,1,1,2,1,2,1,1,2,1,2,1,1,2,1,2,$
$1,1,2,1,1,2,1,1,1,2,1,1,1,2,1,1,1,2,1,1,1,2,1,1,1,5,1,2,1,1)^{\top}$,

$z(E_6) = 27, \Delta(E_6) = 60413/49152 \approx 1.2291056$.

4 Conclusion

We have presented several families of $(2,3)$-instances having $3k + 1$ items with
a recursive structure. Using them we constructed $(2,3)$-instances with a gap
greater than $7/6 - \varepsilon$ for any $\varepsilon > 0$. It improves the best known construction for
$(2,3)$-instances $73/72 - \varepsilon$.

We have expanded our construction for $(2,3)$-instances to the common case
and constructed CSP instances with gaps greater that $59/48 - \varepsilon$ for any $\varepsilon > $
0. This improves the best known gap from 1.203125 to 1.229166. The further
improvement may be in construction of a better version of S-gadget: Eq. (3)
shows that the trade-off for the sensitivity is a bit expensive.

During our computer search for G_ψ-gadgets we have found much more of
them than presented in this paper. We conjecture that it is possible to construct
G_ψ-gadgets for many other values of ψ, namely

Conjecture 1. For any integer $a > 1$ there exists a G_ψ-gadget with $\psi = \frac{1}{2a}$.

The MIRUP conjecture is still open.

Acknowledgments. The author would like to thank Vadim M. Kartak and Jürgen
Rietz for reading the early version of the manuscript and giving valuable remarks. He
would also like to thank Maxim Mouratov and Damir Akhmetzyanov who helped to
improve English in this paper. This research is supported by RFBR, project 19-07-
00895.

References

1. Baum, S., Trotter, L., Jr.: Integer rounding for polymatroid and branching optimization problems. SIAM J. Algebraic Discrete Methods **2**(4), 416–425 (1981)
2. Delorme, M., Iori, M., Martello, S.: Bin packing and cutting stock problems: mathematical models and exact algorithms. Eur. J. Oper. Res. **255**(1), 1–20 (2016)
3. Dyckhoff, H., Kruse, H.J., Abel, D., Gal, T.: Trim loss and related problems. Omega **13**(1), 59–72 (1985)
4. Fieldhouse, M.: The duality gap in trim problems. SICUP Bul. **5**(4), 4–5 (1990)
5. Gau, T.: Counter-examples to the IRU property. SICUP Bull. **12**(3) (1994)
6. Gilmore, P., Gomory, R.: A linear programming approach to the cutting-stock problem. Oper. Res. **9**(6), 849–859 (1961)
7. Jansen, K., Solis-Oba, R.: A simple OPT+1 algorithm for cutting stock under the modified integer round-up property assumption. Inf. Process. Lett. **111**(10), 479–482 (2011). https://doi.org/10.1016/j.ipl.2011.02.009
8. Kantorovich, L.V.: Mathematical methods of organizing and planning production. Manage. Sci. **6**(4), 366–422 (1960)
9. Kartak, V.M., Ripatti, A.V.: Large proper gaps in bin packing and dual bin packing problems. J. Global Optim. **74**(3), 467–476 (2018). https://doi.org/10.1007/s10898-018-0696-0
10. Kartak, V.M., Ripatti, A.V., Scheithauer, G., Kurz, S.: Minimal proper non-IRUP instances of the one-dimensional cutting stock problem. Discrete Appl. Math. **187**(Complete), 120–129 (2015). https://doi.org/10.1016/j.dam.2015.02.020
11. Marcotte, O.: An instance of the cutting stock problem for which the rounding property does not hold. Oper. Res. Lett. **4**(5), 239–243 (1986)
12. Nitsche, C., Scheithauer, G., Terno, J.: New cases of the cutting stock problem having MIRUP. Math. Methods Oper. Res. **48**(1), 105–115 (1998). https://doi.org/10.1007/PL00020909
13. Rietz, J.: Untersuchungen zu MIRUP für Vektorpackprobleme. Ph.D. thesis, Technische Universität Bergakademie Freiberg (2003)
14. Rietz, J., Dempe, S.: Large gaps in one-dimensional cutting stock problems. Discret. Appl. Math. **156**(10), 1929–1935 (2008)
15. Rietz, J., Scheithauer, G., Terno, J.: Families of non-IRUP instances of the one-dimensional cutting stock problem. Discret. Appl. Math. **121**(1), 229–245 (2002)
16. Rietz, J., Scheithauer, G., Terno, J.: Tighter bounds for the gap and non-IRUP constructions in the one-dimensional cutting stock problem. Optimization **51**(6), 927–963 (2002)
17. Ripatti, A.V., Kartak, V.M.: Bounds for non-IRUP instances of cutting stock problem with minimal capacity. In: Bykadorov, I., Strusevich, V., Tchemisova, T. (eds.) MOTOR 2019. CCIS, vol. 1090, pp. 79–85. Springer, Cham (2019). https://doi.org/10.1007/978-3-030-33394-2_7
18. Ripatti, A.V., Kartak, V.M.: Constructing an instance of the cutting stock problem of minimum size which does not possess the integer round-up property. J. Appl. Ind. Math. **14**(1), 196–204 (2020). https://doi.org/10.1134/S1990478920010184
19. Ripatti, A.V., Kartak, V.M.: Sensitive instances of the cutting stock problem. In: Kochetov, Y., Bykadorov, I., Gruzdeva, T. (eds.) MOTOR 2020. CCIS, vol. 1275, pp. 80–87. Springer, Cham (2020). https://doi.org/10.1007/978-3-030-58657-7_9

20. Scheithauer, G., Terno, J.: About the gap between the optimal values of the integer and continuous relaxation one-dimensional cutting stock problem. In: Gaul, W., Bachem, A., Habenicht, W., Runge, W., Stahl, W.W. (eds.) Operations Research Proceedings 1991, vol. 1991, pp. 439–444. Springer, Heidelberg (1992). https://doi.org/10.1007/978-3-642-46773-8_111
21. Scheithauer, G., Terno, J.: The modified integer round-up property of the one-dimensional cutting stock problem. Eur. J. Oper. Res. **84**(3), 562–571 (1995)
22. Scheithauer, G., Terno, J.: Theoretical investigations on the modified integer round-up property for the one-dimensional cutting stock problem. Oper. Res. Lett. **20**(2), 93–100 (1997)
23. Sweeney, P.E., Paternoster, E.R.: Cutting and packing problems: a categorized, application-orientated research bibliography. J. Oper. Res. Soc. **43**(7), 691–706 (1992)

Stochastic Optimization

On the Resource Allocation Problem to Increase Reliability of Transport Systems

Aleksei Ignatov[(✉)] [iD]

Moscow Aviation Institute, Moscow, Russian Federation
alexei.ignatov1@gmail.com

Abstract. The resource allocation problem to increase reliability of transport systems is considered in the paper. We use probabilities of various undesirable events to describe reliability of transport system elements. Probabilistic and quantile criteria are considered for optimization. The problem with probabilistic criterion is reduced to the integer linear programming problem. We propose the procedure to obtain an approximate solution in the problem with the quantile criterion based on the Chernoff bound and solution of mixed integer nonlinear programming problems.

Keywords: Reliability · Transport systems · Probability ·
Transportation · Quantile · Chernoff bound · Integer programming

1 Introduction

The resource allocation problem is a well-known operations research problem [1]. Its practical application is extremely wide: from portfolio optimization problems to production planning problems [2]. A large number of studies of the resource allocation problem is related to the optimization of traffic on transport [3–8]. Railroad tracks, trains, locomotives, airplanes, trucks act as resources in such problems, and the purpose of optimization is the carriage of goods/passengers in the shortest time/lowest cost of carriage. The loss of the cargo during carriage is possible due to some various undesirable events may occur. For example, such an event at freight trains carriage is a derailment due to a malfunction of a rolling stock or violation of traffic rules. In the transmission of telecommunications signals it is the loss of a signal at the relay station due to equipment malfunctions or insufficient power. In this regard, the problem of increasing reliability of a transport network arises in order to minimize damage/loss of information. Such problem will allow to reduce the quantity of undesirable events.

It is reasonable to consider the probabilistic and quantile criteria for decision-making in such problem. The probabilistic criterion allows to find a strategy with maximal probability that the quantity of undesirable events will not exceed

This work was supported by the Russian Science Foundation, project no. 23-21-00293.

a given level. Unfortunately, it remains unknown, when using a probabilistic criterion, how many undesirable events will occur if we consider a greater level of reliability than the maximal value of the probabilistic criterion. The quantile criterion is used to answer this question.

[9] provides detailed view on the usage of quantile characteristics for reliability analysis. In [10] there was considered the problem to allocate resources between departure vertices of the transport network represented by the directed graph. The optimization goal was to maximize the probability of successful transmitting resources to destination nodes. Capacity of arcs assumed to be random. The reverse problem to [10] was considered in [11]. According terminology of [10,11] considers the resources as fixed in their departure vertices, and there is a control on increasing capacity taking into account that the latter is random. More specifically, [11] considered the problem of the investment fund allocation for devices prohibiting passage through railway crossings in order to reduce the quantity of collisions between vehicles and trains. [11] did not take into account traffic intensity in different parts of the day, which reduces the adequacy of the proposed mathematical model. In addition, it was assumed that even after the collision, the train continues to move. Poisson approximation was used to search for a strategy close to optimal by quantile criterion. Since the calculation did not use the Poisson approximation error, the solution proposed in [11] cannot be considered as a guarantee. More specifically, an estimate of the quantile provided in [11] is neither lower nor upper. These shortcomings are eliminated in the present work.

The present paper modifies and generalizes the problem statement from [11] to the case of a general transport network. As well as in [11], the reliability improvement problems are posed in probabilistic and quantile formulations. The problem with a probabilistic criterion is reduced to the problem of integer linear programming. We propose a searching for an approximate solution procedure for a problem with the quantile criterion. This procedure is based on the use of the Chernoff bound for probability estimation and the subsequent solution of non-linear programming problems. A meaningful example is given.

2   Basic Designations and Assumptions

Let us have I objects, that must traverse the transport network containing M units, on which some undesirable events can occur with objects at time interval $\mathcal{T}$. For example, trains can be interpreted as objects, railway crossings as units. Besides, it is possible to use packages of Internet traffic as objects, routers as units. Also items produced on a conveyor can be interpreted as objects, and units are technical systems that perform manipulations with products. The product may be declared unusable for further use as a result of such manipulations.

Let us divide time interval $\mathcal{T}$ for T parts. Suppose that there are known for i-th object:

- N_i is quantity of units traversed/crossed by object;
- $j_{i,n}$ is the number of n-th unit in series of traversing ($j_{i,n} \in \{1, \ldots, M\}$);

– $t_{i,n}$ is the number of time interval T part when n-th (in series) unit will be crossed by object ($t_{i,n} \in \{1, \ldots, T\}$),

$i = \overline{1, I}$, $n = \overline{1, N_i}$.

Let us group various undesirable events into the one. Probabilities of various undesirable events are transformed to the single one by product rule of probability.

Each unit can be equipped by one or other technical devices allowing to decrease the probability of undesirable event occurrences. Hereinafter such technical devices we will name protection systems to be short. We will consider two sets of technical devices as various protection systems if there is a difference at least in one element of them. For example, we will consider automatic traffic signaling and automatic traffic signaling with automatic barriers in railway industry as different protection systems. We will assume that any unit can be equipped with only one protection system. Also we suppose that there are given for the m-th unit:

– K_m is the quantity of protection systems available to be installed;
– u_m^0 is the number of the (pre)installed protection system ($u_m^0 \in \{1, \ldots, K_m\}$);
– $P_{m,k,t}$ is the probability of undesirable event occurrence with an object when the latter crosses the unit at t-th time interval and when protection system with number k is installed at the unit
– $c_{m,k}$ is cost of the substitution of current protection system for protection system with number k (obviously, $c_{m,u_m^0} = 0$),

$m = \overline{1, M}$, $k = \overline{1, K_m}$, $t = \overline{1, T}$. In practice, probability $P_{m,k,t}$ can be estimated as the frequency. The estimation may be more accurate when there are not only observations for the fact of undesirable event occurrences but for collateral to this event factors. For instance, it can be used the season when a previous object successfully or unsuccessfully crossed a unit as the factor. Such estimate can be obtained by means of logit and probit-regressions.

Let us introduce auxiliary variables $u_{m,k}$ characterizing the installation of protection system with number k at unit with number m: 0 – protection system with number k at unit with number m is not installed, 1 – otherwise, $m = \overline{1, M}$, $k = \overline{1, K_m}$. We have in this notation

$$u_{m,k} \in \{0, 1\}, m = \overline{1, M}, k = \overline{1, K_m}. \tag{1}$$

Only one of available protection systems can be installed for each unit

$$\sum_{k=1}^{K_m} u_{m,k} = 1, m = \overline{1, M}. \tag{2}$$

The total expenses connected with the installation of (new) protection systems can not exceed the investment fund volume C

$$\sum_{m=1}^{M} \sum_{k=1}^{K_m} u_{m,k} c_{m,k} \leq C. \tag{3}$$

We introduce the designation $u \overset{\text{def}}{=} \text{col}(u_{1,1}, \ldots, u_{1,K_1}, \ldots, u_{M,1}, \ldots, u_{M,K_M})$ to be short. The set of admissible strategies U is formed from vectors u satisfying to constraints (1)–(3).

Taking into account (1), (2) and earlier introduced designations, we obtain that probability $P_{m,\cdot,t}(u)$ of the undesirable event occurrence at t-th time interval at unit with number m is defined by formula

$$P_{m,\cdot,t}(u) \overset{\text{def}}{=} \sum_{k=1}^{K_m} u_{m,k} P_{m,k,t}, \quad m = \overline{1,M}, t = \overline{1,T}. \tag{4}$$

Let us introduce random variable $X_i(u)$, which is equal to one when the undesirable event will occur with object with number i at some unit (under chosen strategy of protection systems installation) and is equal to zero otherwise. Probability $q_i(u)$ of occurrence, when the undesirable event will not happen with object with number i, is calculated by formula

$$q_i(u) = \underbrace{(1 - P_{j_{i,1},\cdot,t_{i,1}})}_{\substack{\text{probability that} \\ \text{undesirable event} \\ \text{will not occur} \\ \text{at the first} \\ \text{unit in series}}} \cdot \underbrace{(1 - P_{j_{i,2},\cdot,t_{i,2}})}_{\substack{\text{probability that} \\ \text{undesirable event} \\ \text{will not occur} \\ \text{at the second} \\ \text{unit in series}}} \cdot \ldots \cdot \underbrace{(1 - P_{j_{i,N_i},\cdot,t_{i,N_i}})}_{\substack{\text{probability that no} \\ \text{undesirable event} \\ \text{will not occur} \\ \text{at the} N_i\text{-th} \\ \text{unit in series}}} \tag{5}$$

$$= \prod_{n=1}^{N_i} (1 - P_{j_{i,n},\cdot,t_{i,n}}(u)), \quad i = \overline{1,I}.$$

Let us note, it is true for any random events $A_1, A_2, \ldots, A_n$

$$P(\overline{A}_1 \cdot \overline{A}_2 \cdot \ldots \cdot \overline{A}_n) = 1 - P(A_1 + A_2 + \ldots + A_n) \geq 1 - \sum_{i=1}^{n} P(A_i).$$

Therefore the following is correct

$$q_i(u) \geq \tilde{q}_i(u) \overset{\text{def}}{=} 1 - \sum_{n=1}^{N_i} P_{j_{i,n},\cdot,t_{i,n}}(u), \quad i = \overline{1,I}. \tag{6}$$

As consequence, we have

$$\tilde{p}_i(u) \overset{\text{def}}{=} 1 - \tilde{q}_i(u) \geq 1 - q_i(u) = p_i(u), \tag{7}$$

where $p_i(u)$ is probability of undesirable event occurrence with object with number i.

Obviously, $X_i(u) \sim \text{Bi}(1, p_i(u))$, where $p_i(u) \overset{\text{def}}{=} 1 - q_i(u)$. Let us make an assumption, that random variables $X_1(u)$, $X_2(u)$, ..., $X_N(u)$ are independent. Such assumption causes that units continue to work even in the case of undesirable event occurrence at one or other unit. This assumption can be justified for routers. At the same time such assumption for railroad crossings does not

always refer to the reality: sometimes the liquidation of railroad crossing collision results may take more than the day. But it is very difficult to predict trains
movement changes at other railroad tracks in such situation and, as consequence,
to correct $p_i(u)$, $i = \overline{1, I}$. Though the assumption on independence can not be
justified always, it opens the way to find approximate to optimal strategies of
increasing reliability of transport systems.

We introduce random variable $X(u)$ characterizing the total quantity of undesirable event occurrences:

$$X(u) \stackrel{\text{def}}{=} X_1(u) + X_2(u) + \ldots + X_I(u).$$

Let us consider the probability function

$$P_\varphi(u) \stackrel{\text{def}}{=} \mathcal{P}(X(u) \leq \varphi), \quad \varphi \in \mathbb{Z}_+. \tag{8}$$

Function $P_\varphi(u)$ characterizes probability that no more than φ undesirable event
occurrences will happen at time interval $\mathcal{T}$. In practice, the goal is to maximize
function $P_\varphi(u)$ when $\varphi = 0$. In this case the strategy, at which the probability
that no undesirable event occurrences will happen is maximal, will be found.

Also we consider the quantile function connected with probability function
$P_\varphi(u)$

$$\varphi_\alpha(u) \stackrel{\text{def}}{=} \min\{\varphi : P_\varphi(u) \geq \alpha\}, \alpha \in (0, 1).$$

The quantile function characterizes the maximal quantity of undesirable event
occurring at time interval $\mathcal{T}$ for the predefined reliability level α. The problem
in consideration is a problem to improve the system reliability, that's why we
will consider only $\alpha > 1/2$ in future.

Let us state problems to search for the optimal strategy of improving transport systems reliability

$$u_0^* = \arg\max_{u \in U} P_0(u), \tag{9}$$

$$u_\alpha^* = \arg\min_{u \in U} \varphi_\alpha(u). \tag{10}$$

3   Solving Optimization Problems to Improve Reliability

3.1   Optimization of Probability Function

We have by definition of probability function (8)

$$P_0(u) = \mathcal{P}(X(u) \leq 0).$$

Random variables $X_1(u)$, $X_2(u)$, ..., $X_I(u)$ are nonnegative and independent,
that's why we have

$$P_0(u) = \mathcal{P}(X(u) = 0) = \mathcal{P}(X_1(u) + X_2(u) + \ldots + X_I(u) = 0)$$
$$= \mathcal{P}(\{X_1(u) = 0\} \cdot \{X_2(u) = 0\} \cdot \ldots \cdot \{X_I(u) = 0\})$$
$$= \mathcal{P}(X_1(u) = 0) \cdot \mathcal{P}(X_2(u) = 0) \cdot \ldots \cdot P(X_I(u) = 0) = \prod_{i=1}^{I} q_i(u). \tag{11}$$

We get, substituting (11) in (5),

$$P_0(u) = \prod_{i=1}^{I} \prod_{n=1}^{N_i} (1 - P_{j_{i,n}, \cdot, t_{i,n}}(u)).$$

We have, taking into account (4),

$$P_0(u) = \prod_{i=1}^{I} \prod_{n=1}^{N_i} \left(1 - \sum_{k=1}^{K_{j_{i,n}}} u_{j_{i,n},k} P_{j_{i,n},k,t_{i,n}} \right). \tag{12}$$

Logarithm is monotonously increasing function, so optimal strategy for function $\ln(P_0(u))$ will be coincident with strategy optimal for $P_0(u)$. Based on that, we introduce function $P_0^L(u) \stackrel{\text{def}}{=} \ln(P_0)$. We obtain on the base of (12)

$$P_0^L(u) = \ln \left[\prod_{i=1}^{I} \prod_{n=1}^{N_i} \left(1 - \sum_{k=1}^{K_{j_{i,n}}} u_{j_{i,n},k} P_{j_{i,n},k,t_{i,n}} \right) \right]$$

$$= \sum_{i=1}^{I} \sum_{n=1}^{N_i} \ln \left(1 - \sum_{k=1}^{K_{j_{i,n}}} u_{j_{i,n},k} P_{j_{i,n},k,t_{i,n}} \right). \tag{13}$$

Only one of protection systems can be installed at each unit, that's why

$$\ln \left(1 - \sum_{k=1}^{K_{j_{i,n}}} u_{j_{i,n},k} P_{j_{i,n},k,t_{i,n}} \right) = \sum_{k=1}^{K_{j_{i,n}}} u_{j_{i,n},k} \ln \left(1 - P_{j_{i,n},k,t_{i,n}} \right). \tag{14}$$

Therefore we obtain by substituting (14) in (13) the following

$$P_0^L(u) = \sum_{i=1}^{I} \sum_{n=1}^{N_i} \sum_{k=1}^{K_{j_{i,n}}} u_{j_{i,n},k} \ln \left(1 - P_{j_{i,n},k,t_{i,n}} \right).$$

Based on the above, strategy u_0^* can be found by solving of the following problem

$$u_0^* = \arg \max_{u \in U} P_0^L(u). \tag{15}$$

Thus, the initial integer nonlinear programming problem (9) reduced to integer linear programming problem (15).

3.2   Optimization of the Quantile Function

Since it is difficult to find the distribution law for the random variable $X(u)$, we will search for a solution in problem (10). We carry out simple transformations to do this

$$P_\varphi(u) = \mathcal{P}(X(u) \leq \varphi) = 1 - \mathcal{P}(X(u) > \varphi).$$

The following equalities are true due to $\varphi \in \mathbb{Z}_+$

$$P_\varphi(u) = 1 - \mathcal{P}(X(u) > \varphi) = 1 - \mathcal{P}(X(u) \geq \varphi + 1).$$

We obtain using the Chernoff bound

$$P_\varphi(u) = 1 - \mathcal{P}(X(u) \geq \varphi + 1) \geq 1 - \frac{\mathbf{M}[\exp(\varepsilon X(u))]}{\exp(\varepsilon(\varphi + 1))} \quad \forall \varepsilon \geq 0. \tag{16}$$

Since random variables $X_1(u)$, $X_2(u)$, ..., $X_N(u)$ are independent, we get

$$\mathbf{M}[\exp(\varepsilon X(u))] = \mathbf{M}[\exp(\varepsilon(X_1(u) + X_2(u) + \ldots + X_I(u)))]$$
$$= \mathbf{M}[\exp(\varepsilon X_1(u))]\mathbf{M}[\exp(\varepsilon X_2(u))] \cdot \ldots \cdot \mathbf{M}[\exp(\varepsilon X_I(u))]$$
$$= \prod_{i=1}^{I} \underbrace{\mathbf{M}[\exp(\varepsilon X_i(u))]}_{\substack{\text{moment-generating} \\ \text{function of } X_i(u)}}. \tag{17}$$

Random variable $X_i(u)$ is distributed by Bernoulli law, that's why we obtain according to [12]

$$\mathbf{M}[\exp(\varepsilon X_i(u))] = q_i(u) + p_i(u)\exp(\varepsilon). \tag{18}$$

We have, substituting (18) in (17),

$$\mathbf{M}[\exp(\varepsilon X(u))] = \prod_{i=1}^{I} [q_i(u) + p_i(u)\exp(\varepsilon)]. \tag{19}$$

We get, taking into account (16), (19),

$$P_\varphi(u) \geq 1 - \frac{\prod\limits_{i=1}^{I} [q_i(u) + p_i(u)\exp(\varepsilon)]}{\exp(\varepsilon(\varphi + 1))}. \tag{20}$$

It is needed to optimize the right side of inequality (20). But function $q_i(u)$ contains the multiplying of binary variables that causes difficulties in optimization. Therefore we obtain using (6), (7) the other bound of function $P_\varphi(u)$. Namely,

$$P_\varphi(u) \geq 1 - \frac{\prod\limits_{i=1}^{I} [q_i(u) + p_i(u)\exp(\varepsilon)]}{\exp(\varepsilon(\varphi + 1))} = 1 - \frac{\prod\limits_{i=1}^{I} [1 - p_i(u) + p_i(u)\exp(\varepsilon)]}{\exp(\varepsilon(\varphi + 1))}$$

$$= 1 - \frac{\prod\limits_{i=1}^{I} [1 + p_i(u)(\exp(\varepsilon) - 1)]}{\exp(\varepsilon(\varphi + 1))}.$$

Since $\varepsilon \geq 0$ and $p_i(u) \geq 0$, $i = \overline{1, I}$ by definition, then, taking into account (7), we get

$$\frac{\prod\limits_{i=1}^{I} [1 + p_i(u)(\exp(\varepsilon) - 1)]}{\exp(\varepsilon(\varphi + 1))} \leq \frac{\prod\limits_{i=1}^{I} [1 + \tilde{p}_i(u)(\exp(\varepsilon) - 1)]}{\exp(\varepsilon(\varphi + 1))}.$$

Hence,

$$P_\varphi(u) \geq 1 - \frac{\prod\limits_{i=1}^{I} [1 + \tilde{p}_i(u)(\exp(\varepsilon) - 1)]}{\exp(\varepsilon(\varphi + 1))} = 1 - \frac{\prod\limits_{i=1}^{I} [\tilde{q}_i(u) + \tilde{p}_i(u) \exp(\varepsilon)]}{\exp(\varepsilon(\varphi + 1))}.$$

Let us introduce function

$$\tilde{P}_\varphi(u, \varepsilon) \overset{\text{def}}{=} 1 - \frac{\prod\limits_{i=1}^{I} [\tilde{q}_i(u) + \tilde{p}_i(u) \exp(\varepsilon)]}{\exp(\varepsilon(\varphi + 1))}$$

It is needed to solve problem

$$(\tilde{u}_\varphi^*, \varepsilon_\varphi^*) = \arg \max_{u \in U, \varepsilon \geq 0} \tilde{P}_\varphi(u, \varepsilon) \tag{21}$$

for each $\varphi \in \mathbb{Z}_+$ to find an approximate solution of problem (10). If it is true $\tilde{P}_{\varphi^*}(\tilde{u}_{\varphi^*}^*, \varepsilon_{\varphi^*}^*) \geq \alpha$ for some φ^*, then one should take $\tilde{u}_{\varphi^*}^*$ as the approximate solution of problem (10), and φ^* as the estimate of $\varphi_\alpha(u_\alpha^*)$.

We note that the optimal solution of problem (21) one can find in other way

$$(\tilde{u}_\varphi^*, \varepsilon_\varphi^*) = \arg \min_{u \in U, \varepsilon \geq 0} \frac{\prod\limits_{i=1}^{I} [\tilde{q}_i(u) + \tilde{p}_i(u) \exp(\varepsilon)]}{\exp(\varepsilon(\varphi + 1))}. \tag{22}$$

Eventually, we obtain by taking the logarithm of criterial function in (22) the following optimization problem

$$(\tilde{u}_\varphi^*, \varepsilon_\varphi^*) = \arg \min_{u \in U, \varepsilon \geq 0} \sum_{i=1}^{I} \ln [\tilde{q}_i(u) + \tilde{p}_i(u) \exp(\varepsilon)] - \varepsilon(\varphi + 1).$$

4   Example

Let us have $M = 10$ units, each one of that can be equipped by one of eight protection systems, i.e. $K_1 = K_2 = \ldots = K_M = 8$. We divide the day into 2 parts. We assign probabilities of undesirable event occurrence depending on one or other protection system and part of the day in Table 1.

Table 1. Data on probabilities (multiplied by 10^{-9}) of undesirable event occurrence with object when the latter traverses various units at installation one or other protection system (further shortened as *PS*) in the first | the second part of the day

Unit \ PS	i		ii		iii		iv		v		vi		vii		viii	
1	40	30	25	20	20	15	15	12	13	10	9	8	5	3	0	0
2	40	30	25	20	20	15	15	12	13	10	9	8	5	3	0	0
3	40	30	25	20	20	15	15	12	13	10	9	8	5	3	0	0
4	40	30	25	20	20	15	15	12	13	10	9	8	5	3	0	0
5	40	30	25	20	20	15	15	12	13	10	9	8	5	3	0	0
6	40	30	25	20	20	15	15	12	13	10	9	8	5	3	0	0
7	40	30	25	20	20	15	15	12	13	10	9	8	5	3	0	0
8	40	30	25	20	20	15	15	12	13	10	9	8	5	3	0	0
9	40	30	25	20	20	15	15	12	13	10	9	8	5	3	0	0
10	40	30	25	20	20	15	15	12	13	10	9	8	5	3	0	0

Suppose that PS installation cost is given according to Table 2.

Table 2. Data on protection system installation cost for various units (PS pre(installed) at the unit is highlighted by bold font)

Unit	Available protection system (installation cost, mln. roubles)							
1	i (0)	ii (0,6)	iii (0,7)	iv (0,9)	v (1)	vi (1,2)	vii (1,5)	viii (800)
2	i (0.1)	**ii (0)**	iii (0,1)	iv (0,3)	v (0,4)	vi (0,6)	vii (0,9)	viii (800)
3	i (0)	ii (0,6)	iii (0,7)	iv (0,9)	v (1)	vi (1,2)	vii (1,5)	viii (800)
4	i (0.1)	**ii (0)**	iii (0,1)	iv (0,3)	v (0,4)	vi (0,6)	vii (0,9)	viii (800)
5	i (0)	ii (0,6)	iii (0,7)	iv (0,9)	v (1)	vi (1,2)	vii (1,5)	viii (800)
6	i (0)	ii (0,6)	iii (0,7)	iv (0,9)	v (1)	vi (1,2)	vii (1,5)	viii (800)
7	i (0)	ii (0,6)	iii (0,7)	iv (0,9)	v (1)	vi (1,2)	vii (1,5)	viii (800)
8	i (0.1)	**ii (0)**	iii (0,7)	iv (0,9)	v (1)	vi (1,2)	vii (1,5)	viii (800)
9	i (0.1)	ii (0.1)	iii (0.1)	**iv (0)**	v (0,1)	vi (0,3)	vii (0,6)	viii (800)
10	i (0.1)	ii (0.1)	iii (0.1)	**iv (0)**	v (0,1)	vi (0,3)	vii (0,6)	viii (800)

Let us have $I = 9300$ objects. Let us make an assumption that the traversing of units occurs either in the first part of the day or in the second. Let routes of traversing, i.e. chains of units numbers sequentially traversed by objects, are specified according to Table 3.

Table 3. Traversing routes in the first | the second part of the day

Route of traversing	Quantity of objects
$1 \to 2 \to 3 \to 4 \to 5 \to 6 \to 7 \to 8 \to 9 \to 10$	1000 | 300
$1 \to 2 \to 3 \to 4 \to 5 \to 6$	3000 | 1000
$1 \to 3 \to 5 \to 6 \to 7$	1500 | 1500
$5 \to 6 \to 7 \to 9 \to 10$	500 | 500

We will consider two variants: when the investment fund volume C is 2 and 3 mln. roubles respectively. We will use $\alpha = 0.999$.

It turned out during numerical experiments that for $C = 2$ there is the following: $\varphi^* = 1$, $P_0(u_0^*) = 0.9987$. For $C = 3$ there are results: $\varphi^* = 1$, $P_0(u_0^*) = 0.9989$. Increase of the budget by 1 mln (with respect to case $C = 2$) allowed to increase probability that no undesirable event occurrences will happen. But this increase is not enough to decrease the estimate of an optimal value of the quantile criterion from 1 to zero.

We will show strategies to modify and install PS by probabilistic and quantile criteria in the following table.

Table 4. Data on units and strategies to install PS

Unit	Quantity of objects traversing unit		PS at unit				
	In the first part of the day	In the second part of the day	Pre- installed	Optimal			
				$C = 2$		$C = 3$	
				u_0^*	$\tilde{u}_{\varphi^*}^*$	u_0^*	$\tilde{u}_{\varphi^*}^*$
1	5500	2800	i	i	i	iii	iii
2	4000	1300	ii	iii	iii	iii	ii
3	5500	2800	i	i	i	iii	iii
4	4000	1300	ii	iii	iii	iii	ii
5	6000	3300	i	iv	iv	iii	iv
6	6000	3300	i	iv	iv	iii	iii
7	3000	2300	i	i	i	i	i
8	1000	300	ii	ii	ii	ii	ii
9	1500	800	iv	iv	iv	iv	iv
10	1500	800	iv	iv	iv	iv	iv

Let us comment results in Table 4. As follows from this Table, strategies by quantile and probabilistic criteria does not coincide always. Obtained solutions does not worsen total reliability, as installation of PS with lower (than currently unit has) reliability is not provided. Investments are allocated mostly to units with more objects when $C = 2$. Investments allocation is more uniform when $C = 3$. If we will compare expenses for PS at various units, then for optimal strategy by probabilistic criterion investments allocation is more uniform than investments allocation for the approximate solution by the quantile criterion.

Since solving of mixed integer nonlinear programming problems is needed to search for strategy $\tilde{u}_{\varphi^*}^*$, the computation time of this strategy is a lot more than for strategy optimal by probabilistic criterion. For instance, when $C = 3$ strategy $\tilde{u}_\varphi^*$ was found after 1.5 h. At the same time strategy u_0^* was found after 3 s. But denial from the quantile criterion is not reasonable. It is reasonable to search for the optimal by probabilistic criterion strategy and use the latter as an initial approximation in problem (22). The other way is searching for φ, at which

$$\max_{\varepsilon \geq 0} \tilde{P}_{\varphi}(u_0^*, \varepsilon) \geq \alpha.$$

Such value φ will characterize the quantity of undesirable event occurrences, guaranteed on the level α, when strategy u_0^* has been chosen.

All results were obtained using Opti Toolbox package with SCIP solver in Matlab on the personal computer (Intel Core i5 4690, 3.5 GHz, 16 GB DDR3 RAM).

We additionally note that if quantity of objects will be doubled proportionally, then computation time to find the approximate to optimal strategy by quantile criterion is increased for 25% in the example under consideration. It is very interesting especially for the quantile criterion to analyze computation time with respect to various values of probabilities, quantity of objects and PS, and will be the subject of future research. It should be additionally noted that the speed in obtaining the approximate to optimal strategy by quantile criterion may be faster with using some other solver than the one used in the paper.

5 The Conclusion

In this paper we considered the problem to increase reliability of units in the transport network. Each object, being located in the transport network, traversed/crossed certain units on which undesirable event could occur with the object. One or another protection system could be installed on each of units to prevent the occurrence of undesirable event. Mathematical programming problems were formulated to find the best protection systems for one or other unit. Investment fund volume and reliabilities of various protection systems were taken into account. There were used quantile and probabilistic criteria to formulate mathematical programming problems. The problem with the probabilistic criterion was reduced to the integer linear programming problem, the solution of the problem with the quantile criterion was found approximately with the help of the Chernoff bound on the basis of solving mixed integer nonlinear programming problems. The numerical experiment showed that for a different data, the obtained strategies may be different, may be coincident. At the same time, the solution for the probabilistic criterion is found much faster, and the strategy according to this criterion divides resources (investments) into units more uniformly than the proposed approximate strategy for the quantile criterion.

References

1. Katoh, N., Ibaraki, T.: Resource allocation problems. In: Du, D.-Z., Pardalos, P.M. (eds.) Handbook of Combinatorial Optimization, pp. 905–1006. Springer, New York (1998). https://doi.org/10.1007/978-1-4613-0303-9_14
2. Patriksson, M.: A survey on the continuous nonlinear resource allocation problem. Eur. J. Oper. Res. **185**(1), 1–46 (2008)

3. Caprara, A., Fischetti, M., Toth, P.: Modeling and solving the train timetabling problem. Oper. Res. **50**(5), 851–861 (2002)
4. Kroon, L., Maroti, G., et al.: Stochastic improvement of cyclic railway timetables. Transp. Res. Part B: Methodol. **42**(6), 553–570 (2008)
5. Bosov, A.V., Ignatov, A.N., Naumov, A.V.: Model of transportation of trains and shunting locomotives at a railway station for evaluation and analysis of side-collision probability. Inform.Appl. **12**(3), 107–114 (2018). (in Russian)
6. Pita, J., Barnhart, C., Pais Antunes, A.: Integrated flight scheduling and fleet assignment under airport congestion. Transp. Sci. **47**, 477–492 (2013)
7. Aydinel, M., Sowlati, T., et al.: Optimization of production allocation and transportation of customer orders for a leading forest products company. Math. Comput. Modell. **48**(7–8), 1158–1169 (2008)
8. Ignatov, A.N.: On the scheduling problem of cargo transportation on a railway network segment and algorithms for its solution. Bull. South Ural State Univ. Ser. Mat. Model. Progr. **14**(3), 61–76 (2021)
9. Nair, N., Paduthol, S., Balakrishnan, N.: Quantile-based reliability analysis (2013)
10. Hsieh, C.-C., Lin, M.-H.: Reliability-oriented multi-resource allocation in a stochastic-flow network. Reliab. Eng. Syst. Saf. **81**(2), 155–161 (2003)
11. Kibzun, A.I., Ignatov, A.N.: On the task of allocating investment to facilities preventing unauthorized movement of road vehicles across level crossings for various statistical criteria. Dependability **18**(2), 31–37 (2018)
12. Casella, G., Berger, R.L.: Statistical Inference, 2nd edn. Duxbury, Pacific Grove (2002)

Distributionally Robust Optimization by Probability Criterion for Estimating a Bounded Signal

Konstantin Semenikhin[(✉)][iD] and Alexandr Arkhipov[iD]

Moscow Aviation Institute (National Research University), Volokolamskoye shosse 4, 125993 Moscow, Russia
siemenkv@mail.ru

Abstract. This paper aims at solving a distributionally robust minimax estimation problem to recover a bounded smooth signal from the finite number of measurements with known second-order moment characteristics of the observation noise. The objective functional is the probability that the L2-norm of the estimation error will exceed a given threshold. To take into account the prior uncertainty, the upper bound of the probability functional is considered over the family of noise distributions and the set of signals with bounded second derivative. The goal of the problem is to minimize the worst-case error probability over the class of linear estimators. A specific feature of this problem is a major significance of the bias and its guaranteed bound. To solve the robust optimization problem with the probability objective we follow two methods: 1) direct minimization of the MSE-bound derived from the Markov inequality; 2) applying the explicit multivariate Selberg bound to the problem with quantile criterion. Numerical simulations are performed to compare the two methods applied to the problem of target path recovery.

Keywords: minimax estimation · distributionally robust optimization · error probability · bounded signal · multivariate Selberg bound · cvx

1 Introduction

Distributionally robust optimization covers a great variety of stochastic models to be optimized under the prior uncertainty in probability distributions of random data [12]. This approach has recently received significant attention in both the operations research and statistical learning communities [18]. Basically, the key idea of robust stochastic optimization models is to assign the maximum value of a probabilistic functional (risk/losses/errors) as an objective to be minimized over the class of feasible strategies/controls/estimators when the maximum (or worst-case) value is taken over the set of probability distributions.

As an important example of describing available information on random parameters/disturbances/noises, we first mention the case of constraints on

© The Author(s), under exclusive license to Springer Nature Switzerland AG 2023
M. Khachay et al. (Eds.): MOTOR 2023, LNCS 13930, pp. 181–194, 2023.
https://doi.org/10.1007/978-3-031-35305-5_12

second-order moment characteristics such as expectation vector and covariance matrix [14]. This approach was used in stochastic programming and its applications [6,7,21]. The recent publications devoted to distributionally robust optimization focus on using some metric to measure discrepancy between the true probability and the underlying empirical distribution (e.g., Wasserstein or relative entropy distance) [4,5,17].

In this paper, we consider a distributionally robust optimization problem in an estimation setting such that the accuracy of estimates is measured according to the error probability—the probability of the estimation error exceeding a given threshold. Such an objective is used in many areas: multi-estimation or confidence estimation [1,4,15], minimax classification with quadratic discrimination functions [10], minimax estimation by probability criteria [2,16].

For constructing robust statistical decision rules under uncertainty in expectations and covariance matrices [6,11], the techniques of semidefinite programming and linear matrix inequalities demonstrate successful applications due to the efficient optimization tools such as YALMIP and cvx (see [13] and [8], respectively).

The goal of the estimation problem we study is to find the estimate of a bounded signal in such a way that the worst-case probability of the error falling outside the ball is minimized. To determine this probability objective, we have to take into account that the bias is nonzero. So the ball is not centered at the mean of the estimation error. In this case the well-known Markov inequality does not provide a sharp bound on the worst-case error probability. The sharp bound can be obtained by means of an algorithmic solution via semidefinite programming (SDP) [20]. However, this result cannot be applied directly because of two issues: the bias is uncertain and the SDP problem has a huge dimension. The both issues can be eliminated by using an explicit expression for the multivariate Selberg bound [3]. The desired solution to the distributionally robust estimation problem is obtained by means of reducing to cvx specifications.

The paper is organized as follows: Sect. 2 formulates the distributionally robust optimization problem in the estimation setting; Sect. 3 describes the minimax method based on using mean-square-error criterion; Sect. 4 provides the desired solution of the estimation problem by probability criterion; Sect. 5 describes results of computer simulation experiments; Sect. 6 formulates the key contribution of the paper.

2   Model Description and Problem Statement

Let us consider a real-valued smooth signal $x(t)$, $t \in [0,T]$ that is to be estimated from a finite number of observations in the presence of the following prior information: the initial state $x(0) = x_0$ and derivative $\dot{x}(0) = v_0$ are given exactly whereas the second derivative is a completely unknown but bounded piecewise continuous function, i.e., $|\ddot{x}(t)| \leq w$ at each point $t \in [0,T]$, where w is a specified positive constant.

We assume that the observations $\{x^{obs}(t_k)\}$ are available at the set of points $\mathcal{T} = \{t_1, \ldots, t_N\} \subset (0,T)$ and have additive mean-zero random errors $\{\eta(t_k)\}$

whose joint distribution is uncertain and a scaled identity matrix $\sigma^2 I_N$ with known σ^2 plays the role of the worst-case covariance matrix.

The goal of the distributionally robust estimation problem is to recover the signal $x(t)$ taking into account its worst-case behavior together with the uncertainty in describing the noise distribution. To formulate the problem we use the probability optimization criterion: by choosing an estimate $\widetilde{x}(t)$, minimize the error probability—the probability that $L_2[0,T]$-norm of the estimation error exceeds a given threshold h:

$$\min_{\widetilde{x}} P\{\|\widetilde{x} - x\| > h\}, \quad \text{where} \quad \|\varepsilon\|^2 = \frac{1}{T} \int_0^T \varepsilon^2(t)\, dt. \tag{1}$$

The signal to be recovered is represented in the form

$$x(t) = x^{ref}(t) + \int_0^T g(t - \tau)\, \theta(\tau)\, d\tau, \quad \theta \in \mathsf{K}, \tag{2}$$

using the reference signal $x^{ref}(t) = x(0) + tv(0)$, integral kernel $g(u) = \max\{u, 0\}$, and uncertain function $\theta(t)$ that belongs to the space of piecewise-continuous functions K and satisfies

$$\|\theta\|_\infty \le w, \quad \text{where} \quad \|\theta\|_\infty = \operatorname*{ess\,sup}_{t \in [0,T]} |\theta(t)|. \tag{3}$$

We will consider linear estimates of the form:

$$\widetilde{x}(t) = x^{ref}(t) + \sum_{k=1}^{N} f(t, t_k) Y(t_k), \quad t \in [0, T], \tag{4}$$

where $\{f(t, t_k)\}$ are smooth functions with zero initial conditions

$$f(0, t_k) = 0, \quad \left.\frac{\partial f(t, t_k)}{\partial t}\right|_{t=0} = 0 \tag{5}$$

and $\{Y(t_k)\}$ are the centered observations

$$Y(t_k) = x^{obs}(t_k) - x^{ref}(t_k) = \int_0^T g(t_k - \tau)\, \theta(\tau)\, d\tau + \eta(t_k). \tag{6}$$

Thus the model described above can be written as

$$X = A\theta, \quad \theta \in \Theta, \quad Y = B\theta + \eta, \quad \eta \sim \mathcal{P}(0, \sigma^2 I_N), \tag{7}$$

in terms of the following notation:

Θ is a ball of radius w in the space K, i.e., $\theta \in \Theta$ means (3);

$X = x(\cdot) - x^{ref}(\cdot)$ is a signal to be estimated as an element of the Hilbert space $\mathsf{H} = \mathsf{L}_2[0,T]$;

$Y = \operatorname{col}[Y(t_1), \ldots, Y(t_N)]$ and $\eta = \operatorname{col}[\eta(t_1), \ldots, \eta(t_N)]$ are the observation vector and the noise vector, respectively, from the Euclidean space $\mathbb{R}^N$;

$\eta \sim \mathcal{P}(0, R)$ describes the assumption

$$\mathsf{M}\eta = 0, \qquad \mathrm{cov}\{\eta, \eta\} \preceq R \tag{8}$$

($S \preceq R$ or $R \succeq S$ means that $R - S$ is positive semidefinite);
$A \colon \mathsf{K} \to \mathsf{H}$ and $B \colon \mathsf{K} \to \mathbb{R}^N$ are integral linear operators defined by the rule

$$(A\theta)(t) = \int_0^T g(t - \tau)\,\theta(\tau)\,d\tau, \qquad (B\theta)(t_k) = \int_0^T g(t_k - \tau)\,\theta(\tau)\,d\tau. \tag{9}$$

Therefore we obtain the correct optimization formulation of the distributionally robust estimation problem with probability criterion:

$$\min_{F \in \mathcal{F}} \sup_{\theta \in \Theta} \sup_{\eta \sim \mathcal{P}(0, \sigma^2 I_N)} \mathsf{P}_\theta\{\|FY - X\| > h\}, \tag{10}$$

where $\mathcal{F}$ is a class of linear operators $F \colon \mathbb{R}^N \to \mathsf{H}$ that determine the estimates $\widetilde{x}(t) = x^{ref}(t) + (FY)(t)$ in the form (4).

So the goal of the problem (10) is to find an estimate $\hat{x}(t)$ that has the minimum worst-case error probability if compared with other linear estimates $\widetilde{x}(t)$ from some class. The choice of this class will be discussed later.

3   Mean-Square-Error Approach

Instead of elaborating methods for a direct solution to the infinite-dimensional optimization problem (10), we focus on finding relaxed or approximate formulations that can be solved using standard optimization tools such as cvx designed for disciplined convex programming [8].

First we apply the Markov inequality to write an upper bound on the error probability:

$$\mathsf{P}_\theta\{\|FY - X\| > h\} \leq \frac{\mathsf{M}_\theta\|FY - X\|^2}{h^2}, \tag{11}$$

where the mean-square error (MSE) falls into the squared bias and the total variance:

$$\mathsf{M}_\theta\|FY - X\|^2 = \|(FB - A)\theta\|^2 + \sigma^2 \,\mathrm{tr}[FF^*]. \tag{12}$$

Here $F^* \colon \mathsf{H} \to \mathbb{R}^N$ denotes the adjoint operator:

$$(F^*u)(t_k) = \frac{1}{T} \int_0^T f(\tau, t_k)\,u(\tau)\,d\tau, \qquad u \in \mathsf{H}, \quad k = 1, \dots, N. \tag{13}$$

Since the uncertain element $\theta \in \Theta$ affects the bias, we need to find a guaranteed bound on it. To do this, we use the following evident result.

Lemma 1. *For any integral operator $\Psi \colon \mathsf{K} \to \mathsf{H}$*

$$(\Psi\vartheta)(t) = \int_0^T \psi(t,\tau)\vartheta(\tau)\,d\tau, \quad \vartheta \in \mathsf{K}, \ t \in [0,T], \tag{14}$$

with Borelean bounded kernel $\psi(t,\tau)$, we have

$$\sup_{\theta \in \Theta} \|\Psi\theta\| \leq w\|\psi\|_1 \tag{15}$$

where the $\|\cdot\|_1$-norm is defined by the rule

$$\|\psi\|_1^2 = \frac{1}{T}\int_0^T \left(\int_0^T |\psi(t,\tau)|\,d\tau\right)^2 dt. \tag{16}$$

Although inequality (15) is not tight in general, it provides the explicit expression for an upper bound which turns out to be the best upper estimate in the case of positive $\psi(t,\tau)$.

Therefore we obtain the *Markov bound* on the objective functional in the minimization problem (10):

$$\sup_{\theta \in \Theta} \sup_{\eta \sim \mathcal{P}(0,\sigma^2 I_N)} \mathsf{P}_\theta\{\|FY - X\| > h\} \leq \frac{R(F) + D(F)}{h^2}, \tag{17}$$

where the total variance $D(F)$ and the maximum squared bias $R(F)$ have the form

$$D(F) = \sigma^2 \operatorname{tr}[FF^*] = \frac{\sigma^2}{T}\int_0^T \sum_{k=1}^N f^2(t,t_k)\,dt, \tag{18}$$

$$R(F) = w^2\|\psi\|_1^2, \quad \psi(t,\tau) = \sum_{k=1}^N f(t,t_k)g(t_k - \tau) - g(t - \tau). \tag{19}$$

Here we apply Lemma 1 to the bias

$$R_\theta(F) = \|(FB - A)\theta\|^2 = \frac{1}{T}\int_0^T \left\{\int_0^T \psi(t,\tau)\theta(\tau)\,d\tau\right\}^2 dt \tag{20}$$

and exploit the representation of the integral operator $FF^* \colon \mathsf{H} \to \mathsf{H}$

$$(FF^*u)(t) = \frac{1}{T}\int_0^T \sum_{k=1}^N f(t,t_k)f(\tau,t_k)\,u(\tau)\,d\tau \tag{21}$$

together with the fact that its trace coincides with the integral of the kernel function over the diagonal $\{(t,t)\colon t \in [0,T]\}$.

Now we consider the MSE-minimax estimation problem on the finite-dimensional class of operators $\mathcal{F}_\mathsf{S}$:

$$\min_{F \in \mathcal{F}_\mathsf{S}} \{R(F) + D(F)\}. \tag{22}$$

Here $F \in \mathcal{F}_S$ means that the image of F coincides with the space of splines S. This space is a linear span

$$S = \text{span}\{\beta_p(t): p = 1, \dots, P\} \tag{23}$$

of quadratic B-splines that are determined at the uniform grid of knots

$$T_0 = 0 < T_1 < \dots < T_P = T < T_{P+1} < T_{P+2}.$$

Hence, any estimator $F \in \mathcal{F}_S$ is defined by

$$(FY)(t) = \sum_{k=1}^{N} f(t, t_k) Y(t_k), \quad f(t, t_k) = \sum_{p=1}^{P} \beta_p(t) c_p(t_k) \tag{24}$$

using a $(P \times N)$-matrix of coefficients

$$C = \{c_p(t_k): p = 1, \dots, P, \ k = 1, \dots, N\}. \tag{25}$$

This means that we can parameterize all $F \in \mathcal{F}_S$ by matrices C. So we write $F = F_C$ for the estimator (24).

The variance $D(F_C)$ can be calculated directly

$$D(F_C) = \sigma^2 \sum_{k=1}^{N} \sum_{p,q=1}^{P} c_p(t_k) G_{p,q} c_q(t_k) \tag{26}$$

using inner products of the basis splines

$$G_{p,q} = (\beta_p, \beta_q) = \frac{1}{T} \int_0^T \beta_p(t) \beta_q(t) \, dt. \tag{27}$$

Since they form the Gram matrix (which is positive definite)

$$G = \{G_{p,q}: p, q = 1, \dots, P\}, \tag{28}$$

we can write the variance as a positive definite quadratic form

$$D(F_C) = \sigma^2 \, \text{tr}[C^\top G C] = \sigma^2 \|HC\|_F^2, \tag{29}$$

where $\| \cdot \|_F$ denotes the Frobenius norm and H is any $(P \times P)$ matrix from the factorization

$$G = H^\top H. \tag{30}$$

For the bias we have to use an approximate integration scheme (say, the rectangular rule):

$$R_{\mathcal{S}}(F_C) = \frac{w^2 T^2}{M^3} \sum_{i=1}^{M} \left(\sum_{j=1}^{M} |\psi(s_i, s_j)| \right)^2, \tag{31}$$

where $\mathcal{S} = \{s_1, \ldots, s_M\}$ is the uniform partition of the segment $[0, T]$ and the $(M \times M)$ matrix $\bar{\Psi} = \{\psi(s_i, s_j)\}$ is determined from the linear matrix equality

$$\bar{\Psi} = \bar{\beta} C \bar{B} - \bar{A}. \tag{32}$$

Here we use matrices

$$\bar{\beta} = \{\beta_p(s_i) \colon i = 1, \ldots, M, p = 1, \ldots, P\},$$
$$\bar{B} = \{g(t_k - s_i) \colon k = 1, \ldots, N, i = 1, \ldots, M\},$$
$$\bar{A} = \{g(s_i - s_j) \colon i, j = 1, \ldots, M\}$$

that can be computed before an optimization procedure is launched.

To obtain a solution to the problem

$$\widehat{C} = \arg\min_{F \in \mathcal{F}_{\mathrm{S}}} \{D(F_C) + R_{\mathcal{S}}(F_C)\}, \tag{33}$$

we suggest the following optimization scheme in terms of cvx specifications:

Algorithm 1. Robust MSE-based estimation scheme

Input data: sigma w T P N M H(P,P) beta(M,P) B(N,M) A(M,M)

```
cvx_begin
   variables C(P,N) D R;
   minimize( D + R );
   subject to
      sigma^2*square_pos(norm(H*C,'fro')) <= D;
      w^2*T^2/M^3*sum(square_pos(sum(abs(beta*C*B-A),2))) <= R;
cvx_end
```

Output data: C % Matrix of optimal coefficients $\widehat{C}$
D % Optimal variance $D(F_{\widehat{C}})$
R % Optimal maximum squared bias $R_{\mathcal{S}}(F_{\widehat{C}})$
% (based on the rectangular integration rule)

Finally we obtain the minimax-optimal value of the error probability based on the MSE bound:

$$\bar{\alpha} = \min\left\{\frac{D(F_{\widehat{C}}) + R(F_{\widehat{C}})}{h^2}, 1\right\} \tag{34}$$

4   Method Based on the Multivariate Selberg Bound

It is important to emphasize that the Markov inequality (11) may not provide a sharp bound in the case of a nonzero bias. But this very case describes a major feature of the problem under consideration. So instead of using the conservative MSE-based bound, we need to elaborate a more sophisticated method for evaluating the probability objective.

This probability bound can be computed by means of semidefinite programming [20] when the bias vector and the covariance matrix are fixed. Since these two characteristics are to be chosen within a minimax optimization, this method cannot be applied directly and requires a special study.

Thus we adopt another approach: it based on a simple analytical expression for the tight probability bound of Selberg type (see Corollary 1 from [3]).

Theorem 1 (Multivariate Selberg bound). *For any random vector $\zeta \in \mathbb{R}^n$ and positive constants r, c, and h, the probability bound*

$$\pi_h^{(n)}(r,c) = \sup\{ \, \mathsf{P}(|\zeta + b| > h): $$
$$\zeta \sim \mathcal{P}(0, R), \; b \in \mathbb{R}^n, \; |b| \leq r, \; R \preceq c^2 I_n \} \tag{35}$$

has the following explicit form:

$$\pi_h^{(n)}(r,c) = \begin{cases} \left(1 + \dfrac{\left(\sqrt{nh^2 - (n-1)(nc^2 + r^2)} - r\right)^2}{n^2 c^2}\right)^{-1}, \\ \qquad h^2 \geq (nc^2 + r^2)(1 + c^2/r^2), \hfill \text{(i)} \\[2mm] (nc^2 + r^2)/h^2 < 1, \\ \qquad nc^2 + r^2 < h^2 \leq (nc^2 + r^2)(1 + c^2/r^2), \hfill \text{(ii)} \\[2mm] 1, \qquad h^2 \leq nc^2 + r^2. \hfill \text{(iii)} \end{cases}$$

To apply this result we should take into account that the estimation error $\varepsilon = FY - X$ related to every $F \in \mathcal{F}_\mathsf{S}$ takes values in the infinite-dimensional Hilbert space H, whereas the image of its covariance operator lies in the subspace S of dimension P. Hence we can impose the additional constraint

$$\mathsf{cov}\{\varepsilon, \varepsilon\} \preceq c^2 P_\mathsf{S} \tag{36}$$

using P_S, the projection operator onto S, and c, some coefficient to be further optimized. The condition (36) can be rewritten in the form

$$\sigma^2 GCC^\top G \preceq c^2 G \tag{37}$$

or $\sigma^2 HCC^\top H^\top \preceq c^2 I_P$, which is the same whenever the matrix H forms the factorization (30). Using the spectral norm $\|\cdot\|_S$ we obtain the equivalent constraint

$$\sigma \|HC\|_S \leq c. \tag{38}$$

Therefore, the distributionally robust estimation problem with probability criterion (10) can be reduced to the following relaxed problem:

$$\min_{r,c,C} \pi_h^{(N)}(r,c): \; r, c \in \mathbb{R}, \; C \in \mathbb{R}^{P \times N} \tag{39}$$

subject to

$$r_\mathsf{S}(F_C) \leq r, \quad \sigma \|HC\|_S \leq c, \tag{40}$$

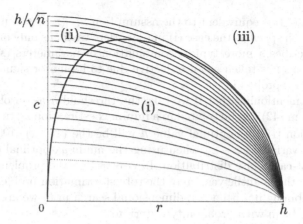

Fig. 1. Contour lines $\pi_h^{(n)}(r,c) = \alpha$ for probability levels $\alpha = 0.05, 0.1, 0.15$ and so on (starting from the lowest curve).

where $r_S(F_C)$ denotes the approximate version of the maximal bias (31), i.e.,

$$r_S(F_C) = \sqrt{R_S(F_C)} = \frac{wT}{M^{3/2}} \left\{ \sum_{i=1}^{M} \left(\sum_{j=1}^{M} |\psi(s_i, s_j)| \right)^2 \right\}^{1/2}. \qquad (41)$$

We call this formulation a *relaxed problem* to stress the following specific feature of its derivation: the linear matrix inequality (36) never holds with equality, because the rank of the covariance operator $\mathrm{cov}\{\varepsilon, \varepsilon\}$ is not greater than the number of observations N which in turn is less than P. This explains why the probability bound $\pi_h^{(N)}(r,c)$ is taken for dimension N.

The bound $\alpha = \pi_h^{(N)}(r,c)$ is quasiconvex as a function of (r,c) (see Fig. 1) but cannot be made convex by any monotonic change of variable α. This is similar to the facts known from the theory of stochastic programming [9].

Nevertheless, let us choose some probability level α and consider a feasibility problem corresponding to (39): find F_C such that $\mathsf{P}\{\|F_C Y - X\| > h\} \leq \alpha$. Then we can represent this problem as a convex program:

$$\min_{r,c,C} r: \; r, c \in \mathbb{R}, \, C \in \mathbb{R}^{P \times N} \qquad (42)$$

subject to

$$(N-1)(Nc^2 + r^2) + \left(r + Nc\sqrt{(1-\alpha)/\alpha} \right)^2 \leq Nh^2, \qquad (43)$$

$$(N+1)c^2 + r^2 + N(c^2/r)^2 \leq h^2, \qquad (44)$$

$$r_S(F_C) \leq r, \quad \sigma\|HC\|_S \leq c. \qquad (45)$$

The constraint (43) is equivalent to the assumption $\pi_h^{(N)}(r,c) \leq \alpha$ under the constraint (44) which specifies the case (i) in (35). Here we focus only on this branch, because it describes a more typical combination of parameters (see Fig. 1). In addition, the bias r is taken as an objective due to its major significance in the robust estimation problem.

It is worth mentioning that if we take the threshold h as an objective function, the problem (42) can be treated in a minimax estimation setting as one with quantile criterion [19]: its optimal value $\hat{h}$ will be the $(1-\alpha) \cdot 100\%$-percentile of the random variable $\|\hat{x} - x\|$ when using the minimax-optimal estimate $\hat{x}(t)$ given the worst-case noise distribution. In contrast to the problem with probability criterion, the quantile version of the robust estimation problem is a convex program. Combining it with a one-dimensional search in h, we are able to solve the original problem with probability criterion.

Below we present a cvx routine (see Algorithm 2) that returns a solution to the distributionally robust estimation problem with the quantile objective.

5   Numerical Experiment

Consider the nonparametric signal $x(t)$ as a target path on the time interval $T = 30$ s with zero initial conditions $x_0 = 0$, $v_0 = 0$, and known bound on the acceleration: $|\ddot{x}(t)| \leq w$, where $w = 1.1 \text{ ms}^{-2}$.

The estimation problem is to recover the target path with the maximum confidence probability $\mathsf{P}\{\|\varepsilon\| \leq h\}$ when h, the threshold for the $\mathsf{L}_2[0,T]$-norm of the error estimation ε, is assigned.

We can also consider the problem with the quantile criterion: find an estimate $\tilde{x}(t)$ that minimizes the radius h of the confidence ball under the constraint on the confidence probability $\mathsf{P}\{\|x - \tilde{x}\| \leq h\} \geq 1 - \alpha$.

The path $x(t)$ is to be estimated from N observations corrupted by zero-mean uncorrelated noises with the standard deviation $\sigma = 1$ m and unknown joint distribution. So the both settings are studied using the distributionally robust optimization approach.

To solve these two problems on the class of spline estimators, we follow:

Algorithm 1—to find a minimax-optimal MSE-estimate with using the Markov bound for evaluating the probability objective (34);

Algorithm 2—to find an estimator that is minimax-optimal with respect to the quantile criterion.

Algorithm 2. Robust quantile-based estimation scheme

```
Input data: alfa sigma w T P N M H(P,P) beta(M,P) B(N,M) A(M,M)
cvx_begin
   variables C(P,N) aux(M,1) h2 r c;
   minimize( h2 );
   subject to
      (N-1)*(N*c^2+r^2)+(r+N*c*sqrt((1-alfa)/alfa))^2 <= N*h2;
      (N+1)*c^2+r^2+N*square_pos(quad_over_lin(c,r)) <= h2;
      sigma*norm(H*C) <= c;
      sum(abs(beta*C*B-A),2) <= aux;
      w*T/M^(3/2)*norm(aux) <= r;
cvx_end
```

Output data: h2	%	Optimal squared quantile: threshold $\hat{h}^2$
C	%	Matrix of optimal coefficients $\widehat{C}$
c^2*N	%	Upper estimate for the optimal variance:
	%	$D(F_{\widehat{C}}) = \sigma^2 \|H\widehat{C}\|_F^2 \leq \hat{c}^2 N$
r	%	Optimal maximum bias $r_S(F_{\widehat{C}})$
	%	(based on the rectangular integration rule)

For $N = 10$ observations obtained at the uniform time grid $\{t_k = Tk/N,\ k = 1, \ldots, N\}$, we take $P = 16$ B-splines and $M = 30$ points for approximate integration by the rectangular rule.

First we present the characteristics of the quantile-based estimator $\hat{F}$ obtained by Algorithm 2 with fixed level of the error probability $\alpha = 0.05$:

the square root of the total variance $\sqrt{D(\hat{F})} = 0.78$ m;

the maximum bias $\sqrt{R_S(\hat{F})} = 1.02$ m;

the worst-case root-MSE $\sqrt{D(\hat{F}) + R_S(\hat{F})} = 1.29$ m;

the minimax-optimal value of the threshold $\hat{h} = 4.84$ m.

Figure 2 shows that the minimax-optimal estimate $\hat{x}(t)$ is not an interpolating spline but provides a tradeoff between making the residuals $|\hat{x}(t_k) - x^{obs}(t_k)|$ small, while keeping the worst-case bias not too big.

Then, we apply Algorithm 1 to find the MSE-based estimator. Its second-order moment characteristics turn out to be almost identical to those for the quantile-based estimator. Furthermore, the plots of the two estimates look very similar. So only one estimate is depicted in Fig. 2. But the error probability obtained via the Markov bound for the same threshold $h = 4.84$ m is significantly greater: $\bar{\alpha} = 0.0704$. To confirm this conjecture, we present Fig. 3: it shows how the error probability α depends on the threshold h for the quantile-based estimators with applying the Selberg-type probability bound (shown as a solid line) and for the MSE-based estimator with applying the Markov bound (shown as a dashed line).

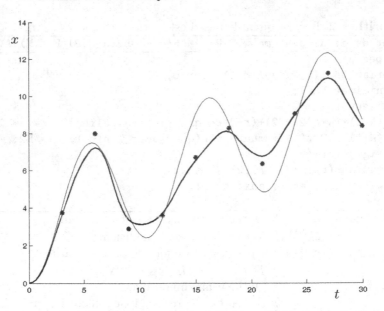

Fig. 2. Target path $x(t)$ (blue line), minimax-optimal estimate $\hat{x}(t)$ (thick black line), and observations $\{x^{obs}(t_k)\}$ (stars) on the time interval $t \in [0, T]$. (Color figure online)

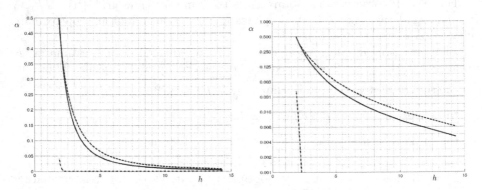

Fig. 3. Error probability α versus threshold h: Selberg-type probability bound (solid lines); Markov bound (dashed lines); Normal probability (dash-dotted lines) on a natural scale (in the left plot) and on a $\log_2$ scale (in the right plot).

It is important to compare these results to those obtained under the assumption that noises are normally distributed (see the dash-dotted line in Fig. 3). For instance, the normal error probability is under 10^{-3} for threshold h near 2.5 and and less than 10^{-6} for h near 5.

Thus, the normal hypothesis leads to wrongly optimistic conclusions. This can be explained by the strong effect of independency, which follows from zero correlation for any jointly Gaussian random variables. In contrast, the hypothesis of uncertain joint distribution does not imply a statistical redundancy of uncorrelated observations.

6   Conclusion

The main contribution of this study is the worst-case probability analysis and optimization of linear estimators in the situation when the true noise distribution is unknown and only expectations and variances are available. More specifically, the main result describes a closed-form convex programming formulation for the distributionally robust problem of estimating a bounded signal by probability and quantile criteria.

References

1. Anan'ev, B.I.: Multistep specific stochastic inclusions and their multiestimates. Autom. Remote. Control. **68**(11), 1891–1899 (2007)
2. Arkhipov, A.S., Semenikhin, K.V.: Minimax linear estimation with the probability criterion under unimodal noise and bounded parameters. Autom. Remote. Control. **81**(7), 1176–1191 (2020)
3. Arkhipov, A.S., Semenikhin, K.V.: A multivariate Chebyshev bound of the Selberg form. Autom. Remote. Control. **83**(8), 1180–1199 (2022)
4. Blanchet, J., Murthy, K., Si, N.: Confidence regions in Wasserstein distributionally robust estimation. Biometrika **109**(2), 295–315 (2022)
5. Chen, R., Paschalidis, I.C.: Distributionally robust learning. Found. Trends Optim. **4**(1–2), 1–243 (2020)
6. Delage, E., Ye, Y.: Distributionally robust optimization under moment uncertainty with application to data-driven problems. Oper. Res. **58**, 595–612 (2010)
7. El Ghaoui, L., Oks, M., Oustry, F.: Worst-case value-at-risk and robust portfolio optimization: a conic programming approach. Oper. Res. **51**(4), 543–556 (2003)
8. Grant, M.C., Boyd, S.P.: The CVX Users' Guide. Release 2.1. CVX Research, Inc. (2018). http://cvxr.com/cvx
9. Kibzun, A.I., Kan, Y.S.: Stochastic Programming Problems with Probability and Quantile Functions. Wiley, London (1996)
10. Kitahara, T., Mizuno, S., Nakata, K.: Quadratic and convex minimax classification problems. J. Oper. Res. Soc. Jpn **51**(2), 191–201 (2008)
11. Kogan, M.M.: Robust estimation and filtering in uncertain linear systems under unknown covariations. Autom. Remote. Control. **76**(10), 1751–1764 (2015). https://doi.org/10.1134/S0005117915100033
12. Lin, F., Fang, X., Gao, Z.: Distributionally robust optimization: a review on theory and applications. Numer. Algebra Control Optim. **12**(1), 159–212 (2022)
13. Löfberg, J.: YALMIP: software for solving convex (and nonconvex) optimization problems. In: Proceedimgs of American Control Conference (ACC 2006). Mineapolis, USA (2006)
14. Marshall, A.W., Olkin, I.: Multivariate Chebyshev inequalities. Ann. Math. Stat. **31**(4), 1001–1014 (1960)
15. Medvedeva, N.V., Timofeeva, G.A.: Comparison of linear and nonlinear methods of confidence estimation for statistically uncertain systems. Autom. Remote. Control. **68**(4), 619–627 (2007)
16. Pankov, A.R., Semenikhin, K.V.: Minimax estimation by probabilistic criterion. Autom. Remote. Control. **68**(3), 430–445 (2007)
17. Parys, B.P.G.V., Esfahani, P.M., Kuhn, D.: From data to decisions: distributionally robust optimization is optimal. arXiv:1704.04118v3 (2019)

18. Rahimian, H., Mehrotra, S.: Frameworks and results in distributionally robust optimization. Open J. Math. Optim. **3**(4), 1–85 (2022)
19. Semenikhin, K.V.: Minimax nature of the linear estimates of the indefinite stochastic vector from the generalized probabilistic criteria. Autom. Remote. Control. **68**(11), 1970–1985 (2007)
20. Vandenberghe, L., Boyd, S., Comanor, K.: Generalized Chebyshev bounds via semidefinite programming. SIAM Rev. **49**(1), 52–64 (2007)
21. Zymler, S., Kuhn, D., Rustem, B.: Distributionally robust joint chance constraints with second-order moment information. Math. Program. **137**, 167–198 (2013)

Scheduling

Approximation Algorithms
for Two-Machine Proportionate Routing
Open Shop on a Tree

Ilya Chernykh[1]([✉]) [ID], Olga Krivonogova[1] [ID], and Anna Shmyrina[2]

[1] Sobolev Institute of Mathematics, Koptyug ave. 4, Novosibirsk 630090, Russia
`idchern@math.nsc.ru`
[2] Novosibirsk State University, Pirogova st. 2, Novosibirsk 630090, Russia

Abstract. The routing open shop problem is a natural generalization of
the open shop problem and the metric traveler salesman problem. Jobs
are located at the nodes of a transportation network, which has to be
traversed by mobile machines in order to process the operations of the
jobs, similar to the classic open shop environment. We consider the proportionate special case of this problem, in which for each job processing
times of its operations are equal. This problem is known to be NP-hard
even for the simplest case with two machines and two nodes.

We present the tight optima localization interval for the two-machine
problem with asymmetric transportation network being arbitrary tree,
yielding a $\frac{7}{6}$-approximation algorithm for this problem. Surprisingly, the
same result holds for a more general problem where travel times are
machine-dependent under the special condition, when one machine is
"not faster" than the other. This stands in contrast to the general routing open shop (without the proportionate condition), for which optima
localization intervals are different for identical travel times and uniform
travel times cases.

Keywords: Proportionate open shop · Routing open shop · Unrelated
travel times · Approximation algorithms · Optima localization

1 Introduction

The routing open shop with proportionate processing times was introduced in
[20]. This paper is a result of direct development of the research, carried out in
[20].

The problem under consideration, namely the routing open shop, is introduced in [1,2] and can be described as follows. A fleet of mobile machines have
to perform operations on each of jobs, located at the nodes of a given transportation network. The processing and travel times are given in advance. Operations

The research was supported by Russian Science Foundation grant N 22-71-10015,
https://rscf.ru/en/project/22-71-10015/.

of every job and every machine can be performed in arbitrary sequences, provided that no two dependent operations (*i.e.* belonging to the same job or the same machine) are processed simultaneously (hence open shop). Machines have to travel along the edges of the network, starting from the common *depot*, and have to come back after performing all operations (hence routing). The goal is to construct a feasible schedule with minimal *makespan*, which is the completion moment of the last machine's activity, which is either traveling back to the depot or performing an operation of job, located at the depot.

This problem is a generalization of the metric traveling salesman problem [11] (the case with zero processing times) and classical open shop scheduling problem to minimize finish time [12] (if the traveling times are zeros, or the transportation network is trivial and consists of a single depot). According to the well-known three-field notation for scheduling problems [15], open shop with m machines is denoted as $Om||C_{max}$, where C_{max} is the maximum completion time of a job, and therefore is the makespan for classical scheduling problems. The open shop problem is introduced in [12] and has a long history of investigation. In the case of two machines ($O2||C_{max}$) it can be solved to the optimum in linear time [12,13,19,22,23], and the optimal makespan coincides with the standard lower bound $\bar{C}$, which is the maximum of machine loads and job durations (see Sect. 2 for details). The three-machine problem ($O3||C_{max}$) is NP-hard [12], although it is not yet known whether it is strongly NP-hard. On the other hand, if the number of machines is the part of an input, then the problem ($O||C_{max}$) is strongly NP-hard even in the case, when processing times do not exceed 2 [24].

In our research we focus on the special case of open shop, namely *proportionate* open shop. In this case processing times of operations are independent on machines. While the commonly used notation for the proportionate open shop is $Om|prpt|C_{max}$, we follow another notation, suggested in [21]: $Om|j\text{-}prpt|C_{max}$, as it helps to distinguish this case from another kind of proportionate open shop, in which processing times are independent on jobs. Some papers (*e.g.* [17,18]) use notation $Om|prpt|C_{max}$ for the latter problem, which can be confusing. Three-machine proportionate open shop ($O3|j\text{-}prpt|C_{max}$) is also NP-hard. The classic prove of NP-hardness of $O3||C_{max}$ from [12] can be easily adapted to the proportionate open shop; formally it was published in [16]. On the other hand, the $O3|j\text{-}prpt|C_{max}$ can be solved in pseudopolynomial time [21], which partly answers the open question, mentioned in the previous paragraph.

We denote the routing open shop with m machines as $ROm||R_{max}$, where the makespan R_{max} is the last completion time of machine's activity (which can differ from C_{max}). Optional notation of form $G = X$ in the second field is used, when the structure of the transportation network is restricted to X (*e.g.*, $G = tree$). The $ROm||R_{max}$ problem is obviously strongly NP-hard even for the case $m = 1$, as is contains the metric TSP as a special case. However, the very restricted special case $RO2|G = K_2|R_{max}$—with two machines on a link—is still NP-hard [2]. Moreover, it was recently proved, that this special case remains NP-hard even with the proportionate restriction [20]. On the other hand, both these cases are not strongly NP-hard, as $RO2|G = K_2|R_{\max}$ and therefore its special case $RO2|j\text{-}prpt, G = K_2|R_{\max}$ admit constructing of FPTAS [14].

We consider the routing open shop problem with *asymmetric* travel times, denoted as $\overrightarrow{RO}||R_{\max}$, as well as its generalization with *machine-dependent* or *individual travel times*. Following [4], the latter is denoted in the second field of the three-field notation as Qtt for *uniform* or *proportional travel times* (when each machine has its own travel speed), or Rtt for *unrelated travel times*. The goal of our investigation is to find so-called *tight optima localization intervals* for various classes of inputs for the problem under consideration, *i.e.* the tightest possible intervals of form $[\bar{R}, \rho\bar{R}]$, guaranteed to contain optimal makespan for each input from the considered class. Here $\bar{R}$ is the standard lower bound for the routing open shop problem (see Sect. 2 for details), which extends the lower bound $\bar{C}$ by adding minimal necessary travel times.

For scheduling problems, the knowledge of tight optima localization interval $[LB, \rho LB]$ (here LB is a corresponding lower bound on the optimal makespan) often leads to an algorithm with linear running time, which obtains a feasible solution with makespan from that interval. Therefore, such algorithm is not only ρ-approximate, but it also has the best theoretically possible approximation ratio in terms of LB. So, although in this paper we focus on finding tight optima localization intervals, the reader should keep in mind, that the by-products of this research are actual linear-time approximation algorithms, which justify the title of this paper.

Previously known results of this type can be found at Table 1. Note that the results from the right part of the table also apply to the problem with uniform travel times (Qtt).

Table 1. Optima localization for the two-machine routing open shop problem with identical and individual travel times.

Problem	Opt. loc.	Ref.	Problem with Qtt/Rtt	Opt. loc.	Ref
$\overrightarrow{RO2}|G = K_2|R_{\max}$	$[\bar{R}, 6/5\bar{R}]$	[1]	$RO2|G = K_2, Rtt|R_{\max}$	$[\bar{R}, 5/4\bar{R}]$	[9]
$RO2|G = K_3|R_{\max}$	$[\bar{R}, 6/5\bar{R}]$	[8]	$RO2|G = K_3, Rtt|R_{\max}$	$[\bar{R}, 5/4\bar{R}]$	[9]
$RO2|G = tree|R_{\max}$	$[\bar{R}, 6/5\bar{R}]$	[6]	$RO2|G = tree, Rtt|R_{\max}$	$[\bar{R}, 5/4\bar{R}]$	[7]
$RO2|j\text{-}prpt, G = K_2|R_{\max}$	$[\bar{R}, 7/6\bar{R}]$	[20]	$RO2|j\text{-}prpt, G = K_2, Rtt|R_{\max}$	?	
$RO2|j\text{-}prpt, G = K_3|R_{\max}$	$[\bar{R}, 7/6\bar{R}]$	[20]	$RO2|j\text{-}prpt, G = K_3, Rtt|R_{\max}$	?	

In this paper we conduct the further investigation of the $\overrightarrow{RO2}||R_{\max}$ and $\overrightarrow{RO2}|Rtt|R_{\max}$ problems. First, we extend the optima localization result from [20] on the $\overrightarrow{RO2}|G = tree|R_{\max}$ problem (Sect. 3). Next, we show, that the same interval $[\bar{R}, 7/6\bar{R}]$ also contains optima for a large subclass of instances of the $\overrightarrow{RO2}|j\text{-}prpt, G = tree, Rtt|R_{\max}$ problem. That class includes in particular all instances of the problem with proportional travel times $\overrightarrow{RO2}|j\text{-}prpt, G = tree, Qtt|R_{\max}$ (Sect. 4). This stands in contrast to the known cases of the $RO||R_{\max}$, where individual travel times (even when travel times are uniform) affect the optima localization interval (see Table 1).

2   Preliminary Notes

2.1   Problem Formulation and Notation

Let us give the formal description of the two-machine proportionate routing open shop problem with asymmetric transportation network $\overrightarrow{RO2}|j\text{-}prpt|R_{\max}$.

The machines M_1 and M_2 and the set of jobs $\mathcal{J} = \{J_1, \ldots, J_n\}$ are given. Each job J_j consists of two operations a_j and b_j with the same processing time p_j, to be performed by the machines M_1 and M_2 respectively. Notation $p_j(I)$ is used when referring to some specific instance I. A transportation network is represented by connected graph $G = \langle V, E \rangle$. For each edge $e = [u, v] \in E$ and for each machine M_i, two non-negative integers $\mathbf{dist}_i(u, v)$ and $\mathbf{dist}_i(v, u)$ are given. For some specific instance I we use notation $\mathbf{dist}_i(I; u, v)$ and $\mathbf{dist}_i(I; v, u)$. The functions $\mathbf{dist}_i$ satisfy the triangle inequality. We distinguish the following cases:

- Identical travel times, $\overrightarrow{RO2}|j\text{-}prpt|R_{\max}$: $\mathbf{dist}_1 \equiv \mathbf{dist}_2$. In this case, we use common notation $\mathbf{dist}$ for both machines.
- Proportional (or uniform) travel times, $\overrightarrow{RO2}|j\text{-}prpt, Qtt|R_{\max}$: $\mathbf{dist}_1 \equiv \alpha\mathbf{dist}_2$. In this case, without loss of generality we assume $\alpha \geqslant 1$.
- Unrelated travel times, $\overrightarrow{RO2}|j\text{-}prpt, Rtt|R_{\max}$: $\mathbf{dist}_1$ and $\mathbf{dist}_2$ are two independent distance functions.

Jobs are distributed among the nodes of G, each node contains at least one job. The machines are mobile and are initially located at the predefined *depot* $v_0 \in V$. Machines have to travel between the nodes, processing jobs in an open shop environment. It takes $\mathbf{dist}_i(u, v)$ time units for machine M_i to reach node v from u. Machines can travel over the same edge simultaneously in any direction, and are allowed to use the same edge multiple times, as well as bypassing nodes. Therefore we assume, that machines take the shortest paths while traversing G, and use the same notation $\mathbf{dist}_i(u, v)$ even for non-adjacent nodes.

A schedule S is defined by specifying the starting times $s(O)$ for each operation O. We also use notation $s_{j1} = s(a_j)$ and $s_{j2} = s(b_j)$. For each job J_j, completion times of its operations a_j and b_j are denoted by $c_{ji} = s_{ji} + p_j$. The *return moment* of machine M_i in S is defined as $R_i(S) = c_{ji} + \mathbf{dist}_i(v, v_0)$, where J_j is located at node v and is the last job, processed by M_i in S. The goal is to minimize the makespan $R_{\max}(S) = \max_i R_i(S)$.

We use the following notation for any instance I of the $\overrightarrow{RO2}|j\text{-}prpt, Rtt|R_{\max}$ problem.

- $\mathcal{J}(I; v)$ – the set of jobs located at v;
- $\ell(I) = \sum\limits_{j=1}^{n} p_j$—the load of both machines M_1 and M_2;
- $d_j(I) = 2p_j$—the duration of job J_j; $d_{\max}(I; v) = \max\limits_{J_j \in \mathcal{J}(I;v)} d_j(I)$—the maximum job duration at node v;
- $T_i^*(I)$—the weight of the optimal cyclic route over the graph G for machine M_i (the corresponding TSP optimum).

– $\Delta(I; v) = \displaystyle\sum_{J_j \in \mathcal{J}(I;v)} d_j(I)$—the load of node v; $\Delta(I) = \displaystyle\sum_{j=1}^{n} d_j(I)$—the total
load of the instance;
– $R^*_{\max}(I)$—the optimal makespan;
– $\overleftrightarrow{\mathbf{dist}}_i(I; u, v) = \mathbf{dist}_i(I; u, v) + \mathbf{dist}_i(I; v, u)$.

We omit I from the notation in case when the context defines the instance clear enough.

For the $\overrightarrow{RO2}|j\text{-}prpt|R_{\max}$ problem we use following standard lower bound, adapted from [1]:

$$R^*_{\max} \geqslant \bar{R} = \max \left\{ \max_i (\ell + T^*_i), \max_{v \in V} \left(d_{\max}(v) + \overleftrightarrow{\mathbf{dist}}(v_0, v) \right) \right\}. \qquad (1)$$

For the $\overrightarrow{RO2}|j\text{-}prpt, Rtt|R_{\max}$ problem it can be generalized as follows:

$$R^*_{\max} \geqslant \bar{R} = \max \left\{ \max_i (\ell + T^*_i), \max_{v \in V} \left(d_{\max}(v) + \overleftrightarrow{\mathbf{dist}}_{\min}(v_0, v) \right) \right\}, \qquad (2)$$

where $\overleftrightarrow{\mathbf{dist}}_{\min}(v, u) = \min\{\mathbf{dist}_1(v, u) + \mathbf{dist}_2(u, v), \mathbf{dist}_2(v, u) + \mathbf{dist}_1(u, v)\}$.

Note that (1) coincides with (2) when $\mathbf{dist}_1 \equiv \mathbf{dist}_2$ and (2) coincides with $\bar{C} = \max\{\ell, \max d_j\}$ in case when all travel distances are zero, or when the transportation network consists of a single node (in this case we have a plain open shop problem).

2.2 Instance Simplification Operations

In our research we use the instance reduction procedure, described in the next subsection. It is based on a couple of known instance simplification operations: job aggregation and terminal edge contraction. Note, that similar approach is used in a series of papers applied to different versions of the routing open shop problem, *e.g.* in [3,5,7–10].

Definition 1. *For instance I of the $\overrightarrow{RO2}|j\text{-}prpt, Rtt|R_{\max}$ problem, let $K \subseteq \mathcal{J}(I; v)$ for some node $v \in V$. Then by* job aggregation *of set K we understand the following instance transformation $I \to I'$:*

$$G(I') = G(I), \mathcal{J}(I'; v) = \mathcal{J}(I;v) \setminus K \cup \{J_K\}, p_K(I') = \sum_{J_j \in K} p_j(I).$$

Job aggregation is obviously a reversible transformation: any schedule of operation of a composite job J_K can be treated as a schedule of respective operations of jobs from set K, processed without any idle time in an arbitrary sequence. However, job aggregation can lead to the growth of the standard lower bound, since it is possible that $d_{\max}(I'; v) = d_K > d_{\max}(I; v)$. Using (2) we obtain the following sufficient and necessary condition for preservation of the standard lower bound $\bar{R}$ during the job aggregation transformation:

$$\bar{R}(I') = \bar{R}(I) \iff \sum_{J_j \in K} d_j(I) \leqslant \bar{R}(I) - \overleftrightarrow{\mathbf{dist}}_{\min}(v_0, v). \qquad (3)$$

Definition 2. *A node v from $G(I)$ of some problem instance I is* overloaded *if*

$$\Delta(I;v) > \bar{R}(I) - \overleftrightarrow{\mathbf{dist}}_{\min}(I; v_0, v).$$

Otherwise the node is referred to as underloaded.

An instance, for which any job aggregation increases the standard lower bound, is called *irreducible*.

Now we describe the *terminal edge contraction* operation for the asymmetric problem with unrelated travel times. This operation was described in [3,7] for the problem with symmetric distances. Here we use notation p_{ji}, $i = 1,2$ to denote the processing times of operations a_j and b_j, respectively. In this case $d_j = p_{j1} + p_{j2}$. Obviously, for the proportionate processing times we have $p_{j1} = p_{j2} = p_j$.

Definition 3. *For instance I of the $\overrightarrow{RO2}|Rtt|R_{\max}$ problem, let $v \neq v_0$ be a terminal node in $G(I)$, and there is a single job $J_j \in \mathcal{J}(I;v)$. Let $e = [u,v]$ be the edge incident to v. Then by the* contraction of edge e *we understand the following instance transformation $I \rightarrow I'$:*

$$\mathcal{J}(I';u) = \mathcal{J}(I;u) \cup \{J_j\}, G(I') = G(I)\backslash\{v\},$$

$$p_{ji}(I') = p_{ji}(I) + \overrightarrow{\mathbf{dist}}_i(u,v), i = 1,2.$$

The transformation described is clearly reversible. Similar to Definition 2 we use the following

Definition 4. *For instance I of the $\overrightarrow{RO2}|Rtt|R_{\max}$ problem, let $v \neq v_0$ be a terminal node in $G(I)$ and there is a single job $J_j \in \mathcal{J}(I;v)$. Let $e = [u,v]$ be an edge incident to v. Then edge e is* overloaded *if*

$$d_j(I) + \overleftrightarrow{\mathbf{dist}}_1(I;u,v) + \overleftrightarrow{\mathbf{dist}}_2(I;u,v) + \overleftrightarrow{\mathbf{dist}}_{\min}(I;v_0,u) > \bar{R}(I),$$

and underloaded *otherwise.*

Overloaded elements make the instance somehow problematic, hindering instance simplification. Fortunately, the number of such elements is rather small. It was proved in [3] that any instance I of the $RO2||R_{\max}$ problem contains at most one overloaded element (either node or edge). In [7] this result was generalized for the problem with unrelated travel times. To adapt this proof to our asymmetric case, it is sufficient to replace $2\mathbf{dist}(u,v)$ with $\overleftrightarrow{\mathbf{dist}}(u,v)$ for identical travel times and with $\overleftrightarrow{\mathbf{dist}}_{\min}(u,v)$ for individual travel times.

2.3   Instance Reduction Procedure

The following instance reduction procedure was described for the $RO2||R_{\max}$ problem in [3], and later adapted to the case with unrelated travel times in [7]. Here we describe this procedure for our problem $\overrightarrow{RO2}|Rtt|R_{\max}$.

Algorithm $\mathcal{A}$.

Input: An instance I of $\overrightarrow{RO2}|Rtt|R_{\max}$ problem.
Output: The reduced instance I'.

1. **For each** underloaded $v \in V$ perform the job aggregation of the set $\mathcal{J}(v)$.
2. **For each** terminal node $v \neq v_0$ with single job and its incident edge $e = [u, v]$,
   If e is underloaded, then
   (a) Perform the contraction of e,
   (b) **If** u is underloaded, **then** perform the job aggregation of the set $\mathcal{J}(u)$.
3. **If** some v is overloaded **then** perform the job aggregations in $\mathcal{J}(v)$ to obtain an irreducible instance (as described, *e.g.* in [8]).

Note that this algorithm can easily be implemented to run $O(n)$ time.

The following lemma describes possible outputs of algorithm $\mathcal{A}$, applied to any instance of the $RO2|G = tree, Rtt|R_{\max}$ problem.

Lemma 1 ([3,7]). *Let I be an instance of the $RO2|G = tree, Rtt|R_{\max}$ and $\tilde{I}$ is obtained from I by the algorithm $\mathcal{A}$. Then $\bar{R}(I) = \bar{R}(\tilde{I})$ and the graph $G(\tilde{I})$ satisfies exactly one of the following conditions:*

1. *$G(\tilde{I})$ has a single node v_0.*
2. *$G(\tilde{I})$ is a chain, connecting v_0 with an overloaded node v, and each node contains only one job except v, which contains two or three jobs.*
3. *$G(\tilde{I})$ is a chain, connecting v_0 with a node v with single job at each node, and the edge incident to v is overloaded.*

The following theorem describes a class of instances of $RO2|G = tree|R_{\max}$ problem such that $R_{\max}(S) = \bar{R}(I)$:

Theorem 1 ([3]). *Let I be an instance of the $RO2|G = tree|R_{\max}$ problem, $\tilde{I}$ is obtained from I by the algorithm $\mathcal{A}$, and one of the following conditions is true:*

1. *$G(\tilde{I})$ has a single node v_0,*
2. *$G(\tilde{I})$ is a chain, connecting v_0 with an overloaded node v, which contains exactly three jobs,*
3. *$G(\tilde{I})$ is a chain, connecting v_0 with a node v with single job at each node, and the edge incident to v is overloaded.*

Then one can in linear time build a feasible schedule $S(\tilde{I})$ such that $R_{\max}(S) = \bar{R}(I)$.

Theorem 1 was adapted to the problem $RO2|Rtt|R_{\max}$ (with symmetric distances) in [7]. Similar to the remark in the previous subsection, it can be also easily generalized for the $\overrightarrow{R}O2|G = tree|R_{\max}$ problem: one just need to substitute $2\mathbf{dist}(v_i, v_j)$ with $\overleftrightarrow{\mathbf{dist}}_{\min}(v_i, v_j)$ and use standard lower bound (2). Note that algorithm $\mathcal{A}$ preserves the proportionate property for the problem with identical travel times, and breaks it for the case of individual travel times.

3   Optima Localization for the Problem with Identical Travel Times

Hereafter we use standard approach of describing schedules via so-called *templates*, given by directed graphs on the set of operations, providing a partial order on them such that, for any job and for any machine its operations are linearly ordered. For each template a case analysis is applied, in order to assess the makespan of corresponding early schedule as a weight of possible critical path. Two auxiliary nodes S and F are used to denote start and finish times, correspondingly. This method is described in details *e.g.* in [7,8,10].

Theorem 2. *For any instance of the $\overrightarrow{R}O2|j\text{-}prpt, G = tree|R_{\max}$ problem a feasible schedule S with $R_{\max}(S) \leqslant \frac{7}{6}\bar{R}$ can be constructed in linear time $O(n)$.*

Proof. According to Lemma 1 and Theorem 1 it is sufficient to prove the claim just for irreducible instances on a chain, connecting depot with an overloaded node, which contains exactly two jobs.

Now consider irreducible instance I of $\overrightarrow{R}O2|j\text{-}prpt, G = chain|R_{\max}$ such that $G(I) = (v_0, \ldots, v)$, there $\mathcal{J}(v_t) = \{J_t\}$ for each $t = 0, \ldots, k$, and $\mathcal{J}(v) = \{J_\alpha, J_\beta\}$ with $T^* = \overleftrightarrow{\mathbf{dist}}(v_0, v)$. Note that

$$2p_\alpha + 2p_\beta + \overrightarrow{\mathbf{dist}}(v_0, v) > \bar{R}, \tag{4}$$

due to the fact that v is overloaded.

Without loss of generality assume

$$p_\alpha \geqslant p_\beta. \tag{5}$$

Note that due to (2)

$$\Delta = 2\ell \leqslant 2(\bar{R} - T^*). \tag{6}$$

Construct an early schedule S_1 according to the partial order of the operations from the template $\mathcal{H}_1$ (see Fig. 1).

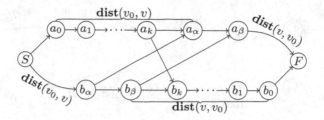

Fig. 1. Template $\mathcal{H}_1$.

Assuming $R_{\max}(S_1) > \bar{R}$ (otherwise the theorem is proved) and using (6), (4) and (5), we have

$$R_{\max}(S_1) = 2p_\alpha + p_\beta + \overleftrightarrow{\mathbf{dist}}(v, v_0) = 2p_\alpha + p_\beta + T^*. \tag{7}$$

Indeed, $2p_\alpha + p_\beta + \overleftrightarrow{\mathbf{dist}}(v, v_0) \geqslant p_\alpha + 2p_\beta + \overleftrightarrow{\mathbf{dist}}(v, v_0)$ due to (5) and

$$\sum_{j=0}^{k} p_k + \sum_{j=k}^{0} p_k + \mathbf{dist}(v_0, v_k) + \mathbf{dist}(v_k, v_0) \leqslant \Delta - d_\alpha - d_\beta + T^* < \bar{R}$$

due to (6) and (4).

Construct an early schedule S_2 according to the partial order of the operations from the template $\mathcal{H}_2$ (see Fig. 2).

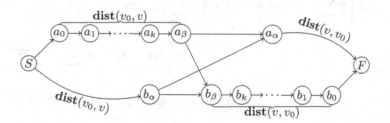

Fig. 2. Template $\mathcal{H}_2$.

Assuming $R_{\max}(S_2) > \bar{R}$ (otherwise the theorem is proved), we have

$$R_{\max}(S_2) = 2\sum_{j=0}^{k} p_j + 2p_\beta + T^*. \tag{8}$$

Similarly, construct schedule S_3 based on the template $\mathcal{H}_3$ (see Fig. 3).

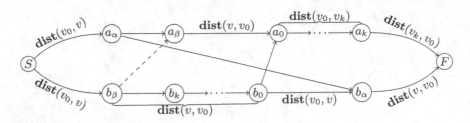

Fig. 3. Template $\mathcal{H}_3$.

Again, assume $R_{\max}(S_3) > \bar{R}$ (otherwise the theorem is proved). Note that, due to (5) $c_{\alpha 1} \geqslant c_{\beta 2}$. Also travel time of M_1 is less than travel time of M_2 by $\overleftrightarrow{\mathrm{dist}}(v_k, v)$. Therefore we have

$$R_{\max}(S_3) = \max\left\{ 2\sum_{j=0}^{k} p_j + p_\beta + 2T^* - \overleftrightarrow{\mathrm{dist}}(v_k, v), \ell + 2T^* \right\}.$$

Case 1. $R_{\max}(S_3) = 2\sum_{j=0}^{k} p_j + p_\beta + 2T^* - \overrightarrow{\mathrm{dist}}(v_k, v)$.
Let $R = \min\{R_{\max}(S_1), R_{\max}(S_2), R_{\max}(S_3)\}$. Then, using (5) we obtain

$$6R \leqslant 3R_{\max}(S_1) + 2R_{\max}(S_2) + R_{\max}(S_3) \leqslant 3(2p_\alpha + p_\beta + T^*)$$
$$+ 2\left(2\sum_{j=0}^{k} p_j + 2p_\beta + T^* \right) + \left(2\sum_{j=0}^{k} p_j + p_\beta + 2T^* \right)$$
$$= 6p_\alpha + 8p_\beta + 6\sum_{j=0}^{k} p_j + 7T^* \leqslant 7\bar{R}.$$

Therefore $R \leqslant \frac{7}{6}\bar{R}$, and the best schedule among S_1, S_2, S_3 satisfy the claim of the theorem.

Case 2. $R_{\max}(S_3) = \ell + 2T^*$.
Again, let $R = \min\{R_{\max}(S_1), R_{\max}(S_2), R_{\max}(S_3)\}$. Then, using (5) we have

$$6R \leqslant 2R_{\max}(S_1) + 3R_{\max}(S_2) + R_{\max}(S_3) = 2(2p_\alpha + p_\beta + T^*)$$
$$+ 3\left(2\sum_{j=0}^{k} p_j + 2p_\beta + T^* \right) + (\ell + 2T^*) \leqslant 7\ell + 7T^* \leqslant 7\bar{R}.$$

Therefore $R \leqslant \frac{7}{6}\bar{R}$.

Note that both Algorithm $\mathcal{A}$ and constructions of early schedules require at most $O(n)$ running time. This concludes the proof.   □

Note that in [20] an instance $\tilde{I}$ of $RO|j\text{-}prpt, G = K_2|R_{\max}$ problem is described, for which $R_{\max}^*(\tilde{I}) = \frac{7}{6}\bar{R}(\tilde{I})$. Therefore, it follows from Theorem 2 that $[\bar{R}, \frac{7}{6}\bar{R}]$ is the tight optima localization interval for the $\overrightarrow{RO}|j\text{-}prpt, G = tree|R_{\max}$ problem, as well as for $RO|j\text{-}prpt, G = K_2|R_{\max}$.

4   Unrelated Travel Times

As noted before, we can't apply terminal edge contraction operation to the problem with individual travel times, as it would violate the $j\text{-}prpt$ property. Instead, in this case we reduce the problem with unrelated travel times to the problem with identical travel times in the following manner: for each $e = [u, v] \in E$ we replace both $\mathbf{dist}_1(u, v)$ and $\mathbf{dist}_2(u, v)$ with the same $\max\{\mathbf{dist}_1(u, v), \mathbf{dist}_2(u, v)\}$. However, such transformation can significantly increase the standard lower bound. To avoid this, we consider a special case of the $\overrightarrow{RO2}|Rtt|R_{\max}$ problem with so-called *comparable distances*.

Definition 5. *Distance functions* $\mathbf{dist}_1$ *and* $\mathbf{dist}_2$ *are called* comparable, *if*

$$\forall u, v \in V \ \ \mathbf{dist}_1(u, v) \geqslant \mathbf{dist}_2(u, v). \tag{9}$$

Theorem 3. *For any instance I of the $\overrightarrow{RO2}|j\text{-}prpt, G = tree, Rtt|R_{\max}$ with comparable distances a feasible schedule S with $R_{\max}(S) \leqslant \frac{7}{6}\bar{R}$ can be built in linear time.*

Proof. Let I be instance with comparable distances. Then by (2)

$$\bar{R}(I) = \max\left\{\ell + T_1^*, \max_{v \in V}(d_{\max}(v) + \overrightarrow{\mathbf{dist}}_{\min}(v_0, v))\right\}. \tag{10}$$

Consider simplified instance I', which is identical to I except $\mathbf{dist}_2(I') = \mathbf{dist}_1(I)$. Clearly I' is the instance of the $\overrightarrow{RO2}|j\text{-}prpt, G = tree|R_{\max}$ problem with identical travel times. Moreover, any schedule, feasible for I', is also feasible for the initial instance I due to (9). By (1) we have

$$\bar{R}(I') = \max\left\{\ell + T_1^*, \max_{v \in V}(d_{\max}(v) + \overleftarrow{\mathbf{dist}}_1(v_0, v))\right\}. \tag{11}$$

Note that, due to (9) $\bar{R}(I') \geqslant \bar{R}(I)$.

Case 1. $\bar{R}(I') = \bar{R}(I)$.
We can apply Theorem 2 to instance I' and build a feasible schedule S with $R_{\max}(S) \leqslant \frac{7}{6}\bar{R}(I') = \frac{7}{6}\bar{R}(I)$ in linear time. Due to the remark above, S is also feasible for the initial instance I. Inequality (9) implies, that the makespan of this schedule for instance I is not greater than $R_{\max}(S)$ for instance I'.

Case 2. $\bar{R}(I') > \bar{R}(I)$.
Let's prove that in this case one can easily build an optimal schedule of makespan $\bar{R}$.

By (10) and (11) inequality $\bar{R}(I') > \bar{R}(I)$ implies $\bar{R}(I') = d_{max}(v) + \overleftarrow{\text{dist}}_1(v_0, v)$ for some $v \in V$. Let $d_{max}(v) = d_1 = 2p_1$.

Without loss of generality assume $\overleftrightarrow{\text{dist}}_{min}(v_0, v) = \text{dist}_1(v_0, v) + \text{dist}_2(v, v_0)$. (Otherwise we change the numeration of machines.) Since $G = tree$, there exists a closed route over G, which goes straight to node v first and uses each edge exactly twice (therefore it is optimal). Enumerate the jobs according to this route, starting with J_1. Now consider an early schedule S, in which machine M_1 processes jobs in direct order $J_1, \ldots, J_n$, while M_2 processes jobs in the opposite order $J_n, \ldots, J_1$ (see Fig. 4).

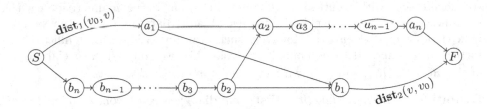

Fig. 4. Template for Theorem 3, Case 2.

Since $d_1 + \text{dist}_1(v_0, v) + \text{dist}_2(v, v_0) > \ell + T_1^*$, in schedule S we have $c_{11} > c_{22}$. By the same assumption machine M_2 finishes last, and

$$R_{max}(S) = 2p_1 + \text{dist}_1(v_0, v) + \text{dist}_2(v, v_0) = d_1 + \overleftrightarrow{\text{dist}}_{min}(v_0, v) = \bar{R}.$$

$\square$

Note that condition (9) automatically holds for any instance of $\overrightarrow{RO2}|G = K_2, Rtt|R_{max}$ and $\overrightarrow{RO2}|Qtt|R_{max}$. Therefore we have the following corollaries.

Corollary 1. *Let I be an instance of the $\overrightarrow{RO2}|j\text{-}prpt, G = K_2, Rtt|R_{max}$. Then $R_{max}^*(I) \leqslant \frac{7}{6}\bar{R}$, and a feasible schedule S with $R_{max}(S) \leqslant \frac{7}{6}\bar{R}$ can be built in linear time.*

Corollary 2. *Let I be an instance of the $\overrightarrow{RO2}|j\text{-}prpt, G = tree, Qtt|R_{max}$. Then $R_{max}^*(I) \leqslant \frac{7}{6}\bar{R}$, and a feasible schedule S with $R_{max}(S) \leqslant \frac{7}{6}\bar{R}$ can be built in linear time.*

5   Conclusion

The main result of this paper is Theorem 2, which generalizes the optima localization interval $[\bar{R}, \frac{7}{6}\bar{R}]$, known from [20] for the $RO2|j\text{-}prpt, G = K_2|R_{max}$ problem on $\overrightarrow{RO}|j\text{-}prpt, G = tree|R_{max}$. Note, that due to the following considerations, routing problem with $G = tree$ also generalizes the problem on a triangular transportation network $G = K_3$.

Consider a triangular network as shown at Fig. 5, a). Add a dummy node v and consider the new network as shown at Fig. 5, b). If we set $x = 1/2(\tau - \mu + \nu)$, $y = 1/2(\tau + \mu - \nu)$ and $z = 1/2(-\tau + \mu + \nu)$, then the star network is metrically equivalent to the triangular one. We may assume, that dummy node v contains dummy job with zero processing times, solve the problem for the star, exclude the dummy job and obtain feasible solution with the same makespan for the initial triangular network. In this sense we may say, that the $\overrightarrow{RO}|j\text{-}prpt, G = tree|R_{\max}$ problem is a generalization of the $RO2|j\text{-}prpt, G = K_3|R_{\max}$ problem.

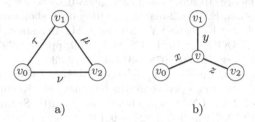

a)   b)

Fig. 5. The connection between triangular (a) and star-like (b) transportation networks.

Theorem 3 provides the optima localization result for a special *comparable* case of the problem with unrelated travel times. Although we still don't know the answer for the $\overrightarrow{RO}|j\text{-}prpt, G = tree, Rtt|R_{\max}$ problem (and even for its special case $RO|j\text{-}prpt, G = K_3, Rtt|R_{\max}$), Corollaries 1 and 2 look curious enough, showing that addition of individual travel times do not change the optima localization interval for those cases. Of course, discovering tight optima localization for the $\overrightarrow{RO}|j\text{-}prpt, G = tree, Rtt|R_{\max}$ problem with incomparable distances would be of most interest for future research.

References

1. Averbakh, I., Berman, O., Chernykh, I.: A 6/5-approximation algorithm for the two-machine routing open-shop problem on a two-node network. Eur. J. Oper. Res. **166**(1), 3–24 (2005). https://doi.org/10.1016/j.ejor.2003.06.050
2. Averbakh, I., Berman, O., Chernykh, I.: The routing open-shop problem on a network: complexity and approximation. Eur. J. Oper. Res. **173**(2), 531–539 (2006). https://doi.org/10.1016/j.ejor.2005.01.034
3. Chernykh, I.D., Lgotina, E.V.: Two-machine routing open shop on a tree: instance reduction and efficiently solvable subclass. Optim. Methods Softw. **36**(4), 821–841 (2020). https://doi.org/10.1080/10556788.2020.1734802
4. Chernykh, I.: Routing open shop with unrelated travel times. In: Kochetov, Y., Khachay, M., Beresnev, V., Nurminski, E., Pardalos, P. (eds.) DOOR 2016. LNCS, vol. 9869, pp. 272–283. Springer, Cham (2016). https://doi.org/10.1007/978-3-319-44914-2_22

5. Chernykh, I.: Two-machine routing open shop: how long is the optimal makespan? In: Pardalos, P., Khachay, M., Kazakov, A. (eds.) MOTOR 2021. LNCS, vol. 12755, pp. 253–266. Springer, Cham (2021). https://doi.org/10.1007/978-3-030-77876-7_17

6. Chernykh, I., Krivonogiva, O.: Optima localization for the two-machine routing open shop on a tree (in Russian) (2020). Submitted to Diskretnyj Analiz i Issledovanie Operacij

7. Chernykh, I., Krivonogova, O.: Efficient algorithms for the routing open shop with unrelated travel times on cacti. In: Jaćimović, M., Khachay, M., Malkova, V., Posypkin, M. (eds.) OPTIMA 2019. CCIS, vol. 1145, pp. 1–15. Springer, Cham (2020). https://doi.org/10.1007/978-3-030-38603-0_1

8. Chernykh, I., Lgotina, E.: The 2-machine routing open shop on a triangular transportation network. In: Kochetov, Y., Khachay, M., Beresnev, V., Nurminski, E., Pardalos, P. (eds.) DOOR 2016. LNCS, vol. 9869, pp. 284–297. Springer, Cham (2016). https://doi.org/10.1007/978-3-319-44914-2_23

9. Chernykh, I., Lgotina, E.: How the difference in travel times affects the optima localization for the routing open shop. In: Khachay, M., Kochetov, Y., Pardalos, P. (eds.) MOTOR 2019. LNCS, vol. 11548, pp. 187–201. Springer, Cham (2019). https://doi.org/10.1007/978-3-030-22629-9_14

10. Chernykh, I., Pyatkin, A.: Irreducible bin packing and normality in routing open shop. Ann. Math. Artif. Intell. **89**(8–9), 899–918 (2021). https://doi.org/10.1007/s10472-021-09759-x

11. Christofides, N.: Graph Theory: An Algorithmic Approach. Academic Press, New York (1975)

12. Gonzalez, T.F., Sahni, S.: Open shop scheduling to minimize finish time. J. ACM **23**(4), 665–679 (1976). https://doi.org/10.1145/321978.321985

13. Khramova, A.P., Chernykh, I.: A new algorithm for the two-machine open shop and the polynomial solvability of a scheduling problem with routing. J. Sched. **24**(4), 405–412 (2021). https://doi.org/10.1007/s10951-021-00694-7

14. Kononov, A.: On the routing open shop problem with two machines on a two-vertex network. J. Appl. Ind. Math. **6**(3), 318–331 (2012). https://doi.org/10.1134/s1990478912030064

15. Lawler, E.L., Lenstra, J.K., Rinnooy Kan, A.H.G., Shmoys, D.B.: Chapter 9. Sequencing and scheduling: algorithms and complexity. In: Logistics of Production and Inventory, Handbooks in Operations Research and Management Science, vol. 4, pp. 445–522. Elsevier (1993). https://doi.org/10.1016/S0927-0507(05)80189-6

16. Liu, C., Bulfin, R.: Scheduling ordered open shops. Comput. Oper. Res. **14**(3), 257–264 (1987). https://doi.org/10.1016/0305-0548(87)90029-3

17. Matta, M.E., Elmaghraby, S.E.: Polynomial time algorithms for two special classes of the proportionate multiprocessor open shop. Eur. J. Oper. Res. **201**(3), 720–728 (2010). https://doi.org/10.1016/j.ejor.2009.03.048

18. Naderi, B., Zandieh, M., Yazdani, M.: Polynomial time approximation algorithms for proportionate open-shop scheduling. Int. Trans. Oper. Res. **21**(6), 1031–1044 (2014). https://doi.org/10.1111/itor.12087

19. Pinedo, M., Schrage, L.: Stochastic shop scheduling: a survey. In: Dempster, M.A.H., Lenstra, J.K., Rinnooy Kan, A.H.G. (eds.) Deterministic and Stochastic Scheduling. NATO Advanced Study Institute Series, vol. 84, pp. 181–196. Springer, Dordrecht (1982). https://doi.org/10.1007/978-94-009-7801-0_9

20. Pyatkin, A.V., Chernykh, I.D.: On complexity of two-machine routing proportionate open shop. Siber. Electron. Math. Rep. **19**(2), 528–539 (2022). https://doi.org/10.33048/semi.2022.19.044 https://doi.org/10.33048/semi.2022.19.044

21. Sevastyanov, S.V.: Some positive news on the proportionate open shop problem. Sibirskie Elektronnye Mat. Izv. **16**, 406–426 (2019). https://doi.org/10.33048/semi.2019.16.023 https://doi.org/10.33048/semi.2019.16.023
22. Soper, A.J.: A cyclical search for the two machine flow shop and open shop to minimise finishing time. J. Sched. **18**(3), 311–314 (2013). https://doi.org/10.1007/s10951-013-0356-7
23. de Werra, D.: Graph-theoretical models for preemptive scheduling. Advances in Project Scheduling, pp. 171–185 (1989). http://infoscience.epfl.ch/record/88562
24. Williamson, D.P., et al.: Short shop schedules. Oper. Res. **45**(2), 288–294 (1997). https://doi.org/10.1287/opre.45.2.288

MIP Heuristics for a Resource Constrained Project Scheduling Problem with Workload Stability Constraints

Igor Vasilyev[1,2] , Ildar Muftahov[2] , and Anton V. Ushakov[1(✉)]

[1] Matrosov Institute for System Dynamics and Control Theory of SB RAS,
134 Lermontov str., 664033 Irkutsk, Russia
{vil,aushakov}@icc.ru
[2] Novosibirsk Research Center, Huawei Russian Research Institute,
6/1 Academician Lavrentiev av., 630090 Novosibirsk, Russia

Abstract. We address a variation of the multi-skill resource-constrained project scheduling problem motivated by the need to find a quality schedule for the construction activities related to rolling-out a large network of mobile stations. The network roll-out (NRO) is a complex industrial project aimed at deploying a set of mobile stations in certain region. This requires attracting a set of contractors, who can provide worker teams that have skills to execute specific project activities depending on their type. The problem also involves some additional specific industrial requirements related to how the contractors' workload should be distributed during the project duration (workload stability constraints). Other important industrial constraints are the precedence redundancy constraints on the project activities. We formulate this problem as a mixed-integer linear program that can be viewed as a disaggregated time-indexed integer programming formulation with pulse variables. To find quality feasible solutions of the problem, we propose a two-phase MIP heuristic consisting of (i) a relax-and-fix heuristic followed by (ii) a large neighborhood search. The effectiveness of the proposed solution approach is demonstrated in extensive computational experiments on real world problem instances.

Keywords: multiple project scheduling · large neighborhood search · relax-and-fix heuristic · network roll-out · resource-constrained project scheduling · relax-and-optimize

1 Introduction

Project management is a complex process aimed at aggregation and application of diverse methods and resources to achieve specific project goals. One of its critically important tasks is project scheduling. Usually, a project consists of a pre-defined set of activities interrelated by precedence constraints. Each activity has a specific execution time and, possibly, requires some amount of specific resources available in limited quantities. The goal is to find the best schedule for activities subject to precedence and resource constraints such that a certain project objective (e.g. makespan) is optimized. Such a problem is known as the resource-constrained project scheduling problem (RCPSP). It has been receiving much attention in the operations research community and has become a standard project scheduling problem [11]. RCPSP was proved to be NP-hard in the strong sense [7] and was first formulated as an integer program in [16]. Note that it is one of the most studied project scheduling problems (for a survey, see [10–12,17,21].

As real world projects usually involve some specific industrial requirements for how activities must be executed and supplied by the resources, there are many variations of RCPSP extending the basic model. One important extension is that the model may involve different types of resources [19]. For example, the basic RCPSP considers only one type of resources, called renewable, which is supposed to be limited only in a given time moment (interval). On the other hand, one may also consider a nonrenewable resource which capacity is limited over the entire project. A nonrenewable resource is obviously useful to model the project budget. The concept of nonrenewable resource naturally arises in a popular extension of RCPSP, called multi-mode RCPSP (for a survey see [23]). Here, there are several modes or alternatives for executing each activity, i.e. one can use different combinations of resources to finish an activity or attract more resources to reduce its execution time. Thus, each mode for an activity consists of a resource combination, whereas the execution time of the activity depends on the mode which it is started in. From the practical point of view, modes may reflect different types of materials or parts feasible to perform an activity. These resources may not only have different price but also affect the activity execution time or/and its quality.

The traditional multi-mode RCPSP, as well as RCPSP, supposes that each particular resource can be utilized in only one specific way. This is especially true if the resources are some raw materials, machines, tools, etc. However, an industrial project often involves human resources (e.g. worker teams) who usually have several skills. In such a problem, each activity requires not a combination of resources but a combination of skills. Each activity may request several skills and several units of a particular skill, but a resource with the skill can contribute to only one skill unit [8]. This extension of RCPSP is known as the multi-skill RCPSP or RCPSP with flexible resources.

Traditional objective for project scheduling problems is to minimize makespan. However, other time-based objectives, like weighted sum of all activities completion times, maximum lateness with respect to some due date, etc., are also possible. In many extensions of RCPSP, one often considers several objectives, e.g. combines a time-based objective with cost-based ones [13,15]. Other variants, called deadline problems, assume that there is a due date to complete a project, hence the objective is to minimize the total cost and/or maximize quality of executed activities.

The research field devoted to variants of the multi-skill RCPSP is quite flourishing (for a survey, see [1]). For the first time, such type of problems was addressed in [5], where employees have skill levels, whereas activities must be executed by an employee who master the required skill at a specified level. The same concept was later used in e.g. [14,25]. The basic multi-skill problem, where each resource is supposed to have several skills, was studied in e.g. [3,8]. As for other project scheduling problems, the multi-skill RCPSP may include various industrial requirements and additional objective functions rather than time-based ones only. For example, in [24] the authors address a variant where the objective function is to maximize the total effectiveness of the resource assignment with respect to skill mastering. Other interesting variations include employee learning effects or the need for additional time to switch between different skills [20].

In this paper we address a variant of the multi-skill RCPSP first considered in [22] and motivated by the need to find a quality schedule for the construction activities related to rolling-out a large network of mobile stations. Note that the network roll-out (NRO) is a complex industrial project aimed at deploying a set of mobile stations in a certain region. To construct a base station, a specific sequence of activities must be accomplished (e.g. construction of different parts of the station, equipment installation, equipment setup, etc.). To roll-out a network, one needs to attract a pre-defined set of contractors. Each contractor can provide a set of worker teams. Each worker team in turn has a certain team type responsible to what skills the team master. Thus, the number of teams of certain type is a renewable resource of the problem. We also suppose that the cost of assigning a team to a particular activity and its execution time are dependent on the team type. Obviously, the construction requires not only worker teams but also materials and tools which are also renewable resources. In addition to the standard multi-skill RCPSP constraints, the problem incorporates some additional specific industrial requirements for contractor workloads.

In this paper, we propose a new effective two-phase heuristic approach to finding high-quality solutions of this complex industrial problem. Our approach is based on the two effective MIP heuristics, relax-and-fix and large neighborhood search procedures. We have implemented and tested the proposed algorithms on several real-world NRO test instances, demonstrating that nearly optimal solutions are obtained in short running times outperforming a state-of-the-art MIP solver.

2   Problem Statement

First of all, let us formulate the aforementioned multi-skill RCPSP as a mixed integer program. The latter can be viewed as a disaggregated time-indexed integer programming formulation with pulse variables. We suppose that there is a time horizon $T = \{0, \dots, |T| - 1\}$ divided into time units $D = \{0, \dots, |D| - 1\}$. The latter can be interpreted as weeks, month, etc. The number of days in a unit $d \in D$ is given as n^d; we also denote the starting time of unit d as τ_d. The project consists in building mobile stations, each of which may have different type or characteristics but requires completing activities of the same types. We suppose that J is a set of activity types, and the number of activities of type $j \in J$

equals n_j. The execution time of an activity is denoted as p_j. Preemption is not allowed. To build mobile stations, a set of contractors C has to be attracted. A contractor $c \in C$ may provide worker teams of several types. Note that each contractor has her own types of worker teams. We denote the set of all available team types as B. As was mentioned above, a team of particular type has a limited number of skills. Each activity requires only one specific skill. Thus, the cost of assigning a team of type b to an activity of type j is c_{jb}. Each team can be assigned to only one activity at the same time. There are precedence relations between activity types. The relations are defined by the immediate predecessor activity type. We assume that an activity cannot start before its predecessor is completed and there is a time lag of f_j between the start time of an activity of some type and the finish time of its predecessor.

In addition, our multi-skill RCPSP problem incorporates some specific industrial requirements. The first one imposes limits on the ratio of the number of completed activities to the number of their completed pre-tasks in one time unit $d \in D$. In other words, at the end of a time unit d, either all the pre-task activities of a task of type j are completed or the number of completed predecessors is larger than the number of completed activities multiplied by risk factor $r_j > 0$. Obviously, the reasoning behind this type of constraints is to reduce the risk of a too long project schedule. Note that these constraints closely relate to feeding precedence relation constraints introduced in [2].

The second specific requirement relates to the workload of each contractor. Specifically, the goal is to assign worker teams to activities in such a way to obtain a specific "smooth" team utilization profile. First, during a time unit $d \in D$, the number of involved teams per interval t must be similar, i.e. the requirements for a contractor's teams should be distributed evenly during interval d. Secondly, the maximal number of the worker teams requested from a contractor per time unit has to be non-decreasing up to some time unit d and then non-increasing. Note that the latter constraints are supposed to be soft.

The additional denotations required for problem statement are the following:

1. $B^c \subset B$ is a subset of team types of contractor $c \in C$.
2. u_t^b is the number of teams of type $b \in B$ available at time t.
3. $J^b \subset J$ is a subset of activity types which a team of type $b \in B$ is able to execute (has the required skill).
4. $B^j \subset B$ is a subset of team types with the skill to execute an activity of type $j \in J$.
5. $\bar{J} \subset J$ is a set of task types, which have a preceding task j_j^* (pre-task), i.e. $J \setminus \bar{J}$ is a set of task types that are the first in construction sequence, and they do not have a pre-task.
6. $j_j^* \in J$ is a predecessor activity type of an activity of type $j \in \bar{J}$.
7. r_j is the risk factor between task of type $j \in J$ and its pre-task.

We suppose that there is a deadline up to which the project must be completed. The problem is to find a feasible schedule such that the total cost of assigning worker teams to activities is minimized subject to the resource, precedence, workload, and availability constraints. The problem can be formulated as a mixed-integer linear program. Let us introduce the integer variables x_{jbt} which equal to the number of activities of type $j \in J$ started by a team of type $b \in B$ at time $t \in T$. We also introduce the binary variables:

$$y_{jd} = \begin{cases} 1, \text{ if not all precedence activities of type } j_j^* \text{ are completed} \\ \quad\ \text{at the end of time unit d,} \\ 0, \text{ otherwise.} \end{cases} \quad j \in \bar{J}, d \in D.$$

With these notations our multi-skill RCPSP is

$$\min \sum_{j \in J} \sum_{b \in B^j} \sum_{t \in T} c_{jb} x_{jbt} + S \sum_{c \in C} \sum_{d \in D} \psi_{cd}, \tag{1}$$

$$\sum_{b \in B^j} \sum_{t \in T} x_{jbt} \geq n_j, \qquad\qquad j \subset J \tag{2}$$

$$\sum_{b \in B^{j_j^*}} \sum_{l=0}^{t-p_{j_j^*}-f_j} x_{j_j^* bl} \geq \sum_{b \in B^j} \sum_{l=0}^{t} x_{jbl}, \qquad j \in \bar{J}, t \in T \tag{3}$$

$$\sum_{j \in J^b} \sum_{l=t}^{t-p_j+1} x_{jbl} \leq u_t^b, \qquad\qquad b \in B, t \in T, \tag{4}$$

$$n_{j_j^*} y_{jd} \geq n_{j_j^*} - \sum_{b \in B^{j_j^*}} \sum_{t=0}^{\tau_{d+1}-p_{j_j^*}} x_{j_j^* bt}, \qquad j \in \bar{J}, d \in D, \tag{5}$$

$$\sum_{b \in B^{j_j^*}} \sum_{t=0}^{\tau_{d+1}-p_{j_j^*}} x_{j_j^* bt} + n_j(1 - y_{jd}) \geq r_j \sum_{b \in B^j} \sum_{t=0}^{\tau_{d+1}-1} x_{jbt}, \ j \in \bar{J}, d \in D, \tag{6}$$

$$y_{jd} \geq y_{jd+1}, \qquad\qquad j \in \bar{J}, d = 1, |D| - 1 \tag{7}$$

$$\psi_{cd} \geq \sum_{b \in B^c} \sum_{j \in J^b} \sum_{l=t-p_j+1}^{t} x_{jbl}, \qquad c \in C, d \in D, t = \tau_d, \dots, \tau_{d+1} - 1, \tag{8}$$

$$\begin{aligned} \psi_{cd} &\leq \psi_{cd+1} + M z_{cd}, \\ \psi_{cd+1} &\leq \psi_{cd} + M(1 - z_{cd}), \qquad c \in C, d = 1, \dots, |D| - 1, \\ z_{cd} &\leq z_{cd+1}, \end{aligned} \tag{9}$$

$$z_{cd} \in \{0, 1\}, \qquad\qquad c \in C, d = 1, \dots, |D| - 1, \tag{10}$$

$$x_{jbt} \in \{0, 1\}, \qquad\qquad j \in J, b \in B^j, t \in T, \tag{11}$$

$$\psi_{cd} \geq 0, \qquad\qquad c \in C, d \in D. \tag{12}$$

The objective function minimizes the overall cost of the schedule and the total sum of the maximal daily workload of contractors, thus making it as evenly distributed as possible during time units. Constraints (2) requires that all activities must be completed. Inequalities (3) impose precedence relations between activities. Indeed, the number of the activities of type j cannot be greater than the number of completed preceded activities. Constraints (4) are resource capacity constraints, i.e. they guarantee that the number of requested teams at time t does not exceed available capacity. Inequalities (5)–(7) ensure that if not all preceded activities of an activity of type j are finished at the end of d, then the number of completed preceded activities must be at least as large as the number of the activities of type j multiplied by a risk factor r_j. In particular, constraints (5) bind variables y_{jd} and x_{jbt}, and constraints (7) are natural valid inequalities. Finally, constraints (8),(9) are responsible for contracts' workload requirements. These are soft constraints and are modeled using additional variables ψ_{cd} and z_{cd}. Specifically, constraints (8) define the maximal daily workload of each contractor in time unit d, while inequalities (9) ensure that the maximal contractor workload per time unit should first be increasing and, after some peak, decreasing. The remained inequalities impose integrality and boundary constraints on decision variables.

3   Solution Algorithms

We can see that the mixed-integer program (1)–(12) is tightly constrained and, due to large size of real world industrial problem instances, may require prohibitively long time to be solved. Moreover, it involves a relatively large number of the so-called big M constraints, e.g. (5), (6) or (9), that render the natural LP relaxation weak. In our particular case, the optimal solution of the LP relaxation turns out to be "very fractional" which makes it difficult to extract some useful information about the optimal integer solution.

Nevertheless, we can leverage our knowledge of the problem structure to develop effective and efficient heuristics based on the mixed integer formulation. Such types of heuristics are often referred to as MIP heuristics and are divided into constructive and improvement ones. In this paper we develop a two-phase heuristic approach consisting of two principle components. The first one is a relax-and-fix heuristic that is used to fast find an initial feasible solution of the problem from scratch. Then, we try to improve this solution with the second component, a large neighborhood search heuristic. Our approach is designed to fast find quality feasible solutions of problem instances of real world size within reasonable running time. In addition to speed, its main benefits is flexibility and simplicity.

Relax-and-Fix is a well-known general heuristic to solve mixed-integer linear programs. Its idea is simple. First, the set of integer decision variables is partitioned into non-overlapping subsets and the order of how they have to be processed is defined. An iteration of a relaxed-and-fix heuristic consists in relaxing the integrality constraints for all but one of these subsets. The obtained sub-problem is then solved and the integer variables are fixed at their current values. This iterative process is repeated for all the subsets of integer variables. If at some iteration the corresponding subproblem is infeasible, then the heuristic is claimed to be failed. Otherwise, the solution found is feasible in the original problem.

A proper strategy of partitioning integer variables into subsets is crucially important for the relax-and-fix heuristic to be effective. For example, when dealing with problems involving time horizon, one often splits the time interval in a given number of subintervals, which introduces a partition of variables. For our problem, we develop a variant of relax-and-fix heuristic that we call *two-stage heuristic*. It employs a strategy specific for our multi-skill RCPSP. First, we note that our problem involves a large number of complex soft constraints, hence the two-stage heuristic starts by excluding them from the problem formulation. The variables x_{jbt} and y_{jd} define two input subsets of the decision variables that are sequentially relaxed and fixed. The outline of the two-stage heuristic is presented in Algorithm 1.

Algorithm 1. Two-stage relax-and-fix heuristic

Stage I: Remove constraints (8)–(10),(12) from the formulation. Relax the integrality constraints for all variables x_{jbt}. Solve the corresponding subproblem.

Stage II: Fix variables y_{jd} at the found binary values $\bar{y}_{jd}$. Return back the integrality of x_{jbt}. Solve the corresponding subproblem and find integer values $\bar{x}_{jbt}$.

Return $(\bar{y}, \bar{x})$. The objective value can be found by solving the original problem (with constraints (8)–(10),(12)) with $x_{jbt} = \bar{x}_{jbt}$ and $y_{jd} = \bar{y}_{jd}$.

Once a feasible solution is found, it is quite natural to try to find a better solution with an improvement heuristic. For our problem, we develop an approach that can be viewed as a large neighborhood search heuristic [4]. It is also often referred to as exchange heuristic, improvement version of relax-and-fix heuristic [18], or fix-and-optimize heuristic [9]. Moreover, it can be considered as a variation of conventional local search. Indeed, given an initial feasible solution (incumbent), the heuristic first defines a subproblem obtained by fixing most of integer variables (except for a small subgroup) at their values in the current incumbent. After that, the subproblem, that is much easier than the original one due to a small number of free variables, is solved with a MIP solver. If the solution found has a better objective value, it replaces the current incumbent, and the procedure is repeated, i.e. other decision variables remained free are selected. As in the relax-and-fix heuristic, the way of selecting the free variables in each iteration and their number are key components that significantly affects the heuristic performance.

In our particular case, we employ the following variable decomposition strategy. We suppose that all variables y_{jd} are free, as they can be easily defined by variables x_{jbt}. In each iteration, to define the required group of free variables x_{jbt}, we select a number of team types $b \in \hat{B}$, $\hat{B} \subset B$, uniformly at random; after that we randomly choose a time period consisting of a specific number of sequential time units $d \in \hat{D} \subset D$. All the variables x_{jbt}, $b \in \hat{B}$, $j \in J^b$, $d \in \hat{D}$ are supposed to be unfixed and the corresponding subproblem is solved with a

MIP solver. Our stopping criterion is the number of iterations (or the number of solved subproblems).

4   Computational Experiments

The proposed solution algorithms were implemented using C++ programming language (compiled with GNU C++ 12.2.0 compiler) and tested on a PC with Intel(R) Core(TM) i7-8665U and 16 GB of RAM. We use SCIP 8.0.3, distributed under Apache 2.0 license [6], as a general MIP solver. Note that it was run sequentially, i.e. using only one thread.

Our test bed consists of 10 problem instances derived from real world data. The problem statistics is presented in Table 1, where column *Name* presents problem names, column $|K|$—the number of contractors, $|B|$—the number of team types, $|J|$ – the number of activity types, $\bar{H}$—the minimal makespan in weeks. The last two columns demonstrate the parameters of our large neighborhood search heuristic that we use in our experiments, i.e. the numbers of randomly chosen team types and time units.

Table 1. Instance details

Name	$|K|$	$|B|$	$|J|$	$\bar{H}$	$|\hat{B}|$	$|\hat{D}|$
Task01	6	27	12	24	7	6
Task02	6	26	21	26	7	5
Task03	8	28	21	33	9	9
Task04	8	21	28	19	5	5
Task05	8	29	28	18	7	5
Task06	7	32	28	15	8	8
Task07	6	26	21	26	6	7
Task08	8	28	21	33	6	8
Task09	8	21	28	20	5	5
Task10	7	32	28	24	8	6

Furthermore, for our constructive heuristic we set a time limit of 300 seconds. For the MIP solver, we define a GAP limit of 1% when solving MIP subproblems in each phase of our approach. In total, we have prepared 10 different test instances. For each of them, we defined 5 scenarios depending on the deadline (makespan). In our experiments, we compare the solutions found by our two-phase solution approach with the solutions obtained with the SCIP solver and with the three-phase heuristic [22]. We did not apply any tuning of solver parameters, i.e. they are default initialized. In the large neighborhood search, we set $|\hat{B}| \approx 0.25 \cdot |B|$ and $|\hat{D}| \approx 0.25 \cdot |H|$. These values may however be increased by one or two depending on the problem complexity. Recall that the size of $\hat{B}$ and $\hat{D}$ defines the number of free x_{jbt} variables which values are supposed to improve. The stopping criterion of the large neighborhood search is the number of iterations that we set to be 120 for all problem instances. The results of computational experiments are reported in Table 2. The first column presents the problem name and the next column $Weeks$ contains the chosen makespan (project deadline in time units). Then, we report the computational results for SCIP and for each component of our two-phase heuristic approach. We run SCIP for 2 h and report GAP (in percent) between the corresponding upper UB and lower LB bounds found. Recall that GAP is computed by SCIP as $GAP = (UB - LB)/LB * 100$. For the relax-and-fix and large neighborhood search heuristics, we report run time and GAP computed using the lower bound found by SCIP. The column $Totaltime$ contains the total time of our two-phase heuristic approach for a particular instance. Besides the run time of heuristics, it also includes the time for reading input data, setting up MIP models using SCIP, etc.

As a result, we received 50 final test runs, among which only one test case 9 for makespan 20 was not solved by our approaches in the defined time and GAP limits. In Table 2 for the other 49 launches, we highlighted the best solutions in bold.

We can see that in many cases the MIP solver and the three-stage heuristic were not capable of finding feasible solutions. Test instances $test5$–$test8$ turned to be challenging for both the MIP solver and the three-stage heuristic. For example, we can observe that SCIP was not able to find even a feasible solution for $test7$ and $test8$ after two hours of run time. The three-stage heuristic found a feasible solution only for $test7$ with the largest deadline of 30 weeks. At the same time, the two-phase heuristic managed to discover quality solutions for these instances in about 5 min. The gap is about 3% for $test7$ but it is larger for $test8$. We can see that the problem $test9$ with deadline of 20 weeks turned out to be much challenging: none of the competing algorithms found any feasible solutions.

Though the relax-and-fix heuristic provides the initial solutions with relatively large gap values, its main benefit is high speed. This property allowed the two-phase approach to run much faster than the three-stage heuristic, especially on hard problem instances. In general, in 41 test runs, the best results in terms of quality and time were demonstrated by our two-phase heuristic. In 6 cases, the three-stage heuristics provided a solution with a minimum gap. In the remaining two cases, no heuristics could show the result better than the SCIP with time limit of 2 h.

Table 2. Computational results and comparison of the proposed MIP heuristics with SCIP

| | | | | | Two-phase heuristic | | | | |
| | | SCIP | Three stage | | Relax and fix | | LNS | | |
Prob.	Weeks	GAP	GAP	Time, s	GAP	Time, s	GAP	Time, s	Total time, s
test1	24	-	-	-	1.91	344	**1.17**	88	432
	25	3.70	2.28	523	3.09	509	**1.63**	150	659
	26	0.55	0.78	276	2.38	415	**0.62**	177	592
	27	-	-	-	3.65	441	**1.17**	137	578
	28	2.44	1.97	472	3.25	179	**1.27**	139	318
test2	26	-	2.28	664	17.64	398	**1.41**	105	503
	27	-	2.94	764	20.44	438	**1.59**	125	563
	28	-	-	-	26.59	485	**4.43**	111	596
	29	-	**2.33**	603	25.22	387	3.90	106	493
	30	6.85	**3.18**	698	29.16	160	5.82	125	285
test3	33	8.35	-	-	43.51	81	**2.88**	203	284
	34	-	5.29	769	45.11	225	**4.29**	159	384
	35	-	-	-	52.95	139	**3.36**	133	272
	36	10.24	-	-	51.16	107	**3.41**	142	249
	37	-	-	-	49.48	100	**2.69**	114	214
test4	19	-	-	-	17.72	104	**2.76**	113	217
	20	-	1.63	482	20.55	119	**1.12**	131	250
	21	5.12	**1.19**	281	13.01	314	1.54	151	465
	22	6.30	**1.31**	577	24.35	72	1.41	153	225
	23	4.50	**2.82**	511	29.65	202	4.77	200	402
test5	18	-	-	-	21.29	603	**4.36**	275	878
	19	8.16	-	-	15.45	146	**1.03**	238	388
	20	13.05	-	-	18.92	273	**2.54**	257	530
	21	6.24	-	-	19.83	107	**0.75**	178	285
	22	6.03	-	-	22.15	607	**4.11**	275	882
test6	15	-	-	-	11.53	249	**1.34**	197	446
	16	-	-	-	16.68	255	**2.37**	220	475
	17	16.70	-	-	17.10	344	**1.62**	225	569
	18	-	-	-	16.77	199	**0.94**	221	420
	19	13.44	6.15	906	22.14	391	**2.51**	254	645
test7	26	-	-	-	19.48	339	**1.16**	134	473
	27	-	-	-	19.23	247	**3.26**	113	360
	28	-	-	-	22.73	276	**2.95**	118	394
	29	-	-	-	21	272	**3.03**	93	365
	30	-	3.47	923	24.98	466	**2.78**	96	562
test8	33	-	-	-	40.37	161	**4.44**	153	314
	34	-	-	-	49.1	266	**2.96**	155	421
	35	-	-	-	52.71	287	**10.03**	115	402
	36	-	-	-	38.63	158	**7.87**	106	264
	37	-	-	-	41.95	150	**8.38**	91	241
test9	20	-	-	-	-	-	-	-	-
	21	9.46	6.24	709	20.93	382	**2.27**	194	576
	22	17.78	2.87	580	21.73	236	**1.43**	146	382
	23	-	-	-	24.07	295	**3.12**	154	449
	24	17.04	-	-	26.89	209	6.03	132	341
test10	24	14.15	**1.47**	601	27.4	71	2.37	124	195
	25	5.75	4.21	692	18.33	193	**1.77**	112	305
	26	**2.67**	-	-	34.07	139	5.4	180	319
	27	**3.32**	-	-	39.5	61	5.14	103	164
	28	8.72	-	-	37.56	60	**6.15**	96	156

5   Conclusion

In this paper we addressed a variation of the multi-skill resource-constrained project scheduling problem that arises in the specific industrial application. The problem includes some specific application-oriented constraints on the workload of the contractors and on the ration between the numbers of completed activities and their predecessors. We have developed an effective two-phase MIP-based solution approach consisting of a relax-and-fix heuristic and a large neighborhood search that demonstrate superior performance on real-life problem instances.

Our future research may be focused on extending the problem formulation by incorporating additional specific industrial requirements. Moreover, one can also try improve the proposed MIP heuristics by employing more sophisticated techniques for selecting the required subsets of decision variables, e.g. machine learning ones.

References

1. Afshar-Nadjafi, B.: Multi-skilling in scheduling problems: a review on models, methods and applications. Comput. Ind. Eng. **151**, 107004 (2021). https://doi.org/10.1016/j.cie.2020.107004
2. Alfieri, A., Tolio, T., Urgo, M.: A project scheduling approach to production planning with feeding precedence relations. Int. J. Prod. Res. **49**(4), 995–1020 (2011). https://doi.org/10.1080/00207541003604844
3. Almeida, B.F., Correia, I., da Gama, F.S.: Modeling frameworks for the multi-skill resource-constrained project scheduling problem: a theoretical and empirical comparison. Int. Trans. Oper. Res. **26**(3), 946–967 (2019). https://doi.org/10.1111/itor.12568
4. Avella, P., D'Auria, B., Salerno, S., Vasil'ev, I.: A computational study of local search algorithms for Italian high-school timetabling. J. Heuristics **13**, 543–556 (2007). https://doi.org/10.1007/s10732-007-9025-3
5. Bellenguez, O., Néron, E.: Lower bounds for the multi-skill project scheduling problem with hierarchical levels of skills. In: Burke, E., Trick, M. (eds.) PATAT 2004. LNCS, vol. 3616, pp. 229–243. Springer, Heidelberg (2005). https://doi.org/10.1007/11593577_14
6. Bestuzheva, K., et al.: The SCIP Optimization Suite 8.0. Technical report, Optimization Online (2021). http://www.optimization-online.org/DB_HTML/2021/12/8728.html
7. Blazewicz, J., Lenstra, J.K., Kan, A.H.G.R.: Scheduling subject to resource constraints: classification and complexity. Discret. Appl. Math. **5**(1), 11–24 (1983). https://doi.org/10.1016/0166-218X(83)90012-4
8. Correia, I., Lourenço, L.L., da Gama, F.S.: Project scheduling with flexible resources: formulation and inequalities. OR Spectr. **34**, 635–663 (2012). https://doi.org/10.1007/s00291-010-0233-0
9. Dorneles, A.P., de Araújo, O.C.B., Buriol, L.S.: A fix-and-optimize heuristic for the high school timetabling problem. Comput. Oper. Res. **52**, 29–38 (2014). https://doi.org/10.1016/j.cor.2014.06.023
10. Hartmann, S., Briskorn, D.: A survey of variants and extensions of the resource-constrained project scheduling problem. Eur. J. Oper. Res. **207**(1), 1–14 (2010). https://doi.org/10.1016/j.ejor.2009.11.005

11. Hartmann, S., Briskorn, D.: An updated survey of variants and extensions of the resource-constrained project scheduling problem. Eur. J. Oper. Res. **297**(1), 1–14 (2022). https://doi.org/10.1016/j.ejor.2021.05.004

12. Herroelen, W., De Reyck, B., Demeulemeester, E.: Resource-constrained project scheduling: a survey of recent developments. Comput. Oper. Res. **25**(4), 279–302 (1998). https://doi.org/10.1016/S0305-0548(97)00055-5

13. Li, H., Womer, K.: Optimizing the supply chain configuration for make-to-order manufacturing. Eur. J. Oper. Res. **221**(1), 118–128 (2012). https://doi.org/10.1016/j.ejor.2012.03.025

14. Lin, J., Zhu, L., Gao, K.: A genetic programming hyper-heuristic approach for the multi-skill resource constrained project scheduling problem. Expert Syst. Appl. **140**, 112915 (2020). https://doi.org/10.1016/j.eswa.2019.112915

15. Maghsoudlou, H., Afshar-Nadjafi, B., Niaki, S.T.A.: A multi-objective invasive weeds optimization algorithm for solving multi-skill multi-mode resource constrained project scheduling problem. Comput. Chem. Eng. **88**, 157–169 (2016). https://doi.org/10.1016/j.compchemeng.2016.02.018

16. Patterson, J.H., Huber, W.D.: A horizon-varying, zero-one approach to project scheduling. Manag. Sci. **20**(6), 990–998 (1974). https://doi.org/10.1287/mnsc.20.6.990

17. Pellerin, R., Perrier, N., Berthaut, F.: A survey of hybrid metaheuristics for the resource-constrained project scheduling problem. Eur. J. Oper. Res. **280**(2), 395–416 (2020). https://doi.org/10.1016/j.ejor.2019.01.063

18. Pochet, Y., Wolsey, L.A.: Production Planning by Mixed Integer Programming. Springer Series in Operations Research and Financial Engineering, Springer, New York (2006). https://doi.org/10.1007/0-387-33477-7

19. Słowiński, R.: Two approaches to problems of resource allocation among project activities - a comparative study. J. Oper. Res. Soc. **31**, 711–723 (1980). https://doi.org/10.1057/jors.1980.134

20. Tian, Y., Xiong, T., Liu, Z., Mei, Y., Wan, L.: Multi-objective multi-skill resource-constrained project scheduling problem with skill switches: model and evolutionary approaches. Comput. Ind. Eng. **167**, 107897 (2022). https://doi.org/10.1016/j.cie.2021.107897

21. Vanhoucke, M.: Resource-Constrained Project Scheduling, pp. 107–137. Springer, Heidelberg (2012). https://doi.org/10.1007/978-3-642-25175-7_7

22. Vasilyev, I., Rybin, D., Kudria, S., Ren, J., Zhang, D.: Multiple project scheduling for a network roll-out problem: MIP formulation and heuristic. In: Pardalos, P., Khachay, M., Mazalov, V. (eds.) MOTOR 2022. LNCS, vol. 13367, pp. 123–136. Springer, Cham (2022). https://doi.org/10.1007/978-3-031-09607-5_9

23. Węglarz, J., Józefowska, J., Mika, M., Waligóra, G.: Project scheduling with finite or infinite number of activity processing modes - a survey. Eur. J. Oper. Res. **208**(3), 177–205 (2011). https://doi.org/10.1016/j.ejor.2010.03.037

24. Yannibelli, V., Amandi, A.: A knowledge-based evolutionary assistant to software development project scheduling. Expert Syst. Appl. **38**(7), 8403–8413 (2011). https://doi.org/10.1016/j.eswa.2011.01.035

25. Zheng, H., Wang, L., Zheng, X.: Teaching–learning-based optimization algorithm for multi-skill resource constrained project scheduling problem. Soft. Comput. **21**(6), 1537–1548 (2015). https://doi.org/10.1007/s00500-015-1866-3

Hybrid Evolutionary Algorithm with Optimized Operators for Total Weighted Tardiness Problem

Yulia Zakharova$^{(\boxtimes)}$ [ID]

Sobolev Institute of Mathematics, Novosibirsk, Russia
kovalenko@ofim.oscsbras.ru

Abstract. A new evolutionary algorithm with optimal recombination is proposed for the total weighted tardiness problem on the single machine. We solve the optimal recombination problem in a crossover operator. The NP-hardness of this problem is proved for various practically important cases. We construct the initial population means of greedy constructive heuristics. The insert and swap local search heuristics are used to improve the initial and the final populations. A computational experiment on the OR-Library instances shows that the proposed algorithm yields results competitive to those of well-known algorithms and confirms that the optimal recombination may be used successfully in evolutionary algorithms.

Keywords: Evolutionary Algorithm · Optimal Recombination · Scheduling · Experiment

1 Introduction

1.1 Problem Statement

We consider the following problem of scheduling a set of jobs $\mathcal{J} = \{1, \dots, n\}$ on a single machine. Job $j \in \mathcal{J}$ is characterized by release date r_j, processing time p_j, positive weight w_j, and due date d_j by which it should be completed. The machine can execute at most one job at a time and preemptions are disallowed. Let C_j denote the completion time of job $j \in \mathcal{J}$, then the tardiness T_j of job j is computed as $\max\{0; C_j - d_j\}$. The goal is to construct a schedule such that the total weighted tardiness $\sum_{j \in \mathcal{J}} w_j T_j$ is minimized. Using the classical three-field notation our problem is denoted as $1|r_j, d_j, w_j| \sum_j w_j T_j$ and called *the one-machine total weighted tardiness problem (TWTP)*. This problem arises in several practical settings, in particular in the chemical industry [28]. In addition, scheduling models with a single machine also have implications for scheduling research involving multiple machines, where the results obtained from single-machine problems can often be applied to more complex scheduling environments, such as parallel machines, flow shops, job shops, and open shops.

Let $\pi = (\pi_1, \dots, \pi_n)$ denote a permutation of the jobs, i.e. π_i is the i-th job on the machine, $i = 1, \dots, n$. Then the completion times $C(\pi_i) := C(\pi_{i-1}) + p_{\pi_i}$,

M. Khachay et al. (Eds.): MOTOR 2023, LNCS 13930, pp. 224–238, 2023.
https://doi.org/10.1007/978-3-031-35305-5_15

and tardiness $T(\pi_i) := \max\{0; C(\pi_i) - d_{\pi_i}\}$ for jobs in positions $i = 1, \ldots, n$ (suppose $C(\pi_0) := 0$). We denote the total weighted tardiness for permutation π by $T(\pi) = \sum_{i=1}^{n} w_{\pi_i} T(\pi_i)$.

The considered problem is NP-hard, so exact techniques are only applicable to small-sized instances in practice. Therefore metaheuristics are appropriate for the problem, in particular evolutionary algorithms (EAs).

Performance of evolutionary algorithms depends on the crossover operator, where the components of parent solutions are combined to build the offspring. *Optimal recombination problem* (ORP) is a subproblem, that consists in finding the best possible offspring as a result of a crossover operator, given two feasible parent solutions under the requirement that the recombination should be respectful and gene transmitting as coined by N. Radcliffe [23].

The optimal recombination problem for $1|r_j, d_j, w_j| \sum_j w_j T_j$ with the position-based representation of solutions is formulated as follows (see, e.g., [13]). Given two parent solutions π^1 and π^2. It is required to find a permutation π' such that:

(I) $\pi'_i = \pi^1_i$ or $\pi'_i = \pi^2_i$ for all $i = 1, \ldots, n$;
(II) π' has the minimum value of objective function $T(\pi')$ among all permutations that satisfy condition (I).

The corresponding recombination operators, where the ORP is solved, are called optimized crossovers. Previously, the computational complexity of optimal recombination of single-machine scheduling problem with setup times [15], permutation flow-shop scheduling problems [20], Travelling Salesmen Problem (TSP) and Boolean programming problems [13] has been investigated. It is known algorithms of solving NP-hard ORPs for permutation problems and their variations (e.g., dynamic programming [30], enumeration of perfect matchings in a special bipartite graph [20], partition of graph-vertices into recombining components [27], an analogue of the recursive algorithm of Eppstein for cubic graphs [16]). Experimental evaluation shows that the optimal recombination can be successfully used in EAs (see, e.g., [4,14,30]).

In this paper we propose new evolutionary algorithm with optimal recombination for the single-machine total weighted tardiness problem. The optimal recombination problem is solved in a crossover operator. The computational complexity of this problem is analysed. We construct the initial population means of greedy constructive heuristics. The insert and swap local search heuristics are used to improve the initial and the final populations. The experimental results show applicability of the proposed algorithm to the the OR-Library instances.

1.2 Previous Research

Here we briefly present the state-of-the-art metaheuristics known for problem $1|r_j, d_j, w_j| \sum_j w_j T_j$: local search methods, ant colony optimization methods, evolutionary algorithms and other population-based methods.

Competitive experimental results for $1|r_j, d_j, w_j| \sum_j w_j T_j$ are demonstrated by single-solution and population-based local search methods [6,9,10,17,18,29,

31]. The reason is the "good" properties of the objective function, which significantly reduce the search area of solutions in the neighbourhoods (see, e.g., [18]).

Ant colony optimization (ACO) is inspired by the behaviour of real ant colonies. The main steps of the ACO are constructing solutions, local improving and updating trails. In [1,2,5], competitive ACO algorithms have been developed for the TWTP.

Hybrid evolutionary algorithms (HEAs) occupy one of the leading positions among the metaheuristics known for the considered problem. One of the first genetic algorithms for the TWTP was presented in [3]. The algorithm uses various greedy heuristics in operators assigning jobs to positions. The authors applied uniform and single-point crossovers, and perturbation techniques. Different crossover operators were compared in [19]. The position and order-based crossovers demonstrated the best results in the canonical genetic algorithm (basic scheme) for the TWTP. L. Chaabane [7] has proposed for TWTP a hybrid genetic algorithm applying simulated annealing at local improving stage, one point crossover and insert mutation. The hybrid method [11] combining evolutionary algorithm with position-based crossover and dynasearch shows the most effective results at this time in terms of both objective value and running time.

In differential evolution (DE) methods [22,25,26] for the considered problem, a solution is encoded by a real vector of length n, and the corresponding permutation is constructed using the smallest position value (SPV) rule, where jobs are sorted in increasing order of vector-component values. Various local search methods (for example, insert and swap) and their combinations are used at iterations to improve results.

The bee colony algorithm is also a population-based method inspired by the nectar-searching behaviours of bees and their activities in the hive. In the BCO algorithm from [24] for the considered problem the solutions are represented by real-valued vectors, and such vectors are decoded into permutations using the smallest position value (SPV) rule. The operators from continuous optimization and local search improving methods are involved in the search mechanism.

One of the aims of this paper is to compare the optimized crossover based on position solution representation with other known techniques, in particular local optimization methods. As far as we know, such results have not been provided in the literature at this time.

2   Computational Complexity of Optimal Recombination Problem

For proving NP-hardness of the ORP for $1|r_j, d_j, w_j| \sum_j w_j T_j$ we will use the following *Restricted Even-Odd Partition Problem* [12]: given ordered set $A = \{a_1, a_2, \ldots, a_{2n_0}\}$ and weight e_i of each element $a_i \in A$, where

$$\sum_{a_i \in A} e_i = 2E, \ e_i > e_{i+1}, \ i = 1, \ldots, 2n_0 - 1, \tag{1}$$

$$e_{2j} > e_{2j+1} + \delta, \ e_{2j-1} \le e_{2j} + \delta, j = 1, \ldots, n_0. \tag{2}$$

Here $\delta = \frac{1}{2}\sum_{i=1}^{n_0}(e_{2i-1} - e_{2i})$. Note that $E = \frac{1}{2}\sum_{i=1}^{2n_0} e_i = \sum_{i=1}^{n_0} e_{2i-1} - \delta = \sum_{i=1}^{n_0} e_{2i} + \delta$. The question is to decide whether set A can be partitioned on two subsets A_1 and A_2 such, that $\sum_{a_i \in A_1} e_i = \sum_{a_i \in A_2} e_i = E$, $|A_1| = |A_2| = n_0$ and subset A_1 includes only one element from each pair a_{2i-1}, a_{2i}.

We prove NP-hardness of the ORP for two cases, when weights of jobs are different and when weights are identical under the condition that all release dates $r_j = 0$. NP-hard case for non-zero release dates will be also provided. Note that condition (2) is not used in proofs of Theorems 1 and 3. An exact method for the ORP and polynomial solvability in "almost all" cases will be presented.

2.1 NP-Hardness

Theorem 1. *Optimal recombination problem (I)-(II) for* $1|r_j = 0, d_j, w_j| \sum_j w_j T_j$ *is NP-hard even in the case when all jobs except two have the same due dates.*

Proof. Let the number of jobs $n = 2n_0 + 2$. Jobs $j \in \{1, \ldots, 2n_0\}$ have processing times $p_j = e_j$, weights $w_j = e_j$, and due dates $d_j = 0$. Two last jobs have duration $p_{2n_0+1} = p_{2n_0+2} = 3E^2 + 1$, weight $w_{2n_0+1} = w_{2n_0+2} = 3(3E^2 + 1 + E)E + 1$, due dates $d_{2n_0+1} = 3E^2 + 1$ and $d_{2n_0+2} = 2(3E^2 + 1) + E$.

Set the parent permutations in ORP as $\pi^1 := (2n_0+1, 1, 3, \ldots, 2n_0-1, 2n_0+2, 2, 4, \ldots, 2n_0)$, $\pi^2 := (2n_0 + 1, 2, 4, \ldots, 2n_0, 2n_0 + 2, 1, 3, \ldots, 2n_0 - 1)$.

We show that permutation π', for which conditions (I) and $\sum_{j=1}^n w_{\pi'_j} T(\pi'_j) \leq 3(3E^2 + 1 + E)E$ hold, exists if and only if the Restricted Even-Odd Partition Problem has a positive answer (see Fig. 1).

Let the Restricted Even-Odd Partition Problem has a positive answer and subsets A_1 and A_2 are such that $\sum_{a_i \in A_1} e_i = \sum_{a_i \in A_2} e_i = E$, $|A_1| = |A_2| = n_0$. Denote by $\pi(A_1)$ and $\pi(A_2)$ permutations of jobs in the order, corresponding to increasing of numbers from A_1 and A_2, respectively. Then for permutation $\pi' = (2n_0 + 1, \pi(A_1), 2n_0 + 2, \pi(A_2))$ we have

$$\sum_{j=1}^n w_{\pi'_j} T(\pi'_j) \leq (3E^2 + 1 + E)E + 2(3E^2 + 1 + E)E = 3(3E^2 + 1 + E)E,$$

as all jobs from $\pi(A_1)$ have completion times no more than $(3E^2 + 1 + E)$, and jobs from $\pi(A_2)$ completes execution on time moment less than $2(3E^2 + 1 + E)$. Moreover the total sum of weights for jobs from $\pi(A_1)$ (similarly from $\pi(A_2)$) is equal to E, and jobs $2n_0 + 1$ and $2n_0 + 2$ are executed in time.

Let permutation π' satisfy conditions (I) and $\sum_{j=1}^n w_{\pi'_j} T(\pi'_j) \leq 3(3E^2 + 1 + E)E$. Then jobs $2n_0 + 1$ and $2n_0 + 2$ are executed in time, as their weights are greater than $3(3E^2 + 1 + E)E$. Denote by W the sum of weights of jobs, executed after job $2n_0 + 2$. Then

$$\sum_j w_j T_j(\pi') \geq (3E^2 + 1)(2E - W) + 2(3E^2 + 1)W = (3E^2 + 1)(2E + W).$$

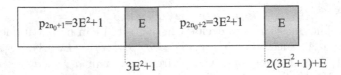

Fig. 1. Illustration of a feasible schedule for the ORP of $1|r_j = 0, d_j, w_j| \sum_j w_j T_j$.

It is evidence that $W \geq E$, since job $2n_0 + 2$ must be completed before time $2(3E^2 + 1) + E$. Suppose that $W \geq E + 1$, then $\sum_{j=1}^{n} w_{\pi'_j} T(\pi'_j) \geq (3E^2 + 1)(2E + E + 1) = 3(3E^2 + 1)E + (3E^2 + 1) > 3(3E^2 + 1 + E)E$, and we have a contradiction. So, n_0 jobs with total weight E are performed between jobs $2n_0 + 1$ and $2n_0 + 2$. Therefore, the Restricted Even-Odd Partition Problem has a positive answer.

Theorem 2. *Optimal recombination problem (I)-(II) for $1|r_j = 0, d_j| \sum_j T_j$ is NP-hard.*

Proof. The presented reduction uses some ideas from [12], but we prove NP-hardness of new ORP of the scheduling problem and identify its property instead of the classic scheduling statement. Let the number of jobs $n = 3n_0 + 1$. The jobs $j \in \{1, \ldots, 2n_0\}$ will be called S-jobs: $S_1, \ldots, S_{2n_0}$, and the rest $n_0 + 1$ will be called B-jobs: $B_1, \ldots, B_{n_0}, B_{n_0+1}$. We denote by b the value $(4n_0 + 1)\delta$.

Processing times of S-jobs are equal to $p(S_i) = e_i$, $i = 1, \ldots, 2n_0$, while durations of B-jobs $p(B_i) = b$, $i = 1, \ldots, n_0 + 1$. Now we provide due dates of jobs (see Fig. 2)

$$d(S_i) = \begin{cases} (j-1)b + \delta + (e_2 + e_4 + \cdots + e_{2j}), & \text{if } i = 2j - 1, \\ d(S_{2j-1}) + 2(n_0 - j + 1)(e_{2j-1} - e_{2j}), & \text{if } i = 2j, \end{cases}$$

$j = 1, \ldots, n_0.$

$$d(B_i) = \begin{cases} ib + (e_2 + e_4 + \cdots + e_{2i}), & \text{if } i = 1, \ldots, n_0, \\ d(B_{n_0}) + \delta + b, & \text{if } i = n_0 + 1. \end{cases}$$

Define the parent permutations in ORP as $\pi^1 := (S_1, B_1, S_3, B_2, \ldots, B_{n_0-1}, S_{2n_0-1}, B_{n_0}, B_{n_0+1}, S_{2n_0}, S_{2n_0-2}, \ldots, S_2)$, $\pi^2 := (S_2, B_1, S_4, B_2, \ldots, B_{n_0-1}, S_{2n_0}, B_{n_0}, B_{n_0+1}, S_{2n_0-1}, S_{2n_0-3}, \ldots, S_1)$.

Show that there is a permutation π' satisfying conditions (I) and $\sum_{j=1}^{n} T(\pi'_j) \leq T_0$, if and only if the Restricted Even-Odd Partition Problem has a positive answer. Here $T_0 := E + n_0 P - n_0(n_0 - 1)b/2 - n_0 \delta - \sum_{i=1}^{n_0}(n_0 - i + 1)(e_{2i-1} + e_{2i})$, where $P := (n_0 + 1)b + 2E$ is the total duration of all jobs.

We start the proof with the statement that the inequality $\sum_{j=1}^{n} T(\pi'_j) \geq T_0$ holds for permutation $\pi' := (S_{1,1}, B_1, S_{2,1}, B_2, \ldots, B_{n_0-1}, S_{n_0,1}, B_{n_0}, B_{n_0+1}, S_{n_0,2}, S_{n_0-1,2}, \ldots, S_{1,2})$ (here $\{S_{i,1}, S_{i,2}\} = \{S_{2i-1}, S_{2i}\}$). Moreover, $\sum_{j=1}^{n} T(\pi'_j) = T_0$ if and only if $\sum_{i=1}^{n_0} p(S_{i,1}) = \sum_{i=1}^{n_0} p(S_{i,2})$.

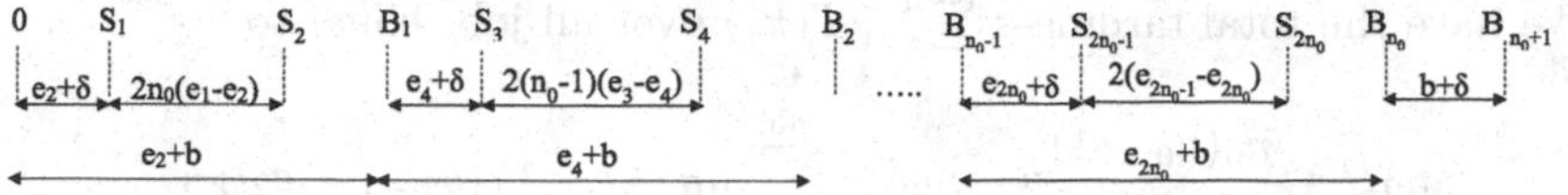

Fig. 2. Illustration of the due dates for the ORP of $1|r_j = 0, d_j| \sum_j T_j$.

Compute the total tardiness for jobs B_i, $i = 1, \ldots, n_0$. Since $e_{2j-1} > e_{2j}$, then $C(B_i) \geq d(B_i)$ for completion times of jobs B_i, $i = 1, \ldots, n_0$. In addition $C(B_i) = ib + \sum_{j=1}^{i} p(S_{i,1})$, $\sum_{i=1}^{n_0} d(B_i) = \frac{n_0(n_0+1)}{2} b + \sum_{i=1}^{n_0} (n_0 - i + 1)e_{2i}$. Therefore

$$\sum_{i=1}^{n_0} (C(B_i) - d(B_i)) = \sum_{i=1}^{n_0} C(B_i) - \sum_{i=1}^{n_0} d(B_i) =$$

$$\frac{n_0(n_0+1)}{2} b + \sum_{i=1}^{n_0} (n_0 - i + 1)p(S_{i,1}) - \frac{n_0(n_0+1)}{2} b - \sum_{i=1}^{n_0} (n_0 - i + 1)e_{2i} =$$

$$\sum_{i=1}^{n_0} (n_0 - i + 1)(p(S_{i,1}) - e_{2i}).$$

Now we calculate the total tardiness for jobs $S_{i,2}$, $i = 1, \ldots, n_0$. Note that we have $C(S_{i,2}) \geq d(S_{i,2})$ for jobs $S_{i,2}$, $i = 1, \ldots, n_0$, because $C(S_{i,2}) \geq (n_0 + 1)b + \sum_{j=1}^{n_0} a_{2j}$, but $d(S_{i,2}) \leq (n_0 - 1)b + (2n_0 + 1)\delta + \sum_{j=1}^{n_0} a_{2j}$ from definitions of due dates and processing times. Moreover $C(S_{i,2}) = P - \sum_{j=1}^{i-1} p(S_{j,2})$. If $S_{i,2} = S_{2i-1}$, then $d(S_{i,2}) = (i - 1)b + \delta + \sum_{j=1}^{i} e_{2j}$, while if $S_{i,2} = S_{2i}$, then $d(S_{i,2}) = (i - 1)b + \delta + \sum_{j=1}^{i} e_{2j} + 2(n_0 - i + 1)(e_{2i-1} - e_{2i})$. Therefore $d(S_{i,2}) = (i - 1)b + \delta + \sum_{j=1}^{i} e_{2j} + (n_0 - i + 1)(e_{2i-1} - e_{2i}) + (n_0 - i + 1)(p(S_{i,1}) - p(S_{i,2}))$. Thus,

$$\sum_{i=1}^{n_0} (C(S_{i,2}) - d(S_{i,2})) =$$

$$n_0 P - \sum_{i=1}^{n_0-1} (n_0 - i)p(S_{i,2}) - \frac{n_0(n_0-1)}{2} b - n_0\delta -$$

$$\sum_{i=1}^{n_0} (n_0 - i + 1)e_{2i-1} - \sum_{i=1}^{n_0} (n_0 - i + 1)(p(S_{i,1}) - p(S_{i,2})) =$$

$$n_0 P - \frac{n_0(n_0-1)}{2} b - n_0\delta - \sum_{i=1}^{n_0} (n_0 - i + 1)e_{2i-1} - \sum_{i=1}^{n_0} (n_0 - i + 1)p(S_{i,1}) + \sum_{i=1}^{n_0} p(S_{i,2}).$$

We have the total tardiness $\sum_{j=1}^{n} T(\pi_j')$ over all jobs equal to

$$n_0 P - \frac{n_0(n_0 - 1)}{2} b - n_0 \delta - \sum_{i=1}^{n_0} (n_0 - i + 1)(e_{2i-1} + e_{2i}) +$$

$$\sum_{i=1}^{n_0} p(S_{i,2}) + \sum_{i=1}^{n_0} T(S_{i,1}) + T(B_{n_0+1}).$$

If $\sum_{i=1}^{n_0} p(S_{i,1}) \leq E = \sum_{i=1}^{n_0} e_{2i} + \delta$, then for any $k = 1, \ldots, n_0$ we have $\sum_{i=1}^{k} p(S_{i,1}) \leq \sum_{i=1}^{k} e_{2i} + \delta$ (proved by contradiction). And form the definition of values $d(S_i)$ and $d(B_{n_0+1})$ we obtain $C(S_{i,1}) \leq d(S_{i,1})$ and $T(S_{i,1}) = 0$; $C(B_{n_0+1}) \leq d(B_{n_0+1})$ and $T(B_{n_0+1}) = 0$. From the definition of T_0 and expression for $\sum_{j=1}^{n} T(\pi_j')$ we conclude that $\sum_{i=1}^{n_0} p(S_{i,2}) \geq E$ and $\sum_{j=1}^{n} T(\pi_j') \geq T_0$. Moreover, equality is possible only when $\sum_{i=1}^{n_0} p(S_{i,1}) = \sum_{i=1}^{n_0} p(S_{i,2}) = E$.

If $\sum_{i=1}^{n_0} p(S_{i,1}) > E$, then

$$T(B_{n_0+1}) = \sum_{i=1}^{n_0} p(S_{i,1}) + (n_0+1)b - \left((n_0 + 1)b + \delta + \sum_{i=1}^{n_0} e_{2i} \right) = \sum_{i=1}^{n_0} p(S_{i,1}) - E.$$

Moreover, we can find index l such that $\sum_{i=1}^{l} p(S_{i,1}) > \sum_{i=1}^{l} e_{2i} + \delta$. Let k is the smallest from such indices, then $S_{k,1} = S_{2k-1}$, $C(S_{k,1}) > d(S_{k,1}) = \sum_{i=1}^{k} e_{2i} + \delta + (k-1)b$ by definition and $T(S_{k,1}) > 0$. As a result we have

$$\sum_{j=1}^{n} T(\pi_j') \geq n_0 P - \frac{n_0(n_0 - 1)}{2} b - n_0 \delta - \sum_{i=1}^{n_0} (n_0 - i + 1)(e_{2i-1} + e_{2i}) +$$

$$\left(\sum_{i=1}^{n_0} p(S_{i,2}) + \sum_{i=1}^{n_0} p(S_{i,1}) - E \right) + T(S_{k,1}) > T_0.$$

So, a permutation π' satisfying conditions (I) and $\sum_{j=1}^{n} T(\pi_j') \leq T_0$ exists if and only if the Restricted Even-Odd Partition Problem has a positive answer.

Theorem 3. *Optimal recombination problem (I)-(II) for $1|r_j, d_j| \sum_j T_j$ is NP-hard even in the case when we have only one job with individual release date and due date.*

The proof is similar to the proof of NP-hardness of $1|r_j, d_j| \sum_j T_j$ (see Theorem 4 in [21]).

2.2 Exact Method

For solving the ORP of $1|r_j, d_j, w_j| \sum_j w_j T_j$ we will use an exact method, similar to that proposed in [20] for solving ORP of the flow-shop scheduling problem.

Construct a bipartite graph $G = (\mathcal{J}_n, \mathcal{J}, U)$ with subsets of vertices $\mathcal{J}_n, \mathcal{J}$ having equal sizes and the set of edges $U = \{(i, j) : i \in \mathcal{J}_n, j = \pi_i^1 \text{ or } j = \pi_i^1\}$, where $\mathcal{J}_n = \{1,\dots,n\}$. There exists a one-to-one correspondence between the set of feasible solutions to the considered ORP and the set of perfect matchings in graph $\bar{G}$: permutation $\pi^m = (j^1,\dots,j^n)$ gives perfect matching $m^\pi = \{\{1,j^1\},\ \dots,\ \{n,j^n\}\}$ and vice versa.

According to terminology [15] an edge $\{i,j\} \in \bar{U}$ is called *special* if $\{i,j\}$ belongs to any perfect matching of G. The maximum connected subgraph of G containing at least two edges represents a cycle. Denote by q the number of cycles in G. Each cycle G contains exactly two maximal (perfect) matchings, and they are edge disjoint. Moreover, any perfect matchings in $\bar{G}$ is uniquely defined by maximal matchings of cycles and special edges. Note that maximal matchings of cycles and special edges may be computed in $O(n)$ time.

Thus, ORP (I)-(II) can be solved by enumeration of perfect matchings in graph G. For each perfect matching m, we construct the corresponding permutation π and calculate the objective value. As a result, we find the optimal solution of the ORP in $O(2^q n)$ time, where $q \leq \frac{n}{2}$. Moreover, "almost all" pairs of parent permutations give graphs G with $q \leq \frac{\ln(n)}{\ln(2)}$ cycles, i.e., ORP (I)-(II) has at most n feasible solutions (see proof in [15]).

In addition to the previous research, we propose here the following speed-up procedure for the considered ORP based on the properties of criterion $\sum_{j\in\mathcal{J}} w_j T_j$. In the searching process, we will guarantee that the maximum matching is changed only in one cycle (for example, using the code of Gray), when we go from one perfect matching to another. Let permutation π (corresponding to the perfect matching m at the previous step) is replaced by permutation $\pi' = (\pi_1,\dots,\pi'_a, \pi'_{a+1},\dots,\pi'_b,\dots,\pi_n)$ by changing the maximum matching in the cycle with the minimum index of the left part equal to a and the maximum index equal to b. Then, the completion times of jobs in positions $a, a+1,\dots,b$ are only changed for the permutation π' in comparison to the permutation π. Only these modifications should be taken into account in computing the objective value for π' using the objective value for π (if r_j is nonzero, then the completion times of jobs in positions $b, b+1,\dots,n$ must be updated after changing even one maximum matching of a cycle).

Solving the Optimal recombination problem in the crossover operator may be considered as a derandomization of the well-known *CX* (Cycle Crossover) operator [19], where maximal matchings are selected randomly in cycles.

3   Hybrid Evolutionary Algorithm

We propose a hybrid evolutionary algorithm with optimal recombination based on the steady-state replacement scheme [16] (see Algorithm 1). Each individual of the evolutionary process corresponds to the tentative solution of the problem.

The number of individuals in the population is denoted by N and remains constant during the search. The initial population includes solutions generated randomly or using constructive heuristics. The main constructive heuristic is the following: a random non-scheduled job is selected at each step; the position is assigned to this job such that gives the best objective for the current partial solution. Also, the population contains permutations, that correspond non-decreasing order of due dates d_j, non-increasing order of w_j/p_j and non-increasing order of $(w_j/p_j) \cdot exp\{-\max\{0, (d_j - p_j)\}/(n * p_{aver})\}$, where $p_{aver} = \sum_{j=1}^{n} p_j/n$.

Individuals of the last population is improved by a local optimization procedure based on the *swap* neighborhood (positions of two jobs are exchanged) and the *insert* neighborhood (a job is inserted in some other position). For reducing the running time of local search heuristics, we use strategies provided in [18] for $1|r_j = 0, d_j, w_j| \sum_j w_j T_j$. In particular, obviously unpromising moves are excluded and only a part of the neighborhood is considered by moving jobs only from positions located at the distance of no more than $20\%n$ from each other.

Algorithm 1. Hybrid Evolutionary Algorithm with Steady State Replacement Scheme.

1: Construct the initial population.
2: Until a termination condition is satisfied, perform steps

    2.1 Select two parent solutions π^1 and π^2.
    2.2 Apply mutation operator to parent solutions.
    2.3 Generate an offspring p' by applying a crossover operator to π^1 and π^2.
    2.4 Offspring p' replaces the worst individual of the population.

3: Perform local improvements of individuals of the last population.

We use two mutation operators that perform a random jump within swap or insert neighborhood [18]. The operators are used for mutation with equal probability. The mutation is applied with probability p_{mut}. In the crossover operator we solve the ORP with probability p_{cross} (this crossover is called *OCX*, Optimized Cycle Crossover), and use the well-known *OX* (Ordered Crossover) [19] with probability $1 - p_{\text{cross}}$.

One of the perspective approaches to solve the considered TWTP problem is local optimization (Variable Neighbourhood Search, Variable Neighbourhood Descent, Tabu Search, Dynasearch and others) [6,9–11,17,18,29,31]. The Optimal recombination may be also considered as a best-improving move in a position-based neighbourhood defined by two parent solutions.

The classic restating rule is used: EA is restarted as soon as the current iteration number becomes twice the iteration number when the record was found.

4   Computational Experiment

In this section, the proposed evolutionary algorithm HEA-OCX is compared to the heuristics which previously demonstrated their competitiveness to the most advanced special-purpose metaheuristics for the TWTP known at that time:

Genetic algorithm (GA) from [3] is the population-based algorithm with local improvements, where initial population is generated randomly; some biased randomly chosen subset of jobs is fixed in the selected solution and local improvements are applied on the rest jobs with adopting priorities at iterations (it is tested on a computer with Pentium II 400 MHz, 96 Mb).

Population-based VNS algorithm (PVNS) [29] uses the scheme of the population-based variable neighbourhood search method with shaking procedure based on insertion and swap neighbourhoods and such local search techniques as Variable depth search, Path-relinking and Tabu search; the initial population is constructed by NEH-heuristic with random sequences of jobs and by deterministic rules (it is tested on a computer with Pentium IV 3.0 GHz, 512 Mb).

Table 1. Results for series with $n = 40$.

Algorithm	Time, sec.	n_{hit}	n_{opt}	Δ_{aver}
GA	29.11	–	100%	0
PVNS	6.19	100%	100%	0
m-VNS	0.908	100%	100%	–
HEA-OCX	0.05	100%	100%	0
HEA-DS	0.003	100%	100%	0

Table 2. Results for series with $n = 50$.

Algorithm	Time, sec.	n_{hit}	n_{opt}	Δ_{aver}
GA	41.02	–	99.2%	0.000
PVNS	12.15	100%	100%	0
m-VNS	1.62	100%	100%	–
HEA-OCX	0.14	100%	100%	0
HEA-DS	0.006	100%	100%	0

Multiple VNS algorithm (m-VNS) [8] is also correspond to the basic principle of VNS, where the initial solution is generated using the dispatching rule or the roulette wheel rule; the classic insertion and swap neighbourhoods and their compound versions with series of independent moves are used in local search, specific disturbing and matching operations are applied for selecting neighbourhoods at iterations (it is tested on a computer with Intel Core 2.3 GHz, 2 Gb).

The hybrid evolutionary algorithm (HEA-DS) from [11] represents the steady-state genetic algorithm, where the initial population is generated randomly, position-based crossover and perturbation by independent swaps compose reproduction operators, dynasearch based on swap moves is applied at the initialization stage and EA iterations for local improving solutions (it is tested on a computer with Xeon E5440 2.83GHz, 16Gb). Recall that dynasearch is a neighbourhood search algorithm, where exponential-sized neighbourhoods are searched using dynamic programming.

We use the TWTP instances from OR-Library (http://people.brunel.ac.uk/mastjjb/jeb/orlib/wtinfo.html). The number of jobs in the series of instances are 40, 50 and 100. Each series contains 125 tests. HEA-OCX was run on each instance for 50 times. The algorithm is set to stop when it obtains an optimal solution or reaches a maximum number of generations – 6000 (such condition is selected in order to compare our results to the ones form [3,8,11,29]). We set $p_{mut} = 0.15$, $p_{cross} = 0.8$, and use the tournament selection with the tournament size $s = 5$. The experiment was carried out on a computer with Intel Core i3-10100F CPU 3.60 GHz, 16 Gb.

Table 3. Results for series with $n = 100$.

Algorithm	Time, sec.	n_{hit}	n_{opt}	Δ_{aver}
GA	118.97	–	66.4%	0.02
PVNS	183.47	100%	100%	0
m-VNS	5.845	99.84%	100%	–
HEA-OCX	1.5	100%	100%	0
HEA-DS	0.032	100%	100%	0

Table 4. Computational comparison of four crossover operators.

Operator	n_{bt}	n_{opt}	Δ_{aver}
$n = 40$			
OCX	–	100%	0
CX	43/125	82.51%	0.0029
OX	20/125	93.73%	0.00078
OX+LD	17/125	94.56%	0.00076
$n = 50$			
OCX	–	100%	0
CX	67/125	68.27%	0.0057
OX	41/125	84%	0.0017
OX+LD	38/125	92.88%	0.0013
$n = 100$			
OCX	–	100%	0
CX	92/125	34.59%	0.0097
OX	90/125	40.56%	0.0055
OX+LD	80/125	62.32%	0.0031

The results of the experiment is provided in Tables 1, 2 and 3. Here n_{hit} is the average percentage number of optimal or best known solution values found for an instance out of the given trial runs, n_{opt} is the percentage of instances, where

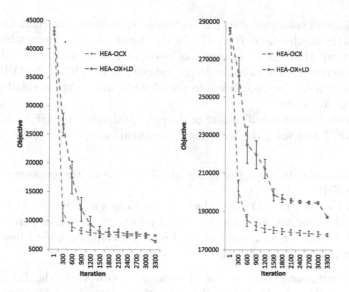

Fig. 3. Dynamics of the objective function over the EA run.

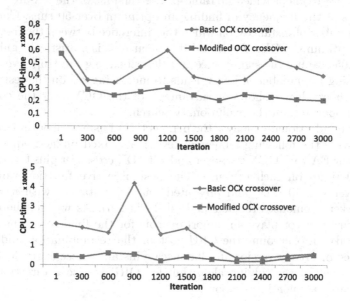

Fig. 4. Dynamics of the utilized CPU time (nanosec.) in ORP solving.

the optimum is found, Δ_{aver} is the average relative deviation from the optimum. We see that our algorithm HEA-OCX demonstrates competitive results, finding the record in all cases within comparable time. Note that we know from the previous research that the dynasearch method based on the swap-neighbourhood demonstrates leading results on the considered here series of instances in the context quality of solutions and time resources. Our algorithm HEA-OCX shows slightly worth results than HEA-DS in the term of time for reaching of the record. But we think that our time is also good from the practical point of view. Our current goal is to estimate the rationality of solving the ORP in EAs. Further research can be undertaken to parallelized and/or approximate methods in optimized crossovers.

We also considered modifications of the proposed algorithm HEA-OCX, where instead of OCX crossover the following recombination operators [19] are used:

- Cycle Crossover (CX) is the randomized version of the crossover based on solving ORP;
- Ordered Crossover (OX) saves absolute positions for some jobs from one parent and the relative order for the rest of jobs from the other parent;
- Ordered Crossover with subsequent Local Descent algorithm based on the insertion neighbourhood (OX+LD).

The computational comparison of various crossovers in the EA under the same given CPU time limit (equal to the time when HEA-OCX reaches the best known record) is provided in Table 4. The versions of the EA are compared on the bases of the frequency of finding an optimum over 50 runs. Column n_{bt} shows the number of instances, on which the difference between the frequency of finding an optimum of HEA-OCX and its concurrent is statistical significant at level $\alpha \leq 0.05$ (we use the significance test described, e.g., in [16]). We conclude that increasing the number of jobs n leads to increasing n_{bt} on the tested series. This experiment clearly shows an advantage of the OCX over the considered randomized operators in the evolutionary search.

In order to additionally analyse the perspective of using the optimized crossover, we provide the following numerical test: For a fixed budget, equal to 4000 iterations, our EA with OCX crossover and with OX crossover plus Local Descent was run 50 times on each instance. The best objective function values were averaged over the 50 runs, and presented for two instances with $n = 100$ in Fig. 3 (whiskers demonstrate the standard deviation). As we see, the choice of the crossover operator plays an important role for the final result, however we must also take into account the CPU cost of the recombination and may be use restricted or greedy versions for large-scale instances (see examples in [16]). Parallel versions of the solving algorithm for ORP is also a direction for the further research as mentioned above.

In this context, we carried out an additional experiment to evaluate the CPU cost of the optimized crossover in the original version and the modified variant with the speed-up procedure proposed in Subsect. 2.2. The dynamics of the utilized CPU time for the 1st and the 15th instances of the series with $n = 100$ is provided in Fig. 4 (results are similar for the rest instances). In general, the speed-up of the algorithm for solving the ORP using the objective properties depends on the generation. The dimension of the ORP decreases with increasing the iteration number in all considered instances and runs. The modified algorithm works approximately 3 times more quickly on average than the original one at the initial generations.

5   Conclusion

We proposed the hybrid evolutionary algorithm with optimized operators for the total weighted tardiness scheduling problem on the single machine. The NP-hardness of the Optimal recombination problem was proved for three practically important cases. An experimental evaluation on OR-Library instances indicates that the proposed approach gives competitive results.

Acknowledgements. The research was supported by Russian Science Foundation grant N 22-71-10015, https://rscf.ru/en/project/22-71-10015/.

References

1. Abitz, D., Hartmann, T., Middendorf, M.: A weighted population update rule for PACO applied to the single machine total weighted tardiness problem. In: GECCO 2020: Proceedings of the 2020 Genetic and Evolutionary Computation Conference, pp. 4–12 (2020)
2. Anghinolfi, D., Paolucci, M.: A new ant colony optimization approach for the single machine total weighted tardiness scheduling problem. Int. J. Oper. Res. **5**(1), 44–60 (2008)
3. Avci, S., Akturk, M., Storer, R.H.: A problem space algorithm for single machine weighted tardiness problems. IIE Trans. **35**, 479–486 (2003)
4. Balas, E., Niehaus, W.: Optimized crossover-based genetic algorithms for the maximum cardinality and maximum weight clique problems. J. Heuristics **4**(2), 107–122 (1998)
5. den Besten, M., Stutzle, T., Dorigo, M.: Ant colony optimization for the total weighted tardiness problem. In: Parallel Problem Solving from Nature (PPSN-2000), LNCS, vol. 1917 (2000)
6. Bilge, U., Kurtulan, M., Kirac, F.: A tabu search algorithm for the single machine total weighted tardiness problem. Eur. J. Oper. Res. **176**, 1423–1435 (2007)
7. Bouchra, D., Chaabane, L.: An enhanced genetic algorithm for the single machine total weighted tardiness problem. Rom. J. Inf. Comput. Sci. **9**(1), 66–77 (2007)
8. Chung, T.P., Fu, Q., Liao, C.J., Liu, Y.T.: Multiple-variable neighbourhood search for the single-machine total weighted tardiness problem. Eng. Optim. **49**(7), 1133–1147 (2016)
9. Chunga, T., Fua, Q., Liaob, C., Liub, Y.: On hybrid evolutionary algorithms for scheduling problem with tardiness criterion. Eng. Optim. **49**(7), 1133–1147 (2017)
10. Congram, R., Potts, C., van de Velde, S.: An iterated dynasearch algorithm for the single-machine total weighted tardiness scheduling problem. INFORMS J. Comput. **14**(1), 52–67 (2002)
11. Ding, J., Lu, Z., Cheng, T., Xu, L.: A hybrid evolutionary approach for the single machine total weighted tardiness problem. Comp. Indust. Eng. **108**, 70–80 (2017)
12. Du, J., Leung, Y.: Minimizing total tardiness on one machine is NP-hard. Math. Oper. Res. **15**(3), 483–495 (1990)
13. Eremeev, A., Kovalenko, Y.: Optimal recombination in genetic algorithms for combinatorial optimization problems. Yugoslav J. Oper. Res. **24**(1–2), 387–401 (2014)
14. Eremeev, A.V., Kovalenko, J.V.: Experimental evaluation of two approaches to optimal recombination for permutation problems. In: Chicano, F., Hu, B., García-Sánchez, P. (eds.) EvoCOP 2016. LNCS, vol. 9595, pp. 138–153. Springer, Cham (2016). https://doi.org/10.1007/978-3-319-30698-8_10

15. Eremeev, A., Kovalenko, Y.: On complexity of optimal recombination for one scheduling problem with setup times. Diskretn. Anal. Issled. Oper. **19**(3), 13–26 (2012)
16. Eremeev, A., Kovalenko, Y.: A memetic algorithm with optimal recombination for the asymmetric travelling salesman problem. Memetic Comp. **12**, 23–36 (2020)
17. Fu, Q., Chung, T.: On hybrid evolutionary algorithms for scheduling problem with tardiness criterion. In: 2016 IEEE International Conference on Industrial Engineering and Engineering Management (IEEM), IEEE, pp. 438–441 (2016)
18. Geiger, M.J.: On heuristic search for the single machine total weighted tardiness problem - some theoretical insights and their empirical verification. Eur. Jour. Oper. Res. **207**, 1235–1243 (2010)
19. Kellegöz, T., Toklu, B., Wilson, J.: Comparing efficiencies of genetic crossover operators for one machine total weighted tardiness problem. Appl. Math. Comput. **199**(2), 590–598 (2008). https://doi.org/10.1016/j.amc.2007.10.013
20. Kovalenko, Y.V.: On complexity of optimal recombination for flowshop scheduling problems. J. Appl. Ind. Math. **10**(2), 220–231 (2016). https://doi.org/10.1134/S1990478916020071
21. Lenstra, J., Rinnooy Kan, A., Brucker, P.: Complexity of machine scheduling problem. In: Studies in Integer Programming. Annals of Discrete Mathematics, pp. 343–362. The Netherlands, North Holland, Amsterdam (1977)
22. Nearchou, A.: A hybrid metaheuristic for the single-machine total weighted tardiness problem. Cybern. Syst. **43**, 651–668 (2012)
23. Radcliffe, N.: The algebra of genetic algorithms. Ann. Math. Artif. Intell. **10**(4), 339–384 (1994)
24. Sharma, N., Sharma, H., Sharma, A.: An effective solution for large scale single machine total weighted tardiness problem using lunar cycle inspired artificial bee colony algorithm. In: IEEE/ACM Transactions on Computational Biology and Bioinformatics, pp. 1573–1581. IEEE (2019)
25. Tasgetiren, M., Liang, Y., Sevkli, M., Gencyilmaz, G.: Particle swarm optimization and differential evolution for the single machine total weighted tardiness problem. Int. J. Prod. Res. **44**(22), 4737–4754 (2006)
26. Tasgetiren, M., Sevkli, M., Liang, Y., Gencyilmaz, G.: Particle swarm optimization algorithm for single machine total weighted tardiness problem. In: 2004 Proceedings of the 2004 Congress on Evolutionary Computation, pp. 1412–1419. IEEE (2004)
27. Tinos, R., Whitley, L.D., Ochoa, G.: A new generalized partition crossover for the traveling salesman problem: tunneling between local optima. Evol. Comput. **28**(2), 255–288 (2020)
28. Wagner, B., Davis, D., Kher, H.: The production of several items in a single facility with linearly changing demand rates. Decis. Sci. **33**(3), 317–346 (2002)
29. Wang, X., Tang, L.: A population-based variable neighborhood search for the single machine total weighted tardiness problem. Comp. Oper. Res. **36**, 2105–2110 (2009)
30. Yagiura, M., Ibaraki, T.: The use of dynamic programming in genetic algorithms for permutation problems. Eur. Jour. Oper. Res. **92**, 387–401 (1996)
31. Yahyaoui, H., Krichen, S., Derbel, B., Talbi, E.: A variable neighborhood descent for solving the single machine total weighted tardiness problem. In: ICMSAO, p. 6. IEEE (2013)

Game Theory

Equilibrium Arrivals to Preemptive Queueing System with Fixed Reward for Completing Request

Julia V. Chirkova[✉] [iD]

Institute of Applied Mathematical Research of Karelian Research Centre of RAS,
Pushkinskaya st. 11, Petrozavodsk, Russia
julia@krc.karelia.ru

Abstract. This paper considers a single-server queueing system in which players send requests into the system with preemptive access and fixed reward for completing request. When a request enters the system, the server immediately starts its service. The current request leaves the system when its service is completed or when the next request comes into the system. We study the non-cooperative game, where each player decides when to arrive at the queueing system within a certain period of time. The objective of the player is to serve his request completely or to maximize the time of receiving service. We investigate Nash equilibrium properties in this game. Finally, we present results of numerical experiments demonstrating the equilibria with different values of the model parameters.

Keywords: Queueing system · Preemptive access · Strategic users · Optimal arrivals · Nash equilibrium · Fixed reward

1 Introduction

We consider a development of the single-server queueing system described in [4]. Players send requests into the system with preemptive access during a fixed time interval $[0, T]$. When a request enters the system, the server immediately starts its service. The current request leaves the system when its service is completed or when the next request comes into the system. The motivation for such service discipline we described with examples in [4]. The service discipline is often found in real life in systems without any access management: animal territory markers, advertising board, graffiti, etc. In transport system we can consider an example where a bus arriving to bus stop needs to leave it when the next one comes. One more interpretation is given in the paper [18], where each of farmers wants to

This research was supported by the Russian Science Foundation: grant No. 22-11-20015, https://rscf.ru/project/22-11-20015/, jointly with support of the authorities of the Republic of Karelia with funding from the Venture Investment Foundation of the Republic of Karelia.

decide the optimal time to put his product on the market until the next harvest season, maximizing the price increasing with time but decreasing with number of sellers. If we assume that buyers choose only last seller with a fresh product and ignore all previous sellers, we obtain the model closed to considered in the current paper.

Usually in queuing theory models the input process is determined by given rate of requests. But game-theoretic approach assumes that the input process of arrivals into the system is strategic ([1,3,6–17]). These arrivals are determined by selfish and independent players choosing the arrival time moments into the system, working on a time interval $[0, T]$, trying to maximize their payoff. The paper [7] is the first work that considers the single-server queue which is formed by selfish strategy of users, trying to minimize their waiting time. A model with several servers and a limited buffer is explored in [10]. A game with the batch service is considered in [8]. The common property in such games is the mixed symmetric Nash equilibrium strategy which is a probability distribution of arrival time moments on a working interval of the system. Further researches develop this area adding variations into the system and payoff functions of players. In [15] we explore a single-server queuing system without buffer where the players payoff includes a time-sensitivity convenience function. A model where the players take into account not only waiting, but also tardiness costs is considered in [13]. In the paper [12] authors combine the tardiness costs, waiting costs, and restrictions on the opening and closing times. The paper [2] presents a queueing system with the last-come first-served discipline and preemptive-resume. Our paper [5] is devoted a single-server queueing system with retrievals.

This work is the development of the model in [4], where players also choose a time instant to enter the system. The queuing system is the same as presented in [4], the difference of the model is in the payoff function. In [4] each player's aim is to maximize the probability to perform his request completely, obtaining a reward 1 only in case of successful completing service of his request. In this paper we assume that each player tries to serve his request completely or to maximize a service time for his request. He obtains a fixed reward α in case of completing request service, or a value 1 for each unit of a service time which his request get. Similarly to [4] we concentrate on the symmetric equilibria at this paper, considering each player to be independent and similar to any another. We suppose that they all act in the same way and don't distinct them. Of course, it is possible to consider some non-symmetric equilibria, defining their structures. It can be interesting problem for further investigations.

The description of the generalized model is provided in Sect. 2. In the Sect. 3 we analyse properties of a symmetric equilibrium. The Sect. 4 presents the analytical solution for the game with two players. Finally, we present results of numerical experiments demonstrating the equilibria with different values of the model parameters. We also try to compare this model with a model given in [4] where players maximize the probability of their request completion.

2   Description of the Model

2.1   Queueing System

Consider a one-server system that receives user requests on a time interval $[0, T]$ and $N + 1$ users who hope to perform their requests. The system has no queues, the server may simultaneously perform only one request. The service time for a request is exponentially distributed with parameter μ. When the request enters the system, it starts to perform immediately. It can leave the system if when completes its service or when the next request arrives. If more then one request arrives into the system at the same time moment, the server chooses equiprobably one of them for further servicing.

At the moment T the system can accept the last request and becomes closed for requests entering after this moment. However, if some request coming into the system at a time moment $t \in [0, T]$ has not finished its service yet till the time moment T, the system continues its service and completes it successfully.

The request discipline is not defined in the system. It is actually formed by the users seeking to maximize the probability of complete servicing for their requests or to maximize the time of receiving service.

2.2   Game Model

Consider the optimal request arrivals problem as a non-cooperative game. A set of players consist of $N + 1$ users each of which chooses the moment to send his request into the system for servicing. In general the number of players it can be fixed or random, in the current paper we consider the case of deterministic number of players. A player sends his request to the system to serve it. The request leaves the system when it is served completely or when the next request comes into the system.

Each player chooses the time to send his request to the system, providing its complete service or a maximum service time for his requests. The reward for the player is a fixed value $\alpha > 0$ in case of successful completion of his request service. Otherwise he obtains 1 for each unit of a service time which his request get. We don't lose generality because both values can be divided by profit from each service time unit. This value α can be considered to be proportional to a ticket price in a transport interpretation.

Here we don't consider the case $\alpha \leq 0$ which can be interpreted as follows. Each request performing in the system can be only changed by the next one or die with completing its service. α is the cost of request loss in this case.

Consider a player i. His pure strategy is the arrival time t_i of his request in the system. The mixed strategy is the distribution function $F_i(t)$ with the density $f_i(t)$ of the arrival times in the system on the time interval $[0, T]$.

We concentrate on the symmetric equilibria at this paper, considering each player to be independent and similar to any another. We suppose that they all act in the same way and don't distinct them. Players act selfishly without cooperation and can achieve some symmetric Nash equilibrium. Their the strategies are the same, i.e., $F_i(t) = F(t)$ for all i.

The payoff for each player is his expected reward which he obtains sending and performing his request in the system.

Definition 1. *A distribution function $F(t)$ of the arrival times t in the system is a symmetric Nash equilibrium if there exists a constant C such that at any time $t \in [0, T]$ the payoff does not exceed C and is equal to C on the support of $F(t)$.*

Lemma 1. *The support of the equilibrium strategy possess the following properties. The point $t = T$ is an atom, i.e. there is a strictly positive probability $p > 0$ to arrive at the time moment T. Also there is a time interval (t_e, T) when no requests arrive into the system.*

Proof. Let's imagine that no requests arrive into the system at time T in equilibrium, that is $p = 0$. Then any request can deviate and arrive into the system at the instant T receiving servicing completely and obtaining α with probability 1. Let's imagine that this value is less than S, which is the expected payoff at any another time instant of the strategy support.

So, if $p = 0$, a player coming at time $T-$ obtains α, which is not satisfied value, that's why he does not come at $T-$. The same is true for $(T-)-$ and so on. Then there is some interval without arrivals $[t, T]$ and some moment t when requests stop to arrive. Then any request coming at $t-$ obtains α too. It means that the interval without arrivals is $(0, T]$. In this case all arrive at the instant 0 and obtain $\frac{\alpha}{N+1}$ which is less than α. That the point T is an atom in the equilibrium strategy.

Denote X_p a random variable which is the number of requests coming at the last time moment T. The expected service time for each request coming at the moment T is

$$\alpha E \left[\frac{1}{X_p + 1} \right] = \alpha \left(P(X_p = 0) + E \left[\frac{1}{X_p + 1} \Big| X_p > 0 \right] \right).$$

Assume that there are no arrivals after the time instant t till the last time moment T, i.e. on the interval (t, T). If the request arrives at the instant t, its payoff is

$$\alpha \left(P(X_p = 0) + P(X_p > 0)(1 - e^{-\mu(T-t)}) \right) + P(X_p > 0)e^{-\mu(T-t)}(T - t)$$
$$= \alpha - P(X_p > 0)e^{-\mu(T-t)}(\alpha - (T - t)).$$

It decreases by t when $t > T - \alpha - \frac{1}{\mu}$ due to

$$\left(e^{-\mu(T-t)}(\alpha - (T - t)) \right)'_t = e^{-\mu(T-t)}\left(\alpha - \frac{1}{\mu}(T - t)\right)$$

is positive for $t > T - \alpha - \frac{1}{\mu}$.

If the request arrives at $t = T-$ immediately before the last moment T, its payoff is

$$\lim_{t \to T-} \alpha \left(P(X_p = 0) + P(X_p > 0)(1 - e^{-\mu(T-t)}) \right) + P(X_p > 0)e^{-\mu(T-t)}(T - t)$$
$$= \alpha P(X_p = 0).$$

It is less than the payoff for the request coming at the instant T.

Thus, the player's payoff decreases on the interval $(T - \alpha - \frac{1}{\mu}, T)$ and remains less than at the moment T. That's why no requests arrive into the system on the time interval $[t_e, T)$. $\qquad \square$

Assume that p is a positive known equilibrium probability. The player's payoff is a decreasing function on the interval $(T - \alpha - \frac{1}{\mu}, T)$. Hence, if there exists an instant $t_e > T - \alpha - \frac{1}{\mu}$ such that payoffs at the instants t_e and T are equal, it satisfies to an equation

$$\alpha E\left[\frac{1}{X_p + 1}\right] = \alpha - P(X_p > 0)e^{-\mu(T - t_e)}(\alpha - (T - t_e))). \qquad (1)$$

Since $E\left[\frac{1}{X_p+1}\right]$ is a probability to be the one winner among $X_p + 1$ uniform players, it does not exceed 1. The Eq. (1) can be transformed as

$$\frac{\alpha(1 - E\left[\frac{1}{X_p+1}\right])}{P(X_p > 0)} = e^{-\mu(T - t_e)}(\alpha - (T - t_e)).$$

The left part of the equation is positive for at least two players. It means that if a root t_e of the equation exists, it is such that $\alpha - (T - t_e) > 0$. So, we obtained the following lemma.

Lemma 2. t_e is a root of the Eq. (1) iff $\alpha - (T - t_e) > 0$.

Further we assume that the Eq. (1) has a root. Note also that $\alpha - (\alpha - (t_2 - t_1))e^{-\mu(t_2 - t_1)} = \alpha(1 - e^{-\mu(t_2 - t_1)}) + (t_2 - t_1)e^{-\mu(t_2 - t_1)} \geq 0$ for any $t_2 \geq t_1 \geq 0$, since $e^{-\mu(t_2 - t_1)} \leq 1$.

Lemma 3. *If $0 < t_e$, then there is a strictly positive equilibrium density function $f(t) > 0$ of the arrival moments at the interval $[0, t_e]$. This interval has no atoms or discontinuities.*

Proof. We show first that the interval $[0, t_e]$ has no discontinuities, that is the equilibrium distribution density function is strictly positive over the entire interval. Assume that there is is some interval $(t_1, t_2) \in [0, t_e]$ such that the equilibrium distribution density function is zero on this interval. If some request arrives at the moment t_1, its payoff is

$$\alpha - (\alpha - (t_2 - t_1))e^{-\mu(t_2 - t_1)} + \int_{t_2}^{T} (\alpha - (\alpha - (\theta - t_1))e^{-\mu(\theta - t_1)})dP(\theta),$$

where $dP(\theta)$ is a probability that another request arrives to the system at the instant θ. The same request could arrive also at the moment t_2, providing its payoff to be equal to

$$\int_{t_2}^{T} (\alpha - (\alpha - (\theta - t_2))e^{-\mu(\theta - t_2)})dP(\theta),$$

but it is less than payoff at the moment t_1. It contradicts with the demand of constant payoff on the equilibrium strategy support.

Now, show that the strategy on the interval $[0, t_e]$ has no atoms. Assume there is a point $t \in [0, t_e]$ which is an atom for the strategy and the probability to come at the time moment t is $p > 0$. Denote $t+$ the moment, which is immediately after the moment t. Consider the request arriving at the moment t. Denote X_p the random value which is the number of another requests coming at the same moment, it is positive since p is positive. represent the number of his opponents whose requests entered the system at the instant t. Due to the strict positivity of the probability p, this random variable must also be positive. The payoff for request at the moment t is

$$E \frac{1}{X_p + 1} \int_{t+}^{T} (\alpha - (\alpha - (\theta - t+))e^{-\mu(\theta - t+)})dP(\theta).$$

At the moment $t+$ immediately after t the payoff is

$$\int_{t+}^{T} (\alpha - (\alpha - (\theta - t+))e^{-\mu(\theta - t+)})dP(\theta).$$

So, if there is an atom at the moment t, it is better to come immediately after this moment.   □

3   The Nash Equilibrium in the Queueing Game

Let's consider one of $N + 1$ players sending requests into the system. This player shares the service with N opponents. We assume that at time $t = T$ each of N opponents sends his request to the system with probability p. Let the random value X_p with binomial distribution $Bin(N, p)$ be the number of requests arriving into the system at time T. Then, for our player the payoff at time T is

$$S(T) = \alpha E\left[\frac{1}{X_p + 1}\right] = \alpha \sum_{i=0}^{N} \binom{N}{i} p^i (1-p)^{N-i} \frac{1}{i+1} = \alpha \frac{1 - (1-p)^{N+1}}{p(N+1)}. \quad (2)$$

If the same player sends his request at the instant $t_e < T$ in case there is no customers arriving at the interval (t_e, T), his payoff is defined by

$$S(t_e) = \alpha - (1 - (1-p)^N)(\alpha - (T - t_e))e^{-\mu(T - t_e)}. \quad (3)$$

In the equilibrium both payoffs are equal, so p and t_e satisfy the equation

$$\frac{1 - (1-p)^{N+1}}{p(N+1)} = 1 - (1 - (1-p)^N)(1 - \frac{1}{\alpha}(T - t_e))e^{-\mu(T - t_e)}. \quad (4)$$

Lemma 4. *If $N = 1$, the Eq. (4) defines a constant t_e that is independent of p. If $N > 1$ the Eq. (4) defines a function $t_e(p)$ that strictly increases in p.*

Proof. Let $N = 1$. The Eq. (4) becomes

$$(1 - p) \cdot 1 + p\frac{1}{2} = 1 - pe^{-\mu(T-t_e)}(1 - \frac{1}{\alpha}(T - t_e)).$$

which is transformed to

$$e^{\mu(T-t_e)} = 2(1 - \frac{1}{\alpha}(T - t_e)) \tag{5}$$

So, the solution t_e doesn't depend on p.

Let $N = n > 1$. The Eq.(4) has the form

$$\sum_{k=0}^{n} \frac{1}{k+1} \binom{n}{k} p^k (1-p)^{n-k} = 1 - (1 - (1-p)^n)(1 - \frac{1}{\alpha}(T - t_e))e^{-\mu(T-t_e)}. \tag{6}$$

Expressing the exponent from (6) we obtain

$$e^{-\mu(T-t_e)} = \frac{1 - \sum_{k=0}^{n} \frac{1}{k+1} \binom{n}{k} p^k (1-p)^{n-k}}{(1 - (1-p)^n)(1 - \frac{1}{\alpha}(T - t_e))}. \tag{7}$$

We differentiate (7) by p and obtain

$$\mu e^{-\mu(T-t_e)} \frac{dt_e}{dp} = \frac{(1 - (1-p)^n)^2 - n^2 p^2 (1-p)^{n-1}}{(n+1)(1 - (1-p)^n)^2 p^2 (1 - \frac{1}{\alpha}(T - t_e))}. \tag{8}$$

The expression $1 - \frac{1}{\alpha}(T - t_e)$ is positive due to Lemma 2.

Further we use Cauchy inequality

$$\frac{\sum_{i=0}^{n-1} (1-p)^i}{n} \geq \left(\prod_{i=0}^{n} (1-p)^i \right)^{\frac{1}{n}} = (1-p)^{\frac{n-1}{2}}.$$

It yields

$$(1 - (1-p)^n)^2 = p^2 \left(\sum_{i=0}^{n-1} (1-p)^i \right)^2 \geq n^2 p^2 (1-p)^{n-1},$$

which provides a positivity of the right side of (8) for $0 < p \leq 1$, and, so, of the derivative dt_e/dp.

The Lemma 4 yields that the higher the probability p of coming at the last moment T, the larger the interval $[0, t_e]$ where requests arrive with a positive density. Then if $t_e(1) \leq 0$, then the equilibrium strategy is pure and all requests come into the system at time $t = T$ with probability 1. Further, we assume that $t_e(1) > 0$.

It is necessary to find the equilibrium density function $f(t)$ for the arrival time in the system on the interval $[0, t_e]$. As the queuing system is the same as in

[4], we use defined there Markov process with system states (i) at each instant $t \in [0, t_e]$, where $i \in \{0, \ldots, N\}$ indicates the number of players who have sent their requests to the system before the time t. The state probabilities for this process are [4]

$$p_i(t) = \binom{N}{i} F(t)^i (1 - F(t))^{N-i} \text{ for } i = 0, \ldots, N. \tag{9}$$

Since to the moment $t = 0$ the system is empty, the initial state probabilities are $p_0(0) = 1$ and $p_i(0) = 0$ for $i = 1, \ldots, N$.

Then the payoff function of player arriving at the instant $t \in [0, t_e]$ is

$$S(t) = \sum_{i=0}^{N-1} p_i(t) S_{N-i}(t) + \alpha p_N(t),$$

where $S_j(t)$ is a conditional payoff of player coming at the instant $t \in [0, t_e]$, provided that j requests have not arrived into the system yet before the instant t.

Let $j = 1$ and τ_1 is an instant when opponent's request arrives to the system. Then

$$
\begin{aligned}
S_1(t) &= \alpha \left(E(1 - e^{-\mu(\tau_1 - t)} | t \le \tau_1 \le t_e) + P(\tau_1 = T)(1 - e^{-\mu(T-t)}) \right) \\
&\quad + E((\tau_1 - t)e^{-\mu(\tau_1 - t)} | t \le \tau_1 \le t_e) + P(\tau_1 = T)e^{-\mu(T-t)}(T - t) \\
&= E(\alpha - e^{-\mu(\tau_1 - t)}(\alpha - (\tau_1 - t)) | t \le \tau_1 \le t_e) + P(\tau_1 = T)(\alpha - e^{-\mu(T-t)}(\alpha - (T - t))) \\
&= \tfrac{1}{1-F(t)} \left[\int_t^{t_e} \alpha - e^{-\mu(\tau - t)}(\alpha - (\tau - t)) dF(\tau) \right. \\
&\quad \left. + p \left(\alpha - e^{-\mu(T-t)}(\alpha - (T - t)) \right) \right].
\end{aligned}
$$

For $j = 2$ let τ_1, τ_2 be the time moments when two opponents' requests come into the system. Hence

$$
\begin{aligned}
S_2(t) &= \alpha \left(2E(1 - e^{-\mu(\tau_1 - t)} | t \le \tau_1 \le \tau_2 \le t_e) + 2E(1 - e^{-\mu(\tau_1 - t)} | t \le \tau_1 \le t_e, \tau_2 = T) \right. \\
&\quad \left. + P(\tau_1 = T, \tau_2 = T)(1 - e^{-\mu(T-t)}) \right) \\
&\quad + \left(2E((\tau_1 - t)e^{-\mu(\tau_1 - t)} | t \le \tau_1 \le \tau_2 \le t_e) + 2E((\tau_1 - t)e^{-\mu(\tau_1 - t)} | t \le \tau_1 \le t_e, \tau_2 = T) \right. \\
&\quad \left. + P(\tau_1 = T, \tau_2 = T)((T - t)e^{-\mu(T-t)}) \right) \\
&= 2E(\alpha - e^{-\mu(\tau_1 - t)}(\alpha - (\tau_1 - t)) | t \le \tau_1 \le \tau_2 \le t_e) \\
&\quad + 2E(\alpha - e^{-\mu(\tau_1 - t)}(\alpha - (\tau_1 - t)) | t \le \tau_1 \le t_e, \tau_2 = T) \\
&\quad + P(\tau_1 = T, \tau_2 = T)(\alpha - e^{-\mu(T-t)}(\alpha - (T - t))) \\
&= \tfrac{1}{(1-F(t))^2} \left(2 \int_t^{t_e} dF(t_1) \int_{t_1}^{t_e} dF(t_2)(\alpha - e^{-\mu(t_1 - t)}(\alpha - (t_1 - t))) \right. \\
&\quad + 2p \int_t^{t_e} dF(t_1)(\alpha - e^{-\mu(t_1 - t)}(\alpha - (t_1 - t))) \\
&\quad \left. + p^2(\alpha - e^{-\mu(T-t)}(\alpha - (T - t))) \right) \\
&= \tfrac{1}{(1-F(t))^2} \left(2 \int_t^{t_e} dF(t_1)(1 - F(t_1))(\alpha - e^{-\mu(t_1 - t)}(\alpha - (t_1 - t))) \right. \\
&\quad \left. + p^2(\alpha - e^{-\mu(T-t)}(\alpha - (T - t))) \right).
\end{aligned}
$$

Similarly we obtain for $k = 2, \ldots, N$

$$
\begin{aligned}
S_k(t) &= kE(\alpha - e^{-\mu(\tau_1 - t)}(\alpha - (\tau_1 - t)) | t \le \tau_1 \le t_e, \tau_1 \le \tau_j, j = 2, \ldots, N) + \\
&\quad P(\tau_j = T, j = 1, \ldots, N)(\alpha - e^{-\mu(T-t)}(\alpha - (T - t))) = \\
&\quad \tfrac{1}{(1-F(t))^k} \left(k \int_t^{t_e} dF(t_1)(1 - F(t_1))^{k-1}(\alpha - e^{-\mu(t_1 - t)}(\alpha - (t_1 - t))) \right. \\
&\quad \left. + p^k(\alpha - e^{-\mu(T-t)}(\alpha - (T - t))) \right).
\end{aligned}
$$

Substituting all $S_j(t)$ into $S(t)$ we obtain

$$S(t) = F(t)^N + \sum_{i=0}^{N-1} \binom{N}{i} F(t)^i p^{N-i} (\alpha - e^{-\mu(T-t)}(\alpha - (T-t))) +$$
$$\sum_{i=0}^{N-1} \binom{N}{i} F(t)^i (N-i) \int_t^{t_e} (\alpha - e^{-\mu(s-t)}(\alpha - (s-t)))(1 - F(s))^{N-i-1} dF(s).$$

The first sum is

$$(F(t)+p)^N (\alpha - e^{-\mu(T-t)}(\alpha - (T-t))) - F(t)^N (\alpha - e^{-\mu(T-t)}(\alpha - (T-t))).$$

Now we simplify the second sum

$$\sum_{i=0}^{N-1} \binom{N}{i} F(t)^i (N-i) \int_t^{t_e} (\alpha - e^{-\mu(s-t)}(\alpha - (s-t)))(1 - F(s))^{N-i-1} dF(s) =$$
$$N \sum_{i=0}^{N-1} \binom{N-1}{i} F(t)^i \int_t^{t_e} (\alpha - e^{-\mu(s-t)}(\alpha - (s-t)))(1 - F(s))^{N-i-1} dF(s) =$$
$$N \int_t^{t_e} (\alpha - e^{-\mu(s-t)}(\alpha - (s-t)))(F(t) + 1 - F(s))^{N-1} dF(s).$$

Finally, we get

$$S(t) = (F(t)+p)^N (\alpha - e^{-\mu(T-t)}(\alpha - (T-t))) + F(t)^N e^{-\mu(T-t)}(\alpha - (T-t)) +$$
$$N \int_t^{t_e} (\alpha - e^{-\mu(s-t)}(\alpha - (s-t)))(F(t) + 1 - F(s))^{N-1} dF(s).$$

$$(10)$$

The equilibrium payoff is a constant on the interval $[0, t_e]$, so the distribution $F(t)$ must satisfy the equation $S(t) = S(t_e)$ for $t \in [0, t_e]$, that is

$$(F(t)+p)^N (\alpha - e^{-\mu(T-t)}(\alpha - (T-t))) + F(t)^N e^{-\mu(T-t)}(\alpha - (T-t)) +$$
$$N \int_t^{t_e} (\alpha - e^{-\mu(s-t)}(\alpha - (s-t)))(F(t) + 1 - F(s))^{N-1} dF(s) = \qquad (11)$$
$$\alpha - (1 - (1-p)^N)(\alpha - (T-t_e))e^{-\mu(T-t_e)}.$$

Now we need to determine p, which is the probability to come at the moment $t = T$. It can be found from the normalization condition

$$\int_0^{t_e} dF(t) + p = 1. \qquad (12)$$

So, we obtain the following:

Theorem 1. *The symmetric Nash equilibrium in the $N+1$-person queueing game with preemptive access and fixed reward for completing request is described by the distribution function $F(t)$ on the interval $[0, T]$, which has the following properties.*

1. *There is a strictly positive probability p for a request to arrive into the system at the instant T.*
2. *There is the interval $[t_e, T)$, where t_e is determined by (4), when the requests do not enter the service system.*
3. *If the solution of Eq. (4) is negative for $p = 1$, then requests arrive into the system at instant T. Otherwise, $p < 1$, and t_e is greater than 0; in addition, the density function $f(t)$ on the support $[0, t_e]$ is determined from Eq. (11).*
4. *The probability p to arrive into the system at the instant T is determined from Eq. (12).*
5. *In equilibrium, the expected reward which a player receives sending and performing his request in the system is equal to $S(T) = \alpha \frac{1-(1-p)^{N+1}}{\mu p(N+1)}$.*

4   Two-Players Game

Consider the game with two symmetric players ($N = 1$) in details. In this case t_e can be found from the Eq. (5) and does not depend on p. The payoff at the instant T is

$$S(T) = \alpha \frac{2-p}{2}.$$

For $0 \leq t \leq t_e$ the payoff has the form

$$S(t) = \alpha \left(p(1 - (1 - \frac{1}{\alpha}(T - t))e^{-\mu(T-t)}) + \int_0^t dF(\theta) + \int_t^{t_e} 1 - (1 - \frac{1}{\alpha}(\theta - t))e^{-\mu(\theta-t)} dF(\theta) \right).$$

As it must be constant on the strategy support, $\frac{dS(t)}{dt} \equiv 0$ for $0 \leq t \leq t_e$. It yields the following differential equation

$$p(\frac{\mu}{\alpha}(T - t) - \mu - \frac{1}{\alpha})e^{-\mu T} + f(t)e^{-\mu t} + \int_t^{t_e} (\frac{\mu}{\alpha}(\theta - t) - \mu - \frac{1}{\alpha})e^{-\mu\theta} f(\theta)d\theta = 0.$$

Denoting $g(t) = f(t)e^{-\mu t}$ we obtain the following equation

$$p(\frac{\mu}{\alpha}(T - t) - \mu - \frac{1}{\alpha})e^{-\mu T} + g(t) + \int_t^{t_e} (\frac{\mu}{\alpha}(\theta - t) - \mu - \frac{1}{\alpha})g(\theta)d\theta \equiv 0. \quad (13)$$

Differentiating it twice we get

$$g'(t) + g(t)(\mu + \frac{1}{\lambda}) - \frac{\mu}{\lambda} \int_t^{t_e} g(\theta)d\theta = p\frac{\mu}{\lambda}e^{-\mu T} \quad (14)$$

and

$$g''(t) + g'(t)(\mu + \frac{1}{\lambda}) + \frac{\mu}{\lambda} = 0.$$

Consider now two cases: $\lambda = \frac{1}{\mu}$ and $\lambda <> \frac{1}{\mu}$.

4.1   Case $\alpha = \frac{1}{\mu}$

In this case a reward for complete service equals to expected service time in the absence of a competition.

The solution for the Eq. (13) is $g(t) = (c_1 + c_2 t)e^{-\mu t}$ where c_1 and c_2 are obtained from (13) and (14), using (5), and are equal to

$$c_1 = p\mu e^{-\mu(T-t_e)}(2 - \mu T + \mu t_e(2 - \mu(T - t_e))) = p\mu(e^{-\mu(T-t_e)} + \frac{\mu t_e + 1}{2}),$$

$$c_2 = -p\mu^2 e^{-\mu(T-t_e)}(1 - \mu(T - t_e)) = -\frac{p\mu^2}{2}.$$

Thus the equilibrium strategy is linear for $0 \le t \le t_e$:

$$f(t) = p\mu(e^{-\mu(T-t_e)} + \frac{1}{2}(\mu t_e - \mu t + 1)).$$

The equilibrium probability to send request at the time moment T can be found from normalization condition (12), that is

$$p\mu t_e(e^{-\mu(T-t_e)} + \frac{\mu t_e}{4} + \frac{1}{2}) = 1 - p.$$

The probability is

$$p = \frac{1}{1 + \mu t_e(e^{-\mu(T-t_e)} + \frac{\mu t_e}{4} + \frac{1}{2})}.$$

4.2   Case $\alpha <> \frac{1}{\mu}$

In this case the solution for the Eq. (13) is $g(t) = c_1 e^{\mu t} + c_2 e^{-\frac{1}{\alpha}t}$ where c_1 and c_2 are obtained from (13) and (14), using (5), and are equal to $c_1 = pk_1$ and $c_2 = pk_2$, where

$$k_1 = -\frac{e^{-\mu(T-t_e)}}{1 - \alpha\mu}\mu^2(\alpha - (T - t_e)) = -\frac{\mu^2\alpha}{2(1 - \alpha\mu)},$$

$$k_2 = \frac{e^{-(\mu T - \frac{t_e}{\alpha})}}{\alpha(1 - \alpha\mu)}(1 - \mu(T - t_e)).$$

Thus the equilibrium strategy for $0 \le t \le t_e$ is as follows

$$f(t) = c_1 + c_2 e^{-(\frac{1}{\alpha} - \mu)t}.$$

The equilibrium probability to send request at the time moment T can be found from normalization condition (12), that is

$$p(k_1 t_e + k_2 \frac{\alpha}{1 - \alpha\mu}(1 - e^{-(\frac{1}{\alpha} - \mu)t_e})) = 1 - p.$$

The probability is

$$p = \frac{1}{1 + k_1 t_e + k_2 \frac{\alpha}{1 - \alpha\mu}(1 - e^{-(\frac{1}{\alpha} - \mu)t_e})}.$$

5   Numerical Examples

Let $T = 12$, $\mu = 0.25$. The equilibrium optimal values for different N and α are presented in Table 1. The shapes of density functions for the equilibrium arrivals at the interval $[0, t_e]$ are shown at Fig. 1. We give visual illustrations for $\alpha = 2$ and $\alpha = 4$ only, since the case for $\alpha = 6$ looks like the case $\alpha = 4$.

Table 1. Optimal p, t_e and payoff for $T = 12$.

	$\alpha = 2$			$\alpha = \frac{1}{\mu} = 4$			$\alpha = 6$		
N	$S(t)$	p	t_e	$S(t)$	p	t_e	$S(t)$	p	t_e
2	1.90668	0.04741	11.22394	3.29235	0.18879	10.79619	4.57380	0.26028	10.53862
5	1.58973	0.09276	11.27590	2.46270	0.19945	10.96604	3.30488	0.24776	10.79655
10	1.16216	0.11629	11.37912	1.71094	0.19223	11.18377	2.25358	0.22799	11.08822
20	0.73663	0.12059	11.53625	1.05661	0.17728	11.44830	1.37644	0.20594	11.41092
100	0.17470	0.11334	11.88061	0.24812	0.15962	11.87296	0.32147	0.18479	11.86842

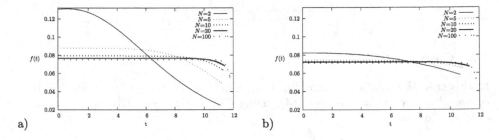

Fig. 1. Equilibrium PDF $f(t)$ for $T = 12$: a) for $\alpha = 2$, b) for $\alpha = 4$.

Now we explore how the value of a reward α influences on the probability to complete request performing in the equilibrium. From the paper [4] we know an expression for this probability

$$C(t) = (F(t) + p)^N (1 - e^{-\mu(T-t)}) + F(t)^N e^{-\mu(T-t)} +$$
$$N \int_t^{t_e} (1 - e^{-\mu(s-t)})(F(t) + 1 - F(s))^{N-1} dF(s),$$

where $F(\cdot)$ is a distribution function of players' arrival instants on the interval $[0, T]$. The Fig. 2 allows to observe how the probability is changed on the working interval for different α. It's interesting that if reward for complete service is small, the equilibrium probability to perform request completely is less at the begin and more at the end of working interval compared with the situation where α is about (and more than) $\frac{1}{\mu}$.

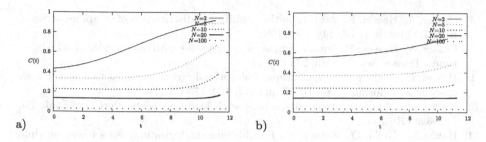

Fig. 2. Equilibrium $C(t)$ for $T = 12$: a) for $\alpha = 2$, b) for $\alpha = 4$.

6 Conclusion

We explored a game-theoretic model for a single-server queueing system where players send their requests into the system according to preemptive access on a time interval $[0, T]$. We shown that the symmetric equilibrium possesses the following features. The strictly positive density function of the request arrival time instants is defined at the time interval $[0, t_e]$. On a time interval (t_e, T) players don't send requests. At the last time moment T the players send their requests to the system with a certain positive probability p. At this moment each player has a chance to obtain its performing winning one place among other players, who also send their requests at the moment T. If it wins, it exactly obtains its complete service, because no one sends any request after the moment T. We present results of numerical experiments demonstrating the equilibria with different values of the model parameters. Also we explore numerically how the value of a reward influences on the probability to complete request performing.

References

1. Altman, E., Shimkin, N.: Individually optimal dynamic routing in a processor sharing system. Oper. Res. **46**, 776–784 (1998)
2. Breinbjerg, J., Platz, T.T., Østerdal, L.P.: Equilibrium Arrivals to a Last-come First-served Preemptive-resume Queue, Working Papers 17–2020, Copenhagen Business School, Department of Economics (2020). https://ideas.repec.org/p/hhs/cbsnow/2020_017.html. Accessed 28 Apr 2022
3. Chirkova, Y.V.: Optimal arrivals in a two-server random access system with loss. Autom. Remote. Control. **78**(3), 557–580 (2017). https://doi.org/10.1134/S0005117917030146
4. Chirkova, J., Mazalov, V.: Optimal arrivals to preemptive queueing system. In: Pardalos, P., Khachay, M., Mazalov, V. (eds.) MOTOR 2022. LNCS, vol. 13367, pp. 169–181. Springer, Cham (2022). https://doi.org/10.1007/978-3-031-09607-5_12
5. Chirkova, J., Mazalov, V., Morozov, E.: Equilibrium in a queueing system with retrials. Mathematics **10**(3), 428 (2022)
6. Dimitriou, I.: A queueing system for modeling cooperative wireless networks with coupled relay nodes and synchronized packet arrivals. Perform. Eval. **114**(C), 16–31 (2017)

7. Glazer, A., Hassin, R.: ?/M/1: on the equilibrium distribution of customer arrivals. Eur. J. Oper. Res. **13**, 146–150 (1983)
8. Glazer, A., Hassin,: R. Equilibrium arrivals in queues with bulk service at scheduled times. Transp. Sci. **21**, 273–278 (1987)
9. Hassin, R., Kleiner, Y.: Equilibrium and optimal arrival patterns to a server with opening and closing times. IIE Trans. **43**, 164–175 (2011)
10. Haviv, M.: When to arrive at a queue with tardiness costs. Perform. Eval. **70**, 387–399 (2013)
11. Haviv, M., Kella, O., Kerner, Y.: Equilibrium strategies in queues based on time or index of arrival. Prob. Eng. Inform. Sci. **24**, 13–25 (2010)
12. Haviv, M., Ravner, L.: A survey of queueing systems with strategic timing of arrivals. Queueing Syst. **99**, 163–198 (2021)
13. Jane, R., Juneja, S., Shimkin, N.: The concert queueing game: to wait or to be late. Discrct. Evcnt Dyn. Syst. **21**, 103–138 (2011)
14. Mazalov, V., Chirkova, J.: Networking Games. Network Forming Games and Games on Networks, p. 322. Academic Press, Cambridge (2019)
15. Mazalov, V.V., Chuiko, J.V.: Nash equilibrium in the optimal arrival time problem. Comput. Technol. **11**, 60–71 (2006)
16. Haviv, M., Ravner, L.: Strategic timing of arrivals to a finite queue multi-server loss system. Queueing Syst. **81**(1), 71–96 (2015). https://doi.org/10.1007/s11134-015-9453-y
17. Ravner, L., Haviv, M.: Equilibrium and socially optimal arrivals to a single server loss system. In International Conference on Network Games Control and Optimization 2014 (NETGCOOP 2014), Trento, Italy (2014)
18. Teraoka, Y., Hohjo, H.: N-person silent game on sale. Sci. Math. Japonicae **63**(2), 237–240 (2006)

On Optimal Positional Strategies in Fractional Optimal Control Problems

Mikhail I. Gomoyunov[1,2]([✉])[ID]

[1] Krasovskii Institute of Mathematics and Mechanics, Ural Branch of Russian
Academy of Sciences, Ekaterinburg 620108, Russia
[2] Ural Federal University, Ekaterinburg 620002, Russia
m.i.gomoyunov@gmail.com

Abstract. We consider an optimal control problem for a dynamical system described by a Caputo fractional differential equation and a Mayer cost functional. We propose a new method for constructing an optimal positional control strategy, which allows us to generate near optimal controls for any initial system state by using time-discrete recursive feedback control procedures. The method is based on the knowledge of the optimal result functional and uses a special Lyapunov–Krasovskii functional.

Keywords: Fractional differential equation · Optimal control problem · Positional control strategy · Lyapunov–Krasovskii functional

1 Introduction

In this paper, we consider an optimal control problem for a dynamical system described by a nonlinear fractional differential equation

$$(^{C}D^{\alpha}x)(\tau) = f(\tau, x(\tau), u(\tau)), \tag{1a}$$

where $\tau \in [0, T]$ is time, $T > 0$ is a fixed time horizon, $x(\tau) \in \mathbb{R}^n$ and $u(\tau) \in P$ are values of the state and control vectors at a current time τ, respectively, $P \subset \mathbb{R}^{n_u}$ is a compact set, $(^{C}D^{\alpha}x)(\tau)$ is a Caputo fractional derivative of order $\alpha \in (0, 1)$ defined by (see, e.g., [9, Section 2.4] and [2, Chapter 3])

$$(^{C}D^{\alpha}x)(\tau) \doteq \frac{1}{\Gamma(1-\alpha)} \frac{\mathrm{d}}{\mathrm{d}\tau} \int_{0}^{\tau} \frac{x(\xi) - x(0)}{(\tau - \xi)^{\alpha}} \, \mathrm{d}\xi,$$

and Γ is the gamma-function. Given an initial condition

$$x(0) = x_0, \tag{1b}$$

where $x_0 \in \mathbb{R}^n$ is an initial value of the state vector, satisfying the inclusion $x_0 \in B(M) \doteq \{x \in \mathbb{R}^n \colon \|x\| \leq M\}$ with a fixed number $M > 0$, the goal of control is to minimize the value of a Mayer cost functional

$$J \doteq \sigma(x(T)). \tag{1c}$$

This work was supported by the Russian Science Foundation, project 21-71-10070, https://rscf.ru/en/project/21-71-10070/.

We study the question of constructing an optimal positional control strategy (see, e.g., [10,11,14] and also [4–6] for the case of fractional-order systems) in problem (1), which allows us to generate ζ-optimal controls with any specified tolerance $\zeta > 0$ and for any initial value of the state vector $x_0 \in B(M)$ by using the corresponding time-discrete recursive feedback control procedures.

Let us note that such questions are particularly important if the dynamical system is additionally affected by disturbances and we consider a problem of optimization of a guaranteed result (which may be embedded into the corresponding zero-sum differential game). Nevertheless, for brevity, we focus on the optimal control problem (1), leaving behind the fact that the results obtained in the paper admit a straightforward generalization to this more complex class of problems. Let us note also that the results can be directly extended to the case of a Bolza cost functional.

According to [4–6], a position of system (1a) is a pair $(t, w(\cdot))$ consisting of a time $t \in [0, T]$ and a function $w \colon [0, t] \to \mathbb{R}^n$, treated as a motion history on the interval $[0, t]$. The need to take into account not only the current value of the state vector but the entire motion history is related to hereditary nature of the Caputo fractional derivative. This feature is one of the main reasons for using fractional-order systems in mathematical modeling of real processes with memory effects. On the other hand, it indicates that such systems are essentially infinite-dimensional. This circumstance complicates behavioral analysis of fractional-order systems, including development of feedback control schemes, which is the subject of this paper.

Let us recall that two methods for constructing optimal positional control strategies in problem (1) were already suggested in [5,6]. Both of them are based on the knowledge of the value functional (the optimal result functional), and their key elements are Lyapunov–Krasovskii functionals with suitable properties. However, one of the disadvantages of these methods is that the corresponding functionals are quite complex and difficult to deal with. In particular, they do not have differentiability properties useful in practice.

A method for constructing an optimal positional control strategy proposed in this paper goes back to the constructions from, e.g., [1,3] and [14, Section 12.2] (see also, e.g., [12] for the case of time-delay systems). It is generally close to the method from [6], but is devoid of the indicated disadvantage. Indeed, a Lyapunov–Krasovskii functional that forms its basis is relatively simple and continuously differentiable in the sense of fractional coinvariant differentiation (see [4]) and has a simple analog in the classical non-fractional case (i.e., when $\alpha = 1$). This functional was built in [7] and used there in order to prove a uniqueness theorem for viscosity solutions of the Hamilton–Jacobi–Bellman equation associated with problem (1). Let us stress that this Lyapunov–Krasovskii functional, in general, does not satisfy the set of conditions from [6, Section 8] (in particular, the key condition (c)), and, therefore, the results obtained in this paper are not a consequence of [6]. Nevertheless, under an assumption of local Lipschitz continuity of the function σ from (1c), which entails the corresponding local Lipschitz continuity of the value functional [8, Lemma 1], and by appropriately modifying the proof of [6, Lemma 3], we succeed in substantiating the optimality of the constructed positional control strategy.

The paper is organized as follows. In Sect. 2, after formulating basic assumptions, we introduce a space of positions of system (1a), describe motions of this system generated by open-loop controls, and give a definition of an optimal positional control strategy in problem (1). In Sect. 3, we consider the value functional and the Lyapunov–Krasovskii functional and recall some of their properties. In Sect. 4, we describe and substantiate the new method for constructing an optimal positional control strategy. In Sect. 5, we draw some conclusions.

2 Problem Statement

2.1 Basic Assumptions

Throughout the paper, it is assumed that the function $f\colon [0,T] \times \mathbb{R}^n \times P \to \mathbb{R}^n$ is continuous; for any $R > 0$, there exist $\lambda_f > 0$ and $\lambda_\sigma > 0$ such that

$$\|f(\tau, x, u) - f(\tau, x', u)\| \le \lambda_f \|x - x'\|, \quad |\sigma(x) - \sigma(x')| \le \lambda_\sigma \|x - x'\| \qquad (2)$$

for all $\tau \in [0,T]$, x, $x' \in B(R)$, and $u \in P$; and there exists $c_f > 0$ such that

$$\|f(\tau, x, u)\| \le c_f (1 + \|x\|) \quad \forall \tau \in [0,T] \quad \forall x \in \mathbb{R}^n \quad \forall u \in P. \qquad (3)$$

In the above, $\|\cdot\|$ stands for the Euclidean norm in $\mathbb{R}^n$.

2.2 Space of Admissible Positions

Denote by $AC^\alpha([0,T], \mathbb{R}^n)$ the set of all functions $x\colon [0,T] \to \mathbb{R}^n$ each of which can be represented in the following form for some (Lebesgue) measurable and essentially bounded function $f\colon [0,T] \to \mathbb{R}^n$ (see, e.g., [13, Definition 2.3]):

$$x(\tau) = x(0) + \frac{1}{\Gamma(\alpha)} \int_0^\tau \frac{f(\xi)}{(\tau - \xi)^{1-\alpha}}\, d\xi \quad \forall \tau \in [0,T]. \qquad (4)$$

The second term in the right-hand side of this equality is the Riemann–Liouville fractional integral of order α of the function $f(\cdot)$ (see, e.g., [13, Definition 2.1]). In accordance with, e.g., [13, Remark 3.3], the set $AC^\alpha([0,T], \mathbb{R}^n)$ is considered as a subset of the space of all continuous functions from $[0,T]$ to $\mathbb{R}^n$ endowed with the usual supremum norm. Due to, e.g., [13, Theorem 2.4], any function $x(\cdot) \in AC^\alpha([0,T], \mathbb{R}^n)$ has at almost every (a.e.) $\tau \in [0,T]$ a Caputo fractional derivative $(^C D^\alpha x)(\tau)$. In addition, if equality (4) holds for some measurable and essentially bounded function $f(\cdot)$, then $(^C D^\alpha x)(\tau) = f(\tau)$ for a.e. $\tau \in [0,T]$.

Let us introduce the set X of all functions $x(\cdot) \in AC^\alpha([0,T], \mathbb{R}^n)$ such that $x(0) \in B(M)$ and $\|(^C D^\alpha x)(\tau)\| \le c_f(1 + \|x(\tau)\|)$ for a.e. $\tau \in [0,T]$, where M is the number that bounds an initial value of the state vector (see (1b)) and c_f is the number from assumption (3). Then, in accordance with [4–6], we define an admissible position of system (1a) as a pair $(t, w(\cdot))$ such that $t \in [0,T]$ and there exists $x(\cdot) \in X$ for which $w(\cdot) = x_t(\cdot)$, where $x_t(\cdot)$ denotes the restriction of the function $x(\cdot)$ to the interval $[0, t]$. Let G be the set of all admissible positions. We endow the set G with the metric

$$\text{dist}\big((t, w(\cdot)), (t', w'(\cdot))\big) \doteq |t - t'| + \max_{\tau \in [0, T]} \|w(\min\{\tau, t\}) - w'(\min\{\tau, t'\})\| \quad (5)$$

for all $(t, w(\cdot)), (t', w'(\cdot)) \in G$. Let us note that the set X is compact (see, e.g., [5, Assertion 7]), which, in particular, implies compactness of the space G due to continuity of the mapping $[0, T] \times X \ni (t, x(\cdot)) \mapsto (t, x_t(\cdot)) \in G$.

2.3 Open-Loop Controls and System Motions

Let a position $(t, w(\cdot)) \in G$ be given. An admissible open-loop control on the interval $[t, T]$ is a measurable function $u \colon [t, T] \to P$. Let $\mathcal{U}(t, T)$ be the set of all such controls. A motion of system (1a) generated from the position $(t, w(\cdot))$ by a control $u(\cdot) \in \mathcal{U}(t, T)$ is defined as a function $x(\cdot) \in AC^\alpha([0, T], \mathbb{R}^n)$ that satisfies the condition $x_t(\cdot) = w(\cdot)$ and together with the function $u(\cdot)$ satisfies the fractional differential equation (1a) for a.e. $\tau \in [t, T]$. It can be shown (see, e.g., [4, Section 5]) that such a motion $x(\cdot)$ exists, is unique, and belongs to the set X. In what follows, we denote it by $x(\cdot) \doteq x(\cdot \mid t, w(\cdot), u(\cdot))$. Note that this motion $x(\cdot)$ is a unique continuous function that satisfies the condition $x_t(\cdot) = w(\cdot)$ and the nonlinear weakly-singular Volterra integral equation

$$x(\tau) = a(\tau \mid t, w(\cdot)) + \frac{1}{\Gamma(\alpha)} \int_t^\tau \frac{f(\xi, x(\xi), u(\xi))}{(\tau - \xi)^{1-\alpha}} \, d\xi \quad \forall \tau \in [t, T], \qquad (6)$$

where the function $a(\cdot \mid t, w(\cdot))$ is defined by

$$a(\tau \mid t, w(\cdot)) \doteq \begin{cases} w(\tau) & \text{if } \tau \in [0, t], \\ w(0) + \dfrac{1}{\Gamma(\alpha)} \displaystyle\int_0^t \frac{(^C D^\alpha w)(\xi)}{(\tau - \xi)^{1-\alpha}} \, d\xi & \text{if } \tau \in (t, T]. \end{cases} \qquad (7)$$

In other words, $a(\cdot) \doteq a(\cdot \mid t, w(\cdot))$ is a unique function from $AC^\alpha([0, T], \mathbb{R}^n)$ such that $a_t(\cdot) = w(\cdot)$ and $(^C D^\alpha a)(\tau) = 0$ for a.e. $\tau \in [t, T]$.

Note also that initial condition (1b) corresponds to the position $(0, w(\cdot)) \in G$ with $w(0) \doteq x_0$, and, therefore, for any control $u(\cdot) \in \mathcal{U}(0, T)$, we can consider the motion $x(\cdot) \doteq x(\cdot \mid x_0, u(\cdot))$ of system (1a) under initial condition (1b).

2.4 Positional Control Strategies

In accordance with [4–6], we define a positional control strategy in the optimal control problem (1) as a mapping $G_0 \times (0, \infty) \ni ((t, w(\cdot)), \varepsilon) \mapsto U(t, w(\cdot), \varepsilon) \in P$, where $G_0 \doteq \{(t, w(\cdot)) \in G \colon t < T\}$ and ε is an auxiliary accuracy parameter.

Let us fix $\varepsilon > 0$ and choose a partition Δ of the interval $[0, T]$, i.e.,

$$\Delta \doteq \{\tau_j\}_{j \in \overline{1, k+1}}, \quad \tau_1 = 0, \quad \tau_j < \tau_{j+1} \quad \forall j \in \overline{1, k}, \quad \tau_{k+1} = T.$$

Then, given an initial value of the state vector $x_0 \in B(M)$, a positional control strategy U forms a piecewise constant control $u(\cdot) \in \mathcal{U}(0,T)$ and the corresponding motion $x(\cdot) \doteq x(\cdot \mid x_0, u(\cdot))$ of system (1a) by the following time-discrete recursive feedback control procedure: at every time τ_j with $j \in \overline{1,k}$, the history $x_{\tau_j}(\cdot)$ of the motion $x(\cdot)$ on $[0, \tau_j]$ is measured, the value $u_j \doteq U(\tau_j, x_{\tau_j}(\cdot), \varepsilon)$ is computed, and then the constant control $u(\tau) \doteq u_j$ is applied until τ_{j+1}, when a new measurement of the history $x_{\tau_{j+1}}(\cdot)$ is taken. In a short form,

$$u(\tau) \doteq U(\tau_j, x_{\tau_j}(\cdot), \varepsilon) \quad \forall \tau \in [\tau_j, \tau_{j+1}) \quad \forall j \in \overline{1,k}. \tag{8}$$

Formally putting $u(T) \doteq \tilde{u}$ for some fixed $\tilde{u} \in P$, we conclude that the described control procedure generates $u(\cdot)$ and $x(\cdot)$ uniquely. Let us denote the obtained open-loop control by $u(\cdot) \doteq u(\cdot \mid x_0, U, \varepsilon, \Delta)$.

We say that a positional control strategy U° is optimal if, for any number $\zeta > 0$, there exist a number $\varepsilon_* > 0$ and a function $\delta_* \colon (0, \varepsilon_*] \to (0, \infty)$ such that, for any initial value of the state vector $x_0 \in B(M)$, any value of the accuracy parameter $\varepsilon \in (0, \varepsilon_*]$, and any partition $\Delta \doteq \{\tau_j\}_{j \in \overline{1,k+1}}$ with the diameter $\mathrm{diam}(\Delta) \doteq \max_{j \in \overline{1,k}}(\tau_{j+1} - \tau_j) \leq \delta_*(\varepsilon)$, the control $u^\circ(\cdot) \doteq u(\cdot \mid x_0, U^\circ, \varepsilon, \Delta)$ is ζ-optimal in problem (1), i.e., the inequality below takes place:

$$\sigma\big(x(T \mid x_0, u^\circ(\cdot))\big) \leq \inf_{u(\cdot) \in \mathcal{U}(0,T)} \sigma\big(x(T \mid x_0, u(\cdot))\big) + \zeta. \tag{9}$$

The goal of this paper is to propose a new method for constructing an optimal positional control strategy U°.

3 Auxiliary Constructions

We follow the approach that relies on the consideration of the value functional (the optimal result functional) in problem (1), which is defined by

$$\rho(t, w(\cdot)) \doteq \inf_{u(\cdot) \in \mathcal{U}(t,T)} \sigma\big(x(T \mid t, w(\cdot), u(\cdot))\big) \quad \forall (t, w(\cdot)) \in G. \tag{10}$$

According to, e.g., [4, Sects. 6 and 8], the functional $\rho \colon G \to \mathbb{R}$ is continuous, satisfies the boundary condition $\rho(T, x(\cdot)) = \sigma(x(T))$ for all $x(\cdot) \in X$, and has the following property, expressing the dynamic programming principle in problem (1): for any $(t, w(\cdot)) \in G_0$, $t' \in [t, T]$, and $\eta > 0$, there is $u(\cdot) \in \mathcal{U}(t,T)$ such that $\rho(t', x_{t'}(\cdot)) \leq \rho(t, w(\cdot)) + \eta$, where $x(\cdot) \doteq x(\cdot \mid t, w(\cdot), u(\cdot))$. Moreover, by [8, Lemma 1], the functional ρ satisfies a special Lipschitz continuity property with respect to the variable $w(\cdot)$ (see property (ii) formulated below).

Let us introduce a Lyapunov–Krasovskii functional constituting the basis of the method. For any $t \in [0, T]$, let $G(t)$ be the set of all functions $w \colon [0, t] \to \mathbb{R}^n$ such that $(t, w(\cdot)) \in G$. Further, denote by G^* the set of all pairs $(t, w(\cdot))$ such that $t \in [0, T]$ and there exist $w'(\cdot), w''(\cdot) \in G(t)$ for which $w(\cdot) = w'(\cdot) - w''(\cdot)$. We endow the set G^* with a metric dist^* that is defined by analogy with (5).

Let us take $q \doteq 2/(2-\alpha)$, choose $\beta \in (0, \min\{1-\alpha, \alpha/2\})$, and, for any $\varepsilon > 0$, consider a functional $\nu_\varepsilon \colon G^* \to \mathbb{R}$ given by

$$\nu_\varepsilon(t, w(\cdot)) \doteq (\varepsilon^{\frac{2}{q-1}} + \|a(T)\|^2)^{\frac{q}{2}} + \int_0^T \frac{(\varepsilon^{\frac{2}{q-1}} + \|a(\xi)\|^2)^{\frac{q}{2}}}{(T-\xi)^{(1-\alpha-\beta)q}} \, \mathrm{d}\xi - C\varepsilon^{\frac{q}{q-1}} \quad (11)$$

for all $(t, w(\cdot)) \in G^*$, where the function $a(\cdot) \doteq a(\cdot \mid t, w(\cdot))$ is defined in accordance with (7) and $C \doteq 1 + T^{1-(1-\alpha-\beta)q}/(1 - (1-\alpha-\beta)q)$.

Let us briefly recall some properties of the functional ν_ε established in [7, Sect. 5.1]. They will be used in the next Sect. 4.

(i) For any $\varepsilon > 0$, the functional $\nu_\varepsilon \colon G^* \to \mathbb{R}$ is continuous and non-negative. In addition, if $t \in [0, T]$ and $w(\tau) \doteq 0$ for all $\tau \in [0, t]$, then $\nu_\varepsilon(t, w(\cdot)) = 0$.

(ii) There exists $\lambda_\rho > 0$ such that

$$|\rho(t, w(\cdot)) - \rho(t, w'(\cdot))| \le \lambda_\rho (\nu_\varepsilon(t, w(\cdot) - w'(\cdot)) + C\varepsilon^{\frac{q}{q-1}})^{\frac{1}{q}}$$

for all $\varepsilon > 0$, $t \in [0, T]$, and $w(\cdot), w'(\cdot) \in G(t)$.

(iii) For any $R > 0$ and $r > 0$, there is $\varepsilon_* > 0$ such that, for any $\varepsilon \in (0, \varepsilon_*]$, $t \in [0, T]$, and $w(\cdot), w'(\cdot) \in G(t)$ meeting the condition $\nu_\varepsilon(t, w(\cdot) - w'(\cdot))/\varepsilon \le R$, the inequality $\|w(\tau) - w'(\tau)\| \le r$ is valid for all $\tau \in [0, t]$.

(iv) For any $\varepsilon > 0$, there is a continuous mapping $\nabla^\alpha \nu_\varepsilon \colon G_0^* \to \mathbb{R}^n$ such that, for any $\theta \in (0, T)$ and $x(\cdot), x'(\cdot) \in X$, the function $\varkappa(\tau) \doteq \nu_\varepsilon(\tau, x_\tau(\cdot) - x'_\tau(\cdot))$ for all $\tau \in [0, T-\theta]$ is Lipschitz continuous and, for a.e. $\tau \in [0, T-\theta]$,

$$\dot{\varkappa}(\tau) = \langle \nabla^\alpha \nu_\varepsilon(\tau, x_\tau(\cdot) - x'_\tau(\cdot)), ({}^C D^\alpha x)(\tau) - ({}^C D^\alpha x')(\tau) \rangle.$$

In the above, $G_0^* \doteq \{(t, w(\cdot)) \in G^* \colon t < T\}$, $\dot{\varkappa}(\tau) \doteq \mathrm{d}\varkappa(\tau)/\mathrm{d}\tau$, and $\langle \cdot, \cdot \rangle$ stands for the inner product in $\mathbb{R}^n$.

(v) For any $\theta \in (0, T)$, there is $\kappa > 0$ such that, for any $\varepsilon > 0$, $t \in [0, T-\theta]$, and $w(\cdot), w'(\cdot) \in G(t)$, the mapping $\nabla^\alpha \nu_\varepsilon$ from (iv) satisfies the estimate

$$\|\nabla^\alpha \nu_\varepsilon(t, w(\cdot) - w'(\cdot))\| \le \kappa (\nu_\varepsilon(t, w(\cdot) - w'(\cdot)) + C\varepsilon^{\frac{q}{q-1}})^{\frac{q-1}{q}}.$$

It is important to note that the functional ν_ε is continuously differentiable in the sense of coinvariant differentiability of order α and the mapping $\nabla^\alpha \nu_\varepsilon$ is the coinvariant gradient of order α of this functional. For details, the reader is referred to [7, Lemma 5.4], where also an explicit formula for $\nabla^\alpha \nu_\varepsilon$ is presented. In particular, property (iv) is a corollary of [7, Lemma 5.4] and [4, Lemma 9.2]. By virtue of compactness of the space G, property (ii) can be derived from [8, Lemma 1] and [7, Lemma 5.2], while property (iii) follows from [7, Lemma 5.3].

4   Optimal Positional Control Strategy

For any $(t, w(\cdot)) \in G_0$ and $\varepsilon > 0$, taking into account that the value functional ρ and the Lyapunov–Krasovskii functional ν_ε are continuous and that the space G is compact, we consider the value

$$\rho_\varepsilon(t, w(\cdot)) \doteq \min_{z(\cdot) \in G(t)} \left(\rho(t, z(\cdot)) + \frac{\nu_\varepsilon(t, w(\cdot) - z(\cdot))}{\varepsilon} \right) \quad (12)$$

and choose a function $z(\cdot \mid t, w(\cdot), \varepsilon) \in G(t)$ at which the minimum in (12) is attained. Then, we define a positional control strategy U° from the condition

$$U^{\circ}(t, w(\cdot), \varepsilon) \in \underset{u \in P}{\operatorname{argmin}} \left\langle \frac{\nabla^{\alpha} \nu_{\varepsilon}(t, w(\cdot) - z(\cdot \mid t, w(\cdot), \varepsilon))}{\varepsilon}, f(t, w(t), u) \right\rangle \quad (13)$$

for all $(t, w(\cdot)) \in G_0$ and $\varepsilon > 0$.

Theorem 1. *Under the assumptions made in Sect. 2.1, the positional control strategy U° defined by (13) is optimal in problem (1).*

Proof. Let $R_* > 0$ be such that $|\rho(t, w(\cdot))| \leq R_*$ for all $(t, w(\cdot)) \in G$, and let $R^* > 0$ be such that $\nu_{\varepsilon}(t, w(\cdot) - w'(\cdot)) \leq R^*$ for all $t \in [0, T]$, $w(\cdot)$, $w'(\cdot) \in G(t)$, and $\varepsilon \in (0, 1]$. Denote $R \doteq \max\{2R_*, \lambda_{\rho}(R^* + C)^{\frac{1}{q}}\}$, where λ_{ρ}, C, and q are the numbers from property (ii) and definition (11) of ν_{ε}. Using assumption (2), let us take $\lambda_f > 0$ such that, for any $\tau \in [0, T]$, $x(\cdot)$, $x'(\cdot) \in X$, and $u \in P$,

$$\|f(\tau, x(\tau), u) - f(\tau, x'(\tau), u)\| \leq \lambda_f \|x(\tau) - x'(\tau)\|.$$

Let $\zeta > 0$ be fixed. Let us consider $\theta \in (0, T)$ such that

$$|\rho(T, x(\cdot)) - \rho(\tau, x_{\tau}(\cdot))| \leq \frac{\zeta}{4}$$

for all $\tau \in [T - \theta, T]$ and $x(\cdot) \in X$. By this number θ, let us choose $\kappa > 0$ due to property (v). In addition, let us take $r_* > 0$ such that, for any $t \in [0, T]$ and $w(\cdot)$, $w'(\cdot) \in G(t)$ satisfying the condition $\|w(\tau) - w'(\tau)\| \leq r_*$ for all $\tau \in [0, t]$, the inequality below is valid:

$$|\rho(t, w(\cdot)) - \rho(t, w'(\cdot))| \leq \frac{\zeta}{4}.$$

Denote $r \doteq \min\{r_*, \zeta/(8T\lambda_f \kappa(\lambda_{\rho} + C^{\frac{q-1}{q}}))\}$. By the numbers R and r, let us determine $\varepsilon_* \in (0, 1]$ based on property (iii). Finally, for any $\varepsilon \in (0, \varepsilon_*]$, let us choose $\delta_*(\varepsilon) \in (0, \theta/2]$ such that, for any τ, $\tau' \in [0, T - \theta/2]$ with $|\tau - \tau'| \leq \delta_*(\varepsilon)$, any $x(\cdot)$, $x'(\cdot)$, $x''(\cdot) \in X$, and any $u \in P$,

$$\left| \left\langle \frac{\nabla^{\alpha} \nu_{\varepsilon}(\tau, x_{\tau}(\cdot) - x'_{\tau}(\cdot))}{\varepsilon}, f(\tau, x''(\tau), u) \right\rangle \right.$$
$$\left. - \left\langle \frac{\nabla^{\alpha} \nu_{\varepsilon}(\tau', x_{\tau'}(\cdot) - x'_{\tau'}(\cdot))}{\varepsilon}, f(\tau', x''(\tau'), u) \right\rangle \right| \leq \frac{\zeta}{16T}.$$

Let us consider $x_0 \in B(M)$, $\varepsilon \in (0, \varepsilon_*]$, and a partition $\Delta \doteq \{\tau_j\}_{j \in \overline{1, k+1}}$ with $\operatorname{diam}(\Delta) \leq \delta_*(\varepsilon)$ and put $u^{\circ}(\cdot) \doteq u(\cdot \mid x_0, U^{\circ}, \varepsilon, \Delta)$ and $x^{\circ}(\cdot) \doteq x(\cdot \mid x_0, u^{\circ}(\cdot))$. Then, according to (9) and (10), in order to prove the theorem, we need to verify the inequality

$$\sigma(x^{\circ}(T)) \leq \rho(\tau_1, x^{\circ}_{\tau_1}(\cdot)) + \zeta. \quad (14)$$

Let $m \in \overline{2, k}$ be such that $\tau_{m-1} < T - \theta \leq \tau_m$. Note that $\tau_m \leq T - \theta/2$ since $\delta_*(\varepsilon) \leq \theta/2$. For any $j \in \overline{1, m}$, denote $z^{[j]}(\cdot) \doteq z(\cdot \mid \tau_j, x_{\tau_j}^{\circ}(\cdot), \varepsilon)$ for brevity. Due to (12) and the definition of R_*, we derive $\rho_\varepsilon(\tau_m, x_{\tau_m}^{\circ}(\cdot)) \leq \rho(\tau_m, x_{\tau_m}^{\circ}(\cdot)) \leq R_*$. On the other hand, we get $\rho_\varepsilon(\tau_m, x_{\tau_m}^{\circ}(\cdot)) \geq -R_* + \nu_\varepsilon(\tau_m, x_{\tau_m}^{\circ}(\cdot) - z^{[m]}(\cdot))/\varepsilon$ in view of the choice of $z^{[m]}(\cdot)$. Hence, we obtain $\nu_\varepsilon(\tau_m, x_{\tau_m}^{\circ}(\cdot) - z^{[m]}(\cdot))/\varepsilon \leq 2R_*$, which implies that $\|x_{\tau_m}^{\circ}(\tau) - z^{[m]}(\tau)\| \leq r_*$ for all $\tau \in [0, \tau_m]$, and, consequently, $|\rho(\tau_m, x_{\tau_m}^{\circ}(\cdot)) - \rho(\tau_m, z^{[m]}(\cdot))| \leq \zeta/4$. Therefore, we have

$$\rho_\varepsilon(\tau_m, x_{\tau_m}^{\circ}(\cdot)) \geq \rho(\tau_m, z^{[m]}(\cdot)) \geq \rho(\tau_m, x_{\tau_m}^{\circ}(\cdot)) - \frac{\zeta}{4} \geq \rho(T, x^{\circ}(\cdot)) - \frac{\zeta}{2}.$$

Thus, recalling the boundary condition for the value functional ρ, we find that $\rho_\varepsilon(\tau_m, x_{\tau_m}^{\circ}(\cdot)) \geq \sigma(x^{\circ}(T)) - \zeta/2$. Since $\rho_\varepsilon(\tau_1, x_{\tau_1}^{\circ}(\cdot)) \leq \rho(\tau_1, x_{\tau_1}^{\circ}(\cdot))$, we conclude that, in order to verify the desired inequality (14), it remains to show that $\rho_\varepsilon(\tau_m, x_{\tau_m}^{\circ}(\cdot)) - \rho_\varepsilon(\tau_1, x_{\tau_1}^{\circ}(\cdot)) \leq \zeta/2$. To this end, let us take $j \in \overline{1, m-1}$ and prove the inequality

$$\rho_\varepsilon(\tau_{j+1}, x_{\tau_{j+1}}^{\circ}(\cdot)) - \rho_\varepsilon(\tau_j, x_{\tau_j}^{\circ}(\cdot)) \leq \frac{\zeta(\tau_{j+1} - \tau_j)}{2T}. \tag{15}$$

Using the fact that the value functional ρ satisfies the dynamic programming principle, let us choose $u^{[j]}(\cdot) \in \mathcal{U}(\tau_j, T)$ such that

$$\rho(\tau_{j+1}, x_{\tau_{j+1}}^{[j]}(\cdot)) \leq \rho(\tau_j, z^{[j]}(\cdot)) + \frac{\zeta(\tau_{j+1} - \tau_j)}{4T},$$

where $x^{[j]}(\cdot) \doteq x(\cdot \mid \tau_j, z^{[j]}(\cdot), u^{[j]}(\cdot))$. Then, we derive

$$\rho_\varepsilon(\tau_{j+1}, x_{\tau_{j+1}}^{\circ}(\cdot)) - \rho_\varepsilon(\tau_j, x_{\tau_j}^{\circ}(\cdot))$$

$$\leq \rho(\tau_{j+1}, x_{\tau_{j+1}}^{[j]}(\cdot)) + \frac{\nu_\varepsilon(\tau_{j+1}, x_{\tau_{j+1}}^{\circ}(\cdot) - x_{\tau_{j+1}}^{[j]}(\cdot))}{\varepsilon}$$

$$- \rho(\tau_j, z^{[j]}(\cdot)) - \frac{\nu_\varepsilon(\tau_j, x_{\tau_j}^{\circ}(\cdot) - z^{[j]}(\cdot))}{\varepsilon}$$

$$\leq \frac{\nu_\varepsilon(\tau_{j+1}, x_{\tau_{j+1}}^{\circ}(\cdot) - x_{\tau_{j+1}}^{[j]}(\cdot))}{\varepsilon} - \frac{\nu_\varepsilon(\tau_j, x_{\tau_j}^{\circ}(\cdot) - z^{[j]}(\cdot))}{\varepsilon} + \frac{\zeta(\tau_{j+1} - \tau_j)}{4T}. \tag{16}$$

Let us consider the function $\varkappa(\tau) \doteq \nu_\varepsilon(\tau, x_\tau^{\circ}(\cdot) - x_\tau^{[j]}(\cdot))/\varepsilon$ for all $\tau \in [\tau_j, \tau_{j+1}]$. By property (iv), the function $\varkappa(\cdot)$ is Lipschitz continuous and

$$\dot{\varkappa}(\tau) = \left\langle \frac{\nabla^\alpha \nu_\varepsilon(\tau, x_\tau^{\circ}(\cdot) - x_\tau^{[j]}(\cdot))}{\varepsilon}, f(\tau, x^{\circ}(\tau), u^{\circ}(\tau)) - f(\tau, x^{[j]}(\tau), u^{[j]}(\tau)) \right\rangle$$

for a.e. $\tau \in [\tau_j, \tau_{j+1}]$. Hence, taking into account the choice of $\delta_*(\varepsilon)$ and the equality $x_{\tau_j}^{[j]}(\cdot) = z^{[j]}(\cdot)$, we get

$$\dot{\varkappa}(\tau) \leq \left\langle \frac{\nabla^\alpha \nu_\varepsilon(\tau_j, x_{\tau_j}^{\circ}(\cdot) - z^{[j]}(\cdot))}{\varepsilon}, f(\tau_j, x^{\circ}(\tau_j), u^{\circ}(\tau)) \right\rangle$$

$$- \left\langle \frac{\nabla^\alpha \nu_\varepsilon(\tau_j, x_{\tau_j}^{\circ}(\cdot) - z^{[j]}(\cdot))}{\varepsilon}, f(\tau_j, z^{[j]}(\tau_j), u^{[j]}(\tau)) \right\rangle + \frac{\zeta}{8T}$$

for a.e. $\tau \in [\tau_j, \tau_{j+1}]$. According to (8), we have $u^{\circ}(\tau) = U^{\circ}(\tau_j, x^{\circ}_{\tau_j}(\cdot), \varepsilon)$ for all $\tau \in [\tau_j, \tau_{j+1})$, and, therefore, if follows from definition (13) of U° that

$$
\left\langle \frac{\nabla^{\alpha} \nu_{\varepsilon}(\tau_j, x^{\circ}_{\tau_j}(\cdot) - z^{[j]}(\cdot))}{\varepsilon}, f(\tau_j, x^{\circ}(\tau_j), u^{\circ}(\tau)) \right\rangle
$$
$$
= \min_{u \in P} \left\langle \frac{\nabla^{\alpha} \nu_{\varepsilon}(\tau_j, x^{\circ}_{\tau_j}(\cdot) - z^{[j]}(\cdot))}{\varepsilon}, f(\tau_j, x^{\circ}(\tau_j), u) \right\rangle \quad \forall \tau \in [\tau_j, \tau_{j+1}).
$$

Then, based on the choice of λ_f, we derive

$$
\dot{\varkappa}(\tau) \leq \min_{u \in P} \left\langle \frac{\nabla^{\alpha} \nu_{\varepsilon}(\tau_j, x^{\circ}_{\tau_j}(\cdot) - z^{[j]}(\cdot))}{\varepsilon}, f(\tau_j, x^{\circ}(\tau_j), u) \right\rangle
$$
$$
- \left\langle \frac{\nabla^{\alpha} \nu_{\varepsilon}(\tau_j, x^{\circ}_{\tau_j}(\cdot) - z^{[j]}(\cdot))}{\varepsilon}, f(\tau_j, z^{[j]}(\tau_j), u^{[j]}(\tau)) \right\rangle + \frac{\zeta}{8T}
$$
$$
\leq \lambda_f \frac{\| \nabla^{\alpha} \nu_{\varepsilon}(\tau_j, x^{\circ}_{\tau_j}(\cdot) - z^{[j]}(\cdot)) \|}{\varepsilon} \| x^{\circ}(\tau_j) - z^{[j]}(\tau_j) \| + \frac{\zeta}{8T}
$$

for a.e. $\tau \in [\tau_j, \tau_{j+1}]$.

Further, observing that

$$
\rho(\tau_j, x^{\circ}_{\tau_j}(\cdot)) \geq \rho_{\varepsilon}(\tau_j, x^{\circ}_{\tau_j}(\cdot)) = \rho(\tau_j, z^{[j]}(\cdot)) + \frac{\nu_{\varepsilon}(\tau_j, x^{\circ}_{\tau_j}(\cdot) - z^{[j]}(\cdot))}{\varepsilon}
$$

and recalling the definition of λ_{ρ}, we obtain

$$
\frac{\nu_{\varepsilon}(\tau_j, x^{\circ}_{\tau_j}(\cdot) - z^{[j]}(\cdot))}{\varepsilon} \leq \lambda_{\rho}(\nu_{\varepsilon}(\tau_j, x^{\circ}_{\tau_j}(\cdot) - z^{[j]}(\cdot)) + C\varepsilon^{\frac{q}{q-1}})^{\frac{1}{q}}, \tag{17}
$$

which implies the estimate

$$
\frac{(\nu_{\varepsilon}(\tau_j, x^{\circ}_{\tau_j}(\cdot) - z^{[j]}(\cdot)) + C\varepsilon^{\frac{q}{q-1}})^{\frac{q-1}{q}}}{\varepsilon} \leq \lambda_{\rho} + C^{\frac{q-1}{q}}.
$$

Consequently, in view of the choice of κ, we get

$$
\frac{\| \nabla^{\alpha} \nu_{\varepsilon}(\tau_j, x^{\circ}_{\tau_j}(\cdot) - z^{[j]}(\cdot)) \|}{\varepsilon} \leq \kappa(\lambda_{\rho} + C^{\frac{q-1}{q}}).
$$

Moreover, we derive from (17) that $\nu_{\varepsilon}(\tau_j, x^{\circ}_{\tau_j}(\cdot) - z^{[j]}(\cdot))/\varepsilon \leq \lambda_{\rho}(R^* + C)^{\frac{1}{q}}$ since $\varepsilon \leq \varepsilon_* \leq 1$, and, therefore,

$$
\| x^{\circ}(\tau_j) - z^{[j]}(\tau_j) \| \leq \frac{\zeta}{8T\lambda_f\kappa(\lambda_{\rho} + C^{\frac{q-1}{q}})}.
$$

Thus, we conclude that $\dot{\varkappa}(\tau) \leq \zeta/(4T)$ for a.e. $\tau \in [\tau_j, \tau_{j+1}]$. Hence,

$$
\frac{\nu_{\varepsilon}(\tau_{j+1}, x^{\circ}_{\tau_{j+1}}(\cdot) - x^{[j]}_{\tau_{j+1}}(\cdot))}{\varepsilon} - \frac{\nu_{\varepsilon}(\tau_j, x^{\circ}_{\tau_j}(\cdot) - z^{[j]}(\cdot))}{\varepsilon}
$$
$$
= \varkappa(\tau_{j+1}) - \varkappa(\tau_j) \leq \frac{\zeta(\tau_{j+1} - \tau_j)}{4T}. \tag{18}
$$

Combining (16) and (18), we obtain (15) and complete the proof.

5   Conclusion

In the paper, we have considered the fractional optimal control problem (1) and proposed the new method for constructing the optimal positional control strategy U° (see Sect. 4). This strategy U° allows us to generate ζ-optimal controls with any specified tolerance $\zeta > 0$ and for any initial value of the state vector $x_0 \in B(M)$ according to rule (8). The method is based on the knowledge of the value functional ρ (see (10)) and uses the Lyapunov–Krasovskii functional ν_ε (see (11)). Compared to the previous results [5,6], the functional ν_ε is relatively simple and continuously coinvariantly differentiable of order α (see [4]).

Another advantage of the functional ν_ε is that it is defined in terms of the function $a(\cdot \mid t, w(\cdot))$ (see (7)), which takes into account the fractional dynamics (1a). In this regard, it seems possible to further develop the results obtained in the paper for dynamical systems described by Volterra integral equations of a more general form than (6).

References

1. Clarke, F.H., Ledyaev, Y.S., Subbotin, A.I.: Universal feedback control via proximal aiming in problems of control under disturbance and differential games. Proc. Steklov Inst. Math. **224**, 149–168 (1999)
2. Diethelm, K.: The Analysis of Fractional Differential Equations. Springer, Berlin (2010). https://doi.org/10.1007/978-3-642-14574-2
3. Garnysheva, G.G., Subbotin, A.I.: Strategies of minimax aiming in the direction of the quasigradient. J. Appl. Math. Mech. **58**(4), 575–581 (1994). https://doi.org/10.1016/0021-8928(94)90134-1
4. Gomoyunov, M.I.: Dynamic programming principle and Hamilton-Jacobi-Bellman equations for fractional-order systems. SIAM J. Control. Optim. **58**(6), 3185–3211 (2020). https://doi.org/10.1137/19M1279368
5. Gomoyunov, M.I.: Extremal shift to accompanying points in a positional differential game for a fractional-order system. Proc. Steklov Inst. Math. **308**(Suppl. 1), S83–S105 (2020). https://doi.org/10.1134/S0081543820020078
6. Gomoyunov, M.I.: Differential games for fractional-order systems: Hamilton-Jacobi-Bellman-Isaacs equation and optimal feedback strategies. Mathematics **9**(14), art. 1667 (2021). https://doi.org/10.3390/math9141667
7. Gomoyunov, M.I.: On viscosity solutions of path-dependent Hamilton-Jacobi-Bellman-Isaacs equations for fractional-order systems (2021). https://doi.org/10.48550/arXiv.2109.02451
8. Gomoyunov, M.I., Lukoyanov, N.Y.: Differential games in fractional-order systems: inequalities for directional derivatives of the value functional. Proc. Steklov Inst. Math. **315**(1), 65–84 (2021). https://doi.org/10.1134/S0081543821050060
9. Kilbas, A.A., Srivastava, H.M., Trujillo, J.J.: Theory and Applications of Fractional Differential Equations. Elsevier, Amsterdam (2006)
10. Krasovskii, N.N., Krasovskii, A.N.: Control Under Lack of Information. Birkhäuser, Berlin (1995). https://doi.org/10.1007/978-1-4612-2568-3
11. Krasovskii, N.N., Subbotin, A.I.: Game-Theoretical Control Problems. Springer, New York (1988)

12. Lukoyanov, N.Y.: Strategies for aiming in the direction of invariant gradient. J. Appl. Math. Mech. **68**(4), 561–574 (2004). https://doi.org/10.1016/j.jappmathmech.2004.07.009
13. Samko, S.G., Kilbas, A.A., Marichev, O.I.: Fractional Integrals and Derivatives. Theory and Applications. Gordon and Breach Science Publishers, Amsterdam (1993)
14. Subbotin, A.I.: Generalized Solutions of First Order PDEs: The Dynamical Optimization Perspective. Birkhäuser, Basel (1995). https://doi.org/10.1007/978-1-4612-0847-1

On a Single-Type Differential Game
of Retention in a Ring

Igor' V. Izmest'ev[1,2]($\boxtimes$) ⓘ and Viktor I. Ukhobotov[1,2] ⓘ

[1] N.N. Krasovskii Institute of Mathematics and Mechanics,
S. Kovalevskaya Street, 16, 620108 Yekaterinburg, Russia
j748e8@gmail.com, ukh@csu.ru
[2] Chelyabinsk State University, Br. Kashirinykh Street, 129,
454001 Chelyabinsk, Russia

Abstract. In a normed space of finite dimension, we consider a single-type differential game. The vectograms of the players are described by the same ball with different time-dependent radii. The motion is constructed using polygonal lines. The target set is determined by the condition that the norm of the phase vector belongs to a segment with positive ends. In this paper, a set defined by this condition is called a ring. The goal of the first player is to retain the phase vector in the ring until a given time. The goal of the second player is the opposite. We have found the necessary and sufficient conditions of retention and constructed the corresponding controls of the players.

Keywords: Control · Differential game

1 Introduction

The linear differential game with a given duration, using a linear change of variables [4], can be reduced to the form, in which the right side of the new equations of motion contains only the sum of the players' controls with values that belong to time-dependent sets. If the reachability sets of the players are homeomorphic to the same given set, then such games are called «single-type».

An arbitrary linear differential game in which the payoff is determined using the modulus of a linear function is reduced to such games.

In the differential game «isotropic rockets» [2], in its variant in the absence of friction «boy and crocodile» [6] and in the test example of L.S. Pontryagin [6], the vectograms of controls are balls with time-dependent radii. For such differential games, in the case when the terminal set is a ball of a given radius, the alternating integral is constructed in [6]. In [9], optimal positional controls of players are constructed for this problem.

Differential games of pursuit are also relevant, when the pursuer seeks to make the distance to the evader at a fixed time no more than one given number,

The research was supported by a grant from the Russian Science Foundation no. 19-11-00105, https://rscf.ru/project/19-11-00105/.

but no less than another given number. In works [10,11], the set of vectors defined by this condition is called a ring. In [10], a maximal stable bridge is constructed for a single-type differential game with geometric constraints on the player controls, in which the terminal set is a ring. In [11], the corresponding optimal controls of the players are constructed for this problem.

In the present paper, we consider a single-type differential game of retention (see, for example, [1, 7, 8]). The goal of the first player is to retain the phase vector in the ring until a given time. The goal of the second player is the opposite. For this problem, necessary and sufficient termination conditions are found, and the corresponding optimal players controls are constructed.

As an example illustrating the theory, we consider a group pursuit problem in which one pursuer follows another pursuer, retaining the vector of its relative coordinates in the ring.

2 Problem Statement

In the space $\mathbb{R}^n$ with the norm $\|\cdot\|$, the motion occurs according to the equation

$$\dot{z} = -a(t)u + b(t)v, \quad \|u\| \leq 1, \quad \|v\| \leq 1, \quad t \leq p. \tag{1}$$

Here, functions $a(t) \geq 0$ and $b(t) \geq 0$ are summable on each segment of the semiaxis $(-\infty, p]$.

Fix the initial state $t_0 < p$, $z(t_0) \in \mathbb{R}^n$ and numbers $0 < \varepsilon_1 \leq \varepsilon_2$. The goal of the first player, who chooses the control u, is to implement the inclusion

$$z(t) \in S(\varepsilon_1, \varepsilon_2) \quad \text{for} \quad t_0 \leq t \leq p, \tag{2}$$

where

$$S(\varepsilon_1, \varepsilon_2) = \{z \in \mathbb{R}^n : \varepsilon_1 \leq \|z\| \leq \varepsilon_2\}.$$

The goal of the second player, who chooses the control v, is the opposite

Note that the case $\varepsilon_1 = 0$ is considered in the paper [7].

Admissible controls of players are arbitrary functions, which satisfy inequalities

$$\|u(t, z)\| \leq 1, \quad \|v(t, z)\| \leq 1, \quad t \leq p, \quad z \in \mathbb{R}^n. \tag{3}$$

Fix $t_* \in (t_0, p]$. Take partition

$$\omega : t_0 < t_1 < \ldots < t_i < t_{i+1} < \ldots < t_k < t_{k+1} = t_*. \tag{4}$$

with diameter $d(\omega) = \max(t_{i+1} - t_i)$, $i = \overline{0, k}$. Construct polygonal line for (1)

$$z_\omega(t) = z_\omega(t_i) - \left(\int_{t_i}^t a(r)dr \right) u(t_i, z_\omega(t_i)) + \left(\int_{t_i}^t b(r)dr \right) v(t_i, z_\omega(t_i)) \tag{5}$$

for $t_i < t \leq t_{i+1}$. Here, $z_\omega(t_0) = z(t_0)$.

It can be shown that

$$\|z_\omega(\tau) - z_\omega(t)\| \leq \int_t^\tau (a(r) + b(r))dr \quad \text{for} \quad t_0 \leq t < \tau \leq t_*.$$

This equality and absolute continuity theorem on the Lebesgue integral [3, p. 282] imply that the family of these polygonal line (5), which are determined on the segment $[t_0, t_*]$, is uniformly bounded and equicontinuous. By Arzela theorem [3, p. 104], from any sequence of the polygonal lines (5) we can select a subsequence $z_{\omega_m}(t)$ that converges uniformly on the segment $[t_0, t_*]$ to some function $z(t)$.

The motion of the system $z(t)$ on the segment $[t_0, t_*]$ realized with admissible controls (3) from the initial state $z(t_0)$ is defined as any uniform limit of the sequence of the polygonal lines (5), for which diameters of partition tend to zero.

3   Auxiliary Results

Let us present some results for the differential game (1) with a fixed end time p, in which the goal of the first player is to implement the inclusion $z(p) \in S(\varepsilon_1, \varepsilon_2)$.

Denote

$$g(t, \tau) = \int_t^\tau (a(r) - b(r)) dr, \quad t \le \tau.$$

Let us introduce a programmed absorption operator that, for each pair of numbers $t \le \tau$ and for any set $X \subset \mathbb{R}^n$, associates the set $T_t^\tau(X)$ defined by the equalities $T_t^\tau(\emptyset) = \emptyset$ and, if the set is $X \ne \emptyset$, then

$$T_t^\tau(X) = \left(X + \int_t^\tau a(r) dr S \right) \dot{-} \left(\int_t^\tau b(r) dr S \right).$$

Here, $\emptyset$ is the empty set; $S = \{z \in \mathbb{R}^n : \|z\| \le 1\}$; «+» and «$\dot{-}$» denote the Minkowski sum and Minkowski difference, respectively [5].

Lemma 1 [10]. *For any numbers $0 \le \delta \le \varepsilon$ and $t < \tau$ equality*

$$T_t^\tau(S(\delta, \varepsilon)) = \begin{cases} \emptyset & for \quad F_t^\tau(\delta) > \varepsilon + g(t, \tau), \\ S(F_t^\tau(\delta), \varepsilon + g(t, \tau)) & for \quad F_t^\tau(\delta) \le \varepsilon + g(t, \tau) \end{cases}$$

holds.

Here, we denote

$$F_t^\tau(\delta) = \delta - g(t, \tau) \quad for \quad \delta > \int_t^\tau a(r) dr, \quad F_t^\tau(\delta) = 0 \quad for \quad \delta \le \int_t^\tau a(r) dr.$$

In article [10], the operator $T_t^\tau(\cdot)$ is used to construct a solvability set for a differential game with a fixed end time, which is considered in this section.

Lemma 2 [11]. *Let the initial state $t_0 < p$, $z(t_0) \in \mathbb{R}^n$ be such that $\|z(t_0)\| > \varepsilon_2 + g(t_0, p)$. Then the admissible control of the second player $v(t, z) = \phi(z)$ guarantees the inequality $\|z(p)\| > \varepsilon_2$ for any control of the first player and for any realized motion $z(t)$.*

Here,

$$\phi(z) = \frac{z}{\|z\|} \quad \text{for} \quad \|z\| \neq 0 \quad \text{and} \quad \phi(\overline{0}) - \text{any with constraint} \quad \|\phi(\overline{0})\| = 1.$$

$$(6)$$

Lemma 3 [11]. *Let the initial state* t_0, $z(t_0) \in \mathbb{R}^n$ *be such that* $q(\varepsilon_1, p) < t_0 < p$ *and* $\|z(t_0)\| < \varepsilon_1 - g(t_0, p)$. *Then the admissible control of the second player* $v(t, z) = -\phi(z)$ *guarantees the inequality* $\|z(p)\| < \varepsilon_1$ *for any control of the first player and for any realized motion* $z(t)$.

Here,

$$q(\varepsilon_1, p) = \inf \{t \le p : \varepsilon_1 > g(r, p) \quad \text{for all} \quad t < r \le p\}.$$

4   Main Results

4.1   Construction of the Solvability Set

Denote by $W(t, \tau)$ the set of positions $z(t)$ starting from which the first player can guarantee the inclusion $z(r) \in S(\varepsilon_1, \varepsilon_2)$ for $r \in [t, \tau]$ and for any control of the second player.

Let us introduce notations

$$f_1(t) = \varepsilon_1 - \min_{t \le \tau \le p} g(t, \tau), \quad f_2(t) = \varepsilon_2 + \min_{t \le \tau \le p} g(t, \tau); \quad (7)$$

$$t(\varepsilon_1, \varepsilon_2) = \inf \left\{ t < p : \varepsilon_1 - \min_{t \le r \le p} \min_{r \le \tau \le p} g(r, \tau) \le \varepsilon_2 + \min_{t \le r \le p} \min_{r \le \tau \le p} g(r, \tau) \right\}.$$

$$(8)$$

Construct the set $W(t, p)$ for $t(\varepsilon_1, \varepsilon_2) \le t < p$. Take a partition

$$\omega : t = t^{(0)} < t^{(1)} < \ldots < t^{(k)} < t^{(k+1)} = p$$

such that

$$\int_{t^{(i)}}^{t^{(i+1)}} a(r) dr < \varepsilon_1, \quad i = \overline{0, k}.$$

Using Lemma 1 and the procedure from paper [7], we construct the set

$$W(t^{(k)}, p) = \bigcap_{t^{(k)} \le \tau \le p} T_{t^{(k)}}^{\tau} (S(\varepsilon_1, \varepsilon_2))$$

$$= S(\varepsilon_1 - \min_{t^{(k)} \le \tau \le p} g(t^{(k)}, \tau), \varepsilon_2 + \min_{t^{(k)} \le \tau \le p} g(t^{(k)}, \tau)) = S(f_1(t^{(k)}), f_2(t^{(k)})).$$

Next, we will construct set $W(t^{(k-1)}, p)$ as follows

$$W(t^{(k-1)}, p) = W(t^{(k-1)}, t^{(k)}) \bigcap T_{t^{(k-1)}}^{t^{(k)}} (W(t^{(k)}, p)).$$

Let's explain this formula. Based on the results of paper [10], it can be shown that $T^{t^{(k)}}_{t^{(k-1)}}(W(t^{(k)}, p))$ is the set of positions $z(t^{(k-1)})$ starting from which the first player can guarantee the inclusion $z(t^{(k)}) \in W(t^{(k)}, p)$ for any control of the second player and for any realized motion. Intersecting this set and $W(t^{(k-1)}, t^{(k)})$, we get a set of positions $z(t^{(k-1)})$ starting from which inclusions $z(t) \in S(\varepsilon_1, \varepsilon_2)$ for $t \in [t^{(k-1)}, t^{(k)}]$ and $z(t^{(k)}) \in W(t^{(k)}, p)$ are guaranteed. The answer to the question about the existence of the corresponding control u is given below by Theorem 1.

Calculate

$$T^{t^{(k)}}_{t^{(k-1)}}(W(t^{(k)}, p)) = S\left(f_1(t^{(k)}) - g(t^{(k-1)}, t^{(k)}), f_2(t^{(k)}) + g(t^{(k-1)}, t^{(k)})\right),$$

$$W(t^{(k-1)}, t^{(k)}) = S\left(\varepsilon_1 - \min_{t^{(k-1)} \leq \tau \leq t^{(k)}} g(t^{(k-1)}, \tau), \varepsilon_2 + \min_{t^{(k-1)} \leq \tau \leq t^{(k)}} g(t^{(k-1)}, \tau)\right).$$

Then

$$W(t^{(k-1)}, p) = S\left(\max\left(\varepsilon_1 - \min_{t^{(k-1)} \leq \tau \leq t^{(k)}} g(t^{(k-1)}, \tau), \varepsilon_1 - \min_{t^{(k)} \leq \tau \leq p} g(t^{(k-1)}, \tau)\right),\right.$$

$$\left.\min\left(\varepsilon_2 + \min_{t^{(k-1)} \leq \tau \leq t^{(k)}} g(t^{(k-1)}, \tau), \varepsilon_2 + \min_{t^{(k)} \leq \tau \leq p} g(t^{(k-1)}, \tau)\right)\right)$$

$$= S(f_1(t^{(k-1)}), f_2(t^{(k-1)})).$$

Constructing by analogy the sets $W(t^{(i)}, p)$, $i = \overline{0, k-2}$, we get that

$$W(t, p) = S\left(\varepsilon_1 - \min_{t \leq \tau \leq p} g(t, \tau), \varepsilon_2 + \min_{t \leq \tau \leq p} g(t, \tau)\right) = S(f_1(t), f_2(t)).$$

Construct the set $W(t, p)$ for $t < t(\varepsilon_1, \varepsilon_2)$. Formulas (7) and (8) imply equality $f_1(t(\varepsilon_1, \varepsilon_2)) = f_2(t(\varepsilon_1, \varepsilon_2))$ and that there exists $t_* < t(\varepsilon_1, \varepsilon_2)$ such that

$$\int_{t_*}^{t(\varepsilon_1, \varepsilon_2)} a(r)dr < \varepsilon_1 \quad \text{and} \quad g(t, t(\varepsilon_1, \varepsilon_2)) < 0 \quad \text{for} \quad t \in [t_*, t(\varepsilon_1, \varepsilon_2)).$$

Using Lemma 1, we obtain

$$T^{t(\varepsilon_1, \varepsilon_2)}_{t_*}(W(t(\varepsilon_1, \varepsilon_2), p)) = \emptyset,$$

because $f_1(t(\varepsilon_1, \varepsilon_2)) - g(t_*, t(\varepsilon_1, \varepsilon_2)) > f_2(t(\varepsilon_1, \varepsilon_2)) + g(t_*, t(\varepsilon_1, \varepsilon_2))$. Further, $W(t_*, p) = \emptyset$, and $W(t, p) = \emptyset$ for $t \leq t_*$. Letting t_* tend to $t(\varepsilon_1, \varepsilon_2)$, we get $W(t) = \emptyset$ for $t < t(\varepsilon_1, \varepsilon_2)$.

Thus,

$$W(t, p) = S(f_1(t), f_2(t)) \quad \text{for} \quad t(\varepsilon_1, \varepsilon_2) \leq t \leq p, \quad W(t, p) = \emptyset \quad \text{for} \quad t < t(\varepsilon_1, \varepsilon_2).$$

4.2   Solution of Retention Problem

Lemma 4. *Let inequalities $t(\varepsilon_1, \varepsilon_2) \leq t \leq p$ hold. Then*

$$f_1(t) \leq \frac{\varepsilon_1 + \varepsilon_2}{2} \leq f_2(t).$$

Proof. Note that formula (8) implies the inequality

$$f_1(t) \leq f_2(t) \quad \text{for} \quad t(\varepsilon_1, \varepsilon_2) \leq t \leq p. \tag{9}$$

Assume the contrary to the assertion of the Lemma 4, namely, let

$$f_2(t) = \varepsilon_2 + \min_{t \leq \tau \leq p} g(t, \tau) < \frac{\varepsilon_1 + \varepsilon_2}{2}.$$

Then

$$\min_{t \leq \tau \leq p} g(t, \tau) < \frac{\varepsilon_1 - \varepsilon_2}{2}. \tag{10}$$

On the other hand, taking into account formulas (7) and (10), we obtain the system of inequalities

$$f_2(t) < \frac{\varepsilon_1 + \varepsilon_2}{2} = \varepsilon_1 - \frac{\varepsilon_1 - \varepsilon_2}{2} < \varepsilon_1 - \min_{t \leq \tau \leq p} g(t, \tau) = f_1(t).$$

Thus, we have come to a contradiction with (9).
   Assume that

$$\frac{\varepsilon_1 + \varepsilon_2}{2} < f_1(t) = \varepsilon_1 - \min_{t \leq \tau \leq p} g(t, \tau).$$

From here we obtain (10). On the other hand, taking into account (7) and (10), we have a system of inequalities

$$f_1(t) > \frac{\varepsilon_1 + \varepsilon_2}{2} = \varepsilon_2 + \frac{\varepsilon_1 - \varepsilon_2}{2} > \varepsilon_2 + \min_{t \leq \tau \leq p} g(t, \tau) = f_2(t).$$

We have come to a contradiction with (9).

Theorem 1. *Let the initial state $t_0 < p$, $z(t_0)$ be such that the inclusion $z(t_0) \in W(t_0, p)$ holds. Then the control of the first player*

$$u(t, z) = \mathrm{sign} \left(\|z\| - \frac{\varepsilon_1 + \varepsilon_2}{2} \right) \phi(z) \tag{11}$$

guarantees the fulfillment of the condition (2) for any admissible control of the second player and for any realized motion $z(t)$.

Here, we denote sign $x = 1$ for $x \geq 0$ and sign $x = -1$ for $x < 0$.

Proof. Let the second player choose an arbitrary admissible control $v(t, z)$, and $z(t)$ is the realized motion under the chosen controls with the initial condition $z(t_0)$.

Assume the contrary to the assertion of Theorem 1. Let there exists $\tau^* \in (t_0, p]$ such that $z(\tau^*) \notin S(\varepsilon_1, \varepsilon_2)$.

There is a sequence of partitions (4)

$$\omega_m : t_0 = t_0^{(m)} < t_1^{(m)} < \ldots < t_{k_m}^{(m)} < t_{k_m+1}^{(m)} = \tau^*$$

with partition diameters $d(\omega_m) \to 0$ for which the sequence $z_{\omega_m}(t)$ of polygonal lines (5) converges uniformly on the segment $[t_0, \tau^*]$ to $z(t)$. Below we will denote $z_{\omega_m}(t) = z_m(t)$ and $t_i^{(m)} = t_i$.

Case 1. Assume that $\|z(\tau^*)\| > \varepsilon_2$. Then from the condition of Theorem 1 and the form of the function $f_2(t)$ (7) we get that

$$\|z(t_0)\| - \varepsilon_2 - g(t_0, \tau^*) \le \|z(t_0)\| - f_2(t_0) \le 0, \quad \|z(\tau^*)\| - \varepsilon_2 - g(\tau^*, \tau^*) > 0.$$

From here and from the continuity of the functions $\|z(t)\|$ and $g(t, \tau^*)$ it follows that there exists a number $\tau \in [t_0, \tau^*)$ such that

$$\|z(\tau)\| = \varepsilon_2 + g(\tau, \tau_*), \quad \|z(t)\| > \varepsilon_2 + g(t, \tau_*) \quad \text{for} \quad \tau < t \le \tau^*. \tag{12}$$

Fix an arbitrary number $\theta \in (\tau, \tau^*)$. Then it follows from (12) that there exists a number $\sigma > 0$ such that $\|z(t)\| \ge \varepsilon_2 + g(t, \tau^*) + 2\sigma$ for all $\theta \le t \le \tau^*$. From this inequality, using the uniform convergence of the sequence $z_m(t)$ to the realization $z(t)$, we can obtain that for all partitions ω_m with sufficiently small diameters the following inequality holds:

$$\|z_m(t)\| \ge \varepsilon_2 + g(t, \tau^*) + \sigma \quad \text{for all} \quad \theta \le t \le \tau^*. \tag{13}$$

It follows from the absolute continuity theorem for the Lebesgue integral that inequalities

$$\int_t^l a(r)dr < \sigma, \quad \theta \le t < l < t + d(\omega_m), \quad l \le \tau^*. \tag{14}$$

hold for all partitions ω_m with sufficiently small diameters. In this case, we further consider partitions for which inequalities (13) and (14) are satisfied.

Let the number $\theta \in [t_{i-1}, t_i)$. Next, we will prove the inequality

$$\|z_m(t_{s+1})\| \le \|z_m(t_s)\| - \int_{t_s}^{t_{s+1}} (a(r) - b(r))dr, \quad s = i, i+1, \ldots, k. \tag{15}$$

For the polygonal line (5) the following inequality holds:

$$\|z_m(t_{s+1})\| \le \left\| z_m(t_s) - \left(\int_{t_s}^{t_{s+1}} a(r)dr \right) u(t_s, z_m(t_s)) \right\| + \int_{t_s}^{t_{s+1}} b(r)dr.$$

It follows from inequality (13), Lemma 4, and formula (11) that $u(t_s, z_m(t_s)) = \phi(z_m(t_s))$. Let's substitute this control into the previous formula. Then, using formula (6), inequalities (13) and (14), we obtain inequality (15).

Inequality (15) implies that

$$\|z_m(\tau^*)\| \leq \|z_m(t_i)\| - \int_{t_i}^{\tau^*} (a(r) - b(r))dr.$$

In this inequality, we pass to the limit as $d(\omega_m) \to 0$, and then we tend θ to τ. Obtain

$$\|z(\tau^*)\| \leq \|z(\tau)\| - \int_{\tau}^{\tau^*} (a(r) - b(r))dr.$$

This and the first equality in (12) imply that $\|z(\tau^*)\| \leq \varepsilon_2$. We obtain a contradiction with the assumption $\|z(\tau^*)\| > \varepsilon_2$.

Case 2. Let us now assume that $\|z(\tau^*)\| < \varepsilon_1$. Then from the condition of the Theorem 1 and the form of the function $f_1(t)$ (7) we get that

$$\|z(t_0)\| - \varepsilon_1 + g(t_0, \tau^*) \geq \|z(t_0)\| - f_1(t_0) \geq 0, \quad \|z(\tau^*)\| - \varepsilon_1 + g(\tau^*, \tau^*) < 0.$$

Therefore, there exists a number $\tau \in [t_0, \tau^*)$ such that

$$\|z(\tau)\| = \varepsilon_1 - g(\tau, \tau^*), \quad \|z(t)\| < \varepsilon_1 - g(t, \tau^*) \quad \text{for all} \quad \tau < t \leq \tau^*. \tag{16}$$

Fix an arbitrary number $\theta \in (\tau, \tau^*)$. Then it follows from (16) that there exists a number $\sigma > 0$ such that $\|z(t)\| < \varepsilon_1 - g(t, \tau^*) - 2\sigma$ for all $\theta \leq t \leq \tau^*$. From here and from the uniform convergence of $z_m(t)$ to $z(t)$ it follows that

$$\|z_m(t)\| < \varepsilon_1 - g(t, \tau^*) - \sigma \quad \text{for all} \quad \theta \leq t \leq \tau^* \tag{17}$$

and for all partitions ω_m with sufficiently small diameters.

Let the number $\theta \in [t_{i-1}, t_i)$. Next, we will prove the inequality

$$\|z_m(t_{s+1})\| \geq \|z_m(t_s)\| + \int_{t_s}^{t_{s+1}} (a(r) - b(r))dr, \quad s = i, i+1, \ldots, k. \tag{18}$$

For the polygonal line (5) the following inequality holds:

$$\|z_m(t_{s+1})\| \geq \left\| z_m(t_s) - \left(\int_{t_s}^{t_{s+1}} a(r)dr \right) u(t_s, z_m(t_s)) \right\| - \int_{t_s}^{t_{s+1}} b(r)dr.$$

It follows from inequality (17), Lemma 4, and formula (11) that $u(t_s, z_m(t_s)) = -\phi(z_m(t_s))$. We substitute this control into the previous inequality. Taking into account formula (6), we obtain inequality (18).

Inequality (18) implies that

$$\|z_m(\tau^*)\| \geq \|z_m(t_i)\| + \int_{t_i}^{\tau^*} (a(r) - b(r))dr.$$

In this inequality, we pass to the limit as $d(\omega_m) \to 0$, and then we tend θ to τ. Obtain

$$\|z(\tau^*)\| \geq \|z(\tau)\| + \int_\tau^{\tau^*} (a(r) - b(r))dr.$$

This and the first equality in (16) imply that $\|z(\tau^*)\| \geq \varepsilon_1$. We obtain a contradiction with the assumption $\|z(\tau^*)\| < \varepsilon_1$.

4.3 Solution of Evasion Problem

Theorem 2. *Let the initial state $t_0 < p$, $z(t_0)$ be such that $z(t_0) \notin W(t_0, p)$. Then there is a time $\tau \in [t_0, p]$ and a control of the second player that guarantee the fulfillment of inequality $\|z(\tau)\| < \varepsilon_1$ or inequality $\|z(\tau)\| > \varepsilon_2$ for any control of the first player and for any realized motion $z(t)$.*

Proof. Case 1. Let $t(\varepsilon_1, \varepsilon_2) \leq t_0 < p$.

Case 1.1. Let $\|z(t_0)\| > f_2(t_0)$. Then there exists $\tau^* \in [t_0, p]$ such that $\|z(t_0)\| > \varepsilon_2 + g(t_0, \tau^*)$. From here we fall into the condition of Lemma 2 with $p = \tau^*$.

Case 1.2. Let $\|z(t_0)\| < f_1(t_0)$. Then there is $\tau^* \in [t_0, p]$ such that $f_1(t_0) = \varepsilon_1 - g(t_0, \tau^*)$. Therefore, $\|z(t_0)\| < \varepsilon_1 - g(t_0, \tau^*)$.

Let us show that $q(\varepsilon_1, \tau^*) < t_0$. Assume the opposite. Let there exists $\widehat{t} \in [t_0, \tau^*)$ such that $g(\widehat{t}, \tau^*) \geq \varepsilon_1 > 0$. From here we have

$$g(t_0, \tau^*) = g(t_0, \widehat{t}) + g(\widehat{t}, \tau^*) > g(t_0, \widehat{t}).$$

We get a contradiction with optimality of τ^*.

Thus, we fall into the condition of Lemma 3 with $p = \tau^*$.

Case 2. Let $t_0 < t(\varepsilon_1, \varepsilon_2)$. Then (8) implies inequality

$$\varepsilon_1 - \min_{t_0 \leq r \leq p} \min_{r \leq \tau \leq p} g(r, \tau) > \varepsilon_2 + \min_{t_0 \leq r \leq p} \min_{r \leq \tau \leq p} g(r, \tau).$$

Take $t_* \in [t_0, p]$ and $\tau_* \in (t_*, p]$ such that

$$\min_{t_0 \leq r \leq p} \min_{r \leq \tau \leq p} g(r, \tau) = g(t_*, \tau_*).$$

The second player can choose any control for $t < t_*$. At time t_*, one of the following inequalities hold:

$$\|z(t_*)\| > \varepsilon_2 + g(t_*, \tau_*); \tag{19}$$

$$\|z(t_*)\| < \varepsilon_1 - g(t_*, \tau_*). \tag{20}$$

Let inequality (19) hold, then we fall into the condition of Lemma 2 with $t_0 = t_*$ and $p = \tau_*$.

Let inequality (20) hold. Let us show that $q(\varepsilon_1, \tau_*) < t_*$. Assume the opposite. Let there exists $\widehat{t} \in [t_*, \tau_*)$ such that

$$g(\widehat{t}, \tau_*) \geq \varepsilon_1 > 0.$$

Then
$$g(t_*, \tau_*) = g(t_*, \widehat{t}) + g(\widehat{t}, \tau_*) > g(t_*, \widehat{t}).$$

This gives us a contradiction with the optimality of the pair (t_*, τ_*). Thus, we fall into the condition of Lemma 3 with $t_0 = t_*$ and $p = \tau_*$.

Remark 1. The control of the second player, which is described in case 2 of the proof of Theorem 2, is a positional control with memory, since for $t > t_*$ its value depends on the value of the norm $\|z(t_*)\|$ of the phase vector realized at time t_*.

5 Example

Consider the problem of group pursuit in the space $\mathbb{R}^n$:

$$\dot{x}_1 = a_1(t)u, \quad \|u\| \leq 1, \quad \dot{x}_2 = a_2(t)\widehat{u}, \quad \|\widehat{u}\| \leq 1, \quad \dot{y} = b(t)v, \quad \|v\| \leq 1.$$

Here, functions $a_1(t) \geq 0$, $a_2(t) \geq 0$ and $b(t) \geq 0$ are summable on each segment of $\mathbb{R}$. The first pursuer, who chooses the control u, controls point x_1. When control u choosing, the first pursuer uses information about the coordinates of evader y. The second pursuer, who chooses the control $\widehat{u}$, controls point x_2. However, when control $\widehat{u}$ choosing, the second pursuer can use only information about the coordinates of the first pursuer x_1 (the first pursuer leads the second pursuer behind him). The evader, who chooses the control v, controls point y.

Fix the initial state $t_0 < p$, $x_1(t_0) \in \mathbb{R}^n$, $x_2(t_0) \in \mathbb{R}^n$, $y(t_0) \in \mathbb{R}^n$ and numbers $0 < \varepsilon$, $0 < \varepsilon_1 \leq \varepsilon_2$. The goal of the pursuers is to implement the inequalities

$$\|y(p) - x_1(p)\| \leq \varepsilon, \quad \|y(p) - x_2(p)\| \leq \varepsilon + \varepsilon_2 \tag{21}$$

as quickly as possible ($p \to$ min). In addition, condition

$$\|x_1(t) - x_2(t)\| \geq \varepsilon_1 \quad \text{for} \quad t \in [t_0, p] \tag{22}$$

must be satisfied. The goal of the evader is the opposite.

Let's make the change of variables $z_1 = y - x_1$. Consider the differential game of pursuit with a fixed end time $p > t_0$

$$\dot{z}_1 = -a_1(t)u + b(t)v, \quad \|z_1(p)\| \leq \varepsilon.$$

According to [6], necessary and sufficient conditions under which the first pursuer guarantees the fulfillment of inequality $\|z_1(p)\| \leq \varepsilon$ have the form

$$\|z_1(t_0)\| \leq \varepsilon + \int_{t_0}^{p} (a_1(r) - b(r))dr, \quad \varepsilon + \min_{t_0 \leq t \leq p} \int_{t}^{p} (a_1(r) - b(r))dr \geq 0. \tag{23}$$

Denote by p_* the minimum p for which conditions (23) are satisfied for given initial state.

Let's make the change of variables $z_2 = x_1 - x_2$ and consider the differential game of retention in a ring

$$\dot{z}_2 = -a_2(t)\widehat{u} + a_1(t)u, \quad z_2(t) \in S(\varepsilon_1, \varepsilon_2) \quad \text{for} \quad t \in [t_0, p_*].$$

Using the results of Sect. 4, we obtain that sufficient conditions under which the second pursuer guarantees inclusion $z_2(t) \in S(\varepsilon_1, \varepsilon_2)$ for $t \in [t_0, p_*]$ are defined as follows

$$\varepsilon_1 - \min_{t_0 \le \tau \le p_*} \int_{t_0}^{\tau} (a_2(r) - a_1(r))dr \le \|z_2(t_0)\| \le \varepsilon_2 + \min_{t_0 \le \tau \le p_*} \int_{t_0}^{\tau} (a_2(r) - a_1(r))dr,$$
(24)

$$\varepsilon_1 - \min_{t_0 \le r \le p_*} \min_{r \le \tau \le p_*} \int_{r}^{\tau} (a_2(r) - a_1(r))dr \le c_2 + \min_{t_0 \le r \le p_*} \min_{r \le \tau \le p_*} \int_{r}^{\tau} (u_2(r) - a_1(r))dr.$$
(25)

Let us show that conditions (23) with $p = p_*$, (24) and (25) are sufficient for the pursuers to guarantee the fulfillment of inequalities (21) and (22). We make the reverse change of variables and, relying on the results from [9] and Theorem 1, obtain that there are pursuer controls that guarantee

$$\|y(p_*) - x_1(p_*)\| \le \varepsilon, \quad \varepsilon_1 \le \|x_1(t) - x_2(t)\| \le \varepsilon_2 \quad \text{for} \quad t \in [t_0, p_*].$$

Next,

$$\|y(p_*) - x_2(p_*)\| \le \|y(p_*) - x_1(p_*)\| + \|x_1(p_*) - x_2(p_*)\| \le \varepsilon + \varepsilon_2.$$

6   Conclusion

In this paper, we consider an antagonistic differential game. Vectograms of controls are balls with radii that are determined by functions of time. The goal of the first player is to retain the phase vector in a ring until a given time.

We have constructed a solvability set $W(t_0, p)$ in the retention problem. Theorem 1 and Theorem 2 are proven, from which it follows that the set $W(t_0, p)$ defines necessary and sufficient conditions for the possibility of being retained in a ring. Moreover, the corresponding optimal controls of the players are constructed.

As an example, we consider the problem of group pursuit, in which one of the pursuers must retain the vector of relative coordinates of the other pursuer in the ring during the movement.

In the future, we plan to develop a theory for solving differential games of retention, to which group pursuit problems are reduced. For example, to develop an algorithm for conflict-free movement within a group consisting of more than two pursuers.

References

1. Chentsov, A.G.: An abstract confinement problem: a programmed iterations method of solution. Autom. Remote. Control. **65**, 299–310 (2004). https://doi.org/10.1023/B:AURC.0000014727.63912.45
2. Isaacs, R.: Differential Games: A Mathematical Theory with Applications to Warfare and Pursuit, Control and Optimization. Wiley, New York (1965)
3. Kolmogorov, A.N., Fomin, S.V.: Elements of the Theory of Functions and Functional Analysis. Nauka Publ., Moscow (1972). (in Russian)
4. Krasovskii, N.N., Subbotin, A.I.: Positional Differential Games. Nauka Publ., Moscow (1974). (in Russian)
5. Pontryagin, L.S.: Linear differential games. II. Sov. Math. Dokl. **8**, 910–912 (1967)
6. Pontryagin, L.S.: Linear differential games of pursuit. Math. USSR-Sbornik **40**(3), 285–303 (1981). https://doi.org/10.1070/SM1981v040n03ABEH001815
7. Ukhobotov, V.I.: On the construction of a stable bridge in a retention game. J. Appl. Math. Mech. **45**(2), 169–172 (1981). https://doi.org/10.1016/0021-8928(81)90030-7
8. Ukhobotov, V.I.: Domain of indifference in single-type differential games of retention in a bounded time interval. J. Appl. Math. Mech. **58**(6), 997–1002 (1994). https://doi.org/10.1016/0021-8928(94)90115-5
9. Ukhobotov, V.I.: Synthesis of control in single-type differential games with fixed time. Bull. Chelyabinsk Univ. **1**, 178–184 (1996). (in Russian)
10. Ukhobotov, V.I.: Single-type differential game with a terminal set in the form of a ring. In: Batukhtin, V.D. (ed.) Some Problems of Dynamics and Control, Collection of Scientific Papers, pp. 108–123. Chelyabinsk State University, Chelyabinsk (2005). (in Russian)
11. Ukhobotov, V.I., Izmest'ev, I.V.: Single-type differential games with a terminal set in the form of a ring. In: Systems Dynamics and Control Process, Proceedings of the International Conference, Dedicated to the 90th Anniversary of Acad. N.N. Krasovskiy, Ekaterinburg, 15–20 September 2014, pp. 325–332. Publishing House of the UMC UPI, Ekaterinburg (2015). (in Russian)

Harmonic Numbers in Gambler's Ruin Problem

Vladimir Mazalov[1,2] and Anna Ivashko[1,3(✉)]

[1] Institute of Applied Mathematical Research, Karelian Research Center, Russian
Academy of Sciences, ul. Pushkinskaya 11, Petrozavodsk 185910, Russia
{vmazalov,aivashko}@krc.karelia.ru
[2] Saint Petersburg State University, 7/9 Universitetskaya nab.,
Saint Petersburg 199034, Russia
[3] Petrozavodsk State University, ul. Lenina 33, Petrozavodsk 185910, Russia

Abstract. The gambler's ruin problem is studied. At each of n steps,
the probability that the player wins at the next step depends on the
win/lose ratio in previous steps. The player's payoff and the asymptotic
formula for large game durations were determined. The numerical results
of payoff simulation for different n values are reported.

Keywords: Random walk · Gambler's ruin problem · Ruin
probability · Reflection principle

1 Introduction

This paper considers the following multistage model in discrete time. A random
walk related to the ruin problem [1,2] is being monitored. In such problems,
at each step, a particle moves one unit to the right on the integer number line
when the player wins and one unit to the left if the player loses. Accordingly, the
player's capital increases or decreases by a unit depending on whether they win
or lose. The player's initial capital is fixed and when the random walk reaches
this level, it is absorbed, which is considered as the moment of ruin of the player.
In the classical model, the random walk is symmetrical, which corresponds to
equal chances for a player to win and to lose. In this paper, we investigate a
model in which the probability of winning increases with an increase in the total
number of wins, and decreases with an increase in the number of losses. The
player's goal is to augment his/her capital as much as possible without going
ruin.

Models with symmetric random walks were considered in different ways
depending on the player's goal. Shepp [3] studied a problem in which the goal was
to maximize the value of the payoff per unit time. Tamaki [4] solved the problem of maximizing the probability of stopping on any of the last few maximum
values.

Supported by the Russian Science Foundation (No. 22-11-00051, https://rscf.ru/
project/22-11-00051).

Problems related to the sequence of dependent random variables were considered in various urn schemes. Tamaki [5], Mazalov, Tamaki [6] considered variants of the problem of maximizing the probability of stopping at the largest value in an urn scheme, where the probability of transition to the next state depends on the trajectory at the previous steps. Shepp [3], Boyce [7], Ivashko [8] studied setups in which the goal was to maximize the value of the trajectory. Variants of balls-and-bins problem were considered by Tijms [9] and Ivashko [10]. Other extensions of the gambler's ruin problem to the case of multidimensional random walks [11] and the case of two players [12] have been studied.

In this paper, we consider the ruin problem associated with a random walk, where the probability of transition to the next state depends on the ratio of the number of wins and losses at the previous steps: the more wins, the greater the probability of success at the next step.

This paper is structured as follows. Section 2 gives the statement of gambler's ruin problem. Section 3 suggests an analytical solution of the problem for different cases and asymptotic behavior for large values of n. Finally, in Sect. 4, we present the findings and conclusions, and draft plans for the future.

2   Gambler's Ruin Problem

The paper considers the ruin problem of the following form. A time interval n is set at the beginning of the game. A random walk on the integer line starts from 0 and at each step i, $(i = 1, 2, ..., n)$ moves $+1$ to the right or -1 to the left. At the beginning of the game, the player can win or lose with equal probability of $\frac{1}{2}$. In the following, the transition probabilities are calculated based on the following assumptions. Suppose that at step i we are in the state (p, q), where p is the quantity $+1$ and q is the quantity -1, $i = p + q$. We will move to the state $(p + 1, q)$ with the probability $\frac{p + 1}{p + q + 2}$ and to the state $(p, q + 1)$ with the probability $\frac{q + 1}{p + q + 2}$.

Let $X_1, ..., X_n$ be a sequence of random variables that take the values $+1$ or -1 corresponding to the random walk considered above. Then, $S_i = \sum_{j=1}^{i} X_j$, $i = 1, 2, ..., n$ is the difference between $+1$ and -1 during i steps, i.e. the position of the particle at time i, $S_0 = 0$.

Moving over to coordinates on the plane (i, j): $i = p+q$ and $j = p-q$, we find that if $S_i = j$, then X_{i+1} takes the value $+1$ with a probability $\frac{1}{2}\left(1 + \frac{j}{i + 2}\right)$ and the value -1 with a probability $\frac{1}{2}\left(1 - \frac{j}{i + 2}\right)$, $j = \overline{-i, i}$, $i = \overline{0, n}$.

The value sequence $X_1, ..., X_n$ form a certain trajectory on the plane (see, e.g., Fig. 1).

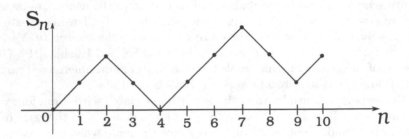

Fig. 1. Example of trajectory S_n for $n = 10$.

Let us describe the ruin problem associated with the specified random walk. We take the case of a cautious gambler who uses the following strategy. He/she leaves the game as soon as the number of losses exceeds the number of wins. Otherwise, he/she continues playing until the end instant n. The payoff when stopping at the time instant $\tau \leq n$ is denoted as S_τ (difference between the number of wins and the number of losses). Then, $S_\tau = -1$ if the random walk goes down to the level -1 or $S_\tau = S_n \geq 0$ if the game continues to the end time instant.

The player's payoff in this problem is

$$V_n = \overline{U}_n + U_n,$$

where

$$\overline{U}_n = \sum_{j \geq 0}^{n} j \cdot \mathbf{P}\{S_1 > -1, S_2 > -1, ..., S_{n-1} > -1, S_n = j\},$$

$$U_n = \sum_{j=1}^{n}(-1)\mathbf{P}\{S_1 > -1, ..., S_{j-1} > -1, S_j = -1\}$$

$$= (-1)\left(1 - \mathbf{P}\{S_1 > -1, ..., S_{n-1} > -1, S_n > -1\}\right).$$

Here, $\overline{U}_n$ is the payoff when stopping at the last step at a non-negative value (see Fig. 2), U_n is the payoff when stopping at -1 (see Fig. 3).

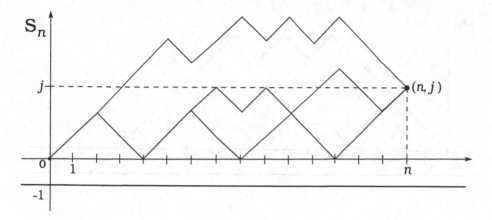

Fig. 2. Payoff when stopping at the last step at a non-negative value.

Remark 1. The probability P_n^r of the player's ruin with this strategy is

$$P_n^r = \mathbf{P}\{\inf_{1 \leq j \leq n} S_j < 0\} = \sum_{j=1}^{n} \mathbf{P}\{S_1 > -1, ..., S_{j-1} > -1, S_j = -1\}$$

$$= 1 - \mathbf{P}\{S_1 > -1, ..., S_{n-1} > -1, S_n > -1\}.$$

To compute the payoff, we need to know the probability of an arbitrary trajectory getting from point $(0,0)$ to point (n,j). Let us find this probability, e.g., for the case where p successes were followed by $n - p$ failures.

$$P(n,j) = P((0,0);(n,j)) = \prod_{i=0}^{p-1} \frac{1}{2}\left(1 + \frac{i}{i+2}\right) \prod_{i=p}^{n-1} \frac{1}{2}\left(1 - \frac{2p-i}{i+2}\right),$$

where $p = \dfrac{n+j}{2}$ is the number of successes.

Simplifying the last expression, we get

$$P(n,j) = \frac{2}{n+j+2} \prod_{i=p}^{n-1} \frac{2i-n-j+2}{2(i+2)} = \frac{1}{p+1} \prod_{i=p}^{n-1} \frac{i-p+1}{i+2} = \frac{1}{\binom{n}{p}(n+1)}.$$

This proves to be valid for any trajectory.

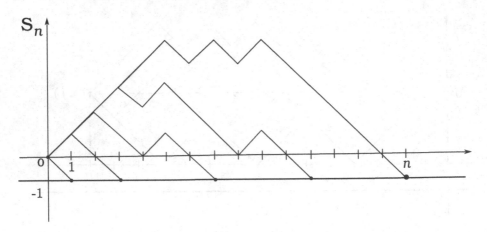

Fig. 3. Payoff when stopping at -1.

Lemma 1. *The probability of getting from point $(0,0)$ to point (n,j), where $n+j$ is even and $-n \leq j \leq n$, via any trajectory does not depend on the travel path and equals*

$$P(n,j) = \frac{1}{\binom{n}{p}(n+1)}, \tag{1}$$

where $p = \dfrac{n+j}{2}$.

Proof. Let us prove the lemma by induction on n. For $n = 1$ and $n = 2$, formula 1 is validated directly. E.g., for $n = 2$ the point $(2,0)$ can be reached by two paths via $(1,1)$ and $(1,-1)$ with equal probability $\dfrac{1}{6}$, which coincides with (1) since $n = 2$, $p = 1$, and $P(2,0) = \dfrac{1}{2 \cdot 3} = \dfrac{1}{6}$.

Assuming this statement has been proved for any $n-1$, let us prove it for n. Let j be such that $n+j$ is an even number and $-n \leq j \leq n$. There are two ways for the trajectory to get to the point (n,j) from the initial point $(0,0)$: via the point $(n-1, j-1)$ and via the point $(n-1, j+1)$ (see Fig. 4). The probability of getting from the point $(n-1, j-1)$ to the point (n,j) is

$$P((n-1, j-1); (n,j)) = \frac{1}{2}\left(1 + \frac{j-1}{n+1}\right),$$

and the probability of getting from the point $(n-1, j+1)$ to the point (n,j) is

$$P((n-1, j+1); (n,j)) = \frac{1}{2}\left(1 - \frac{j+1}{n+1}\right).$$

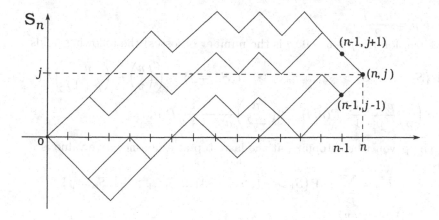

Fig. 4. Trajectories of getting from point $(0,0)$ to point (n,j).

Then, according to the induction proposition for $n-1$, the probability of the trajectory connecting the points $(0,0)$ and (n,j) and traveling via the point $(n-1, j-1)$ will be

$$P_1(n,j) = \frac{1}{2}\left(1 + \frac{j-1}{n+1}\right) \cdot \frac{1}{\binom{n-1}{\frac{n+j-2}{2}} n} = \frac{1}{\binom{n}{\frac{n+j}{2}}(n+1)}.$$

Similarly, the probability of the trajectory connecting the points $(0,0)$ and (n,j) and traveling via the point $(n-1, j+1)$ will be

$$P_2(n,j) = \frac{1}{2}\left(1 - \frac{j+1}{n+1}\right) \cdot \frac{1}{\binom{n-1}{\frac{n+j}{2}} n} = \frac{1}{\binom{n}{\frac{n+j}{2}}(n+1)}.$$

The two probabilities are equal and coincide with the expression (1). The Lemma is proved.

To find $\overline{U}_n$ and U_n, we use the following lemma based on the geometrical principle of trajectory reflection.

Lemma 2 (Feller [1]). *If $a > 0$ and $b > 0$, then the number of paths $(s_1, s_2, ..., s_n)$ such that $s_1 > -b$, $s_2 > -b$, ..., $s_{n-1} > -b$, $s_n = a$ equals $N_{n,a} - N_{n,a+2b}$, where $N_{n,x} = \binom{n}{p}$, $p = \dfrac{n+x}{2}$.*

Applying this lemma, we find that the number of trajectories from point $(0,0)$ to point (n,j) lying in the non-negative half-plane is

$$N_{n,j} - N_{n,j+2} = \binom{n}{p} - \binom{n}{p+1},$$

where $p = \dfrac{n+j}{2}$.

Then, for $j = p - q > 0$ (q is the number of losses) the following holds

$$\mathbf{P}\{S_1 > -1, S_2 > -1, ..., S_{n-1} > -1, S_n = j\} = \left(\binom{n}{p} - \binom{n}{p+1}\right) P(n,j)$$

$$= \binom{n}{p}\frac{p-q+1}{p+1}P(n,j) = 2\binom{n}{\frac{n+j}{2}}\frac{j+1}{n+j+2}P(n,j).$$

The payoff when stopping at the last step at a non-negative value is

$$\overline{U}_n = \sum_{j \geq 0}^{n} j \cdot \mathbf{P}\{S_1 > -1, S_2 > -1, ..., S_{n-1} > -1, S_n = j\}$$

$$= \sum_{j > 0}^{n} j \cdot \mathbf{P}\{S_1 > -1, S_2 > -1, ..., S_{n-1} > -1, S_n = j\}$$

$$= \sum_{p-q>0}^{n} (p-q) \cdot \binom{n}{p}\frac{p-q+1}{p+1} \cdot \frac{1}{\binom{n}{p}(n+1)}$$

$$= \sum_{p-q>0}^{n} \frac{(p-q)(p-q+1)}{(p+1)(n+1)}. \tag{2}$$

The payoff when stopping at the value -1 has the form

$$U_n = -\left(1 - \sum_{p-q>-1}^{n} \frac{p-q+1}{(p+1)(n+1)}\right). \tag{3}$$

3   Payoff for Different Values of n

3.1   The Case of Even Values of n

For even $n = p+q = 2m$ ($j = p-q = 2s$, $p = m+s$, $s = -m, ..., m$) the formulas 2 and 3 take the form

$$\overline{U}_n = \sum_{p-q>0}^{n} \frac{(p-q)(p-q+1)}{(p+1)(n+1)} = \sum_{s=1}^{m} \frac{2s(2s+1)}{(m+s+1)(2m+1)}$$

$$= 2\left(\frac{-m^2+m+1}{2m+1} + (m+1)\left(\sum_{s=1}^{2m}\frac{1}{s} + \sum_{s=1}^{m+1}\frac{1}{s}\right)\right),$$

$$U_n = -\left(1 - \sum_{p-q>-1}^{n} \frac{p-q+1}{(p+1)(n+1)}\right) = -\left(1 - \frac{1}{2m+1}\sum_{s=0}^{m}\frac{2s+1}{m+s+1}\right)$$

$$= -\sum_{s=0}^{m-1}\left(\frac{1}{2s+1} - \frac{1}{2s+2}\right) = -\sum_{s=m+1}^{2m}\frac{1}{s} = -\sum_{s=1}^{2m}\frac{1}{s} + \sum_{s=1}^{m}\frac{1}{s}.$$

Then the payoff in this problem is

$$V_n = \overline{U}_n + U_n$$

$$= 2\left(\frac{-m^2 + m + 1}{2m + 1} + (m + 1)\left(\sum_{s=1}^{2m}\frac{1}{s} + \sum_{s=1}^{m+1}\frac{1}{s}\right)\right) - \sum_{s=1}^{2m}\frac{1}{s} + \sum_{s=1}^{m}\frac{1}{s}$$

$$= -\frac{2m(m + 1)}{2m + 1} + (2m + 1)\left(\sum_{s=1}^{2m}\frac{1}{s} - \sum_{s=1}^{m}\frac{1}{s}\right).$$

3.2   The Case of Odd Values of n

For odd values of $n = 2m + 1$ $(j = 2s + 1, p = m + s + 1, s = -m, ..., m)$, the formulas 2 and 3 have the form

$$\overline{U}_n = \sum_{p-q>0}^{n} (p - q) \cdot \frac{p - q + 1}{p + 1}\frac{1}{n + 1}$$

$$= \sum_{s=0}^{m}\frac{(2s + 1)(s + 1)}{(m + s + 2)(m + 1)} = -m - 1 + (2m + 3)\sum_{s=m+2}^{2m+2}\frac{1}{s},$$

$$U_n = -\left(1 - \sum_{p-q>-1}^{n}\frac{p - q + 1}{(p + 1)(n + 1)}\right) = -\left(1 - \sum_{s=0}^{m}\frac{s + 1}{(m + s + 2)(m + 1)}\right)$$

$$= -\sum_{s=m+2}^{2m+2}\frac{1}{s}.$$

Then the payoff in this problem is

$$V_n = \overline{U}_n + U_n = -m - 1 + (2m + 3)\sum_{s=m+2}^{2m+2}\frac{1}{s} - \sum_{s=m+2}^{2m+2}\frac{1}{s}$$

$$= -m - 1 + (2m + 2)\sum_{s=m+2}^{2m+2}\frac{1}{s} = -m + (2m + 2)\sum_{s=m+2}^{2m+1}\frac{1}{s}.$$

Thus, the following theorem has been proved.

Theorem 1. *In the gambler's ruin problem, the payoff V_n has the form*
   1) for even values of $n = 2m$

$$V_n = -\frac{2m(m + 1)}{2m + 1} + (2m + 1)\left(H_{2m} - H_m\right),$$

   2) for odd values of $n = 2m + 1$

$$V_n = -m + (2m + 2)\left(H_{2m+1} - H_{m+1}\right),$$

where $H_m = \sum_{s=1}^{m}\frac{1}{s}$ is a harmonic number.

The results presented in the theorem show that the payoff in this ruin problem contains harmonic numbers. Interestingly, harmonic numbers occur in optimal stopping problems. To wit, in the best choice problem in [13], the probability of choosing the best item from the set of items N using an optimal strategy k^* is equal to $\dfrac{k^* - 1}{N}\left(H_{N-1} - H_{k^*-2}\right)$.

For large values of $n = 2m$ we have an approximation $H_m \approx \ln(m)$. Then,

$$V_n = -\frac{2m(m+1)}{2m+1} + (2m+1)\left(H_{2m} - H_m\right)$$

$$\approx -\frac{2m(m+1)}{2m+1} + (2m+1)\ln 2 \approx m(\ln 4 - 1) \approx 0.193n.$$

Remark 2. For the probability of gambler's ruin we get an asymptotic estimate

$$P_n^r = H_{2m} - H_{m-1} \approx \ln 2 \approx 0.693.$$

Although the probability of gambler's ruin in the strategy under study is high, his/her average payoff grows without limit with a growing number of game rounds.

Remark 3. For a symmetric random walk, the probability of the gambler's ruin with the given strategy is approximately $1 - \dfrac{1}{\sqrt{\pi m}}$ for even values of $n = 2m$. For greater values of n, the probability of ruin tends to 1, contrary to the suggested random walk.

Table 1 gives the numerical results for payoff values V_n at different values of n.

Table 1. Player's payoff for different values of n

n	2	3	4	5	6	7	8	9	10	100	200	300
V_n	0.167	0.333	0.517	0.7	0.888	1.076	1.266	1.456	1.647	19.010	38.324	57.638

4   Conclusion

We investigated the problem of gambler's ruin during a given time interval n. It is assumed that the probability of the player winning at each next step depends on the ratio of wins and losses in the previous steps. The player continues playing until the number of losses exceeds the number of wins. The player's payoff in this problem was determined for different values of game duration. The payoff is related to harmonic numbers. An asymptotic formula was built for computing the player's payoff at large values of n.

Further studies may focus on other models related to the ruin problem based on the suggested random walk scheme. Here the player's goal is to increase his/her capital without going ruin. Or, in other words, it is required to stop the random walk with the maximum possible value. Note that the payoff function can have a different form, for example, it can be any arbitrary increasing function of the stopped value. In particular, it would be interesting to examine the ruin problem with two players where the game continues until one of the players is ruined.

References

1. Feller, W.: An Introduction to Probability Theory and Its Applications, vol. 1. Wiley, New York (1968)
2. Shiryaev, A.: Probability-1. Springer, New York (2016). https://doi.org/10.1007/978-0-387-72206-1
3. Shepp, L.A.: Explicit solutions to some problems of optimal stopping. Ann. Math. Stat. **40**, 993–1010 (1969)
4. Tamaki, M.: Sum the multiplicative odds to one and stop. J. Appl. Probab. **47**(03), 761–777 (2010)
5. Tamaki, M.: Optimal stopping on trajectories and the ballot problem. J. Appl. Probab. **38**, 946–959 (2001)
6. Mazalov, V.V., Tamaki, M.: Duration problem on trajectories. Stochast. Int. J. Probab. Stochast. Processes **79**(3–4), 211–218 (2007)
7. Boyce, W.M.: Stopping rules for selling bonds. Bell J. Econ. Manage. Sci. **1**(1), 27–53 (1970)
8. Ivashko, A.A.: [Gain maximization problem in the urn scheme] Trudy Karelskogo nauchnogo centra RAN [Transactions of the Karelian Research Centre RAS], 4, 62–66. (2014). (In Russ.)
9. Tijms, H.: One-step improvement ideas and computational aspects. In: Boucherie, R.J., van Dijk, N.M. (eds.) Markov Decision Processes in Practice. ISORMS, vol. 248, pp. 3–32. Springer, Cham (2017). https://doi.org/10.1007/978-3-319-47766-4_1
10. Ivashko, A.A.: Optimal stopping in the balls-and-bins problem. Contrib. Game Theory Manage. **14**, 183–191 (2021)
11. Hussain, A., Cheema, S.A., Hanif, M.: A game based on successive events. Stat **9**, e274 (2020)
12. Lorek, P.: Generalized Gambler's ruin problem: explicit formulas via siegmund duality. Methodol. Comput. Appl. Probab. **19**, 603–613 (2017)
13. Gilbert, J., Mosteller, F.: Recognizing the maximum of a sequence. J. Am. Statist. Assoc. **61**(313), 35–73 (1966)

Exploitation and Recovery Periods in Dynamic Resource Management Problem

Vladimir Mazalov[1,2] and Anna Rettieva[1,2(✉)]

[1] Institute of Applied Mathematical Research, Karelian Research Center of RAS, 11 Pushkinskaya str., Petrozavodsk 185910, Russia
{vmazalov,annaret}@krc.karelia.ru
[2] Saint Petersburg State University, 7/9 Universitetskaya nab., Saint Petersburg 199034, Russia

Abstract. Dynamic game related to resource management problem is considered. The planning horizon is assumed to be divided into the periods of exploitation where many players use a common resource and the periods of recovery where the resource stock is evolving according to the natural growth rule. Both noncooperative and coordinated players' behaviors are investigated. The conditions linking the values of exploitation and recovery periods in order to maintain the sustained resource usage are determined. To illustrate the presented approaches, a dynamic bioresource management problem (harvesting problem) with many players and compound planning horizon is investigated.

Keywords: Dynamic games · Resource management problem · Exploitation period · Recovery period

1 Introduction

Resource management problems (exploitation of common renewable resources) is one of the real-life challenges that are very important for ecology and economics. It encompasses a number of different issues including the phenomenon known as the tragedy of the commons. The commons denotes a natural resource extracted by many individuals and the tragedy means that the participants tend to overexploitation in the absence of regulation [15]. There is an extensive literature on renewable resources management in economics, operations research and optimal control theory.

Real-life problems such as the exploitation processes involve dynamics of the renewable resource and a number of decision makers. Hence, they can be investigated applying the technics of optimal control theory and dynamic games. The game-theoretic approach to resource exploitation was pioneered by Munro

This research was supported by the Russian Science Foundation grant No. 22-11-00051, https://rscf.ru/en/project/22-11-00051/.

[14] and Clark [1], who applied the Nash equilibrium for the models of fisheries. The optimal behavior of players in harvesting problems is investigated by many scientists including Hamalainen et al. [5,8], Kaitala and Lindroos [11], Levhari, Fisher and Mirman [6,7,10], Petrosyan and Zackharov [17], Plou [18], Tolwinski [9,19], Zaccour et al. [2,3]. Several types of optimal behavior regulation have been suggested such as incentive equilibrium [5,12], time-consistent imputation distribution procedure [13,16], moratorium regimes [1,2,14], prohibited for exploitation areas [13] and others.

Some resource management problems involve periodic processes when a natural resource is exploited over a certain period of time, and then a time period for resource recovery or moratorium regime [2] is established. The matter of environmental concernment is the balance between time intervals of exploitation and recovery for sustained resource usage.

Such periodic exploitation processes are wide spread in harvesting problems. For example, in lake Onego in 2004, the moratorium for exploitation of the Shuya salmon population was withdrawn. Salmon fishing has become popular among fishing firms in northwestern Russia. However, in 2010, the size of salmon population that was about a hundred tons of fish according to official statistics, began to decline very sharply. Already by 2014, it reached, according to the data of Karelian harvesting agency, forty tons. The catch, primarily illegal, significantly exceeded reproduction. Therefore, in 2015, a new edition of the fishing rules introduced a new moratorium for salmon exploitation. Similar problems arise the forest management issues and other renewable resource exploitation processes.

The main goal of this paper is to determine the ratio between exploitation and recovery periods depending on the natural environment parameters and the exploitation load in order to maintain the sustainable nature resource evolution. For that purpose a dynamic bioresource management problem (harvesting problem) with many players and compound planning horizon is investigated. Both egoistic (noncooperative) and coordinated players' behaviors are investigated. The obtained conditions linking the values of exploitation and recovery periods allow to establish optimal for renewable resource evolution moratorium regimes.

Further exposition has the following structure. Section 2 describes the main model where the planning horizon is divided into the periods of exploitation and recovery. Two-periods dynamic resource management problem is presented in Sect. 3 while Sect. 4 describes the infinite horizon problem with interchanging periods of exploitation and recovery. The ratio between exploitation and recovery periods is determined for both models in noncooperative and coordinated settings. Finally, Sect. 5 provides the basic results and their discussion.

2   Main Model

Consider a renewable natural resource evolving according to

$$x'(t) = f(x(t)), \quad x(0) = x_0, \quad t \in (0, \infty), \tag{1}$$

where $x(t) \geq 0$ denotes the resource stock at time $t \geq 0$, $f(x)$ is the resource natural growth function.

Let $N = \{1, \ldots, n\}$ players exploit a common resource during time period $t \in [0, T]$. The state dynamics takes the form

$$x'(t) = f(x(t)) - u_1(t) - \ldots - u_n(t), \quad x(0) = x_0, \ t \in [0, T], \qquad (2)$$

where $u_i(t) \geq 0$ gives the exploitation rate of player i at time t, $i \in N$.

Then, the moratorium for exploitation is established during time interval $(T, T + \tau]$ when the resource recovers and evolves according to the dynamics (1). Section 3 considers only two periods of exploitation and recovery while Sect. 4 describes a periodic exploitation process where after the moratorium the players continue resource usage till the next recovery period and so on.

The payoff functions of the players on a finite planning horizon $[0, T]$ have the form

$$J_i(x, u_1, \ldots, u_n) = \int_0^T e^{-\rho t} g_i(x(t), u_1(t), \ldots, u_n(t)) \, dt, \qquad (3)$$

where $g_i(\cdot) \geq 0$, $i \in N$, are the instantaneous payoff functions, $\rho \in (0, 1)$ denotes the discount factor.

For sustainable resource evolution we suppose that after the recovery period the size of the population shouldn't be less than the initial one x_0[1]. The main goal of this paper is to determine the ratio between exploitation and recovery periods (T and τ) depending on the environmental and economical parameters.

Linear natural resource growth function as well as quadratic instantaneous payoff functions are considered in the next sections. The noncooperative (egoistic) case where each player wishes to maximize individual payoff (3) and the coordinated one where the players combine their exploitation rates and optimize the joint payoff are investigated.

3   Two-Periods Model

We begin with two-periods model where players exploit the resource during time interval $[0, T]$ and the resource recovers during time interval $(T, T + \tau]$. The main goal is to determine the conditions linking the values of exploitation and recovery periods (T and τ) in order to maintain the sustainable resource usage. The relations for noncooperative and coordinated cases are constructed and compared.

Let $N = \{1, \ldots, n\}$ players exploit a natural renewable resource during time period $t \in [0, T]$. The evolution of the resource takes the form

$$x'(t) = \varepsilon x(t) - u_1(t) - \ldots - u_n(t), \quad x(0) = x_0, \ t \in [0, T], \qquad (4)$$

---

[1] The results obtained in the paper are naturally extended for the case when the desired size of the population has the form kx_0 for any $k \in (0, \infty)$. The dependence on x_0 is induced by the fact that the only information that the regulator possesses at the beginning of the planning period is the initial size of the population.

where $x(t) \geq 0$ denotes the resource stock at time $t \geq 0$, $\varepsilon \geq 1$ denotes the natural birth rate and $u_i(t) \geq 0$ gives the exploitation rate of player i at time t, $i \in N$.

During the recovery period a renewable resource evolves according linear dynamics

$$x'(t) = \varepsilon x(t), \quad x(0) = x_0, \quad t \in (T, T + \tau]. \tag{5}$$

Each player wishes to maximize the revenue from resource sales and to minimize the exploitation costs. Assume that the players have the same market prices and costs that depend quadratically on the exploitation rate. The payoff functions of the players take the forms

$$J_i(x, u_1, \ldots, u_n) = \int_0^T e^{-\rho t} [p u_i(t) - h u_i(t)^2] \, dt, \tag{6}$$

where $p \geq 0$ is the market price of the resource, $h \geq 0$ indicates the exploitation cost, and $\rho \in (0,1)$ denotes the discount factor. Further assume that $\varepsilon \geq n\rho$ (see Proposition 1).

Due to the symmetry of the instantaneous payoff functions dividing (6) by p and denoting $c = h/p$ gives the players payoff functions in the form

$$J_i(x, u_1, \ldots, u_n) = \int_0^T e^{-\rho t} [u_i(t) - c u_i(t)^2] \, dt. \tag{7}$$

3.1 Noncooperative Behavior

First, we consider noncooperative behavior and construct the Nash equilibrium strategies u_i^N that satisfy Nash inequalities

$$J_i(x, u_1^N, \ldots, u_{i-1}^N, u_i, u_{i+1}^N, \ldots, u_n^N) \leq J_i(x, u_1^N, \ldots, u_n^N) \ \forall u_i \in U_i = [0, \infty), \ i \in N.$$

Note, that the Nash equilibrium will be constructed in the feedback form $u_i(t) = u_i(x(t))$, $i \in N$.

Proposition 1. *The Nash equilibrium strategies in problem (4), (5), (7) have the form*

$$u_i^N(x) = \frac{2\varepsilon - \rho}{2n - 1} x - \frac{\varepsilon - n\rho}{2\varepsilon c(2n - 1)} \tag{8}$$

and the resource size is given by

$$x^N(t) = \frac{n}{2\varepsilon c} + \left(x_0 - \frac{n}{2\varepsilon c}\right) e^{-\frac{\varepsilon - n\rho}{2n - 1} t}, \quad t \in [0, T]. \tag{9}$$

Proof. To construct an equilibrium we apply dynamic programming principle and construct Hamilton-Jacobi-Bellman (HJB) equations [4]. Since players are symmetric assume, that all the players except the player i use feedback strategies $u_j(x) = \phi(x)$, $j \in N$, $j \neq i$ and find the player i's optimal behavior. To maximize the individual payoff (7) the player i's value function $V_i(x, t)$ satisfy the next HJB equation:

$$\rho V_i(x, t) - \frac{\partial V_i}{\partial t} = \max_{u_i}\left\{u_i - cu_i^2 + \frac{\partial V_i}{\partial x}(\varepsilon x - u_i - (n - 1)\phi(x))\right\}. \qquad (10)$$

The solution of (10) will be constructed in quadratic form

$$V_i(x, t) = V_i(x) = A_i x^2 + B_i x + D_i.$$

Substitution to (10) gives

$$\rho A_i x^2 + \rho B_i x + \rho D_i = \max_{u_i}\left\{u_i - cu_i^2 + (2A_i x + B_i)(\varepsilon x - u_i - (n-1)\phi(x))\right\} \quad (11)$$

which yields

$$u_i(x) = -\frac{A_i}{c}x + \frac{1 - B_i}{2c}. \qquad (12)$$

The symmetry of the problems lead to the symmetry of the strategies $u_i(x) = \phi(x)$, $i \in N$. Substituting to (11), the system to define parameters becomes

$$\begin{cases} \rho A_i = \left(\frac{2n-1}{c}A_i^2 + 2\varepsilon A_i\right), \\ \rho B_i = \left(\frac{2n-1}{c}A_i B_i - \frac{n}{c}A_i + \varepsilon B_i\right), \\ \rho D_i = \frac{(B_i - 1)((2n-1)B_i - 1)}{4c}, \end{cases}$$

which yields

$$A_i = -\frac{c(2\varepsilon - \rho)}{2n - 1}, \quad B_i = \frac{n(2\varepsilon - \rho)}{\varepsilon(2n - 1)}, \quad D_i = \frac{(\varepsilon - n\rho)(n(2\varepsilon - \rho) - \varepsilon)}{4\varepsilon^2 c\rho(2n - 1)}.$$

As $A_i \le 0$, to have the nonnegative payoff for player i it is necessary the top of the parabola $A_i x^2 + B_i + D_i$ to lie above the X-line. The abscissa of the top is equal $\bar{x} = \frac{n}{c\varepsilon}$, hence the ordinate takes the form

$$\bar{y} = A_i \frac{n^2}{c^2\varepsilon^2} + B_i \frac{n}{c\varepsilon} + D_i = D_i.$$

It yields that D_i should be nonnegative that is the fact when $\varepsilon \ge n\rho$.

The noncooperative equilibrium strategies become

$$u_i^N(x) = \frac{2\varepsilon - \rho}{2n - 1}x - \frac{\varepsilon - n\rho}{2\varepsilon c(2n - 1)}, \quad i \in N.$$

Substituting (8) into the state dynamics corresponding to the exploitation regime (4), we get

$$x'(t) = -\frac{\varepsilon - n\rho}{2n - 1}x(t) + \frac{n(\varepsilon - n\rho)}{2\varepsilon c(2n - 1)}, \quad x(0) = x_0. \qquad (13)$$

The solution of (13) takes the form

$$x^N(t) = \frac{n}{2\varepsilon c} + \left(x_0 - \frac{n}{2\varepsilon c}\right)e^{-\frac{\varepsilon - n\rho}{2n - 1}t}, \quad t \in [0, T].$$

3.2   Coordinated Behavior

To coordinate the resource usage the players combine their exploitation rates and optimize the joint payoff function that takes the form

$$J(x, u_1, \ldots, u_n) = \int_0^T e^{-\rho t}\Big[\sum_{j=1}^n u_j(t) - c\Big(\sum_{j=1}^n u_j(t)\Big)^2\Big]\,dt. \tag{14}$$

Proposition 2. *The coordinated strategies in problem (4), (5), (14) have the form*

$$u_i^c(x) = \frac{2\varepsilon - \rho}{n}x - \frac{\varepsilon - \rho}{2n\varepsilon c} \tag{15}$$

and the resource size is given by

$$x^c(t) = \frac{1}{2\varepsilon c} + \Big(x_0 - \frac{1}{2\varepsilon c}\Big)e^{-(\varepsilon - \rho)t}, \quad t \in [0, T]. \tag{16}$$

Proof. Applying HJB equation similar to Proposition 1 we construct coordinated behavior. Note that the joint payoff function takes the form

$$V(x, t) = V(x) = Ax^2 + Bx + D,$$

where

$$A = -c(2\varepsilon - \rho), \ B = \frac{(2\varepsilon - \rho)}{\varepsilon}, \ D = \frac{(\varepsilon - \rho)^2}{4\varepsilon^2 c\rho}.$$

3.3   Sustainable Resource Exploitation

For sustainable resource evolution after the recovery period the size of the population shouldn't be less than the initial one. Denote the resource size achieved at the end of the moratorium regime lasting for τ^N time steps in noncooperative case or τ^c steps in the coordinated one as $x^N(T + \tau^N)$ or $x^c(T + \tau^c)$. According to the condition for sustained usage mentioned above the next equality should be fulfilled:

$$x^N(T + \tau^N) = x^c(T + \tau^c) = x_0. \tag{17}$$

Since during the recovery period the population evolves according to dynamics (5) the condition (17) becomes

$$x^N(T)e^{\varepsilon\tau^N} = x^c(T)e^{\varepsilon\tau^c} = x_0. \tag{18}$$

Consider the sustainable condition for noncooperative case. According to (9) it takes the form

$$\Big(\frac{n}{2\varepsilon c} + \Big(x_0 - \frac{n}{2\varepsilon c}\Big)e^{-\frac{\varepsilon - n\rho}{2n-1}T}\Big)e^{\varepsilon\tau^N} = x_0 \tag{19}$$

which yields that the time period for resource recovery shouldn't be less that

$$\tau^N = \frac{1}{\varepsilon}\ln\Big(\frac{2c\varepsilon x_0}{(2c\varepsilon x_0 - n)e^{-\frac{\varepsilon - n\rho}{2n-1}T} + n}\Big). \tag{20}$$

As the dynamics (4) and (5) can describe real fish population possess the huge sizes we consider the condition (20) for large x_0. Under this assumption it gives the ratio between exploitation and recovery periods in the form

$$\frac{\tau^N}{T} \approx \frac{\varepsilon - n\rho}{\varepsilon(2n-1)}. \tag{21}$$

Now consider the sustainable condition for the coordinated case. According to (16) it takes the form

$$\left(\frac{1}{2\varepsilon c} + \left(x_0 - \frac{1}{2\varepsilon c}\right)e^{-(\varepsilon-\rho)T}\right)e^{\varepsilon\tau^c} = x_0 \tag{22}$$

which yields that the time period for resource recovery shouldn't be less that

$$\tau^c = \frac{1}{\varepsilon}\ln\left(\frac{2c\varepsilon x_0}{(2c\varepsilon x_0 - 1)e^{-(\varepsilon-\rho)T} + 1}\right). \tag{23}$$

Again, for large x_0 the condition (23) gives the ratio between exploitation and recovery periods in the form

$$\frac{\tau^c}{T} \approx \frac{\varepsilon - \rho}{\varepsilon}. \tag{24}$$

Theorem 1. *The time period for resource recovery in noncooperative case is less than in the coordinated one.*

Proof. Comparing (21) and (24) observe that

$$\tau^c - \tau^N = \frac{(2\varepsilon - \rho)(n-1)}{\varepsilon(2n-1)}T \geq 0.$$

Hence, the noncooperative behavior is better for population state. This observation differs from the cooperation preference in the "fish wars" model as the population density $0 \leq x \leq 1$ instead of the size of the population was investigated there.

4   Model with Many Periods

Now, consider the case when the periods of exploitation and recovery are repeated many times. As above, the state dynamics (4) correspond to the period of exploitation, while (5) – to the period of recovery.

In this model, the sequence of events is as follows: players exploit the resource for time interval $t \in [0, T]$. Then, the recovery period is implemented for $t \in (T, T + \tau]$, during which the players get zero payoffs. At $t = T + \tau$, the stock level is back to the desired level x_0, and the players can again exploit the resource. Hence, the players' planning horizon is infinite and the payoff functions take the form

$$J_i(x_0, u) = \int_0^\infty e^{-\rho t}(\hat{u}_i(t) - c\hat{u}_i(t)^2)\,dt, \tag{25}$$

where

$$\hat{u}_i(t) = \begin{cases} u_i(t), & t \in [k(T+\tau), (k+1)T + k\tau], \\ 0, & t \in ((k+1)T + k\tau, (k+1)(T+\tau)], \ k = 0, 1, \dots \end{cases}$$

Since after the recovery period the resource size is equal to the initial one x_0 player i's payoff in the game where the moratorium regime is firstly applied at $t = T$ is

$$J_i(x_0, u) = \int_0^T e^{-\rho t}\left(u_i(t) - cu_i(t)^2\right) dt + e^{-\rho(T+\tau)} J_i(x_0, u),$$

which yields

$$J_i(x_0, u) = \frac{1}{1 - e^{-\rho(T+\tau)}} \int_0^T e^{-\rho t}\left(u_i(t) - cu_i(t)^2\right) dt \qquad (26)$$

with state dynamics (4) for $t \in [0, T]$ and (5) for $t \in (T, T+\tau]$. In noncooperative case player $i \in N$ maximizes (26) with respect to u_i and under coordination the players wish to maximize the joint payoff with combined exploitation rates.

As before, we consider both types of players' behavior and construct the equilibrium strategies in feedback form $u_i(t) = u_i(x(t))$, $i \in N$.

4.1   Noncooperative Behavior

First, define the Nash equilibrium in the feedback strategies when each player i maximizes individual payoff (26).

Proposition 3. *The Nash equilibrium strategies in problem (4), (5), (26) have the form*

$$u_i^N(x) = \frac{2\varepsilon - \rho(1 - e^{-\rho(T+\tau^N)})}{2n - 1} x - \frac{\varepsilon - n\rho(1 - e^{-\rho(T+\tau^N)})}{2\varepsilon c(2n - 1)} \qquad (27)$$

and the resource size is given by

$$x^N(t) = \frac{n}{2\varepsilon c} + \left(x_0 - \frac{n}{2\varepsilon c}\right) e^{-\frac{\varepsilon - n\rho(1 - e^{-\rho(T+\tau^N)})}{2n-1} t}, \quad t \in (0, \infty). \qquad (28)$$

Proof. To construct an equilibrium we apply HJB equation again. Since players are symmetric assume, that all the players except the player i use feedback strategies $u_j(x) = \phi(x)$, $j \in N$, $j \neq i$ and find the player i's optimal behavior. To maximize the individual payoff (26) the player i's value function $V_i(x, t)$ satisfy the next HJB equation:

$$\rho V_i(x, t) - \frac{\partial V_i(x, t)}{\partial t}$$

$$= \frac{1}{1 - e^{-\rho(T+\tau^N)}} \max_{u_i}\left\{u_i - cu_i^2 + \frac{\partial V_i(x, t)}{\partial x}(\varepsilon x - u_i - (n-1)\phi(x))\right\}. \quad (29)$$

As before we seek the value function in quadratic form

$$V_i(x,t) = V_i(x) = A_i x^2 + B_i x + D_i \,.$$

Similarly to Proposition 1 we get

$$u_i(x) = -\frac{A_i}{c} x + \frac{1 - B_i}{2c}$$

and

$$A_i = -\frac{c(2\varepsilon - \rho(1 - e^{-\rho(T+\tau^N)}))}{2n - 1} \,, \quad B_i = \frac{n(2\varepsilon - \rho(1 - e^{-\rho(T+\tau^N)}))}{\varepsilon(2n - 1)} \,,$$

$$D_i = \frac{-(n-1)^2 c^2 \mid n^2 (c \quad \rho(1 \quad e^{-\rho(T+\tau^N)}))}{4c\rho\varepsilon^2(1 - e^{-\rho(T+\tau^N)})} \,.$$

Hence, the noncooperative equilibrium strategies become

$$u_i^N(x) = \frac{2\varepsilon - \rho(1 - e^{-\rho(T+\tau^N)})}{2n - 1} x - \frac{\varepsilon - n\rho(1 - e^{-\rho(T+\tau^N)})}{2\varepsilon c(2n - 1)} \,.$$

Substituting (27) into the state dynamics corresponding to the exploitation regime (4) we get

$$x'(t) = -\frac{\varepsilon - n\rho(1 - e^{-\rho(T+\tau^N)})}{2n - 1} x(t) + \frac{n(\varepsilon - n\rho(1 - e^{-\rho(T+\tau^N)}))}{2\varepsilon c(2n - 1)} \,, \quad x(0) = x_0 \,.$$
(30)

The solution of (30) takes the form

$$x^N(t) = \frac{n}{2\varepsilon c} + \left(x_0 - \frac{n}{2\varepsilon c}\right) e^{-\frac{\varepsilon - n\rho(1 - e^{-\rho(T+\tau^N)})}{2n - 1} t} \,.$$

4.2   Coordinated Behavior

Now, define the coordinated equilibrium in the feedback strategies when players maximizes the joint payoff with combined exploitation rates:

$$J(x_0, u) = \frac{1}{1 - e^{-\rho(T+\tau)}} \int_0^T e^{-\rho t} [\sum_{i=1}^n u_i(t) - c\left(\sum_{i=1}^n u_i(t)\right)^2] \, dt \,.$$
(31)

Proposition 4. *The coordinated strategies in problem (4), (5), (31) have the form*

$$u_i^c(x) = \frac{2\varepsilon - \rho(1 - e^{-\rho(T+\tau^c)})}{n} x - \frac{\varepsilon - \rho(1 - e^{-\rho(T+\tau^c)})}{2n\varepsilon c}$$
(32)

and the resource size is given by

$$x^c(t) = \frac{1}{2\varepsilon c} + \left(x_0 - \frac{1}{2\varepsilon c}\right) e^{-(\varepsilon - \rho(1 - e^{-\rho(T+\tau^c)}))t} \,, \quad t \in (0, \infty) \,.$$
(33)

Proof. Applying HJB equation similar to Proposition 3 we construct coordinated behavior.

Note that the joint payoff function takes the form

$$V(x,t) = V(x) = Ax^2 + Bx + D,$$

where

$$A = -c(2\varepsilon - \rho(1 - e^{-\rho(T+\tau^c)})), \quad B = \frac{2\varepsilon - \rho(1 - e^{-\rho(T+\tau^c)})}{\varepsilon},$$

$$D = \frac{(\varepsilon - \rho(1 - e^{-\rho(T+\tau^c)}))^2}{4\varepsilon^2 c\rho(1 - e^{-\rho(T+\tau^c)})}.$$

4.3   Sustainable Resource Exploitation

As before assume that for sustainable resource evolution after the recovery period the size of the population should be equal to the initial one x_0. For both types of players' behavior the corresponding condition take the form

$$x^N(T)e^{\varepsilon \tau^N} = x^c(T)e^{\varepsilon \tau^c} = x_0. \tag{34}$$

Consider the sustainable condition for noncooperative case. According to (28) it takes the form

$$\left(\frac{n}{2\varepsilon c} + \left(x_0 - \frac{n}{2\varepsilon c}\right)e^{-\frac{\varepsilon - n\rho(1 - e^{-\rho(T+\tau^N)})}{2n-1}T}\right)e^{\varepsilon \tau^N} = x_0 \tag{35}$$

which yields for large x_0 the ratio between exploitation and recovery periods in the form

$$\frac{\tau^N}{T} \approx \frac{\varepsilon - n\rho}{\varepsilon(2n-1)} + \frac{1}{\rho T}W\left(\frac{\rho^2 nT}{\varepsilon(2n-1)}e^{-\frac{\rho n(2\varepsilon-\rho)T}{\varepsilon(2n-1)}}\right), \tag{36}$$

where $W(\cdot)$ is the Lambert function.

Now consider the sustainable condition for the coordinated case. According to (33) it takes the form

$$\left(\frac{1}{2\varepsilon c} + \left(x_0 - \frac{1}{2\varepsilon c}\right)e^{-(\varepsilon - \rho(1 - e^{-\rho(T+\tau^c)}))T}\right)e^{\varepsilon \tau^c} = x_0 \tag{37}$$

which yields (for large x_0) the ratio between exploitation and recovery periods in the form

$$\frac{\tau^c}{T} \approx \frac{\varepsilon - \rho}{\varepsilon} + \frac{1}{\rho T}W\left(\frac{\rho^2 T}{\varepsilon}e^{-\frac{\rho(2\varepsilon-\rho)T}{\varepsilon}}\right). \tag{38}$$

Theorem 2. *The time period for resource recovery in noncooperative case is less than in the coordinated one for the model with many periods.*

Proof. Let us compare (36) and (38). From (35) and (37) (for large x_0) we get

$$\tau^c = T(\frac{\varepsilon - \rho}{\varepsilon} + \frac{\rho}{\varepsilon}e^{-\rho(T+\tau^c)}),$$

$$\tau^N = T(\frac{\varepsilon - n\rho}{\varepsilon(2n-1)} + \frac{\rho n}{\varepsilon(2n-1)}e^{-\rho(T+\tau^N)}),$$

which yields

$$\tau^c - \tau^N = \frac{(2\varepsilon - \rho)(n-1)}{\varepsilon(2n-1)}T + \frac{T\rho e^{-\rho(T+\tau^N)}}{\varepsilon(2n-1)}((2n-1)e^{-\rho(\tau^c-\tau^N)} - n). \quad (39)$$

The solution of (39) can be also obtained via Lambert function in the next form

$$\tau^c - \tau^N = \frac{T}{\varepsilon(2n-1)}\left((2\varepsilon - \rho)(n-1) - \rho n e^{-\rho(T+\tau^N)}\right)$$

$$+ \frac{1}{\rho}W\left(\frac{\rho^2 T}{\alpha}e^{-\rho(T+\tau^N)}e^{-\frac{\rho((2\varepsilon-\rho)(n-1)-\rho n e^{-\rho(T+\tau^N)})}{\varepsilon(2n-1)}}\right). \quad (40)$$

As the Lambert function is positive for nonnegative argument and $(2\varepsilon - \rho)(n-1) - \rho n e^{-\rho(1+\tau^N)}$ is larger that $\rho n(1 - e^{-\rho(1+\tau^N)}) > 0$ (40) yields that

$$\tau^c - \tau^N \geq 0.$$

5   Conclusions

Dynamic game related to resource management problem (renewable resource exploitation process) is considered. The evolution of the resource and exploitation processes are assumed to be periodic. Namely, the periods of extraction of the renewable resource are interchanged with recovery periods in order to maintain the sustained resource usage.

The desired resource size after the recovery period is assumed to be equal to the initial one for long-term exploitation. First, the model with one extraction and one recovery periods is considered. Then, the extension with many rotated exploitation periods and moratorium regimes is presented. Both egoistic (noncooperative) and coordinated players' behaviors are investigated. The conditions linking the values of exploitation and recovery periods are derived analytically. It is shown that the time period needed for resource recovery in noncooperative case is less than in the coordinated one. The obtained ratio between exploitation and recovery periods allow to establish optimal for renewable resource evolution moratorium regimes.

Note that the solutions are obtained under assumption that the value of exploitation period T is externally given. The problem of optimal extraction period size determination in order to maintain sustained exploitation process is planned for near future work.

References

1. Clark, C.W.: Bioeconomic Modelling and Fisheries Management. Wiley, New York (1985)
2. Dahmouni, I., Parilina, E.M., Zaccour, G.: Great fish war with moratorium. Technical report, Les Cahiers du GERAD G-2022-07. GERAD, HEC Montreal, Canada (2022)
3. Dahmouni, I., Vardar, B., Zaccour, G.: A fair and time-consistent sharing of the joint exploitation payoff of a fishery. Nat. Resour. Model. **32**, 12216 (2019)
4. Dockner, E.J., Jorgensen, S., Long, N.V., Sorger, G.: Differential Games in Economics and Management Science. Cambridge University Press, Cambridge, UK (2000)
5. Ehtamo, H., Hamalainen, R.P.: A cooperative incentive equilibrium for a resource management problem. J. Econ. Dyn. Control **17**, 659–678 (1993)
6. Fisher, R.D., Mirman, L.J.: Strategic dynamic interaction: fish wars. J. Econ. Dyn. Control **16**, 267–287 (1992)
7. Fisher, R.D., Mirman, L.J.: The complete fish wars: biological and dynamic interactions. J. Environ. Econ. Manag. **30**, 34–42 (1996)
8. Hamalainen, R.P., Kaitala, V., Haurie, A.: Bargaining on whales: a differential game model with Pareto optimal equilibria. Oper. Res. Lett. **3**(1), 5–11 (1984)
9. Haurie, A., Tolwinski, B.: Acceptable equilibria in dynamic games. Large Scale Syst. **6**, 73–89 (1984)
10. Levhari, D., Mirman, L.J.: The great fish war: an example using a dynamic Cournot-Nash solution. Bell J. Econ. **11**(1), 322–334 (1980)
11. Lindroos, M., Kaitala, V.T., Kronbak, L.G.: Coalition games in fishery economics. In: Advances in Fishery Economics. Blackwell (2007)
12. Mazalov, V.V., Rettieva, A.N.: Incentive equilibrium in discrete-time bioresource sharing model. Dokl. Math. **78**(3), 953–955 (2008)
13. Mazalov, V.V., Rettieva, A.N.: Fish wars and cooperation maintenance. Ecol. Model. **221**, 1545–1553 (2012)
14. Munro, G.R.: The optimal management of transboundary renewable resources. Can. J. Econ. **12**(8), 355–376 (1979)
15. Ostrom, E.: Governing the Commons: The Evolution of Institutions for Collective Action. Cambridge University Press, Cambridge, UK (1990)
16. Petrosjan, L.A., Danilov, N.N.: Stable solutions of nonantogonostic differential games with transferable utilities. Viestnik Leningrad Univ. **1**, 52–59 (1979)
17. Petrosyan, L.A., Zakharov, V.V.: Mathematical Models in Ecology. Saint-Petersburg University Press, Saint-Petersburg, Russia (1997)
18. Plourde, C.G., Yeung, D.: Harvesting of a transboundary replenishable fish stock: a noncooperative game solution. Mar. Resour. Econ. **6**, 57–70 (1989)
19. Tolwinski, B., Haurie, A., Leitmann, G.: Cooperative equilibria in differential games. J. Math. Anal. Appl. **119**, 182–202 (1986)

Trade-Off Mechanism to Sustain Cooperation in Pollution Reduction

Shimai Su$^{(\boxtimes)}$ (ID) and Elena M. Parilina (ID)

Saint Petersburg State University, St. Petersburg, Russia
st073379@student.spbu.ru, e.parilina@spbu.ru

Abstract. We consider an asymmetric differential game of pollution control with two players: developed and developing countries. The vulnerable tolerance to the pollution problem distinguishes a developed country from a developing one. This internal characteristic pushes the former to persuade the latter to decrease the polluting production or activities through a contract since the developed country deals with pollution reduction alone in the noncooperative setting. However, concerning the necessary and actual participation of the developing country in a pollution disposal, in this paper, we introduce a trade-off mechanism by trading partial workload of a pollution disposal between the developed country and the developing one. In return, the developed country shares its profit with the developing country to make the trade work. Finally, the numerical example demonstrates the explicit performance of the proposed trade-off mechanism. We also investigate the efficiency of the trade-off mechanism compared with the fully cooperative and noncooperative cases.

Keywords: Differential game · Pollution control · Trade-off mechanism · Partial workload · Pareto optimality

1 Introduction

For interpreting how trade-off mechanism [1] works in a two-player differential game, it's inevitable to decipher the rationale in a supply chain due to the critical importance to our understanding of it. Generally, there are two prominent supply chain models in use: forward supply chain and closed-loop supply chain. Forward supply chain, namely, the flow of the product is one-directional through the chain. Meanwhile, closed-loop supply chain (CLSC), in which the used product can be recycled and sold again after remanufacturing, is another popular model. Moreover, contrary to forward supply chains, CLSC has its inherent characteristics—closed-loop, which naturally makes it implement in an environment-friendly and profitable way [1].

However, no matter which model is used, various pollution control policies or constraints have been leveraged concerning the exacerbating environment problems, such as carbon tax [16,17], cap-and-trade [5,9,19], green supply chain

M. Khachay et al. (Eds.): MOTOR 2023, LNCS 13930, pp. 300–313, 2023.
https://doi.org/10.1007/978-3-031-35305-5_21

management [6,20], consumers' low-carbon preference [5,7,16,18], low-carbon subsidy [17,21], contract design [1], etc. In the paper [15], a repeated game between countries polluting the atmosphere is examined. The authors propose to implement mechanism of transfers between players to reduce a pollution level at each stage of the game. Under given conditions, the profile of constructed strategies is subgame perfect equilibrium realizing Pareto-optimal payoffs in each period of the game. In comparison with this work, we do not consider conditions of subgame perfectness of the proposed solution and suggest to implement a trade-off mechanism to reduce the emission level and to obtain Pareto-optimal payoffs in the whole game.

We consider a two-player differential game, when the players are vulnerable and invulnerable countries creating emissions by their production activities [3,10]. A vulnerable country takes care of emission stock by bearing the costs equivalent to the quadratic function of the emission stock, while an invulnerable player has not such costs in his objective function. We consider cooperative and noncooperative scenarios as in [3,10], and propose a trade-off mechanism, which is similar to the contract design, where players sign a contract and behave following the contract rules over time. In our model, the players are not symmetric participating in a trade-off mechanism, and we examine the conditions when both players benefit adopting a trade-off mechanism in comparison with the Nash equilibrium, i.e. we find the parameters of the contract when it improves the Nash equilibrium payoffs in the sense of Pareto optimality principle. We also show that among such Pareto-improving contracts there exists such under which the pollution stock is less than within a cooperative scenario while earlier the latter scenario is considered as the best from environmental perspective.

In addition, speaking of the role of members in the investigated supply chain model in the previous papers, where either the manufacture is acting as the bellwether or the retailer is dominating, such a configuration is recognized as power structure [9]. Noticeably, in our paper, we do not implement the Stackelberg model and two players choose their strategies simultaneously. This also distinguishes the trade-off mechanism from the cost-revenue sharing contract [1] because the former does not require coordination of players' order in decision making. The trade-off mechanism is also different from a fully cooperative scenario, in which players completely coordinate their actions to maximize the total profit. This scenario requires a total control of players' actions along the cooperative trajectory while in a trade-off mechanism, once the contract is signed, the players act individually and adopt the Nash equilibrium in a redefined differential game. Therefore, there is no need to adopt any allocation mechanism [4,11–13] along the state trajectory.

The paper is organized as follows. In Sect. 2, we specify a two-player differential game in pollution control problem and pinpoint the main target. Section 3 characterizes the equilibria or optimal feedback strategies in noncooperative, cooperative and trade-off mechanism scenarios. The comparisons between a noncooperative scenario and trade-off mechanism, a cooperative scenario and trade-off mechanism are defined in Sect. 4. In Sect. 5, a numerical example is provided to explicitly demonstrate the performance of trade-off mechanism compared with two other scenarios. Section 6 concludes the paper.

2 Model

The set of players $N = \{1, 2\}$ consists of two countries: developed and developing. The asymmetry between two countries is represented by their nonequivalent vulnerabilities to the pollution problem, i.e., the players are of two types: player 1 is a vulnerable player (developed country), and 2 is an invulnerable player (developing country).

Following the model represented in [3, 10], the dynamics of the pollution stock S are given by

$$\dot{S}(t) = \mu \sum_{i \in N} e_i(t) - \varepsilon S(t), \ S(0) = S_0, \tag{1}$$

where $e_i(t)$ denotes the quantity of emissions generated by player i, $\mu > 0$ is the marginal influence on pollution accumulation of the players' emissions, and $\varepsilon > 0$ is the nature's absorption rate.

Vulnerable and invulnerable players vary in their attitudes to the pollution reduction policy. The reactions to different attitudes are indicated in the model with adjustment of damage-cost item, i.e., an invulnerable player maximizes her payoff given by

$$\max_{e_2 > 0} W_2 = \int_0^\infty e^{-\rho t} (\alpha_2 e_2 - \frac{1}{2} e_2^2) dt, \tag{2}$$

whereas the objective function of a vulnerable player goes as

$$\max_{e_1 > 0} W_1 = \int_0^\infty e^{-\rho t} (\alpha_1 e_1 - \frac{1}{2} e_1^2 - \frac{1}{2} \beta_1 S^2) dt, \tag{3}$$

where $\rho > 0$ is a discount rate, and α_i, β_i are positive constants. The damage cost $\frac{1}{2} \beta_1 S^2$ is missed in (2), which conveys that an invulnerable player does not make efforts to reduce the negative impact of pollution reduction on nature.

We consider three possible scenarios in terms of players' behavior/willingness to cooperation. In a noncooperative scenario, both players individually maximize their profits. Such a behavior is not environmentally friendly, i.e., does not overcome the pollution issue. In a cooperative scenario, players maximize their joint profit, which allows to address environmental problem, i.e. to reduce the pollution stock and to achieve the largest joint payoff. There are several problems concerning the realization of a cooperative scenario, and among them: (i) how to allocate the joint profit fairly, and (ii) how to achieve a fully cooperative behavior, especially, when full coordination of the players' behavior is questionable. In the paper, we consider the third scenario, in which the cooperation is different from a fully cooperative scenario, but it is carried through a trade-off mechanism (see [1]), usually used in supply chain coordination [2, 8]. This mechanism is a form of cooperative behavior proposed to find an efficient solution to mitigate the pollution damage, but it does not require full coordination of players' behavior over time. In the proposed trade-off mechanism, although two players are still acting by maximizing their own profits, there is a trade of pollution disposal between them. A vulnerable player compensates an invulnerable player's costs on involving in the production reduction by transferring the share of her profits to the latter.

3   Equilibria Under Different Scenarios

In this section, we examine the Nash equilibria in a two-player differential game under a noncooperative scenario, find the solution of the players' joint optimization problem in a cooperative scenario and obtain the Nash equilibrium strategies in a trade-off mechanism scenario.

3.1   Noncooperaitve Scenario

Under a noncooperative scenario, the two players behave as singletons individually maximizing their profits given by (2) and (3) subject to the state dynamics (1).

Proposition 1. *In a noncooperative scenario, the feedback-Nash equilibrium in two-player differential game defined by objective functions (2) and (3) s.t. (1), is given by*

$$e_1^{nc}(t) = \alpha_1 + \mu(x_{nc}S^{nc}(t) + y_{nc}),$$
$$e_2^{nc}(t) = \alpha_2,$$

where

$$x_{nc} = \frac{\rho + 2\varepsilon - \sqrt{(\rho + 2\varepsilon)^2 + 4\mu^2\beta_1}}{2\mu^2} < 0,$$

$$y_{nc} = \frac{\mu(\alpha_1 + \alpha_2)x_{nc}}{\rho + \varepsilon - \mu^2 x_{nc}} < 0,$$

$$z_{nc} = \frac{(\alpha_1 + \mu y_{nc})^2 + 2\mu y_{nc}\alpha_2}{2\rho}.$$

The corresponding equilibrium state trajectory is

$$S^{nc}(t) = \frac{\mu(\alpha_1 + \alpha_2) + \mu^2 y_{nc}}{\mu^2 x_{nc} - \varepsilon}(e^{(\mu^2 x_{nc} - \varepsilon)t} - 1) + e^{(\mu^2 x_{nc} - \varepsilon)t}S_0.$$

The steady state stock of emissions is

$$S_\infty^{nc} = \frac{\mu(\alpha_1 + \alpha_2)(\rho + \varepsilon)}{(\varepsilon - \mu^2 x_{nc})(\rho + \varepsilon - \mu^2 x_{nc})},$$

which is globally asymptotically stable when $\mu^2 x_{nc} - \varepsilon < 0$.

The Nash equilibrium players' payoffs are

$$V_1^{nc} = \frac{1}{2}x_{nc}S_0^2 + y_{nc}S_0 + z_{nc},$$

$$V_2^{nc} = \frac{\alpha_2^2}{2\rho}.$$

Proof. See [3].

3.2 Cooperative Scenario

In a cooperative scenario, the two players jointly maximize their total payoff, i.e., they solve the following problem:

$$\max_{\substack{e_i \geq 0 \\ i \in N}} \sum_{i \in N} W_i(e_1, e_2),$$

subject to the state dynamics (1), and the players' payoff functions are given by (2) and (3).

Proposition 2. *In a cooperative scenario, the optimal feedback strategies in two-player differential game defined by objective functions (2) and (3) s.t. (1), are given by*

$$e_i^c(t) = \alpha_i + \mu(xS^c(t) + y), \qquad i = 1, 2,$$

where

$$x = \frac{2\varepsilon + \rho - \sqrt{(2\varepsilon + \rho)^2 + 8\mu^2\beta_1}}{4\mu^2} < 0,$$

$$y = \frac{(\alpha_1 + \alpha_2)\mu x}{\rho + \varepsilon - 2x\mu^2} < 0,$$

$$z = \frac{(\alpha_1 + \mu y)^2 + (\alpha_2 + \mu y)^2}{2\rho}.$$

The corresponding cooperative state trajectory is

$$S^c(t) = \frac{\mu(\alpha_1 + \alpha_2) + 2\mu^2 y}{2\mu^2 x - \varepsilon}(e^{(2\mu^2 x - \varepsilon)t} - 1) + e^{(2\mu^2 x - \varepsilon)t}S_0.$$

The steady state stock of emissions is

$$S_\infty^c = \frac{\mu(\alpha_1 + \alpha_2)(\rho + \varepsilon)}{(\varepsilon - 2\mu^2 x)(\rho + \varepsilon - 2\mu^2 x)},$$

which is globally asymptotically stable when $2\mu^2 x - \varepsilon < 0$.
The joint players' payoff is

$$V_{12}^c = \frac{1}{2}xS_0^2 + yS_0 + z.$$

Proof. See [3].

3.3   Trade-Off Mechanism Scenario

In this section, we represent the third scenario, in which players cooperate by making an agreement about the trade-off mechanism of payments/costs over time. The mechanism assumes that players agree on two parameters: (i) the compensation coefficient $0 < \tau < 1$ showing the profit share given by a vulnerable player to an invulnerable player for persuading the latter to deal with pollution problem, (ii) the cost coefficient $0 < \theta < 1$ indicating the magnitude of pollution amount that an invulnerable player should be responsible for. In this case, an invulnerable player's payoff function takes the form:

$$\max_{e_2>0} W_2 = \int_0^\infty e^{-\rho t}\left(\alpha_2 e_2(t) + \tau\alpha_1 e_1(t) - \frac{1}{2}e_2^2(t) - \frac{1}{2}\beta_1\theta S^2(t)\right)dt, \quad (4)$$

while a vulnerable player's payoff function is

$$\max_{e_1>0} W_1 = \int_0^\infty e^{-\rho t}\left((1-\tau)\alpha_1 e_1(t) - \frac{1}{2}e_1^2(t) - \frac{1}{2}\beta_1(1-\theta)S^2(t)\right)dt. \quad (5)$$

The parameters (τ, θ) can be interpreted as a contract between two players and can be negotiated. In the paper, we do consider these parameters as exogenously given, but one can assume them as decision variables of the players in the process of negotiations. Obviously, the feedback-Nash equilibrium significantly depends on the values of (τ, θ).

In the trade-off mechanism scenario, the two players individually maximize their own profits similar to what we have described in a noncooperative scenario. However, the objective functions (4) and (5) are dependent on not only the state variable but the decision variables e_1, e_2.

Proposition 3. *In a trade-off mechanism scenario, the feedback-Nash equilibrium in a two-player differential game defined by objective functions (4) and (5) s.t. (1), is given by*

$$e_1^{ToM}(t) = \alpha_1(1-\tau) + \mu(x_1 S^{ToM}(t) + y_1),$$
$$e_2^{ToM}(t) = \alpha_2 + \mu(x_2 S^{ToM}(t) + y_2),$$

where x_1, x_2, y_1, and y_2 are the solutions of the system of the equations (12) given in the proof.

The corresponding equilibrium state trajectory is

$$S^{ToM}(t) = \frac{\mu B + \mu^2 y_{12}}{\mu^2 x_{12} - \varepsilon}(e^{(\mu^2 x_{12}-\varepsilon)t} - 1) + e^{(\mu^2 x_{12}-\varepsilon)t} S_0, \quad (6)$$

where $x_{12} = x_1 + x_2$, $y_{12} = y_1 + y_2$, and $B = \alpha_1(1-\tau) + \alpha_2$.

The steady state stock of emissions is

$$S_\infty^{ToM} = \frac{\mu B + \mu^2 y_{12}}{\varepsilon - \mu^2 x_{12}}, \quad (7)$$

which is globally asymptotically stable when $\mu^2 x_{12} - \varepsilon < 0$.

The Nash equilibrium players' payoffs are

$$V_1^{ToM} = \frac{1}{2} x_1 S_0^2 + y_1 S_0 + z_1,$$

$$V_2^{ToM} = \frac{1}{2} x_2 S_0^2 + y_2 S_0 + z_2,$$

where z_1 *and* z_2 *are defined in the proof.*

Proof. The optimization problem for each player is

$$W_1^{ToM} = \int_0^\infty e^{-\rho t} \left(\alpha_1 e_1(t)(1-\tau) - \frac{1}{2} e_1^2 - \frac{1}{2} \beta_1 (1-\theta) S^2(t) \right) dt \rightarrow \max_{e_1 \geq 0}, \quad (8)$$

$$W_2^{ToM} = \int_0^\infty e^{-\rho t} \left(\alpha_2 e_2(t) + \tau(\alpha_1 e_1(t)) - \frac{1}{2} e_2^2(t) - \frac{1}{2} \beta_1 \theta S^2(t) \right) dt \rightarrow \max_{e_2 \geq 0}. \quad (9)$$

Assuming the linear-quadratic form of the value functions $V_1(S) = \frac{1}{2} x_1 S^2 + y_1 S + z_1$ and $V_2(S) = \frac{1}{2} x_2 S^2 + y_2 S + z_2$, we write down the HJB equations for (8) and (9):

$$\rho V_1(S) = \max_{e_1} \left\{ \alpha_1 e_1 (1-\tau) - \frac{1}{2} e_1^2 - \frac{1}{2} \beta_1 (1-\theta) S^2 + V_1'(S)[\mu(e_1 + e_2) - \varepsilon S] \right\}, \quad (10)$$

$$\rho V_2(S) = \max_{e_2} \left\{ \alpha_2 e_2 + \tau(\alpha_1 e_1) - \frac{1}{2} e_2^2 - \frac{1}{2} \beta_1 \theta S^2 + V_2'(S)[\mu(e_1 + e_2) - \varepsilon S] \right\}. \quad (11)$$

Maximizing the expression in RHS in (10), we obtain that $e_1 = \alpha_1 + \mu V_1'(S)$, and maximizing the expression in RHS in (11), we obtain that $e_2 = \alpha_2 + \mu V_2'(S)$. Taking into account the derivatives $V_1'(S) = x_1 S + y_1$, $V_2'(S) = x_2 S + y_2$, and substituting these expressions into (10), we obtain an equation:

$$\rho \left(\frac{1}{2} x_1 S^2 + y_1 S + z_1 \right) = \alpha_1 (1-\tau)[\alpha_1 (1-\tau) + \mu(x_1 S + y_1)]$$

$$- \frac{1}{2} [\alpha_1 (1-\tau) + \mu(x_1 S + y_1)]^2 - \frac{1}{2} \beta_1 (1-\theta) S^2$$

$$+ (x_1 S + y_1) \left(\mu[\alpha_1 (1-\tau) + \alpha_2 + \mu(x_1 S + y_1 + x_2 S + y_2)] - \varepsilon S \right).$$

Taking into account the derivative $V_2'(S) = x_2 S + y_2$, and substituting the expressions into (11), we obtain an equation:

$$\rho \left(\frac{1}{2} x_2 S^2 + y_2 S + z_2 \right) = \alpha_2 [\alpha_2 + \mu(x_2 S + y_2)]$$

$$+ \tau \alpha_1 [\mu(x_1 S + y_1) + \alpha_1 (1-\tau)] - \frac{1}{2} [\alpha_2 + \mu(x_2 S + y_2)]^2 - \frac{1}{2} \beta_1 \theta S^2$$

$$+ (x_2 S + y_2) \left(\mu[\mu(x_1 S + y_1) + \alpha_1 (1-\tau) + \alpha_2 + \mu(x_2 S + y_2)] - \varepsilon S] \right).$$

By identification, two linear quadratic equations containing x_1, x_2 can be written as

$$\mu^2 x_1^2 + 2\mu^2 x_1 x_2 - 2\varepsilon x_1 - \rho x_1 - \beta_1(1-\theta) = 0,$$
$$\mu^2 x_2^2 + 2\mu^2 x_1 x_2 - 2\varepsilon x_2 - \rho x_2 - \beta_1 \theta = 0.$$

Rewriting these equations which should be solved to find x_1 and x_2, and summarizing with the rest of equations, we obtain the system:

$$\begin{cases} 3\mu^4 x_1^4 - 4\mu^2(2\varepsilon + \rho)x_1^3 + ((2\varepsilon + \rho)^2 + 6\mu^2\beta_1\theta - 2\mu^2\beta_1)x_1^2 - (1-\theta)^2\beta_1^2 = 0, \\[2mm] 3\mu^4 x_2^4 - 4\mu^2(2\varepsilon + \rho)x_2^3 + ((2\varepsilon + \rho)^2 - 6\mu^2\beta_1\theta + 4\beta_1\mu^2)x_2^2 - \beta_1^2\theta^2 = 0, \\[2mm] y_1 = \dfrac{\mu^3 x_1[(x_2 B + \tau\alpha_1 x_1)A - \mu^2 x_1 x_2 B]}{A(A^2 - \mu^4 x_1 x_2)} - \dfrac{\mu x_1 B}{A}, \\[3mm] y_2 = \dfrac{\mu^3 x_1 x_2 B - \mu(x_2 B + \tau\alpha_1 x_1)A}{A^2 - \mu^4 x_1 x_2}, \\[3mm] z_1 = \dfrac{2\mu y_1 B + \alpha_1^2(1-\tau)^2 + \mu^2 y_1^2 + 2\mu^2 y_1 y_2}{2\rho}, \\[3mm] z_2 = \dfrac{2\mu y_2 B + \alpha_2^2 + 2\alpha_1^2\tau(1-\tau) + \mu^2 y_2^2 + 2\mu^2 y_1 y_2 + \tau\alpha_1\mu y_1}{2\rho}, \end{cases}$$

$$\tag{12}$$

where $A = \mu^2 x_1 + \mu^2 x_2 - \rho - \varepsilon$ and $B = \alpha_1(1-\tau) + \alpha_2$.

In the system (12), we need to solve the first two equations, then substituting x_1 and x_2 into the rest four equations we find y_1, y_2, z_1, and z_2. We should notice that we require that x_1, x_2 be negative to prove the stability of the steady state.

The expression of the equilibrium stock $S^{ToM}(t)$ is obtained as a solution of equation (1) and it is given by (6). If t tends to infinity in (1), we obtain the steady state of emission stock given by (7), which globally asymptotically stable when $\mu^2 x_{12} - \varepsilon < 0$.

4   Comparison of Scenarios

In this section, we investigate the performance of trade-off mechanism with various set values (τ, θ) by comparing the pollution stock, the players' strategies and payoffs under this scenario with noncooperative and cooperative scenarios. We are interested in finding a set of values (τ, θ) which both players are interested in terms of accepting the trade-off mechanism, i.e., their profits in this scenario are larger than they could obtain in a noncooperative case.

4.1   Non-cooperative Scenario vs Trade-Off Mechanism

In the differential game described above, the set (τ, θ), where $\tau \in (0,1)$, $\theta \in (0,1)$, defining the trade-off mechanism can be classified into five types by comparing players' payoffs and the steady-state emission level in noncooperative and trade-off mechanism scenarios: (i) profitable for invulnerable player when only invulnerable player gets a larger payoff in a trade-off mechanism scenario than in a noncooperative one, and the steady-state stock in a trade-off mechanism scenario is lower than in a noncooperative one; (ii) profitable for vulnerable player, when only vulnerable player benefits from a trade-off mechanism scenario with respect to noncooperation, and the steady-state stock with the trade-off mechanism is lower than in noncooperation; (iii) profitable for both players, i.e., both players will obtain higher payoffs adopting the trade-off mechanism; (iv) nonprofitable for both players, namely, this set is not profitable for both players, but the steady-state stock with the trade-off mechanism is lower than in noncooperation; (v) not acceptable, i.e., the two players generate more pollution than in a noncooperative scenario.

Apparently, both players accept the set (τ, θ) if and only if both of them will benefit from it and the steady-state pollution stock is less than in a noncooperative scenario. The types of (τ, θ) and the corresponding inequalities for the profits and the steady state are given in Table 1.

Table 1. Types of the set (τ, θ) (trade-off mechanism vs noncooperation).

	Vul. Player 1	Invul. Player 2	Steady State
(i) Profitable for invulnerable player	$W_1^{ToM} < W_1^{nc}$	$W_2^{ToM} \geq W_2^{nc}$	$S_\infty^{ToM} < S_\infty^{nc}$
(ii) Profitable for vulnerable player	$W_1^{ToM} \geq W_1^{nc}$	$W_2^{ToM} < W_2^{nc}$	$S_\infty^{ToM} < S_\infty^{nc}$
(iii) Profitable for both players	$W_1^{ToM} \geq W_1^{nc}$	$W_2^{ToM} \geq W_2^{nc}$	$S_\infty^{ToM} < S_\infty^{nc}$
(iv) Nonprofitable for both players	$W_1^{ToM} < W_1^{nc}$	$W_2^{ToM} < W_2^{nc}$	$S_\infty^{ToM} < S_\infty^{nc}$
(v) Not acceptable	—	—	$S_\infty^{ToM} \geq S_\infty^{nc}$

It is impossible to verify the inequalities in Table 1 in a general-form game, thus we demonstrate these types on a numerical example in Sect. 5.

4.2   Cooperative Scenario vs Trade-Off Mechanism

In this section, we compare the players' payoffs and steady-state pollution stock in the trade-off mechanism and cooperative scenarios. Repeating the same classification given in Sect. 4.1, we present five types of the set (τ, θ) in Table 2, in which we compare the trade-off mechanism and cooperative scenarios. We again are interested in the subset of (τ, θ) such that both players are beneficial from the trade-off mechanism, but we expect that we cannot find such a subset comparing this scenario with the cooperative one.

We verify the inequalities given in Table 2 in a numerical example in Sect. 5.

5   Numerical Example

In this section, we present a numerical example to illustrate the performance of a trade-off mechanism with respect to the values of (τ, θ). The parameters of the game are

Table 2. Types of the set (τ, θ) (trade-off mechanism vs cooperation).

	Vul. Player 1	Invul. Player 2	Steady State
(i) Profitable for invulnerable player	$W_1^{ToM} < W_1^c$	$W_2^{ToM} \geq W_2^c$	$S_\infty^{ToM} < S_\infty^c$
(ii) Profitable for vulnerable player	$W_1^{ToM} \geq W_1^c$	$W_2^{ToM} < W_2^c$	$S_\infty^{ToM} < S_\infty^c$
(iii) Profitable for both players	$W_1^{ToM} \geq W_1^c$	$W_2^{ToM} \geq W_2^c$	$S_\infty^{ToM} < S_\infty^c$
(iv) Nonprofitable for both players	$W_1^{ToM} < W_1^c$	$W_2^{ToM} < W_2^c$	$S_\infty^{ToM} < S_\infty^c$
(v) Not acceptable	—	—	$S_\infty^{ToM} \geq S_\infty^c$

$$\beta_1 = 1, \ \alpha_1 = 9, \ \alpha_2 = 4,$$
$$\varepsilon = 0.4, \ \mu = 0.35, \ \rho = 0.1, \ S_0 = 1.$$

As shown in Fig. 1, the set (τ, θ) is divided into five areas corresponding to the particular types described in Table 1. The black area represents the subset of (τ, θ) under which the players in a trade-off mechanism pollute more than in a noncooperative scenario. The red (blue) area corresponds to the subset of (τ, θ) when only vulnerable (involnurable) player performs better in a trade-off mechanism and polluting less (in total) than in a noncooperative scenario while the green area gives the Pareto-improving values of (τ, θ), when both players benefit from adopting a trade-off mechanism vs noncooperative scenario. The yellow area indicates that both players are not interested in a trade-off mechanism scenario while they reduce the steady-state pollution stock.

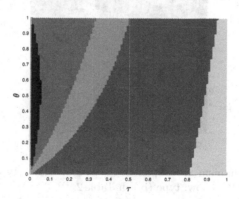

Fig. 1. Trade-off mechanism vs noncooperative scenario. Black: type (v) in Table 1; red: type (ii) in Table 1; blue: type (i) in Table 1; yellow: type (iv) in Table 1. Pareto-improving pairs of values (τ, θ) are green area corresponding to type (iii) in Table 1.

We should notice that for any $\tau \in (0, 0.49)$ there exists a nonempty interval for θ such that the values (τ, θ) define a Pareto-improving trade-off mechanism for the players, i.e. both players are interested in adopting it. Although it is clear that both players are beneficial by choosing (τ, θ) from the green area in Fig. 1, the question is this: how much can they improve their payoffs by a trade-off mechanism? We can calculate the players' benefits (in percentage) obtained by adopting a trade-off mechanism in comparison with a noncooperative scenario using a coefficient

$$M_i = \left| \frac{W_i^{nc} - W_i^{ToM}}{W_i^{nc}} \right| \times 100\%, \, i = 1, 2.$$

We represent some values of Pareto optimal set (τ, θ) (from the green area in Fig. 1) and the improvement coefficients M_1 and M_2 for vulnerable and invulnerable players respectively in Table 3. In a numerical example, the maximal percentage of improvement for player 1 (vulnerable player) is 75.61% and for player 2 (invulnerable player) is 87.8%. It is expected that player 1 is more beneficial with low τ and high θ, while player 2 is interested in high τ.

Table 3. Benefits from adopting trade-off mechanism vs noncooperative scenario for the Pareto-improving values of (τ, θ).

(τ, θ)	$(0.31, 0.89)$	$(0.49, 0.96)$	$(0.4, 0.97)$	$(0.34, 0.89)$	$(0.2, 0.23)$
M_1	75.61%	0.08%	40.23%	60.22%	1.44%
M_2	0.72%	87.8%	46.33%	21.05%	47.35%

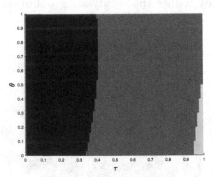

Fig. 2. Trade-off mechanism vs cooperative scenario. Black: type (v) from Table 2; blue: type (i) in Table 2; yellow: type (iv) in Table 2.

We should notice that the trade-off mechanism is useful to outperform a cooperative scenario in terms of the pollution level (the pollution stock in the trade-off mechanism is lower than in a cooperative scenario) as indicated in Fig. 2. In this figure we can see the presence of three areas including a large black area in which the pollution in the trade-off mechanism is larger than in a cooperative scenario. In the blue area, the trade-off mechanism is beneficial only for an invulnerable player. But constructing Fig. 2, in a cooperative scenario we do not use any profit-allocation mechanism to redefine players' payoffs. Therefore, the areas may change after adopting any cooperative payoff allocation rule.

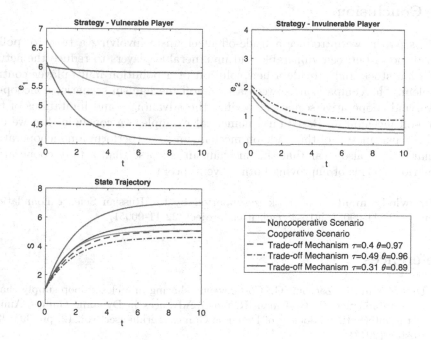

Fig. 3. Strategies of vulnerable and invulernable players, and state trajectory (pollution stock) under different scenarios: noncooperative, cooperative and three trade-off mechanisms with parameters $(\tau, \theta) \in \{(0.31, 0.89), (0.49, 0.96), (0.4, 0.97)\}$.

Apart from the benefit brought to the players when they adopt the trade-off mechanism, the comparisons of players' strategies and pollution stock under different scenarios are also of a particular interest. As demonstrated in Fig. 3, we select three representative sets $(0.31, 0.89), (0.49, 0.96), (0.4, 0.97)$ from the Pareto optimal set of pairs (τ, θ). The first (second) pair of parameters benefits player 1 (player 2) at the highest level, while the third one gives more or less the same improvement to both players (40.23% for player 1 and 46.33% for player 2) as shown in Table 3. In Fig. 3, we observe that after applying trade-off mechanism proposed above, the emission quantity or the strategy of a vulnerable player becomes almost constant (relatively horizontal line), and for an invulnerable player, the quantity of emissions has dropped a lot in comparison with the noncooperative level. More importantly, the pollution stock indicates that trade-off mechanism is capable of reducing pollution even more than a cooperative scenario (see Fig. 3) and brings benefits to both players in comparison with the Nash equilibrium. For instance, when the players adopt a trade-off mechanism with parameters $(\tau, \theta) = (0.49, 0.96)$, then the emission stock is less than in a cooperative scenario, and the players' payoffs increase by 0.08% and 87.8% for player 1 and 2 respectively.

6   Conclusion

In this paper, we introduce a trade-off mechanism involving a trade of pollution disposal between vulnerable and invulnerable players to reduce the actual pollution stock and provide a new solution of a pollution multi-player control problem. The comparison between a trade-off mechanism scenario, noncooperative, and cooperative scenarios specifies the advantages and limitations of the trade-off mechanism. Based on a numerical example, we conclude that we can find the parameters of the trade-off mechanism to outperform a noncooperative scenario. It is also clear that this mechanism is worse than a fully cooperative scenario in terms of improving both players' profits.

Acknowledgements. The work is supported by the Russian Science Foundation, grant no. 22-11-00051, https://rscf.ru/en/project/22-11-00051/

References

1. De Giovanni, P., Zaccour, G.: Cost-revenue sharing in a closed-loop supply chain. In: Cardaliaguet, P., Cressman, R. (eds.) Advances in Dynamic Games. Annals of the International Society of Dynamic Games, Birkhäuser, vol. 12, pp. 395–421. Boston (2013)
2. De Giovanni, P., Zaccour, G.: A selective survey of game-theoretic models of closed-loop supply chains. Ann. Oper. Res. 1–40 (2022). https://doi.org/10.1007/s10479-021-04483-5
3. Fanokoa, P.S., Telahigue, I., Zaccour, G.: Buying cooperation in an asymmetric environmental differential game. J. Econ. Dyn. Control **35**(6), 935–946 (2010)
4. Gromova, E., Plekhanova, K.: A differential game of pollution control with participation of developed and developing countries. Contrib. Game Theory Manag. **8**, 64–83 (2015)
5. Ghosh, S.K., Seikh, M.R., Chakrabortty, M.: Analyzing a stochastic dual-channel supply chain under consumers' low carbon preferences and cap-and-trade regulation. Comput. Ind. Eng, 149 (2020)
6. Herrmann, F.F., Barbosa-Povoa, A.P., Butturi, M.A., Marinelli, S., Sellitto, M.A.: Green Supply Chain management: conceptual framework and models for analysis. Sustainability **13**(15), 8127 (2021). https://doi.org/10.3390/su13158127
7. Ji, J., Zhang, Z., Yang, L.: Carbon emission reduction decisions in the retail-/dual-channel supply chain with consumers' preference. J. Cleaner Prod. 141, 852–867 (2016). https://doi.org/10.1016/j.jclepro.2016.09.135
8. Kuchesfehani, E.K., Parilina, E.M., Zaccour, G.: Revenue and cost sharing contract in a dynamic closed-loop supply chain with uncertain parameters. Ann. Oper. Res. (2022). https://doi.org/10.1007/s10479-022-05055-x
9. Li, C., Gao, J., Guo, J., Wang, J.: Low-carbon supply chain decisions considering carbon emissions right pledge financing in different power structures **15**(15), 5721 (2022). https://doi.org/10.3390/en15155721

10. Masoudi, N., Zaccour, G.: A differential game of international pollution control with evolving environmental costs. Environ. Dev. Econ. **18**(6), 680–700 (2013)
11. Petrosjan, L.A., Danilov, N.A.: Time-consistent solutions of non-antagonistic differential games with transferable payoffs. Vestnik of Leningrad State University, vol 1, pp. 46–54 (1979). (in Russian)
12. Petrosjan, L., Zaccour, G.: Time-consistent shapley value allocation of pollution cost reduction. J. Econ. Dyn. Control **27**(3), 381–398 (2003)
13. Shapley, L.: Notes on the n-Person Game - II: The Value of an n-Person Game. Santa Monica, California (1951)
14. Souza, G.: Closed-loop supply chains: a critical review, and future research. Decision Sci., vol. 44, pp. 7–38 (2013). https://doi.org/10.1111/j.1540-5915.2012.00394.x
15. Vasin, A.A., Divtsova, A.G.: A game-theoretic model of agreement on limitation of transboundary air pollution. Autom. Remote Control, vol. 80, pp. 1164–1176 (2019). https://doi.org/10.1134/S0005117919060134
16. Wang, L., Xu, T., Qin, L.: A study on supply chain emission reduction level based on carbon tax and consumers' low-carbon preferences under stochastic demand. Math. Probl. Eng. vol. 2019, pp. 1–20 (2019). https://doi.org/10.1155/2019/1621395
17. Wu, H., Sun, Y., Su, Y., Chen, M., Zhao, H., Li, Q.: Which is the best supply chain policy: carbon tax, or a low-carbon subsidy? Sustainability **14**(10), 6312, (2022). https://doi.org/10.3390/su14106312
18. Ye, T., Guan, Z., Tao, J., Qu, Y.: Dynamic optimization and coordination about joint emission reduction in a supply chain considering consumer preference to low carbon and reference low-carbon level effect. Chin. J. Manag. Sci **25**, 52–61 (2017)
19. Zhang, G., Zhang, X., Sun, H., Zhao, X.: Three-echelon closed-loop supply chain network equilibrium under cap-and-trade regulation. Sustainability **13**(11), 6472 (2021). https://doi.org/10.3390/su13116472
20. Zhao, R., Neighbour, G., Han, J., McGuire, M., Deutz, P.: Using game theory to describe strategy selection for environmental risk and carbon emissions reduction in the green supply chain. J. Loss Prev. Process Ind. **25**(6), 927–936 (2012). https://doi.org/10.1016/j.jlp.2012.05.004
21. Zhu, Q., Dou, Y.: A game model for green supply chain management based on government subsidies. J. Manag. Sci. China **14**, 86–95 (2011)

Communication Restriction-Based Characteristic Function in Differential Games on Networks

Anna Tur$^{(\boxtimes)}$ and Leon Petrosyan

St. Petersburg State University, 7/9, Universitetskaya nab.,
Saint-Petersburg 199034, Russia
{a.tur,l.petrosyan}@spbu.ru

Abstract. A class of differential games is considered. It is assumed that the players connected by a network interact along paths on this network. It is also supposed that players have the ability to restrict other players' communication. To construct cooperative solutions, a special type of characteristic function is proposed that takes into account the network structure of the game and various types of communication restrictions. The properties of the characteristic function are investigated. The formula for the Shapley value is obtained.

Keywords: Differential game · Cooperative game · Network game · Characteristic function · The Shapley value

1 Introduction

The methods of differential game theory are widely applicable to conflict-controlled dynamical systems. Often, the structure of interaction of players in the system is non-trivial, and is determined by a network or a graph. In this case, the methods of the network game theory are also relevant [1–4]. Players in such models are identified with network nodes, and the way nodes interact is specified by additional rules. In some models, only neighboring players on the network are expected to interact, in others, any connected nodes can communicate. It is also important to determine whether the network changes over time and whether players can influence its structure.

Thus, in papers [3,5], a problem was considered in which players have a new type of strategies with possibility of cutting links with neighboring players during the game. In cooperative differential games on network this new type of strategies has led to the possibility of construction a novel form for the cooperative-trajectory characteristic function [5]. It was proposed to measure the coalition's worth without considering the actions of players who are not members of this coalition.

The work of the second author was supported by the Russian Science Foundation grant No. 22-11-00051, https://rscf.ru/en/project/22-11-00051/.

In this paper, in order to adjust the model to the real process, we add some new abilities for players to manage the network. We assume that any player at any time can restrict communication with any player connected with him. And the players lying on the path between other players are also able to restrict communication of these nodes, thereby reducing their payoff from interacting with each other.

The paper is structured as follows. The definition of the cooperative differential game on a network is given in Sect. 2. In Sect. 3, the properties of the characteristic function based on cooperative strategies are investigated. Based on defined characteristic function, the Shapley value is constructed and an illustrative example is considered in Sect. 4.

2   Problem Formulation

Consider a class of n-person differential games with prescribed duration T. Let $N = \{1; 2; \ldots; n\}$ be the set of players. Players are connected in a network system. A pair (N, L) is called a network, where N is a set of nodes, and $L \subset N \times N$ is a given set of links. The players are represented by nodes. If pair $(i, j) \in L$, there is a link connecting players $i \in N$ and $j \in N$. It is supposed that all links are undirected.

Let $p(i, j)$ be the shortest path between players i and j, $d(i, j)$ – the length of $p(i, j)$ (number of links in path $p(i, j)$). Two vertices i and j are called connected if (N, L) contains a path from i to j.

Denote by $K^m(i) = \{j \in N : d(i, j) = m\}$ – the set of players connected with player $i \in N$ by the shortest path containing exactly m edges.

The system dynamics is given by

$$\dot{x}_i(\tau) = f_i(x_i(\tau); u_i(\tau)); \quad x_i(t_0) = x_i^0; \text{ for } \tau \in [t_0; T] \text{ and } i \in N. \quad (1)$$

Here $x_i(t) \subset R^m$ is the state variable of player $i \in N$ at time t, and $u_i(t) \in U_i \subset R^k$ – the control variable of player $i \in N$. Functions $f_i(x_i; u_i)$ are continuously differentiable in x_i and u_i.

Assume $h_i(x_i(\tau); u_i(\tau))$ is the instantaneous gain that player i can obtain by itself, $\delta^{m-1} h_i^j(x_i(\tau); x_j(\tau); u_i(\tau); u_j(\tau))$ is the instantaneous gain that player i can obtain through the interaction with player $j \in K^m(i)$, $h_i^j(x_i; x_j; u_i; u_j)$ are non-negative and $\delta \in (0, 1)$. The multiplier δ^{m-1} shows that the more distant in network players are from player i, the less their behavior affects his payoff. The presence of this multiplier explains the use of shortest paths for players to interact with each other, since using a longer path for interaction invariably leads to a decrease in their payoffs.

Any player $i \in N$ at any time $\tau \in [t_0, T]$ can restrict communication with any player j connected with him. In this case, for $t \geq \tau$ in the period of time $[t, T]$ if $i \in K^m(j)$, player j can get only $\beta \delta^{m-1} \int_t^T h_j^i(x_j; x_i; u_j; u_i) d\tau$ through the interaction with i. And player i can get only $\beta \int_t^T \delta^{m-1} h_i^j(x_i; x_j; u_i; u_j) d\tau$ through the interaction with j, here $\beta \in (0, 1)$.

Players, lying on the path between a pair of any players p, q and not coinciding with them, at any time $\tau \in [t_0, T]$ can restrict communication p and q. In this case, if $p \in K^m(q)$, player p can get only $\alpha \delta^{m-1} \int_t^T h_p^q(x_p; x_q; u_p; u_q) d\tau$ through the interaction with q. And player q can get only $\alpha \int_t^T \delta^{m-1} h_q^p(x_q; x_p; u_q; u_p) d\tau$ through the interaction with p, for $t \in [\tau, T]$, here $\alpha \in (\beta, 1)$.

Let's also assume that $\alpha > \delta$ and (N, L) contains no cycles of even cardinality. This is necessary in order to avoid a situation where, after the appearance of a communication restrictions for a pair of players, they have a more profitable path for interaction, than the shortest one. Indeed, if (N, L) contains no cycles of even cardinality, then there is a unique shortest path between any pair of players. And if $\alpha > \delta$, then

$$\alpha \delta^{m-1} h_i^j(x_i; x_j; u_i; u_j) > \delta^{m-1+k} h_i^j(x_i; x_j; u_i; u_j), \ \forall \ k > 0.$$

So it is not profitable for players to interact along a path longer than the shortest one, even in conditions of restricted communication.

Then the payoff of player i is given as

$$H_i(x^0; u_1, \ldots, u_n) = \int_{t_0}^T h_i(x_i(\tau); u_i(\tau)) d\tau$$

$$+ \sum_{m=1}^{n-1} \delta^{m-1} \sum_{j \in K^m(i)} \int_{t_0}^T \min\{a_i^j(\tau), b_i^j(\tau)\} h_i^j(x_i(\tau); x_j(\tau); u_i(\tau); u_j(\tau)) d\tau. \quad (2)$$

Here

$$a_i^j(t) = \begin{cases} 1, & t \in [t_0, \tau_1), \\ \alpha, & t \in [\tau_1, \infty), \end{cases}$$

τ_1 – the moment when some of the intermediate players of the path $p(i, j)$ restrict the communication of players i and j. And

$$b_i^j(t) = \begin{cases} 1, & t \in [t_0, \tau_2), \\ \beta, & t \in [\tau_2, \infty), \end{cases}$$

τ_2 – the moment when player j restricts his communication with players i.

A simple interpretation of such problem formulation can be a model in which the nodes represent localities (countries) and the edges represent the roads between them. The two countries invest in the development of road connections between them, and all the localities that lie on the path between them benefit from it. On the other hand, all the nodes along this path can restrict movement through them.

We denote by $x^0 = (x_1^0; x_2^0; \ldots; x_n^0)$ the vector of initial conditions.

Consider the game $\Gamma(x^0, T - t_0)$ if the network (N, L) is defined, the system dynamics (1) and the sets of feasible controls U_i, $i \in N$ are given, and the players' payoffs are determined by (2). Each player, choosing a control variable u_i from his set of feasible controls, steers his own state according to the differential

Eq. (1) and seeks to maximize his objective functional (2). It is supposed that certain assumptions are satisfied in the game under consideration.

Suppose that players can cooperate in order to achieve the maximum joint payoff

$$\sum_{i \in N} \left(\int_{t_0}^{T} h_i(x_i(\tau); u_i(\tau)) d\tau + \sum_{m=1}^{n-1} \delta^{m-1} \sum_{j \in K^m(i)} \int_{t_0}^{T} h_i^j(x_i(\tau); x_j(\tau); u_i(\tau); u_j(\tau)) d\tau \right)$$

(3)

subject to dynamics (1).

Note that players, maximizing the joint payoff, will not restrict any communication (i.e. $a_i^j = b_i^j = 1$, since $h_i^j \geq 0$).

The optimal cooperative strategies of players $\overline{u}(t) = (\overline{u}_1(t), \ldots, \overline{u}_n(t))$, for $t \in [t_0; T]$ are defined as follows

$$\overline{u}(t) = arg \max_{u_1, \ldots, u_n} \sum_{i \in N} H_i(x^0; u_1, \ldots, u_n).$$

(4)

The trajectory corresponding to the optimal cooperative strategies $(\overline{u}_1(t), \ldots, \overline{u}_n(t))$ is the optimal cooperative trajectory $\overline{x}(t) = (\overline{x}_1(t); \overline{x}_2(t); \ldots; \overline{x}_n(t))$. The maximum joint payoff can be expressed as:

$$\sum_{i \in N} \left(\int_{t_0}^{T} h_i(\overline{x}_i(\tau); \overline{u}_i(\tau)) d\tau \right.$$

$$\left. + \sum_{m=1}^{n-1} \delta^{m-1} \sum_{j \in K^m(i)} \int_{t_0}^{T} h_i^j(\overline{x}_i(\tau); \overline{x}_j(\tau); \overline{u}_i(\tau); \overline{u}_j(\tau)) d\tau \right)$$

(5)

subject to dynamics

$$\dot{\overline{x}}_i = f_i(\overline{x}_i(\tau); \overline{u}_i(\tau)); \quad \overline{x}_i(t_0) = x_i^0; \text{ for } \tau \in [t_0; T] \text{ and } i \in N.$$

(6)

Usually the characteristic function is used to determine how to allocate the maximum total payoff among the players under an agreeable scheme [6].

In [5], the new type of the characteristic function for differential games on network was defined. It was supposed that players from S use cooperative strategies under the condition that connections with players from $N \setminus S$ are cut off. The characteristic function constructed in this way is easier to compute and possesses some advantageous properties. In this paper, we will modify this approach according to assumptions of the problem under consideration.

Let $S \subset N$ be a subset of vertices and L_S denote the set of all edges between vertices from S in L. A pair (S, L_S) is called a subgraph induced by S. For player $i \in S$ denote by $K_S^m(i) = \{j \in S : d(i,j) = m, p(i,j) \in S\}$ – the set of players connected with player i by the path in (S, L_S) containing exactly m edges. Let $P_S^m(i) = \{j \in S : d(i,j) = m, p(i,j) \notin S\}$ be the set of players in S connected with player i by the path not lying in (S, L_S) and containing exactly m edges,

$P_S^m(i) = (S \cap K^m(i)) \setminus K_S^m(i)$. Let $B_S^m(i) = K^m(i) \setminus S$ be the set of players in $N \setminus S$ connected with player i by the path containing exactly m edges. Note, that $K^m(i) = K_S^m(i) \cup P_S^m(i) \cup B_S^m(i)$.

For simplicity, we denote

$$h_i(\overline{x}_i(t); \overline{u}_i(t)) = \overline{h}_i(t), \quad h_i^j(\overline{x}_i(t); \overline{x}_j(t); \overline{u}_i(t); \overline{u}_j(t)) = \overline{h}_i^j(t),$$

$$\overline{h}_i^j(t) + \overline{h}_j^i(t) = \overline{h}_{i,j}(t).$$

Let's define a way to construct a characteristic function, taking into account the ability of players to restrict communication. The worth of coalition S in the game, as in [5], is evaluated along the cooperative trajectory

$$V(S; x_0, T - t_0) = \sum_{i \in S} \int_{t_0}^T \overline{h}_i(\tau)d\tau + \sum_{i \in S} \sum_{m=1}^{n-1} \delta^{m-1} \left(\sum_{j \in K_S^m(i)} \int_{t_0}^T \overline{h}_i^j(\tau)d\tau \right.$$

$$\left. + \alpha \sum_{j \in P_S^m(i)} \int_{t_0}^T \overline{h}_i^j(\tau)d\tau + \beta \sum_{j \in B_S^m(i)} \int_{t_0}^T \overline{h}_i^j(\tau)d\tau \right), \quad (7)$$

where $\overline{x}_i(t)$ and $\overline{u}_i(t)$ are the solutions obtained in (4) and (6). So, we assume that players from coalition $N \setminus S$, minimizing the payoff of coalition S, restrict communication with players from S and restrict communication between them if players from $N \setminus S$ are intermediate on the paths between players from S.

In the paper [7], it was assumed that any player i at any instant of time can cut a connection with any other players from $K^1(i)$. Then the worst thing that coalition $N \setminus S$ could do against coalition S was to cut all connections with them. In this case, the characteristic function had the form (7) with $\alpha = 0$, $\beta = 0$.

3   Properties of the Characteristic Function

Characteristic function $V(S; x_0, T - t_0)$ is called *convex* (or supermodular) if for any coalitions $S_1, S_2 \subseteq N$ the following condition holds: $V(S_1 \cup S_2; x_0, T - t_0) + V(S_1 \cap S_2; x_0, T - t_0) \geq V(S_1; x_0, T - t_0) + V(S_2; x_0, T - t_0)$. A game is called convex if its characteristic function is convex.

Proposition 1. *In the game $\Gamma(x^0, T - t_0)$, the characteristic function defined in (7) is convex if $\alpha > \delta$ and (N, L) contains no cycles of even cardinality.*

Proof. According to [8], the convexity of characteristic function $V(S; x_0, T - t_0)$ is equivalent to the fulfillment of conditions

$$V(S_1 \cup \{i\}; x_0, T - t_0) - V(S_1; x_0, T - t_0) \geq V(S_2 \cup \{i\}; x_0, T - t_0) - V(S_2; x_0, T - t_0),$$

for all $S_2 \subset S_1 \subset N \setminus i$.

Here

$$V(S_1 \cup \{i\}; x_0, T - t_0) - V(S_1; x_0, T - t_0)$$

$$= \int_{t_0}^{T} \overline{h}_i(\tau) d\tau + \sum_{m=1}^{n-1} \delta^{m-1} \bigg(\sum_{j \in K_{S_1 \cup i}^m(i)} \int_{t_0}^{T} \overline{h}_i^j(\tau) d\tau + (1 - \beta) \sum_{j \in K_{S_1 \cup i}^m(i)} \int_{t_0}^{T} \overline{h}_j^i(\tau) d\tau$$

$$+ \alpha \sum_{j \in P_{S_1 \cup i}^m(i)} \int_{t_0}^{T} \overline{h}_i^j(\tau) d\tau + (\alpha - \beta) \sum_{j \in P_{S_1 \cup i}^m(i)} \int_{t_0}^{T} \overline{h}_j^i(\tau) d\tau + \beta \sum_{j \in B_{S_1 \cup i}^m(i)} \int_{t_0}^{T} \overline{h}_i^j(\tau) d\tau$$

$$+ (1 - \alpha) \sum_{\{k,l\} \in D_{S_1 \cup i}^m(i)} \int_{t_0}^{T} \overline{h}_{k,l}(\tau) d\tau \bigg), \qquad (8)$$

here $D_S^m(i) = \{k, l \in S : \ k \in K_S^m(l), \ i \in p(k,l), \ i \neq k, i \neq l\}$ – the set of pairs of vertices from S such that the distance between them equals m, all vertices of the path between them belong to S, and i lies on this path (denote $D_N^m(i)$ as $D^m(i)$).

For coalition S_2:

$$V(S_2 \cup \{i\}; x_0, T - t_0) - V(S_2; x_0, T - t_0)$$

$$= \int_{t_0}^{T} \overline{h}_i(\tau) d\tau + \sum_{m=1}^{n-1} \delta^{m-1} \bigg(\sum_{j \in K_{S_2 \cup i}^m(i)} \int_{t_0}^{T} \overline{h}_i^j(\tau) d\tau + (1 - \beta) \sum_{j \in K_{S_2 \cup i}^m(i)} \int_{t_0}^{T} \overline{h}_j^i(\tau) d\tau$$

$$+ \alpha \sum_{j \in P_{S_2 \cup i}^m(i)} \int_{t_0}^{T} \overline{h}_i^j(\tau) d\tau + (\alpha - \beta) \sum_{j \in P_{S_2 \cup i}^m(i)} \int_{t_0}^{T} \overline{h}_j^i(\tau) d\tau + \beta \sum_{j \in B_{S_2 \cup i}^m(i)} \int_{t_0}^{T} \overline{h}_i^j(\tau) d\tau$$

$$+ (1 - \alpha) \sum_{\{k,l\} \in D_{S_2 \cup i}^m(i)} \int_{t_0}^{T} \overline{h}_{k,l}(\tau) d\tau \bigg). \qquad (9)$$

Since $\alpha > \delta$ and (N, L) contains no cycles of even cardinality, players always interact along the shortest path and there is a unique shortest path between any pair of players. Then $K_{S_2 \cup i}^m(i) \subset K_{S_1 \cup i}^m(i)$, $D_{S_2 \cup i}^m(i) \subset D_{S_1 \cup i}^m(i)$, and

$$\sum_{m=1}^{n-1} \delta^{m-1} \bigg(\sum_{j \in K_{S_1 \cup i}^m(i)} \int_{t_0}^{T} \overline{h}_i^j(\tau) d\tau + (1 - \beta) \sum_{j \in K_{S_1 \cup i}^m(i)} \int_{t_0}^{T} \overline{h}_j^i(\tau) d\tau \bigg) \geq$$

$$\sum_{m=1}^{n-1} \delta^{m-1} \bigg(\sum_{j \in K_{S_2 \cup i}^m(i)} \int_{t_0}^{T} \overline{h}_i^j(\tau) d\tau + (1 - \beta) \sum_{j \in K_{S_2 \cup i}^m(i)} \int_{t_0}^{T} \overline{h}_j^i(\tau) d\tau \bigg),$$

$$\sum_{m=1}^{n-1} \delta^{m-1}\left((1-\alpha)\sum_{\{k,l\}\in D_{S_1\cup i}^m(i)}\int_{t_0}^{T}\overline{h}_{k,l}(\tau)d\tau\right) \ge$$

$$\sum_{m=1}^{n-1}\delta^{m-1}\left((1-\alpha)\sum_{\{k,l\}\in D_{S_2\cup i}^m(i)}\int_{t_0}^{T}\overline{h}_{k,l}(\tau)d\tau\right).$$

$P_{S_2\cup i}^m(i) \not\subset P_{S_i\cup i}^m(i)$, but for all $j \in P_{S_2\cup i}^m(i)\setminus P_{S_1\cup i}^m(i)$, we have $\delta^{m-1}(\int_{t_0}^{T}\overline{h}_i^j(\tau)d\tau + (1-\beta)\int_{t_0}^{T}\overline{h}_i^j(\tau)d\tau)$ in (8) and $\delta^{m-1}(\alpha\int_{t_0}^{T}\overline{h}_i^j(\tau)d\tau + (\alpha - \beta)\int_{t_0}^{T}\overline{h}_i^j(\tau)d\tau)$ in (9).

$B_{S_2\cup i}^m(i) \not\subset B_{S_i\cup i}^m(i)$, but for all $j \in B_{S_2\cup i}^m(i)\, B_{S_1\cup i}^m(i)$, we have $\delta^{m-1}\beta\overline{h}_i^j(\tau)d\tau$ in (9) and at least $\delta^{m-1}(\alpha\int_{t_0}^{T}\overline{h}_j^i(\tau)d\tau + (\alpha-\beta)\int_{t_0}^{T}\overline{h}_j^i(\tau)d\tau)$ in (8).

Then

$$V(S_1\cup\{i\};x_0,T-t_0)-V(S_1;x_0,T-t_0) \ge V(S_2\cup\{i\};x_0,T-t_0)-V(S_2;x_0,T-t_0),$$

which proves the convexity of the characteristic function.

4   The Shapley Value

The Shapley value [9] is a classic cooperative solution. According to this solution, each player receives an expected marginal contribution in the game with respect to a uniform distribution over the set of all permutations on the set of players.

The Shapley value is defined by:

$$Sh_i(x_0,T-t_0) = \sum_{S:\, i\in S}\frac{(s-1)!(n-s)!}{n!}(V(S;x_0,T-t_0) - V(S\setminus i;x_0,T-t_0)).$$

Proposition 2 gives the formula for the Shapley value in the game $\Gamma(x^0,T-t_0)$ in the explicit form.

Proposition 2. *The Shapley value in the game* $\Gamma(x^0,T-t_0)$ *has the following form*

$$Sh_i(x_0,T-t_0) = \int_{t_0}^{T}\overline{h}_i(\tau)d\tau + \sum_{m=1}^{n-1}\delta^{m-1}\left(\left(\frac{\alpha}{2}+\frac{1-\alpha}{m+1}\right)\sum_{j\in K^m(i)}\int_{t_0}^{T}\overline{h}_{i,j}(\tau)d\tau\right.$$

$$\left. +\frac{\beta}{2}\sum_{j\in K^m(i)}\int_{t_0}^{T}(\overline{h}_i^j(\tau) - \overline{h}_j^i(\tau))d\tau + \frac{1-\alpha}{m+1}\sum_{\{k,l\}\in D^m(i)}\int_{t_0}^{T}\overline{h}_{k,l}(\tau)d\tau\right). \quad (10)$$

Proof.

$$V(S; x_0, T - t_0) - V(S \setminus i; x_0, T - t_0)$$

$$= \int_{t_0}^{T} \overline{h}_i(\tau)d\tau + \sum_{m=1}^{n-1} \delta^{m-1} \left(\sum_{j \in K_S^m(i)} \int_{t_0}^{T} \overline{h}_i^j(\tau)d\tau + (1 - \beta) \sum_{j \in K_S^m(i)} \int_{t_0}^{T} \overline{h}_j^i(\tau)d\tau \right.$$

$$+ \alpha \sum_{j \in P_S^m(i)} \int_{t_0}^{T} \overline{h}_i^j(\tau)d\tau + (\alpha - \beta) \sum_{j \in P_S^m(i)} \int_{t_0}^{T} \overline{h}_j^i(\tau)d\tau + \beta \sum_{j \in B_S^m(i)} \int_{t_0}^{T} \overline{h}_i^j(\tau)d\tau$$

$$+ (1 - \alpha) \sum_{\{k,l\} \in D_S^m(i)} \int_{t_0}^{T} \overline{h}_{k,l}(\tau)d\tau \right), \quad (11)$$

Consider two fixed players i and j with $d(i,j) = m$. These two vertices (players), together with all vertices (players) from the shortest path between them, are included in C_{n-m-1}^{s-m-1} coalitions of power s. Then

$$\sum_{S:\, i \in S} \frac{(s-1)!(n-s)!}{n!} \sum_{j \in K_S^m(i)} \left(\int_{t_0}^{T} \overline{h}_i^j(\tau)d\tau + (1 - \beta) \int_{t_0}^{T} \overline{h}_j^i(\tau)d\tau \right)$$

$$= \frac{1}{m+1} \sum_{j \in K^m(i)} \left(\int_{t_0}^{T} \overline{h}_i^j(\tau)d\tau + (1 - \beta) \int_{t_0}^{T} \overline{h}_j^i(\tau)d\tau \right).$$

The same for players k and l from $S \setminus i$, such that $k \in K^m(l)$ and $i \in p(k,l)$:

$$\sum_{S:\, i \in S} \frac{(s-1)!(n-s)!}{n!} \sum_{\{k,l\} \in D_S^m(i)} (1 - \alpha) \int_{t_0}^{T} \overline{h}_{k,l}(\tau)d\tau$$

$$= \frac{1}{m+1} \sum_{\{k,l\} \in D^m(i)} (1 - \alpha) \int_{t_0}^{T} \overline{h}_{k,l}(\tau)d\tau.$$

For two fixed players i and j with $d(i,j) = m$, consider such coalitions S, that $j \in P_S^m(i)$. We need to find the number of such coalitions of power s. For the case, when $s < m + 1$, this number is equal to C_{n-2}^{s-2}. If $s \geq m + 1$, we can find this number as $(C_{n-2}^{s-2} - C_{n-m-1}^{s-m-1})$. Then

$$\sum_{S:\ i\in S} \frac{(s-1)!(n-s)!}{n!} \sum_{j\in P_S^m(i)} \left(\alpha \int_{t_0}^{T} \overline{h}_i^j(\tau)d\tau + (\alpha-\beta)\int_{t_0}^{T} \overline{h}_j^i(\tau)d\tau \right)$$

$$= \left(\frac{1}{2} - \frac{1}{m+1}\right) \sum_{j\in K^m(i)} \left(\alpha \int_{t_0}^{T} \overline{h}_i^j(\tau)d\tau + (\alpha-\beta)\int_{t_0}^{T} \overline{h}_j^i(\tau)d\tau \right).$$

For two fixed players i and j, $d(i,j) = m$, consider such coalitions S, that $j \in B_S^m(i)$. There are C_{n-2}^{s-2} such coalitions of power s. Then

$$\sum_{S:\ i\in S} \frac{(s-1)!(n-s)!}{n!} \sum_{j\in B_S^m(i)} \beta\int_{t_0}^{T} \overline{h}_i^j(\tau)d\tau = \frac{\beta}{2} \sum_{j\in K^m(i)} \int_{t_0}^{T} \overline{h}_i^j(\tau)d\tau.$$

Finally, for the Shapley value we obtain

$$Sh_i(x_0, T-t_0) = \int_{t_0}^{T} \overline{h}_i(\tau)d\tau + \sum_{m=1}^{n-1} \delta^{m-1}\left(\left(\frac{\alpha}{2} + \frac{1-\alpha}{m+1}\right) \sum_{j\in K^m(i)} \int_{t_0}^{T} \overline{h}_{i,j}(\tau)d\tau \right.$$

$$\left. + \frac{\beta}{2} \sum_{j\in K^m(i)} \int_{t_0}^{T} (\overline{h}_i^j(\tau) - \overline{h}_j^i(\tau))d\tau + \frac{1-\alpha}{m+1} \sum_{\{k,l\}\in D^m(i)} \int_{t_0}^{T} \overline{h}_{k,l}(\tau)d\tau \right).$$

This concludes the proof.

Note that in the case where $\alpha = 0$ and $\beta = 0$, formula (10) coincides with the result presented in [7]. Such parameter values can be interpreted as an opportunity for players to cut links with neighboring players. Then, according to the Shapley value, for any pair of players i, j $(j \in K^m(i))$, the players' gain $\delta^{m-1}\int_{t_0}^{T} \overline{h}_{i,j}(\tau)d\tau$ from the interaction with each other is divided equally between all the nodes lying on the path between i and j.

As can be seen from formula (10), the presence of the parameter $\alpha \neq 0$ leads to the fact that players i and j get more from $\delta^{m-1}\int_{t_0}^{T} \overline{h}_{i,j}(\tau)d\tau$ than the intermediate players of the path from i to j.

The presence of the parameter $\beta \neq 0$ leads to the fact that if $\int_{t_0}^{T} \overline{h}_i^j(\tau)d\tau > \int_{t_0}^{T} \overline{h}_j^i(\tau)d\tau$ then player i will get more from $\delta^{m-1}\int_{t_0}^{T} \overline{h}_{i,j}(\tau)d\tau$ than player j.

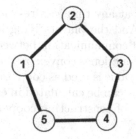

Fig. 1. The network structure of the game

4.1   Example

To illustrate the solution obtained consider the game of 5 players, connected by the network presented on Fig. 1.

An example of construction of $V(S; x_0, T - t_0)$ is demonstrated in (12) for $S = \{1, 2, 4\}$.

$$V(\{1, 2, 4\}; x_0, T - t_0)$$

$$= \int_{t_0}^{T} \Bigg(\overline{h}_1(\tau) + \overline{h}_2(\tau) + \overline{h}_4(\tau) + \overline{h}_{1,2}(\tau) + \alpha\delta(\overline{h}_{1,4}(\tau) + \overline{h}_{2,4}(\tau))$$

$$+ \beta(\overline{h}_1^5(\tau) + \overline{h}_2^3(\tau) + \overline{h}_4^3(\tau) + \overline{h}_4^5(\tau)) +$$

$$+ \beta\delta(\overline{h}_1^3(\tau) + \overline{h}_2^5(\tau)) \Bigg) d\tau. \quad (12)$$

An example of the Shapley value construction is illustrated by (13) for player 1.

$$Sh_1$$

$$= \int_{t_0}^{T} \Bigg(\overline{h}_1(\tau) + (\frac{\alpha}{2} + \frac{1-\alpha}{2})(\overline{h}_{1,2}(\tau) + \overline{h}_{1,5}(\tau)) + \delta(\frac{\alpha}{2} + \frac{1-\alpha}{3})(\overline{h}_{1,3}(\tau) + \overline{h}_{1,4}(\tau))$$

$$+ \frac{\beta}{2}(\overline{h}_1^2(\tau) - \overline{h}_2^1(\tau) + \overline{h}_1^5(\tau) - \overline{h}_5^1(\tau) + \delta(\overline{h}_1^4(\tau) - \overline{h}_4^1(\tau) + \overline{h}_1^3(\tau) - \overline{h}_3^1(\tau)))$$

$$+ \delta\frac{1-\alpha}{3}\overline{h}_{2,5}(\tau) \Bigg) d\tau. \quad (13)$$

5   Conclusion

One class of differential games on networks with communication restrictions is considered. It is assumed that communication on the network is possible between connected players, any player at any time can restrict communication with any player connected with him. And the players lying on the path between other players are also able to restrict communication between them. It is shown that in such games the characteristic function is convex on the networks without cycles of even cardinality. The Shapley value is used as cooperative optimality principle. It is shown that the Shapley value can be calculated in explicit form. An illustrative example demonstrating a way of constructing cooperative optimality principles is considered.

Acknowledgements. The work was supported by the Russian Science Foundation grant No. 22-11-00051, https://rscf.ru/en/project/22-11-00051/.

References

1. Gao, H., Pankratova, Ya.: Cooperation in dynamic network games. Contrib. Game Theory Manage. **10**, 42–67 (2017)
2. Mazalov, V., Chirkova, J.V.: Networking Games: Network Forming Games and Games on Networks. Academic Press, Cambridge (2019)
3. Petrosyan, L., Yeung, D.: Shapley value for differential network games: theory and application. J. Dyn. Games **8**(2), 151–166 (2021)
4. Petrosyan, L., Yeung, D.W.K., Pankratova, Y.: Cooperative differential games with partner sets on networks. Trudy Instituta Matematiki i Mekhaniki UrO RAN **27**(3), 286–295 (2021). https://doi.org/10.21538/0134-4889-2021-27-3-286-295
5. Petrosyan, L., Yeung, D.: Construction of dynamically stable solutions in differential network games. In: Tarasyev, A., Maksimov, V., Filippova, T. (eds.) Stability, Control and Differential Games. LNCISP, pp. 51–61. Springer, Cham (2020). https://doi.org/10.1007/978-3-030-42831-0_5
6. Von Neumann, J., Morgenstern, O.: Theory of Games and Economic Behavior. Princeton University Press, Princeton (1953)
7. Tur, A., Petrosyan, L.: The core of cooperative differential games on networks. In: Pardalos, P., Khachay, M., Mazalov, V. (eds.) Mathematical Optimization Theory and Operations Research, MOTOR 2022. Lecture Notes in Computer Science, vol. 13367, pp. 295–314. Springer, Cham (2022). https://doi.org/10.1007/978-3-031-09607-5_21
8. Shapley, L.S.: Cores of convex games. Int. J. Game Theory **1**, 11–26 (1971)
9. Shapley, L.S.: A value for n-person games. In: Kuhn, H., Tucker, A., (eds.) Contributions to the Theory of Games II, pp. 307–317. Princeton University Press, Princeton (1953). https://doi.org/10.1515/9781400881970-018

Optimal Control and Mathematical Economics

Guaranteed Expectation of the Flock Position with Random Distribution of Items

B. I. Ananyev[(✉)] [iD]

N.N. Krasovskii Institute of Mathematics and Mechanics UB of RAS,
Yekaterinburg 620108, Russian Federation
abi@imm.uran.ru

Abstract. We consider estimation problems for a flock of linear systems with random matrices and additive uncertain disturbances. Because of uncertainty we may only assume that the state of each flock's item belongs to some random information set that can be built according to measurement equations. Our goal is to define and investigate a mean value of random information sets. All the sets are described by its support functions. Some theorems on the approximation of the sets by simpler objects are given. Particular cases, where matrices of the systems or probability spaces have special formes, are elaborated. In conclusion we review some examples.

Keywords: guaranteed estimation · information set · set of attainability · averaged controllability and observability

1 Introduction and Basic Definitions

In this paper, we consider properties of averaged estimation for linear finite-dimensional systems with random coefficients and additive uncertain disturbances. In our analysis, the notions of averaged observability and controllability are used. These notions were introduced by E. Zuazua in [1], and afterwards generalised in PDE setting in [2,3]. On the other hand, the estimation methods that use set-membership description of uncertainty have become widely known in last decades (see [4–6]). We unite two techniques to study continuous-time linear systems described by the state equations with random coefficients and observation:

$$\dot{x}(t) = A(\omega)x(t) + b(\omega)v(t), \quad x(t) \in \mathbb{R}^n, \quad t \in [0, T], \tag{1}$$

$$y(t) = G(\omega)x(t) + cv(t), \quad y(t) \in \mathbb{R}^m, \tag{2}$$

where $\omega \in \Omega$ and $(\Omega, \mathcal{F}, P)$ is a probability space. In (1), (2) we have random matrices $A \in \mathbb{R}^{n \times n}$, $b \in \mathbb{R}^{n \times q}$, and $G \in \mathbb{R}^{m \times n}$ with $\mathcal{F}$-measurable elements. The matrix c being nonrandom has a full rank, i.e. $\text{rank}(c) = m \wedge q$, where $m \wedge q = \min\{m, q\}$. In what follows the stochastic variable ω is often omitted.

M. Khachay et al. (Eds.): MOTOR 2023, LNCS 13930, pp. 327–344, 2023.
https://doi.org/10.1007/978-3-031-35305-5_23

An unknown function $v(t) \in L_2^q[0,T]$ and initial state x_0 are restricted by the constraints

$$\int_0^T |v(t)|^2 dt \leq 1, \quad x_0 \in X_0, \tag{3}$$

where X_0 is a convex compact set. As an auxiliary constraint, we have in mind the following:

$$|x_0 - \hat{x}_0|_{P_0}^2 + \int_0^T |v(t)|^2 dt \leq 1, \quad \hat{x}_0 \in \mathbb{R}^n \text{ is fixed,} \tag{4}$$

where matrix P_0 satisfies the condition of $P_0' = P_0 \geq 0$, and the symbol of $|x|_P^2$ equals $x'Px$. Here and further the symbol of $'$ means the transposition. The symbol of $|B| = \max\{|Bx|/|x| \mid x \neq 0\}$ is used also for the norm of arbitrary matrix B; $|x|$ is the Euclidean norm of vector $x \in \mathbb{R}^n$. The elements of vector x_0 and vector-function $v(t)$ can be both nonrandom and random in dependence on the problem formulation. Constraints (3) or (4) must be fulfilled almost surely (a.s.) in the random case.

One can see that the variables x and y from equations (1) and (2) are bound with each other by means of the function $v(t)$. Let us present the system in the equivalent form. Consider the pseudoinverse c^+ matrix to c, [7]. It is known that $c^+ c$ is the orthogonal projection onto subspace $\operatorname{im} c' = \{v : v = c'y, y \in \mathbb{R}^m\}$. Introduce the matrix $C_1 = I_q - c^+ c$ that is the orthogonal projection onto nullsubspace $\ker c = \{v : cv = 0\}$. Then $v(t) = c^+ cv(t) + C_1 v(t)$ and $cv(t) = y(t) - Gx(t)$. If we introduce a notation

$$\mathbf{b} = bc^+, \quad \mathbf{A} = A - bG, \tag{5}$$

and substitute the orthogonal expansion of $v(t)$ into (1), this equation is converted to the following one

$$\dot{x}(t) = \mathbf{A}x(t) + \mathbf{b}y(t) + bC_1 v(t). \tag{6}$$

Constraints (3), (4) may be rewritten as

$$\int_0^T \left(|y(t) - Gx(t)|_C^2 + |v(t)|_{C_1}^2\right) dt \leq 1, \quad C = (c^+)'c^+, \quad x_0 \in X_0, \tag{7}$$

$$|x_0 - \hat{x}_0|_{P_0}^2 + \int_0^T \left(|y(t) - Gx(t)|_C^2 + |v(t)|_{C_1}^2\right) dt \leq 1 \tag{8}$$

In the case $\operatorname{rank}(c) = m$, we have $c^+ = c'(cc')^{-1}$ and $C = (cc')^{-1}$. In other case $\operatorname{rank}(c) = q$, we obtain $c^+ = (c'c)^{-1}c'$ and $C_1 = O_q$, i.e. zero matrix. In the last case we deal with the unique uncertain element x_0.

Let us formulate the first problem that is investigated in the paper. Suppose that the norm $|e^{At}|$ of matrix exponent belongs to the space $L_\infty(\Omega, \mathcal{F}, P)$ for all numbers $t \in [0,T]$ and the norms $|b|$ and $|G|$ belongs to the Hilbert space $L_2(\Omega, \mathcal{F}, P)$ which is separable for simplicity. We introduce some definitions.

Definition 1. *Let the signal $y(t)$ be generated by* (1), (2) *with the help of unknown random pair $(x_0^*, v^*(\cdot))$ satisfying constraints* (3) *or* (4) *a.s. A random pair $(x_0, v(\cdot))$ is called* compatible *with the measured signal $y(t)$ on $[0, T]$ if the solution $x(t)$ of equation* (6) *and the function $v(t)$ satisfy relations* (7) *or* (8) *a.s., respectively.*

Definition 2. *The set $X(T, y)$ is called the* random information set *(shortly RIS) if it consists of all random vectors $x(T)$ for each of which there exists a generating compatible pair $(x_0, v(\cdot))$ on $[0, T]$ satisfying constraints* (7) *a.s. If constraints* (7) *is changed to* (8), *the RIS is denoted by $X(t, y, P_0)$ and $X(t, y, P_0)$, respectively.*

Our goal to give an analytical description of the *averaged information set* (shortly AIS) $EX(T, y)$, that we denote by $\mathbf{X}(T, y)$, and analyse its dynamic behavior on $[0, T]$. Hereinafter, E is the expectation. Methods of convex analysis [8] is widely used. Let us remind that the support function of the convex closed set $X \subset \mathbb{R}^n$ is defined by the expression $\rho(l|X) = \sup\{l'x|x \in X\} \ \forall l \in \mathbb{R}^n$. It is known that the correspondence between convex closed sets and their support functions is one-to-one. Therefore, we have to find $\rho(l|\mathbf{X}(t, y))$ in order to describe AIS.

We can define other informational sets. For example, assume that the pairs $(x_0, v(\cdot))$ in (3), (4) are nonrandom. Let the signal

$$z(t) = \overline{G}\overline{x}(t) + cv(t), \quad t \in [0, T], \tag{9}$$

be observed instead of (2).

Definition 3. *Let the signal $z(t)$ be generated by* (1), (9) *with the help of unknown nonrandom pair $(x_0^*, v^*(\cdot))$ satisfying constraints* (3) *or* (4). *A nonrandom pair $(x_0, v(\cdot))$ is called* compatible *with the measured signal $z(t)$ on $[0, T]$ if the output* (9) *generated by the pair coincides with function $z(t)$ almost everywhere (a.e.) in t on $[0, T]$ and the pair $(x_0, v(\cdot))$ satisfyes constraints* (3) *or* (4), *respectively.*

Definition 4. *A set $\overline{X}(T, y)$ is said to be the* information set of averaged vectors *(shortly ISAV) if it consists of all vectors $\overline{x}(T) = Ex(T)$ for each of which there exists a generating nonrandom and compatible with $z(\cdot)$ pair $(x_0, v(\cdot))$ on $[0, T]$ satisfying inequality* (3). *If constraint* (3) *is changed to* (4), *the designation $\overline{X}(t, y, P_0)$ is used.*

We intend to give also the analytical description of $\overline{X}(t, y)$ and to compare the entered sets. After that we consider particular cases, where the matrices of the system or probability space have special formes, and give some examples. Of course, we specify conditions about matrices in order that the corresponding expectations existed.

Note that our problems differ from considered in [9]. First of all, the fact that random disturbances are taking in account and not just determined.

2   Main Results

2.1   Description of AIS

We will assume AIS $EX(T, y) = \mathbf{X}(T, y)$ as the family $\{\overline{x}(T) : E x(T)\}$ where $x(T)$ belongs to the RIS $X(T, y)$. Using the matrix exponents $e^{\mathbf{A}t}$, e^{At} and taking the expectation for the solution of (6), we obtain the following equalities

$$\overline{x}(T) = E e^{\mathbf{A}T} x_0 + \int_0^T E e^{\mathbf{A}(T-t)} \left(\mathbf{b} y(t) + b C_1 v(t) \right) dt,$$
$$y(t) = G \left(e^{At} x_0 + \int_0^t e^{A(t-s)} b v(s) ds \right) + c v(t). \tag{10}$$

Further we consider random processes $v(t)$ as elements of the Hilbert space $L_2(\Omega, \mathcal{F}, P; L_2^q[0, T])$, i.e. $E \int_0^T |v(t)|^2 dt < \infty$. From now on, such spaces are denoted by $\mathcal{L}_2^q[0, T]$. The space $\mathcal{L}_2^q[0, T]$ is separable thanks to our assumption about separability of $L_2(\Omega, \mathcal{F}, P)$ (see [10, Problem II.12]). For the existence of expectations and due to the technical requirements, we make

Assumption 1. The matrices A, b, and G satisfy the conditions

$$\left| e^{At} \right|, \ \left| e^{\mathbf{A}t} \right| \in L_\infty(\Omega, \mathcal{F}, P) \ \forall t \in [0, T]; \ |b|, \ |G| \in L_\infty(\Omega, \mathcal{F}, P). \tag{11}$$

Lemma 1. *Under Assumption 1, the vector $\overline{x}(T)$ exists and $y(\cdot) \in \mathcal{L}_2^q[0, T]$ for any $x_0 \in L_2^n(\Omega, \mathcal{F}, P)$ and $v(\cdot) \in \mathcal{L}_2^q[0, T]$.*

Proof. Note that requirements (11) are much stronger, than those which were in Sect. 1. These assumptions are not needed for the existence of $\overline{x}(T)$, but further we use them. Let us establish that $E |y(t)|^2 < \infty$ and $E \int_0^T |y(t)|^2 dt < \infty$. Using Cauchy-Schwarz inequality, we have

$$|y(t)|^2 = \left| G e^{At} x_0 + \int_0^t G e^{A(t-s)} b v(s) ds + c v(t) \right|^2 \leq 2 \Big(\left| G e^{At} x_0 \right|$$
$$+ \left| \int_0^t G e^{A(t-s)} b v(s) ds \right| \Big)^2 + 2|c v(t)|^2 \leq 2 \left(|x_0| \varkappa_1 + \int_0^T |v(s)| ds \varkappa_2 \right)^2 + 2|c v(t)|^2$$
$$\leq 4 \left(|x_0|^2 \varkappa_1^2 + \int_0^T |v(s)|^2 ds T \varkappa_2^2 \right) + 2|c v(t)|^2,$$

where $\varkappa_1, \varkappa_2 \in L_\infty(\Omega, \mathcal{F}, P)$. Hence, the function $|y(t)|^2$ is integrable due to Assumption 1 and $y(\cdot) \in \mathcal{L}_2^q[0, T]$. For the first equality in (10), we obtain $|\overline{x}(T)| \leq \varkappa \sqrt{E|x_0|^2} + \varkappa_1 \sqrt{E \int_0^T |y(t)|^2 dt} + \varkappa_2 \sqrt{E \int_0^T |v(t)|^2 dt}$, where $\varkappa$, $\varkappa_i$ are constants. In particular, equalities (10) are valid for any compatible pair.   □

To reduce records we enter the notation

$$\widetilde{y}(t) = y(t) - \int_0^t G e^{A(t-s)} \mathbf{b} y(s) ds, \quad \widetilde{y}_0(t) = \widetilde{y}(t) - G e^{At} x_0, \quad K(t)$$
$$= G e^{At} b, \quad x_y^T = \int_0^T e^{\mathbf{A}(T-t)} \mathbf{b} y(t) dt, \quad x_0^T = e^{\mathbf{A}T} x_0 + x_y^T, \quad M(t) = e^{\mathbf{A}t} b. \tag{12}$$

It follows from (2), (12) that the integral inequality in (7) has the form

$$\int_0^T \left(\left| \widetilde{y}_0(t) - \int_0^t K(t-s) C_1 v(s) ds \right|_C^2 + |v(t)|_{C_1}^2 \right) dt \leq 1. \tag{13}$$

Remark 1. If rank$(c) = m$ and $m < q$, there is an opportunity to lower the dimension of unknown function $v(t)$. Namely, rank $C_1 = q - m$, as ker $c \oplus$ im $c' = \mathbb{R}^q$, im $C_1 =$ ker c, and dim (im c') $= m$. According to the theorem from linear algebra we have the equality $C_1 = T\widetilde{C}_1 T'$, where T is an orthogonal matrix, $TT' = T'T = I_q$, and $\widetilde{C}_1$ is an diagonal matrix with with zero and units, as $C_1^2 = C_1$. We eliminate m zero columns from $\widetilde{C}_1$ m and designate the received matrix through $\widetilde{D}_1 \in \mathbb{R}^{q \times (q-m)}$. Then $\widetilde{C}_1 = \widetilde{D}_1 \widetilde{D}_1'$ and $C_1 = D_1 D_1'$, where $D_1 = T\widetilde{D}_1$. Further, we define vector-functions $u(t) = D_1' v(t) \in \mathbb{R}^{q-m}$ and receive the equality

$$C_1 v(t) = D_1 u(t), \quad D_1 \in \mathbb{R}^{q \times (q-m)}, \quad \text{rank}(D_1) = q - m.$$

Therefore, we can use the function $D_1 u(t)$ instead $C_1 v(t)$ in relations (6), (7), (8), (10), and (13). Note that $D_1' D_1 = I_{q-m}$ is the unit matrix of dimension $(q - m) \times (q - m)$.

Remark 2. If $q \le m$, then $C_1 = O_q$ is the zero matrix. So, we have $\dot{x} = \mathbf{A}x + \mathbf{b}y$ from (6). The initial vector x_0 is compatible iff $x_0 \in X_0$ and inequality (13) is fulfilled, where C_1 is zero.

Let the symbol $\langle U(\cdot), V(\cdot) \rangle$ mean the integral $\int_0^T U'(t)V(t)dt \in \mathbb{R}^{q \times r}$, where $U(t) \in \mathbb{R}^{p \times q}$, $V(t) \in \mathbb{R}^{p \times r}$, and the elements of matrices belongs to the space $L_2[0,T]$. We use symbol $\langle \cdot, \cdot \rangle$ as the inner product in the spaces like $L_2^{q-m}[0,T]$. If we define the linear self-adjoint operator $\mathbf{K}$ and the function $\mathbf{y}_0(s)$ by the relations

$$\begin{aligned} \mathbf{K}u(s) &= \int_s^T D_1' K'(t-s)C \int_0^t K(t-r)D_1 u(r)drdt + u(s), \\ \mathbf{y}_0(s) &= \int_s^T D_1' K'(t-s)C\widetilde{y}_0(t)dt, \end{aligned} \tag{14}$$

then inequality (13) may be written as

$$\begin{aligned} \|\widetilde{y}_0(\cdot)\|_C^2 - 2\langle \mathbf{y}_0(\cdot), u(\cdot) \rangle + \|u(\cdot)\|_{\mathbf{K}}^2 &\le 1, \quad \text{or} \\ \|\widetilde{y}_0(\cdot)\|_C^2 - \|\mathbf{y}_0(\cdot)\|_{\mathbf{K}^{-1}}^2 + \|u(\cdot) - \mathbf{K}^{-1}\mathbf{y}_0(\cdot)\|_{\mathbf{K}}^2 &\le 1. \end{aligned} \tag{15}$$

The operator $\mathbf{K}^{-1}$ exists as the $\mathbf{K}$ is coercive. The symbol like $\|u(\cdot)\|_{\mathbf{K}}^2$ means $\langle u(\cdot), \mathbf{K}u(\cdot) \rangle$. Now using (10), (12), we are able to calculate the random expression

$$\begin{aligned} R(l, T, x_0, y) &= \max_{u(\cdot)} l'x(T) = l'x_0^T + \max_{u(\cdot)} \langle D_1' M'(T - \cdot)l, u(\cdot) \rangle \\ &= l'x_0^T + \langle D_1' M'(T - \cdot)l, \mathbf{K}^{-1}\mathbf{y}_0(\cdot) \rangle \\ &+ \sqrt{\left(1 + \|\mathbf{y}_0(\cdot)\|_{\mathbf{K}^{-1}}^2 - \|\widetilde{y}_0(\cdot)\|_C^2\right)} \|D_1' M'(T - \cdot)l\|_{\mathbf{K}^{-1}}. \end{aligned} \tag{16}$$

After that we get the support function for RIS:

$$\rho(l|X(T, y)) = \max \left\{ R(l, T, x_0, y) \middle| x_0 \in X_0, \|\widetilde{y}_0(\cdot)\|_C^2 - \|\mathbf{y}_0(\cdot)\|_{\mathbf{K}^{-1}}^2 \le 1 \right\}. \tag{17}$$

Denote by $\mathcal{E}(c, P)$ a set with the support function $\rho(l \mid \mathcal{E}(c, P)) = l'c + \sqrt{l'Pl}$, where the matrix $P \in \mathbb{R}^{n \times n}$ satisfies the conditions $P' = P \geq 0$. The set $\mathcal{E}(c, P)$ is called n-dimensional ellipsoid. It follows from (16) that the set $X(T, y)$ is an ellipsoid if $X_0 = \{x_0\}$. In general, this is not the case. By definition we set

$$\mathbf{X}(T, y) = \{Ex(T) \mid x(T) \in X(T, y)\}. \tag{18}$$

So, we come to the conclusion.

Theorem 1. *The AIS $\mathbf{X}(T, y)$ is a convex compact set with the support function defined by relation*

$$\rho(l \mid \mathbf{X}(T, y)) = E\rho(l \mid X(T, y)) \quad \forall l \in \mathbb{R}^n, \tag{19}$$

where parameters are given in (14), (15), (16) *and* (17).

Proof. Consider the expression $ER(l, T, x_0, y)$. The item $El'x_0^T$ exists by Lemma 1. Since $\mathbf{K} \geq \mathrm{Id}$, we have $\mathbf{K}^{-1} \leq \mathrm{Id}$ and $\|\mathbf{K}^{-1}\| \leq 1$ a.s. Here the Id is the identical operator in $L_2^{q-m}[0, T]$. Let us evaluate the second item in (16). We have
$$\left| \left\langle D_1' b' e^{\mathbf{A}'(T-\cdot)} l, \mathbf{K}^{-1} \mathbf{y}_0(\cdot) \right\rangle \right| \leq \varkappa_1 \|\mathbf{y}_0(\cdot)\| \leq \varkappa_2 \|\widetilde{y}_0(\cdot)\| \leq \varkappa_2(\varkappa_3 \|y(\cdot)\| + \varkappa_4 |x_0|),$$
where $\varkappa_i \in L_\infty((\Omega, \mathcal{F}, P))$, $i \in 1 : 4$. The item is integrable with respect to measure P.

The last item in (16) can be estimated by $\varkappa_1 (1 + \|\mathbf{y}_0(\cdot)\|) \leq \varkappa_1 + \varkappa_2 \|\widetilde{y}_0(\cdot)\| \leq \varkappa_1 + \varkappa_2 (\varkappa_3 \|y(\cdot)\| + \varkappa_4 |x_0|)$ that is integrable as above. As the function $R(l, T, x_0, y)$ is continuous in x_0 and the set X_0 is weakly compact in $L_2^n(\Omega, \mathcal{F}, P)$, we get $\max \{ER(l, T, x_0, y) | x_0 \in X_0, \|\widetilde{y}_0(\cdot)\|_C^2 - \|\mathbf{y}_0(\cdot)\|_{\mathbf{K}^{-1}}^2 \leq 1\} = E \max\{R(l, T, x_0, y) | x_0 \in X_0, \|\widetilde{y}_0(\cdot)\|_C^2 - \|\mathbf{y}_0(\cdot)\|_{\mathbf{K}^{-1}}^2 \leq 1\}$. $\square$

The set $\mathbf{X}(T, y)$ may not have any internal points. In order to exclude such opportunity for any initial sets we give

Definition 5. *System* (1), (2) *with constraints* (3) *is called* simultaneously controllable *on* $[0, T]$ *if the symmetric matrix* $W_T = \int_0^T M(T-t)D_1 \mathbf{K}^{-1}(D_1' M'(T-t))dt$ *from* (16) *has a.s. full rank, i.e.* $W_T > 0$ *a.s. The matrix* D_1 *is defined in Remark* 1.

The Definition 5 means that the condition $D_1' b' e^{\mathbf{A}'(T-t)} l = 0$ $\forall t \in [0, T]$ a.s. implies $l = 0$ if $l \in L_2^n(\Omega, \mathcal{F}, P)$. Of course, the stronger condition of *exact simultaneous controllability* introduced in [11] provides simultaneous controllability. But the rank criterium for this very strong controllability is unknown. In addition, the system can never be exactly simultaneously controllable if the space $L_2(\Omega, \mathcal{F}, P)$ is infinite dimensional (see [11, Proposition 3.1.]). At last, note that the simultaneous controllability is stronger than the averaged controllability. It follows from Jensen's inequality $0 \leq \|EB(\cdot)l\| \leq E\|B(\cdot)l\|$ that is fulfilled for any integrable matrix $B(t)$. A rank condition for the averaged controllability is given in [1].

Now we pass to constraint (4). Let us rewrite inequality (15) with the added term $|x_0 - \hat{x}_0|^2_{P_0}$ in the left side as

$$\|u(\cdot)\|^2_{\mathbf{K}} + 2\langle u(\cdot), U(\cdot)x_0\rangle + |x_0|^2_{\widetilde{P}} + |\hat{x}_0|^2_{P_0} + \|\widetilde{y}(\cdot)\|^2_C - 2\langle u(\cdot), \mathbf{y}(\cdot)\rangle \\ -2x'_0 Y_0 \le 1, \tag{20}$$

where $\mathbf{y}(s) = \displaystyle\int_s^T D'_1 K'(t-s)C\widetilde{y}(t)dt, \quad Y_0 = P_0\hat{x}_0 + \langle L(\cdot), C\widetilde{y}(\cdot)\rangle,$

$$L(t) = Ge^{\mathbf{A}t}, \widetilde{P} = P_0 + N_T, \quad N_T = \langle L(\cdot), CL(\cdot)\rangle, \tag{21}$$

$$U(s) = \int_s^T D'_1 K'(t-s)CL(t)dt.$$

We would like to analyze relations (20) and (21). Note that the condition $N_T > 0$ *a.s.* means the following: if the output $Ge^{\mathbf{A}t}x_0 = 0$ a.s., where $x_0 \in L^n_2(\Omega, \mathcal{F}, P)$, then $x_0 = 0$. This property is called *initial simultaneous observability* for the system $\dot{x} = \mathbf{A}x, \quad y = Gx$. It is stronger condition than *averaged initial observability*. Hereinafter, we suppose that $\widetilde{P} > 0$ a.s. Now, consider the linear self-adjoint operator $\mathbb{K} : L^{q-m}_2[0,T] \times \mathbb{R}^n \to L^{q-m}_2[0,T] \times \mathbb{R}^n$ defined by the relation

$$\mathbb{K}(u(\cdot), x_0) = \Big(\mathbf{K}u(\cdot) + U(\cdot)x_0, \widetilde{P}x_0 + \langle U(\cdot), u(\cdot)\rangle\Big). \tag{22}$$

It can be seen that P_0 may be zero if the system is initially simultaneously observable. It corresponds to the case when restrictions on the initial state x_0 are absent.

If $\widetilde{P} > 0$, the operator $\mathbb{K}$ has a bounded inverse one. It follows from equalities (12) and relations (20), (21) that

$$\rho(l|X(T, y, P_0)) = l'x^T_y + \max_{u(\cdot),x_0}\{\langle D'_1 M'(T-\cdot)l, u(\cdot)\rangle + l'e^{\mathbf{A}T}x_0\}$$

$$= l'x^T_y + \Big\langle\Big(D'_1 M'(T-\cdot)l, e^{\mathbf{A}'T}l\Big), \mathbb{K}^{-1}(\mathbf{y}(\cdot), Y_0)\Big\rangle_{L^{q-m}_2[0,T]\times\mathbb{R}^n} \tag{23}$$

$$+\sqrt{1 - |\hat{x}_0|^2_{P_0} - \|\widetilde{y}(\cdot)\|^2_C + \|(\mathbf{y}(\cdot), Y_0)\|^2_{\mathbb{K}-1}}\left\|\Big(D'_1 M'(T-\cdot)l, e^{\mathbf{A}'T}l\Big)\right\|_{\mathbb{K}-1}.$$

We intend to clarify the item like $\|(a(\cdot), b)\|^2_{\mathbb{K}-1}$. To this end, one has to minimize the term

$$\|u(\cdot)\|^2_{\mathbf{K}} + 2\langle u(\cdot), U(\cdot)x_0\rangle + |x_0|^2_{\widetilde{P}} - 2\langle u(\cdot), a(\cdot)\rangle - 2x'_0 b.$$

The minimizer $(u^*(\cdot), x^*_0)$ equals $\mathbb{K}^{-1}(a(\cdot), b)$ and the minimum is equal to the value $-\|(a(\cdot), b)\|^2_{\mathbb{K}-1}$. Omitting routine operations, we come to relations

$$u^*(\cdot) = \widetilde{\mathbf{K}}^{-1}\widetilde{a}(\cdot), \quad \widetilde{a}(\cdot) = a(\cdot) - U(\cdot)\widetilde{P}^{-1}b, \quad x^*_0 = \widetilde{P}^{-1}(b - \langle U(\cdot), u^*(\cdot)\rangle),$$

$$\widetilde{\mathbf{K}}u(s) = \mathbf{K}u(s) - U(s)\widetilde{P}^{-1}\langle U(\cdot), u(\cdot)\rangle. \tag{24}$$

The value $\|(a(\cdot), b)\|^2_{\mathbb{K}-1} = |b|^2_{\widetilde{P}-1} + \|\widetilde{a}(\cdot)\|^2_{\widetilde{\mathbf{K}}-1}$. Note that the operator $\widetilde{\mathbf{K}}$ is coercive and invertible. The norm $\|\widetilde{\mathbf{K}}^{-1}\| \le 1$. If $P_0 > 0$ we have $\widetilde{\mathbf{K}} =$

$\mathrm{Id} + k^* \left(cc' + LP_0^{-1}L^*\right)^{-1} k$. Here Id is the identical operator in $L_2^{q-m}[0,T]$; the operator $k : L_2^{q-m}[0,T] \rightarrow L_2^m[0,T]$ has the form $ku(t) = \int_0^t K(t-s)D_1 u(s)ds$; the operator $L : \mathbb{R}^n \rightarrow L_2^m[0,T]$ is $Lx = L(\cdot)x$.

Summarizing the above reasonings we obtain

Theorem 2. *The AIS* $\mathbf{X}(t,y,P_0)$ *is a convex compact set under constraint* (4). *Its interior is nonempty under the additional assumption* $\widetilde{P} > 0$. *The support function of* $\mathbf{X}(t,y,P_0)$ *is given by the formula*

$$\rho(l \mid \mathbf{X}(T,y,P_0)) = \mathrm{E}\rho(l \mid X(T,y,P_0)) \quad \forall l \in \mathbb{R}^n, \tag{25}$$

with parameters from (20), (21), (22), *and* (23). *The inverse of the operator* $\mathbb{K}$ *can be calculated by formulas* (24).

The proof of the Theorem 2 is similar to the proof of the Theorem 1.

In conclusion of the section we specify one special case of systems (1) and (2). Suppose that

$$b = [B, O_{n \times m}], \quad c = [O_{m \times r}, I_m], \quad B \in \mathbb{R}^{n \times r}, \quad q = r + m. \tag{26}$$

Then $C = I_m$, $bc' \equiv 0$, $bD_1 = B$, and $bC_1 \equiv b$. In (26) and further we use designations from MATLAB where $[A; B]$ means the vertical concatenation of matrices A, B and $[A, B]$ means the horizontal concatenation.

2.2   Description of ISAV

In this subsection, assume that the compatible pairs $(x_0, v(\cdot))$ are nonrandom. We can write the following relations

$$\mathcal{G}^{-1}\overline{x}(t) = \mathcal{A}(t)x_0 + \int_0^t \mathcal{M}(t-s)\left(c'Cz(s) + D_1 u(s)\right)ds,$$

$$\mathcal{G}^{-1}x(t) = x(t) + \int_0^t \mathcal{M}(t-s)c'C\overline{G}x(s)ds, \quad \mathcal{A}(t) = \mathrm{E}e^{At}, \tag{27}$$

$$\mathcal{M}(t) = \mathrm{E}e^{At}b, \quad \overline{G} = \mathrm{E}G,$$

$$\int_0^T \left(\left|z(t) - \overline{G}\overline{x}(t)\right|_C^2 + |u(t)|^2\right) \leq 1, \tag{28}$$

for any compatible pair $(x_0, u(\cdot))$. Receiving equations (27), (28) is quite similar to system (6), (7). In the case of constraints (4) we have

$$|x_0 - \hat{x}_0|_{P_0}^2 + \int_0^T \left(\left|z(t) - \overline{G}\overline{x}(t)\right|_C^2 + |u(t)|^2\right) \leq 1. \tag{29}$$

Note that the equation $\mathcal{G}^{-1}x(t) = f(t)$ in (27) is the Volterra integral equation of convolution type. It has the unique solution $x(t) = \mathcal{G}f(t) = f(t) - \int_0^t R(t-s)f(s)ds$, where $R(t)$ is the continuous matrix resolvent, [12, sec. 7.1].

The subsequent reasonings is similar to previous ones. To reduce records we enter the notation

$$\tilde{z}(t) = z(t) - \overline{G} \int_0^t \overline{\mathcal{M}}(t-s)c'Cz(s)ds, \quad \tilde{z}_0(t) = \tilde{z}(t) - \overline{G}\overline{\mathcal{A}}(t)x_0, \quad \overline{x}_z^T$$
$$= \int_0^T \overline{\mathcal{M}}(T-s)c'Cz(s)ds, \quad \overline{\mathcal{M}}(t-s) = \mathcal{M}(t-s) - \int_s^t R(t-r)\mathcal{M}(r-s)dr,$$
$$\overline{x}_0^T = \overline{x}_z^T + \overline{\mathcal{A}}(T)x_0, \quad \overline{\mathcal{A}}(t) = \mathcal{A}(t) - \int_0^t R(t-s)\mathcal{A}(s)ds.$$
$$(30)$$

We see that inequality (28) has the form

$$\int_0^T \left(\left| \tilde{z}_0(t) - \overline{G} \int_0^t \overline{\mathcal{M}}(t-s)D_1u(s)ds \right|_C^2 + |u(t)|^2 \right) \le 1. \tag{31}$$

We define the linear self-adjoint operator $\mathcal{K}$ and the function $\mathbf{z}_0(s)$ by the relations

$$\mathcal{K}u(s) = \int_s^T D_1'\overline{\mathcal{M}}'(t-s)\overline{G}'C\overline{G}\int_0^t \overline{\mathcal{M}}(t-r)D_1u(r)drdt + u(s),$$
$$\mathbf{z}_0(s) = \int_s^T D_1'\overline{\mathcal{M}}'(t-s)\overline{G}'C\tilde{z}_0(t)dt. \tag{32}$$

We rewrite (31) as

$$\|\tilde{z}_0(\cdot)\|_C^2 - 2\langle \mathbf{z}_0(\cdot), u(\cdot) \rangle + \|u(\cdot)\|_{\mathcal{K}}^2 \le 1. \tag{33}$$

Let us calculate the expression

$$S(l,T,x_0,y) = \max_{u(\cdot)} l'\overline{x}(T) = l'\overline{x}_0^T + \max_{u(\cdot)} \left\langle D_1'\overline{\mathcal{M}}'(T-\cdot)l, u(\cdot) \right\rangle = l'\overline{x}_0^T$$
$$+ \left\langle D_1'\overline{\mathcal{M}}'(T-\cdot)l, \mathcal{K}^{-1}\mathbf{z}_0(\cdot) \right\rangle + \sqrt{(1 + \|\mathbf{z}_0(\cdot)\|_{\mathcal{K}^{-1}}^2 - \|\tilde{z}_0(\cdot)\|_C^2)} \tag{34}$$
$$\times \left\| D_1'\overline{\mathcal{M}}'(T-\cdot)l \right\|_{\mathcal{K}^{-1}}.$$

After that we obtain

$$\rho(l|\overline{X}(T,y)) = \max \left\{ S(l,T,x_0,y) \middle| x_0 \in X_0, \|\tilde{z}_0(\cdot)\|_C^2 - \|\mathbf{z}_0(\cdot)\|_{\mathcal{K}^{-1}}^2 \le 1 \right\}. \tag{35}$$

We come to the conclusion.

Theorem 3. *Let* $|e^{At}| \in L_\infty(\Omega, \mathcal{F}, P)$ $\forall t \in [0,T]$ *and* $|b|, |G| \in L_1(\Omega, \mathcal{F}, P)$. *Then ISAV* $\overline{X}(T,y)$ *is a convex compact set with the support function defined by relation (35), where parameters are given in (30), (31), (32), and (34). It is an ellipsoid if* $X_0 = \{x_0\}$.

If the set $\overline{X}(T,y)$ is always to be with the nonempty interior for arbitrary initial set X_0, it is enough to assume the *averaged controllability* in the following sense.

Definition 6. *System* (1), (9) *with constraints* (3) *is called* controllable *in average on* $[0,T]$ *if the symmetric matrix* $\overline{W}_T = \int_0^T \overline{\mathcal{M}}(T-t)D_1\mathcal{K}^{-1}(D_1'\overline{\mathcal{M}}'(T-t))dt$ *from (34) has a.s. full rank, i.e.* $\overline{W}_T > 0$ *a.s.*

Actually, Definition 6 means that the condition $D_1'\overline{\mathcal{M}}'(T-t)l = 0 \; \forall t \in [0,T]$ implies $l = 0$ if $l \in \mathbb{R}^n$. This is equivalent to the fact that the image of the operator $\int_0^T \overline{\mathcal{M}}(T-t)D_1u(t)dt$ equals $\mathbb{R}^n$. In other words, the integral equation (see also (27))

$$\mathcal{G}^{-1}x(t) = \int_0^t \mathcal{M}(t-s)D_1u(s)ds \qquad (36)$$

must be controllable from zero to any vector $x(T) \in \mathbb{R}^n$. Unfortunately, no rank controllability criterions for system (36) exist. Moreover, the expectation of the solution of differential equation (6), where $y(t) = z(t)$, does not coincide with the solution of the integral equation in (27).

Under constraint (4) we rewrite inequality (33) with the added term $|x_0 - \hat{x}_0|_{P_0}^2$ in the left side as

$$\|u(\cdot)\|_{\mathcal{K}}^2 + 2\left\langle u(\cdot), \overline{U}(\cdot)\right\rangle x_0 + |x_0|_{\overline{P}}^2 + |\hat{x}_0|_{P_0}^2 + \|\tilde{z}(\cdot)\|_C^2 - 2\left\langle u(\cdot), \mathbf{z}(\cdot)\right\rangle \\ -2x_0'\overline{Y}_0 \leq 1, \qquad (37)$$

where $\mathbf{z}(s) = \int_s^T D_1'\overline{\mathcal{M}}'(t-s)\overline{G}'C\tilde{z}(t)dt, \quad \overline{Y}_0 = P_0\hat{x}_0 + \left\langle \overline{G\mathcal{A}}(\cdot), C\tilde{z}(\cdot)\right\rangle,$

$$\overline{P} = P_0 + \overline{N}_T, \quad \overline{N}_T = \left\langle \overline{G\mathcal{A}}(\cdot), C\overline{G\mathcal{A}}(\cdot)\right\rangle,$$

$$\overline{U}(s) = \int_s^T D_1'\overline{\mathcal{M}}'(t-s)\overline{G}'C\overline{G\mathcal{A}}(t)dt. \qquad (38)$$

We note that the condition $\overline{N}_T > 0$ means the following: if the output $\overline{G\mathcal{A}}(t)x_0 = 0$, where $x_0 \in \mathbb{R}^n$, then $x_0 = 0$. This property is called *initial observability in average* for the system (27). Hereinafter, we suppose that $\overline{P} > 0$. Now, consider the random linear self-adjoint operator $\overline{\mathcal{K}}$ defined by the relation $\overline{\mathcal{K}}(u(\cdot), x_0) = \left(\mathcal{K}u(\cdot) + \overline{U}(\cdot)x_0, \overline{P}x_0 + \left\langle \overline{U}(\cdot), u(\cdot)\right\rangle\right)$. The operator $\overline{\mathcal{K}}$ has a bounded inverse one. It follows from Eq. (27) and relations (30), (37), and (38) that

$$\rho(l|\overline{X}(T, y, P_0)) = l'\overline{x}_z^T + \max_{u(\cdot), x_0} \left\{ \left\langle D_1'\overline{\mathcal{M}}'(T-\cdot)l, u(\cdot)\right\rangle + l'\overline{\mathcal{A}}(T)x_0 \right\}$$

$$= l'\overline{x}_z^T + \left\langle \left(D_1'\overline{\mathcal{M}}'(T-\cdot)l, \overline{\mathcal{A}}'(T)l\right), \overline{\mathcal{K}}^{-1}\left(\mathbf{z}(\cdot), \overline{Y}_0\right)\right\rangle_{L_2^{q-m}[0,T]\times\mathbb{R}^n}$$

$$+ \sqrt{1 - |\hat{x}_0|_{P_0}^2 - \|\tilde{z}(\cdot)\|_C^2 + \|(\mathbf{z}(\cdot), \overline{Y}_0)\|_{\overline{\mathcal{K}}^{-1}}^2}\left\| \left(D_1'\overline{\mathcal{M}}'(T-\cdot)l, \overline{\mathcal{A}}'(T)l\right)\right\|_{\overline{\mathcal{K}}^{-1}}. \qquad (39)$$

The inverse operator $\overline{\mathcal{K}}^{-1}$ can be found similarly to (24). Hence, we obtain

Theorem 4. *The ISAV $\overline{X}(t, y, P_0)$ is a nondegenerate ellipsoid under constraint (4) and with the additional condition $\overline{P} > 0$. The support function of $\overline{X}(t, y, P_0)$ is given by formula (39) with parameters from (37), (38).*

Consider the special case (26). Then relations (27), (28) have the form

$$\overline{x}(t) = \mathcal{A}(t)x_0 + \int_0^t \mathcal{M}(t-s)v(s)ds, \quad \mathcal{A}(t) = Ee^{At}, \quad \mathcal{M}(t) = Ee^{At}B, \quad (40)$$

$$\int_0^T \left(\left|z(t) - \overline{G}\overline{x}(t)\right|^2 + |v(t)|^2 \right) \le 1. \quad (41)$$

Here the integral equation is absent and notions of controllability in average and initial observability in average coincide with introduced early. Relations (32) and (33) has the form $\mathcal{K}v(s) = \int_s^T \mathcal{M}'(t-s)\overline{G}'\overline{G} \int_0^t \mathcal{M}(t-r)v(r)drdt + v(s)$, $\mathbf{z}_0(s) = \int_s^T \mathcal{M}'(t-s)\overline{G}'z_0(t)dt$, $z_0(t) = z(t) - \overline{G}\mathcal{A}(t)x_0$, $\|z_0(\cdot)\|^2 - 2\langle\mathbf{z}_0(\cdot), v(\cdot)\rangle + \|v(\cdot)\|_{\mathcal{K}}^2 \le 1$. Further we can repeat the reasonings in this section. We see that the only difficulty here is to find the inverse operators like $\mathcal{K}^{-1}$ and $\overline{K}^{-1}$. It may be done numerically.

3   Evolution of AIS

3.1   Ellipsoidal RIS

We begin with the set $\mathbf{X}(t, y, P_0)$ under conditions of the Theorem 2. Keeping in mind formula (25), we describe the evolution of parameters for the set $X(t, y, P_0)$. It is a random ellipsoid. We can use formulas (21), (22), (23), but these formulas not so good for evolution description. Let us use the method from [13]. Namely, consider the constraints

$$J(T, x_T, u, y) = |x_0 - \hat{x}_0|_{P_0}^2 + \int_0^T \left(|y(s) - Gx(s)|_C^2 + |u(s)|^2 \right) \le 1, \quad x(T) = x_T,$$
$$(42)$$

for equation (6), where $P_0' = P_0 \ge 0$ and $\hat{x}_0 \in L_2(\Omega, \mathcal{F}, P)$ is the fixed initial vector. The RIS for constraints (42) is denoted by $X(T, y, P_0)$ as before. We can replace T with arbitrary $t \in (0, T]$ and obtain the inequality with the functional $J(t, x_t, u, y)$. It is clear that the random vector $x_t \in X(t, y, P_0)$ iff $\min_{u(\cdot)} J(t, x_t, u, y) \le 1$.

Denote by $V(t, x) = \min_{u(\cdot)} J(t, x, u, y)$ the Bellman function for the minimization problem. We have the equation

$$V_t = \min_u \left\{ -V_x \left(\mathbf{A}x + \mathbf{b}y(t) + bD_1 u \right) + |y(t) - Gx|_C^2 + |u|^2 \right\}, \quad (43)$$
$$V(0, x) = |x - \hat{x}_0|_{P_0}^2.$$

The minimizer is equal to $u_{op}(t) = D_1'b'V_x'/2$, and we set $V(t, x) = x'P(t)x - 2x'd(t) + e(t)$. For parameters we obtain the equations

$$\dot{P} = -PA - A'P + G'CG - PbC_1b'P = -PA - A'P + (G + cb'P)'C(G$$
$$+cb'P) - Pbb'P; \quad \dot{d} = P(t)\mathbf{b}y(t) - A'd + G'Cy(t) - PbC_1b'd = -A'd$$
$$+(G + cb'P)'C(y(t) + cb'd) - Pbb'd; \quad \dot{e} = 2d'(t)\mathbf{b}y(t) \quad (44)$$
$$+|y(t)|_C^2 - d'(t)bC_1b'd(t); \quad P(0) = P_0, \quad d(0) = P_0\hat{x}_0, \quad e(0) = |\hat{x}_0|_{P_0}^2.$$

These equations describe the evolution of RIS and

$$X(t, y, P_0) = \left\{ x \mid |x|^2_{P(t)} - 2x'd(t) + e(t) \le 1 \right\}. \tag{45}$$

From the theory of Riccati equations, it is known that the condition $\widetilde{P} = N_T + P_0 > 0$ a.s. (see (21)) implies $P(t) > 0$ a.s. $\forall t \in (0, T]$. We suppose $\widetilde{P} > 0$ as before. Then it is more convenient to rewrite the formula for $X(t, y, P_0)$ as

$$X(t, y, P_0) = \left\{ x \mid |x - \hat{x}(t)|^2_{P(t)} + h(t) \le 1 \right\}, \tag{46}$$

where parameters $\hat{x} = P^{-1}d$, $h = e - |\hat{x}|^2_P$ satisfy the differential equations

$$\dot{\hat{x}} = \mathbf{A}\hat{x} + \mathbf{b}y(t) + P^{-1}(t)G'C(y(t) - G\hat{x}) = A\hat{x} + (bc' \\ + P^{-1}(t)G')C(y(t) - G\hat{x}), \quad \dot{h} = |y(t) - G\hat{x}(t)|^2_C. \tag{47}$$

Here $\hat{x}(0) = \hat{x}_0$ and $h(0) = 0$ if $P_0 > 0$. Otherwise, $h(0) = 0$ as usual, but $\hat{x}(0)$ is unknown. In that case, it is necessary to use Eqs. (44) and (45). The support function of set (46) is given by the equality

$$\rho(l|X(t, y, P_0)) = l'\hat{x}(t) + \sqrt{1 - h(t)}|l|_{P^{-1}(t)} \ \forall t \in (0, T], \tag{48}$$

which coincides with (23) if $t = T$, but looks more compact.

Let us consider one more limit case. Suppose that $X_0 = \{\hat{x}_0\}$. Surely, in that case $x_0^* = \hat{x}_0$. We take $P_0 = rI_n$ and pass r to ∞. Then we get the limit equation

$$\dot{P}^{-1} = \mathbf{A}P^{-1} + P^{-1}\mathbf{A}' - P^{-1}G'CGP^{-1} + bC_1b' \\ = AP^{-1} + P^{-1}A' - (GP^{-1} + cb')'C(GP^{-1} + cb') + bb', \quad P^{-1}(0) = 0, \tag{49}$$

with Eq. (47) which have the initial states $\hat{x}(0) = \hat{x}_0$ and $h(0) = 0$. Summarizing the above, we can append the Theorem 2.

Theorem 5. *Let the assumptions of the Theorem 2 be valid. Then the support function of AIS $\mathbf{X}(t, y, P_0)$ is given by the expression $\rho(l|\mathbf{X}(t, y, P_0)) = \mathbb{E}\rho(l|X(t, y, P_0)) \ \forall t \in (0, T]$, where the support function of the RIS $X(t, y, P_0)$ is defined by relation (48) along with equations (44), (45), (46), and (47). In the case $X_0 = \{\hat{x}_0\}$, one needs to use limit equation (49). Formula (48) coincides with (23) if $t = T$.*

3.2 Approximation of Non-Ellipsoidal AIS

The evolution of AIS with constraints (3) is more complicated due to max-operation in relations of the type (17). But we can approximate such sets by $X(t, y, P_0)$. Let us use here the following lemma.

Lemma 2. *Let two convex and closed sets of $M \subset \mathbb{R}^n$ and $A \subset \mathbb{R}^n$ be given such that $M \cap A = \varnothing$ while M is compact. Then there exists an ellipsoid $\mathcal{E}(Q, c)$, $Q > 0$, such that $\mathcal{E}(Q, c) \supset M$ and $\mathcal{E}(Q, c) \cap A = \varnothing$. If the set M is centrally symmetric with respect to zero, the vector $c = 0$.*

A proof of the Lemma 2 is given in [14].

For the initial set X_0, we take an arbitrary ellipsoid $\mathcal{E}(P_0^{-1}, \hat{x}_0) \supset X_0, P_0 > 0$, and consider the constraint

$$I(T, x_0, v, y, \alpha) = (1 - \alpha)|x_0 - \hat{x}_0|_{P_0}^2 + \alpha \int_0^T |v(t)|^2 dt \leq 1, \quad \alpha \in (0, 1),$$
$$\text{or} \quad J(T, x_T, u, y, \alpha) = (1 - \alpha)|x_0 - \hat{x}_0|_{P_0}^2 \qquad\qquad (50)$$
$$+\alpha \int_0^T \left(|y(s) - Gx(s)|_C^2 + |u(s)|^2\right) \leq 1, \quad x(T) = x_T.$$

Denote by $X_T^{\mathcal{E}, \alpha}(y)$ the RIS for constraint (50). For mentioned ellipsoids, by definition we have the inclusions

$$X_T^{\mathcal{E}, \alpha}(y) \supset X(T, y) \quad \text{and} \quad \mathbf{X}_T^{\mathcal{E}, \alpha}(y) \supset \mathbf{X}(T, y).$$

The RIS $X_T^{\mathcal{E}, \alpha}(y)$ can be represented by the equations like (46), (47):

$$X_T^{\mathcal{E}, \alpha}(y) = \left\{ x \in \mathbb{R}^n \mid |x - \hat{x}(T)|_{P(T)}^2 + h(T) \leq 1 \right\}; \quad \dot{P} = -PA - A'P$$
$$+\alpha G'CG - PbC_1 b'P/\alpha = -PA - A'P + ((\alpha G + cb'P)'C(\alpha G$$
$$+cb'P) - Pbb'P)/\alpha; \qquad\qquad (51)$$
$$\dot{\hat{x}} = A\hat{x} + \left(bc' + \alpha P^{-1}(t)G'\right)C(y(t) - G\hat{x}); \quad \dot{h}(t) = \alpha |y(t) - G\hat{x}(t)|_C^2;$$
$$h(0) = 0, \quad \hat{x}(0) = \hat{x}_0, \quad P(0) = (1 - \alpha)P_0.$$

Denote by $\mathbb{E}$ the family of ellipsoids $\mathcal{E}(P_0^{-1}, \hat{x}_0)$ with property $\mathcal{E}(P_0^{-1}, \hat{x}_0) \supset X_0$. The following lemma is valid.

Lemma 3. *Let us define the set* $\mathbf{X}_T^{\mathcal{E}}(y) = \bigcap_{\alpha \in (0,1)} \mathbf{X}_T^{\mathcal{E}, \alpha}(y)$. *Then*

$$\rho(l|\mathbf{X}_T^{\mathcal{E}}(y)) = \mathrm{E} \max \left\{ R(l, T, x_0, y) \Big| x_0 \in \mathcal{E}, \;\; \|\tilde{y}_0(\cdot)\|_C^2 - \|\mathbf{y}_0(\cdot)\|_{\mathbf{K}^{-1}}^2 \leq 1 \right\} \qquad (52)$$
$$\forall l \in \mathbb{R}^n.$$

Proof. If vector $\overline{x}_T \in \mathbf{X}_T^{\mathcal{E}}(y)$ then $\overline{x}_T \in \mathbf{X}_T^{\mathcal{E}, \alpha_k}(y)$, where the sequence

$$\alpha_k = \begin{cases} 1 - 1/(k+1), & \text{if } k \text{ odd}, \\ 1/k, & \text{if } k \text{ even}. \end{cases}$$

The corresponding random vector x_T is realized with some function $v_k(\cdot) \in \mathcal{L}_2^q[0, T]$ that belongs to a weakly compact set. For every α_k there exists a randomly compatible pair $(x_0^k, v_k(\cdot))$ (or $(x_0^k, u_k(\cdot))$) from weakly compact set such that it generates x_T and

$$I(T, x_0^k, v_k, y, \alpha_k) \leq 1 \quad \text{or} \quad J(T, x_T^k, u_k, y, \alpha_k) \leq 1, \qquad (53)$$

as $x_T \in X_T^{\mathcal{E}, \alpha_k}(y)$. Hence, thanks to separability of $\mathcal{L}_2^q[0, T]$ and without limiting generality, we can assume that the sequence $v_k(\cdot)$ (and $u_k(\cdot)$) weakly converges to a function $v^*(\cdot)$ ($u_k(\cdot) \to u^*(\cdot)$ weakly). Corresponding initial state x_0^k weakly converges to a vector x_0^* in the space $L_2^n(\Omega, \mathcal{F}, P)$. Passing to the limit by odd k in inequality (53), we obtain that $\|u^*(\cdot)\|^2 \leq 1$ a.s. due to lower semicontinuity of the norm $\|\cdot\|$. Similarly, passing to the limit by even k in inequality (53), we get that $|x_0^* - \hat{x}_0|_{P_0}^2 \leq 1$ a.s. or $x_0^* \in \mathcal{E}(P_0^{-1}, \hat{x}_0)$ a.s. So, for any ellipsoid $\mathcal{E} \in \mathbb{E}$ there exists the randomly compatible pair $(x_0^*, u^*(\cdot))$ generating x_T and such that inequality (15) is fulfilled with $x_0 = x_0^*$ and $u(\cdot) = u^*(\cdot)$. Therefore,

$$l'x_T \leq \max \left\{ R(l, T, x_0, y) \big| x_0 \in \mathcal{E}, \quad \|\widetilde{y}_0(\cdot)\|_C^2 - \|\mathbf{y}_0(\cdot)\|_{\mathbf{K}^{-1}}^2 \leq 1 \right\} \quad \forall l \in \mathbb{R}^n.$$

This means that the vector $\overline{x}_T$ belongs to the set defined by the right side of equality (52). On the other hand, if vector $\overline{x}_T$ is contained in this set, we have the maximization like in (16), (17) with $X_0 = \mathcal{E}$. Hence, $x_T \in X_T^{\mathcal{E}, \alpha}(y)$ and $\overline{x}_T \in \mathbf{X}_{l'}^{\mathcal{E}, \alpha}(y)$ for all $\alpha \in (0, 1)$. □

Theorem 6. *Assume that the probability space has a discrete type, i.e.* $\Omega = \mathbb{N}$, $\mathcal{F} = 2^{\mathbb{N}}$, $P(i) = p_i \geq 0$, *and* $\sum_{i=1}^{\infty} p_i = 1$. *The following equality is valid:* $\mathbf{X}(T, y) = \bigcap_{\mathcal{E} \in \mathbb{E}} \mathbf{X}_T^{\mathcal{E}}(y)$.

Proof. Define the sets $\mathcal{X}_0^i = \left\{ x_0^i \in \mathbb{R}^n \big| \|\widetilde{y}_0(\cdot)\|_C^2 - \|\mathbf{y}_0(\cdot)\|_{\mathbf{K}^{-1}}^2 \leq 1 \right\}$. Suppose that $\{x_0^i | x_0^i \in X_0\} \bigcap \{x_0^i | x_0^i \in \mathcal{X}_0^i\} = \varnothing$. Then thanks to the Hahn-Banach theorem we get $\sum_{i=1}^{\infty} p_i \rho(l_i | X_0) < d \leq \sum_{i=1}^{\infty} p_i l_i x_0^i$, $\forall x_0^i \in \mathcal{X}_0^i$, where $\sum_{i=1}^{\infty} p_i |l_i|^2 < \infty$ and at most one $|l_i| \neq 0$. By Lemma 2 there exist ellipsoids $\mathcal{E}_i \in \mathbb{E}$ such that $\rho(l_i | X_0) \leq \rho(l_i | \mathcal{E}_i) < d$. But for every $\mathcal{E}_j$ there exists a sequence $x_0^{ji} \in \mathcal{X}_0^i$ such that $\sum_{i=1}^{\infty} p_i l_i x_0^{ji} \leq \sum_{i=1}^{\infty} p_i \rho(l_i | \mathcal{E}_j)$. So, we come to contradiction: $\sum_{i=1}^{\infty} p_i \rho(l_i | X_0) \leq \sum_{i=1}^{\infty} p_i \rho(l_i | \mathcal{E}_i) < d \leq \sum_{i=1}^{\infty} p_i l_i x_0^{ii} \leq \sum_{i=1}^{\infty} p_i \rho(l_i | \mathcal{E}_i)$. Hence, for every vector $\overline{x}_T \in \bigcap_{\mathcal{E} \in \mathbb{E}} \mathbf{X}_T^{\mathcal{E}}(y)$ there exists a random compatible pair $(x_0, u(\cdot))$ such that $x_0 \in X_0$ and inequality (15) is valid. This means $\overline{x}_T \in \mathbf{X}(T, y)$. □

Remark 3. Using theorems on measurable selections we can prove the equality $\mathbf{X}(T, y) = \bigcap_{\mathcal{E} \in \mathbb{E}} \mathbf{X}_T^{\mathcal{E}}(y)$ in the general case. Indeed, if the set

$$\mathcal{X}_0 = \left\{ x_0 \in L_2^n(\Omega, \mathcal{F}, P) \big| \|\widetilde{y}_0(\cdot)\|_C^2 - \|\mathbf{y}_0(\cdot)\|_{\mathbf{K}^{-1}}^2 \leq 1 \, a.s. \right\}$$

does not intersect the set $X_0 = \left\{ x_0 \in L_2^n(\Omega, \mathcal{F}, P) | x_0 \in X_0 \, a.s. \right\}$, then thanks to the Hahn-Banach theorem we get $\int_{\Omega} \rho(l(\omega) | X_0) P(d\omega) < d \leq \int_{\Omega} l(\omega) x_0(\omega) \, \forall x_0 \in \mathcal{X}_0$. By Lemma 2 there exist a vector $c(\omega)$ and a matrix $P(\omega) > 0$ such that $\mathcal{E}(\omega) \in \mathbb{E}$ and $\rho(l(\omega) | X_0) \leq \rho(l(\omega) | \mathcal{E}(\omega)) = l'(\omega) c(\omega) + \sqrt{l'(\omega) P(\omega) l(\omega)} < d$. The vector and the matrix can be choosen $\mathcal{F}$-measurable. For every $\mathcal{E}(\omega)$ there exists measurable vector $x_0 \in \mathcal{X}_0$ such that $l'(\omega) x_0(\omega) \leq \rho(l(\omega) | \mathcal{E}(\omega))$. So, we get the contradiction $\int_{\Omega} \rho(l(\omega) | X_0) P(d\omega) \leq \int_{\Omega} \rho(l(\omega) | \mathcal{E}(\omega) P(d\omega) < d \leq \int_{\Omega} l'(\omega) x_0(\omega) P(d\omega) \leq \int_{\Omega} \rho(l(\omega) | \mathcal{E}(\omega)) P(d\omega)$.

3.3   Evolution of ISAV

Unfortunately, we cannot tell anything about evolution of ISAV in the general case. There is no natural ordinary differential equation describing the average

$\overline{x}(t) = Ex(t)$, except when the matrix A is independent of ω for which we have: $\dot{\overline{x}} = A\overline{x} + \overline{b}v(t)$, where $\overline{b} = Eb$ and $v(\cdot)$ is nonrandom. Then the problem is equivalent to the nonrandom one for systems

$$\dot{\overline{x}} = A\overline{x} + \overline{b}v(t), \quad \overline{z}(t) = \overline{G}\,\overline{x}(t) + cv(t), \quad \overline{G} = EG.$$

Corresponding deterministic theory was elaborated in [14].

In spite of that every nonrandom compatible pair $\{x_0, v(\cdot)\}$ is randomly compatible simultaneously, the sets AIS and ISAV are absolutely different. Actually, the ISAV unlike the AIS is built by deterministic rules.

There are other information sets. For example, assume that we can measure $\overline{y}(t)$ along with $y(t)$ and the pairs $\{x_0, v(\cdot)\}$ are nonrandom. Then we obtain one more set of compatible vectors $\overline{x}(T)$.

4   Examples and Special Cases

Example 1. Given is the one-dimensional system

$$\dot{x} = -ix + bv(t), \quad y(t) = x(t)/i + w(t), \quad \text{where} \quad \int_0^T \left(v^2(t) + w^2(t)\right) dt \le 1, \quad x_0 = 0.$$

Suppose that $\Omega = \mathbb{N}$ and $\sum_{i=1}^{\infty} p_i = 1$, $p_i \ge 0$, i.e. the probability space has a discrete type. Here we have the special cases of (26) and (49). Let us write limit equations (47) and (49) for the case $b = 1$:

$$\dot{\hat{x}} = -i\hat{x} + P^{-1}(y(t) - \hat{x}/i)/i, \quad \hat{x}(0) = 0; \quad \dot{h} = (y(t) - \hat{x}/i)^2, \quad h(0) = 0;$$
$$\dot{P}^{-1} = -2iP^{-1} - P^{-2}/i^2 + 1; \quad P^{-1}(0) = 0.$$

It may be verified that the function $P^{-1}(t) = i\sqrt{1 + i^4}\tanh(t\sqrt{1 + i^4}/i + s_i) - i^3$, where $s_i = \log(\sqrt{1 + i^4} + i^2)$, is the solution. Let the $y(t)$ be realized under nonrandom data: $v^*(t) = \sin(t)/\sqrt{2}$ and $w^*(t) = \cos(t)/\sqrt{2}$. Then $y(t) = (e^{-it} - \cos(t) + i\sin(t))/(i\sqrt{2}(i^2 + 1)) + \cos(t)/\sqrt{2}$. For the numerical calculation of AIS $\mathbf{X}(T, y)$, we set $T = 2$, $p_i = 1/16$ for $i \in 1 : 15$. Other p_i are arbitrary, but their sum equals $1/16$. We get $\mathbf{X}(T, y) \approx [-0.0652, 0.0800]$ with calculation error 0.0010 due to the unknown p_i, $i > 15$.

To avoid procedures of the numerical solution of integral equations, we calculate the set $\overline{X}(T, y)$ for this example in more simple form. Namely, assume that $b = 0$, but x_0 is unbounded. We use relations (40), (41). Here $\mathcal{A}(t) =\approx \sum_{i=1}^{15} e^{-it}/16$, $\overline{G} \approx \sum_{i=1}^{15}(16i)^{-1}$. The error is small. Let $T = 2$ and the signal $z(t)$ be generated by the pair $(1/2, w^*(t) \equiv 1/\sqrt{T})$. We get $\overline{X}(T, y) \approx [-0.8399, 1.0416]$.

4.1   The Finite Dimensional Probability Space

In this case, relations (1), (2) have the form

$$\dot{x}_i(t) = A_i x_i(t) + b_i v(t), \quad x_i(t) \in \mathbb{R}^n, \quad t \in [0, T], \tag{54}$$
$$y_i(t) = G_i x_i(t) + cv(t), \quad y_i(t) \in \mathbb{R}^m, \tag{55}$$

where $i \in 1 : d$. Here $\Omega = 1 : d$, $\mathcal{F} = 2^d$, $P(i) = p_i \geq 0$, and $\sum_{i=1}^d p_i = 1$. Now, Assumption 1 is fulfilled automatically. Thanks to the theorem of Cayley-Hamilton, one needs to verify the conditions of averaged controllability and observability only for finite number of objects (see also [15]). For the sake of example, we restrict the analysis to the case of $d = 2$ and $p_1 = p_2 = 1/2$.

Example 2. Consider the two-dimensional system:

$$\dot{x}_i = A_i x_i + bv(t), \quad y_i(t) = G_i x_i + cv(t), \quad A_1 = [0,1;0,0], \quad A_2 = [0,1;-1,0],$$

$$G_1 = [1,0], \quad G_2 = [0,1], \quad b = [0,0;1,0], \quad c = [1,1]; \quad x_i \in \mathbb{R}^2, \quad v(t) \in \mathbb{R}^2.$$

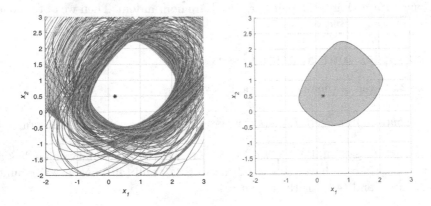

Fig. 1. Intersection of the sets $\mathbf{X}_T^{\mathcal{E},\alpha}(y)$ and the true AIS.

Constraints (3) are valid with the initial set $X_0 = \left\{ x \in \mathbb{R}^2 \mid |x^1| \leq 1, |x^2| \leq 1 \right\}$. Notation (5) has the form $\mathbf{b} = bc'C = [0;1/2]$, $\mathbf{A}_1 = A_1 - \mathbf{b}G_1 = [0,1;-1/2,0]$, $\mathbf{A}_2 = A_2 - \mathbf{b}G_2 = [0,1;-1,-1/2]$. We have $D_1 = [1;-1]/\sqrt{2}$. Let us illustrate the Theorem 6 choosing the signal for the first system generated by $x_0^* = [.5;-.5]$, $v^*(t) = .9[\sin(t);\cos(t)]/\sqrt{T}$, $T = 1$. For the second system the sygnal is generated by $x_0^* = [-.5;.5]$, $v^*(t) = .9[\cos(t);\sin(t)]/\sqrt{T}$, $T = 1$. We take the one-parametric family of ellipses with diagonal matrix $P_0 = [a,0;0,1-a]$ where $a \in (0,1)$. Then any ellipse $\mathcal{E} = \{x \mid x'P_0x \leq 1\}$ contains the square X_0 of initial states. Choose one more parameter $\alpha \in (0,1)$ and consider constraint (50), where $\hat{x}_0 = 0$. The RIS $X_T^{\mathcal{E},\alpha}(y)$ contains the set $X(T,y)$ for any signal. By (51), we have

$$X_T^{\mathcal{E},\alpha,i}(y) = \left\{ x \in \mathbb{R}^n \mid |x - \hat{x}_i(T)|_{P_i(T)}^2 + h_i(T) \leq 1 \right\},$$

$$\dot{P}_i = -P_i\mathbf{A}_i - \mathbf{A}_i'P_i + \alpha G_i'G_i/2 - P_ibC_1b'P_i/\alpha, \quad P_i(0) = (1-\alpha)P_0;$$

$$\dot{\hat{x}}_i = A_i\hat{x}_i + \left(bc' + \alpha P_i^{-1}(t)G_i'\right)(y(t) - G_i\hat{x}_i)/2, \quad \hat{x}_i(0) = 0;$$

$$\dot{h}_i(t) = \alpha |y(t) - G_i\hat{x}_i(t)|^2/2, \quad h_i(0) = 0.$$

The intersection $\bigcap \mathbf{X}_T^{\mathcal{E},\alpha}(y) \approx \mathbf{X}(T,y)$ and the true AIS are shown on Fig. 1. The pictures are obtained with the help of 25×25 grid for the set $\{(a,\alpha) \in (0,1) \times (0,1)\}$. The star is the mean value of the true state at the instant $T = 1$.

Conclusion

- In this work, we consider estimation of the guaranteed expectation for a flock of linear systems with random matrices and additive uncertain disturbances.
- The notions of averaged observability and controllability are used and the paper unites two techniques to study continuous-time linear systems described by the state equations with random coefficients and observation.
- The results of the paper may be used in robotics and the theory of control of aircrafts.
- We are going to extend some results of the paper to nonlinear systems and systems with the distributed parameters.

The work was performed as part of research conducted in the Ural Mathematical Center with the financial support of the Ministry of Science and Higher Education of the Russian Federation (Agreement number 075-02-2023-913).

References

1. Zuazua, E.: Averaged control. Automatica **50**(12), 3077–3087 (2014)
2. Lü, Q., Zuazua, E.: Averaged controllability for random evolution partial differential equations. J. Math. Pures Appl. **105**(3), 367–414 (2016)
3. Lohéac, J.: Zuazua, enrique: averaged controllability of parameter dependent conservative semigroups. J. Differ. Eqn. **262**(3), 1540–1574 (2017)
4. Gollamudi, S., Nagaraj, S., et al.: Set-membership state estimation with optimal bounding ellipsoids. Presented to Symposium on Inform. Theory and its Appl., Victoria, B.C., Canada (1996)
5. Kurzhanski, A., Varaiya, P.: Dynamics and Control of Trajectory Tubes: Theory and Computation, p. 445. Birkhäuser, SCFA, Boston (2014)
6. Liu, Y., Zhao, Y., Wu, F.: Extended ellipsoidal outer-bounding set-membership estimation for nonlinear discrete-time systems with unknown-but-bounded disturbances. In: Discrete Dynamics in Nature and Society Volume, pp. 1–11 (2016)
7. Campbell, S.L., Meyer, C.D. Jr.: Generalized Inverses of Linear Transformations. SIAM, pp. 269 (2009)
8. Rockafellar, R.T.: Convex Analysis. Princeton University Press, pp. xviii+451 (1970)
9. Ananyev, B., Yurovskikh, P.: Averaged estimation of the system flock with uncertainty. In: 16th International Conference on Stability and Oscillations of Nonlinear Control Systems (Pyatnitskiy's Conference - 2022). IEEE Xplore, P. 1–4 (2022)
10. Reed, M., B., Simon, B.: Methods of Modern Mathematical Physics. V. 1, Functional Analysis. Academic Press, p. 357 (1972)
11. Lohéac, Jérôme, Zuazua, Enrique: From averaged to simultaneous controllability. Ann. Fac. Sci. Toulouse, Math. (6), 25(4), 785–828 (2016)
12. Burton, T.A.: Volterra Integral and Differential Equations. Math. in Sci. and Eng., V. 202, 2nd edn. Elsevier, p. 353 (2005)
13. Bertsecas, D.P., Rhodes, I.B.: Recursive state estimation for a set-memberschip description of uncertainty. IEEE Trans. on Auto. Control, AC-16, no. 2, 117–128 (1971)

14. Ananyev, B.I., Yurovskikh, P.A.: On Estimation Problem with Separate Constraints on Initial States and Disturbances. Proc. of Inst. of Math. and Mech., UB of RAS, 28(1), 27–39 (2022)
15. Lazar, M.: Lohéac, Jérôme: output controllability in a long-time horizon. Automatica **113**, 108762 (2020)

Method for Solving a Differential Inclusion with a Subdifferentiable Support Function of the Right-Hand Side

Alexander Fominyh$^{(\boxtimes)}$ (iD)

Institute for Problems in Mechanical Engineering, Russian Academy of Sciences,
Saint Petersburg State University, Saint Petersburg, Russia
`alexfomster@mail.ru`

Abstract. The paper studies differential inclusions such that the support function of the set on the right-hand side considered at each time moment is subdifferentiable. It is required to find a trajectory satisfying the differential inclusion as well as the boundary conditions and some phase constraints. The applied problems are given where such systems arise. The problem is reduced to minimizing the functional. The superdifferentiability of this functional is proved, minimum conditions (in terms of superdifferential) are obtained. The superdifferential descent method is described; the convergence of the method to a stationary point of the functional is proved. An illustrative example is given.

Keywords: Differential inclusion · Support function · Superdifferential

1 Introduction

Differential inclusions are a powerful instrument for modeling dynamical systems. In this paper the differential inclusions of some special structure are explored. More specifically: the support function of the set on the right-hand side of an inclusion is a subdifferentiable (in phase variables) function of a special form. Such differential inclusions arise from systems of differential equations with discontinuous right-hand sides (see [1–3]) and from other practical problems (see an example below). It is required to find a trajectory of the differential inclusion satisfying the boundary conditions as well as some phase constraints.

Let us briefly enumerate the novelties of the paper explicitly: 1) a new algorithm for the differential inclusion considered is developed: the algorithm is not based on any kind of direct discretization of the initial problem, so it is a "continuous" method [4] that is qualitatively different compared to the majority of the existing "discrete" ones, 2) two main constructive ideas are combined in order to explore the problem stated: a) using support functions apparatus in order to

---

The main results of this paper (Sects. 5, 7 and 8) were obtained in IPME RAS and supported by Russian Science Foundation (project 20-71-10032).

M. Khachay et al. (Eds.): MOTOR 2023, LNCS 13930, pp. 345–361, 2023.
https://doi.org/10.1007/978-3-031-35305-5_24

consider the restriction in the form of differential inclusion [5], b) artificial "separation" of phase trajectories and their derivatives [6] in order to overcome some technical difficulties while calculating the supperdifferential required. Although these ideas were introduced earlier in the author's work applied to other problems, to the best of the author's knowledge their combination is implemented in the literature with regard to considered differential inclusions for the first time.

2   Basic Definitions and Notations

In the paper we will use the following notations. Let $C_n[0,T]$ denote the space of n-dimensional continuous on $[0,T]$ vector-functions. Let also $P_n[0,T]$ denote the space of piecewise continuous and bounded on $[0,T]$ n-dimensional vector-functions. We will also require the space $L_n^2[0,T]$ of square-summable on $[0,T]$ n-dimensional vector-functions. If X is some normed space, then $||\cdot||_X$ denotes its norm and X^* denotes the space conjugate to the space given.

For some arbitrary set $F \subset \mathcal{R}^n$ define the support function of the vector $\psi \in \mathcal{R}^n$ as $c(F,\psi) = \sup_{f \in F} \langle f, \psi \rangle$. Hereinafter in the paper we will use $F(x)$ to denote the right-hand side of the differential inclusion; so for the fixed trajectory $x(t)$ and for the fixed time moment t the set F is some subset in $\mathcal{R}^n$. Denote by S_n a unit sphere in $\mathcal{R}^n$ with the center in the origin, also let $B_r(c)$ be a ball with the radius $r \in \mathcal{R}$ and the center $c \in \mathcal{R}^n$. Let the vectors $\mathbf{e}_i$, $i = \overline{1,n}$, form the standard basis in $\mathcal{R}^n$. If $\varphi(x) = \min_{i=\overline{1,M}} f_i(x)$, where $f_i(x) : \mathcal{R}^n \to \mathcal{R}, i = \overline{1,M}$, are some functions, then we call the function $f_{\bar{i}}(x)$, $\bar{i} \in \{1,\dots,M\}$, an active one at the point $x_0 \in \mathcal{R}^n$, if $\bar{i} \in R(x_0) = \{i \in \{1,\dots,M\} \mid f_i(x_0) = \varphi(x_0)\}$.

Consider the space $\mathcal{R}^n \times \mathcal{R}^n$ with the standard norm. Let $d = [d_1, d_2] \in \mathcal{R}^n \times \mathcal{R}^n$ be an arbitrary vector. Suppose that at the point (x,z) there exists such a convex compact set $\overline{\partial}h(x,z) \subset \mathcal{R}^n \times \mathcal{R}^n$ that

$$\frac{\partial h(x,z)}{\partial d} = \lim_{\alpha \downarrow 0} \frac{1}{\alpha}\big(h(x + \alpha d_1, z + \alpha d_2) - h(x,z)\big) = \min_{w \in \overline{\partial}h(x,z)} \langle w, d \rangle. \quad (1)$$

In this case the function $h(x,z)$ is called superdifferentiable at the point (x,z), and the set $\overline{\partial}h(x,z)$ is called its superdifferential at this point.

Consider the space $C_n[0,T] \times P_n[0,T]$ with the norm $L_n^2[0,T] \times L_n^2[0,T]$. Let $g = [g_1, g_2] \in C_n[0,T] \times P_n[0,T]$ be an arbitrary vector-function. Suppose that at the point (x,z) there exists such a convex weakly* compact (compact in the weak-"star" topology) set $\overline{\partial}I(x,z) \subset \big(C_n[0,T] \times P_n[0,T], ||\cdot||_{L_n^2[0,T] \times L_n^2[0,T]}\big)^*$ that

$$\frac{\partial I(x,z)}{\partial g} = \lim_{\alpha \downarrow 0} \frac{1}{\alpha}\big(I(x + \alpha g_1, z + \alpha g_2) - I(x,z)\big) = \min_{w \in \overline{\partial}I(x,z)} w(g). \quad (2)$$

In this case the functional $I(x,z)$ is called superdifferentiable at the point (x,z) and the set $\overline{\partial}I(x,z)$ is called its superdifferential at this point.

3 Statement of the Problem

Consider the differential inclusion

$$\dot{x} \in F(x) \tag{3}$$

with the initial point

$$x(0) = x_0 \tag{4}$$

and with the desired endpoint

$$x_j(T) = x_{Tj}, \quad j \in J. \tag{5}$$

The differential inclusion is considered on the given finite time interval $[0, T]$. We assume that $x(t) \in C_n[0, T]$ is a piecewise continuously differentiable vector-function, so we may say that $\dot{x}(t) \in P_n[0, T]$. In formula (4) $x_0 \in \mathcal{R}^n$ is a given vector; in formula (5) x_{Tj} are given (these are the coordinates of the state vector which are fixed at the right endpoint), $J \subset \{1, \ldots, n\}$ is an index set.

We consider differential inclusions with the special structure, namely possessing both the following properties: 1) the set $F(x)$ on the right-hand side of (3) may be "splitted" into the sets $F_1(x), \ldots F_n(x)$ (such that $\dot{x}_i \in F_i(x), i = \overline{1, n}$) and each of the sets $F_1(x), \ldots F_n(x)$ is convex and compact at every $x \in R^n$; 2) the support functions of the corresponding sets $F_i(x)$, $i = \overline{1, n}$, on the right-hand side of (3) are of the form

$$c(F_i(x), \psi_i) = \sum_{j=1}^{r} \max \left\{ f_{i,j_1}(x)\psi_i, \ldots, f_{i,j_{k(j)}}(x)\psi_i \right\}$$

where $f_{i,j_1}(x), \ldots, f_{i,j_{k(j)}}(x)$, $i = \overline{1, n}$, $j = \overline{1, r}$ (for simplicity of presentation we suppose that r is the same one for each $i = \overline{1, n}$), are continuously differentiable functions. As it was noted in Introduction, such differential inclusions may arise while considering the movement of a discontinuous system along the "discontinuity" surface (we don't give details here for brevity); one meaningful practical problem of another type (leading to the similar structure of a differential inclusion), is considered in the section An example.

Let us also take into consideration the surface

$$s(x) = \mathbf{0_m} \tag{6}$$

where $s(x)$ is a known continuously differentiable m-dimensional vector-function.

We formulate the problem as follows: it is required to find such a trajectory $x^* \in C_n[0, T]$ (with the derivative $\dot{x}^* \in P_n[0, T]$) moving along surface (6), satisfying differential inclusion (3) while $t \in [0, T]$ and meeting boundary conditions (4), (5). Assume that there exists such a solution.

Remark 1. Instead of trajectories from the space $C_n[0, T]$ with derivatives from the space $P_n[0, T]$ one may consider absolutely continuous trajectories on the interval $[0, T]$ with measurable and almost everywhere bounded derivatives on $[0, T]$ what is more natural for differential inclusions. The choice of the solution space in the paper is explained by the possibility of its practical construction.

4   Reduction to a Variational Problem

Insofar as for all $x \in \mathcal{R}^n$ the set $F_i(x)$, $i = \overline{1,n}$, is a convex compact set in $\mathcal{R}^n$ by assumption, then inclusion (3) may be rewritten as follows [7]:

$$\dot{x}_i(t)\psi_i(t) \leq c(F_i(x(t)), \psi_i(t)) \quad \text{for all } \psi_i(t) \in S_1, \quad \text{for all } t \in [0,T], \quad i = \overline{1,n}.$$

Denote $z(t) = \dot{x}(t)$, $z \in P_n[0,T]$, then from (4) one has

$$x(t) = x_0 + \int_0^t z(\tau)d\tau. \tag{7}$$

Let us now realize the following idea (see [6]). "Forcibly" consider the points z and x to be "independent" variables. Since, in fact, there is relationship (7) between these variables (which naturally means that the vector-function $z(t)$ is a derivative of the vector-function $x(t)$), let us take this restriction into account by using the functional

$$v(x,z) = \frac{1}{2}\int_0^T \left(x(t) - x_0 - \int_0^t z(\tau)d\tau \right)^2 dt.$$

Besides, it is seen that condition (4) on the left endpoint is also satisfied if $v(x,z) = 0$.

For $i = \overline{1,n}$ put

$$\ell_i(\psi_i, x, z) = \langle z_i, \psi_i \rangle - c(F_i(x), \psi_i),$$

$$h(x,z) = (h_1(x,z), \ldots, h_n(x,z))', \quad h_i(x,z) = \max_{\psi_i \in S_1} \max\{0, \ell_i(\psi_i, x, z)\},$$

and construct the functional

$$\varphi(x,z) = \frac{1}{2}\int_0^T h^2(x(t), z(t))dt.$$

It is not difficult to check that inclusion (3) takes place iff $\varphi(x,z) = 0$. Introduce the functional

$$\chi(z) = \frac{1}{2}\sum_{j \in J} \left(x_{0j} + \int_0^T z_j(t)dt - x_{Tj} \right)^2.$$

We see that if $v(x,z) = 0$, then condition (5) on the right endpoint is satisfied iff $\chi(z) = 0$.

Introduce the functional

$$\omega(x) = \frac{1}{2}\int_0^T s^2(x(t))dt.$$

Obviously, the trajectory $x(t)$ belongs to surface (6) at each $t \in [0,T]$ iff $\omega(x) = 0$.

Finally construct the functional

$$I(x, z) = \varphi(x, z) + \chi(z) + \omega(x) + \upsilon(x, z). \tag{8}$$

So the original problem has been reduced to minimizing functional (8) on the space $C_n[0, T] \times P_n[0, T]$. Denote by (x^*, z^*) a global minimizer of this functional. Then $x^*(t) = x_0 + \int_0^t z^*(\tau)d\tau$ is a solution of the initial problem.

Despite the fact that the dimension of the functional $I(x, z)$ arguments is n more the dimension of the initial problem (i.e. the dimension of the desired point x^*), the structure of its superdifferential (in the space $C_n[0, T] \times P_n[0, T]$ as a normed space with the norm $L_n^2[0, T] \times L_n^2[0, T]$), as will be seen from what follows, has a rather simple form. This fact will allow us to construct a numerical method for solving the original problem.

5   Minimum Conditions of the Functional $I(x, z)$

By superdifferential calculus rules [8] one may put

$$\overline{\partial}\chi(z) = \left\{ \sum_{j \in J} \left(x_{0j} + \int_0^T z_j(t)dt - x_{Tj} \right) \mathbf{e_j} \right\}, \tag{9}$$

$$\overline{\partial}\omega(x) = \left\{ \sum_{i=1}^m s_i(x(t)) \frac{\partial s_i(x(t))}{\partial x} \right\}, \tag{10}$$

$$\overline{\partial}\upsilon(x, z) = \left\{ \left(\begin{array}{c} x(t) - x_0 - \int_0^t z(\tau)d\tau \\ -\int_t^T \left(x(\tau) - x_0 - \int_0^\tau z(s)ds \right) d\tau \end{array} \right) \right\}. \tag{11}$$

Explore the differential properties of the functional $\varphi(x, z)$. Let the value ψ_i^* be such that $\max\{0, \ell_i(\psi_i^*(x, z), x, z)\} = \max_{\psi_i \in S_1} \max\{0, \ell_i(\psi_i, x, z)\}$, $i = \overline{1, n}$. In [5] it is shown that if $h_i(x, z) > 0$, then $\psi_i^*(x, z)$, $i = \overline{1, n}$, is unique and continuous in (x, z).

For simplicity, consider the case $n = 2$, $r = 1$, $k(1) = 2$, and only the functions $\ell_1(\psi_1, x, z)$ and $h_1(x, z)$ (here we denote them by $\ell(\psi_1, x_1, x_2, z_1)$ and $h(x_1, x_2, z_1)$ respectively) and the time interval $[t_1, t_2] \subset [0, T]$ of nonzero length; the general case may be considered in a similar way. Put $f_1 := f_{1,1_1}$, $f_2 := f_{1,1_2}$, then we have $\ell(\psi_1, x_1, x_2, z_1) = z_1 \psi_1 - \max\{f_1(x)\psi_1, f_2(x)\psi_1\}$. Fix some point $(x_1, x_2) \in \mathcal{R}^2$. Let $f_1(x(t))\psi_1(t) = f_2(x(t))\psi_1(t)$ at $t \in [t_1, t_2]$; other cases may be studied in a completely analogous fashion.

a) Suppose that $h_1(x, z) > 0$, i.e. $h_1(x, z) = \max_{\psi_1 \in S_1} \ell_1(\psi_1, x, z) > 0$.

Our aim is to apply the corresponding theorem on a directional differentiability from [9]. The theorem of this book considers the inf-functions so we will apply this theorem to the function $-\ell(\psi_1, x_1, x_2, z_1)$. Check that the function $h(x_1, x_2, z_1)$ satisfies the following conditions:

i) the function $\ell(\psi_1, x_1, x_2, z_1)$ is continuous on $S_1 \times \mathcal{R}^2 \times \mathcal{R}$;

ii) there exist a number β and a compact set $C \in \mathcal{R}$ such that for every $(\overline{x}_1, \overline{x}_2, \overline{z}_1)'$ in the vicinity of the point $(x_1, x_2, z_1)'$ the level set

$$lev_\beta(-\ell(\cdot, \overline{x}_1, \overline{x}_2, \overline{z}_1)) = \{\psi_1 \in S_1 \mid -\ell(\psi_1, \overline{x}_1, \overline{x}_2, \overline{z}_1) \leq \beta\}$$

is nonempty and is contained in the set C;

iii) for any fixed $\psi_1 \in S_1$ the function $\ell(\psi_1, \cdot, \cdot, \cdot)$ is directionally differentiable at the point $(x_1, x_2, z_1)'$;

iv) if $d = [d_1, d_2] \in \mathcal{R}^2 \times \mathcal{R}$, $\gamma_n \downarrow 0$ and ψ_{1_n} is a sequence in C, then ψ_{1_n} has a limit point $\overline{\psi}_1$ such that

$$\lim_{n \to \infty} \sup \frac{-\ell(\psi_{1_n}, x + \gamma_n d_1, z + \gamma_n d_2) - (-\ell(\psi_{1_n}, x, z))}{\gamma_n} \geq \frac{\partial(-\ell(\overline{\psi}_1, x, z))}{\partial d}$$

where $\dfrac{\partial \ell(\overline{\psi}_1, x_1, x_2, z_1)}{\partial d}$ is the derivative of the function $\ell(\overline{\psi}_1, x_1, x_2, z_1)$ at the point $(x_1, x_2, z_1)'$ in the direction d.

The verification of conditions i), ii) is obvious.

In order to verify the condition iii), it is sufficient to observe that for the fixed $\psi_1 \in S_1$ the function $-\max\{f_1(x)\psi_1, f_2(x)\psi_1\}$ is superdifferentiable [8] (hence, it is differentiable in directions) at the point $(x_1, x_2)'$; herewith, its superdifferential at this point is $\text{co}\left\{\left(\psi_1 \dfrac{\partial f_1(x)}{\partial x_1}, \psi_1 \dfrac{\partial f_1(x)}{\partial x_2}\right)', \left(\psi_1 \dfrac{\partial f_2(x)}{\partial x_1}, \psi_1 \dfrac{\partial f_2(x)}{\partial x_2}\right)'\right\}$. An explicit expression of this function derivative at the point $(x_1, x_2)'$ in the direction d_1 is $-\max\left\{\left\langle \psi_1 \dfrac{\partial f_1(x)}{\partial x}, d_1 \right\rangle, \left\langle \psi_1 \dfrac{\partial f_2(x)}{\partial x}, d_1 \right\rangle\right\}$.

Finally, check condition iv). Let $[d_1, d_2] \in \mathcal{R}^2 \times \mathcal{R}$, $\gamma_n \downarrow 0$ and ψ_{1_n} is some sequence from the set C. Calculate

$$\lim_{n \to \infty} \sup \frac{-\ell(\psi_{1_n}, x_1 + \gamma_n d_{1,1}, x_2 + \gamma_n d_{1,2}, z_1 + \gamma_n d_2) - (-\ell(\psi_{1_n}, x_1, x_2, z_1))}{\gamma_n}$$

$$= \lim_{n \to \infty} \sup \frac{1}{\gamma_n}\left(-(z_1 + \gamma_n d_2)\psi_{1_n}\right.$$

$$+ \max\left\{f_1(x_1 + \gamma_n d_{1,1}, x_2 + \gamma_n d_{1,2})\psi_{1_n}, f_2(x_1 + \gamma_n d_{1,1}, x_2 + \gamma_n d_{1,2})\psi_{1_n}\right\}$$

$$\left. + z_1\psi_{1_n} - \max\left\{f_1(x_1, x_2)\psi_{1_n}, f_2(x_1, x_2)\psi_{1_n}\right\}\right)$$

$$= \lim_{n \to \infty} \sup \frac{1}{\gamma_n}\left(-\gamma_n d_2\psi_{1_n} + \max\left\{\left[f_1(x) + \gamma_n\left\langle \frac{\partial f_1(x)}{\partial x}, d_1\right\rangle + o_1(\gamma_n, x, d)\right]\psi_{1_n},\right.\right.$$

$$\left. \left[f_2(x) + \gamma_n\left\langle \frac{\partial f_2(x)}{\partial x}, d_1\right\rangle + o_2(\gamma_n, x, d)\right]\psi_{1_n}\right\}$$

$$\left. - \max\left\{f_1(x_1, x_2)\psi_{1_n}, f_2(x_1, x_2)\psi_{1_n}\right\}\right)$$

$$\geq \lim_{n\to\infty} \sup \frac{1}{\gamma_n}\left(-\gamma_n d_2\psi_{1_n} + \gamma_n \max\left\{\left\langle \frac{\partial f_1(x)}{\partial x}, d_1\right\rangle\psi_{1_n}, \left\langle\frac{\partial f_2(x)}{\partial x}, d_1\right\rangle\psi_{1_n}\right\}\right.$$

$$\left. + \min\left\{o_1(\gamma_n, x, d)\psi_{1_n}, o_2(\gamma_n, x, d)\psi_{1_n}\right\}\right)$$

where the last inequality follows from the assumption $f_1(x)\psi_1 = f_2(x)\psi_1$ (on the time interval considered) and from the corresponding property of the maximum of two functions [10] in the case when this assumption is satisfied.

Let $\overline{\psi}_1$ be a limit point of the sequence ψ_{1_n}. Then by the directional derivative definition we have

$$\frac{\partial(-\ell(\overline{\psi}_1, x, z))}{\partial d} = -d_2\overline{\psi}_1 + \max\left\{\left\langle \overline{\psi}_1\frac{\partial f_1(x)}{\partial x}, d_1\right\rangle, \left\langle\overline{\psi}_1\frac{\partial f_2(x)}{\partial x}, d_1\right\rangle\right\}.$$

From the last two relations one concludes that condition iv) is fulfilled.

Thus, the function $h(x_1, x_2, z_1)$ satisfies conditions i)-iv) so it is differentiable in directions at the point (x_1, x_2, z_1) [9] and its derivative in the direction d at this point is expressed by the formula

$$\frac{\partial h(x_1, x_2, z_1)}{\partial d} = \sup_{\psi_1 \in \mathcal{S}(x_1, x_2, z_1)} \frac{\partial \ell(\psi_1, x_1, x_2, z_1)}{\partial d}$$

where $\mathcal{S}(x_1, x_2, z_1) = \arg\max_{\psi_1 \in S_1} \ell(\psi_1, x_1, x_2, z_1)$. However, as shown above, in the problem considered the set $\mathcal{S}(x_1, x_2, z_1)$ consists of the only element $\psi_1^*(x_1, x_2, z_1)$, hence

$$\frac{\partial h(x_1, x_2, z_1)}{\partial d} = \frac{\partial \ell(\psi_1^*(x_1, x_2, z_1), x_1, x_2, z_1)}{\partial d}.$$

Finally, recall that by the directional derivative definition one has the equality

$$\frac{\partial(-\ell(\psi_1^*, x, z))}{\partial d} = -d_2\psi_1^* + \max\left\{\left\langle\psi_1^*\frac{\partial f_1(x)}{\partial x}, d_1\right\rangle, \left\langle\psi_1^*\frac{\partial f_2(x)}{\partial x}, d_1\right\rangle\right\}$$

where we have put $\psi_1^* := \psi_1^*(x_1, x_2, z_1)$.

From last two expressions we finally conclude that the function $h(x_1, x_2, z_1)$ is superdifferentiable at the point $(x_1, x_2, z_1)'$ but it is also positive in the case considered so the function $h^2(x_1, x_2, z_1)$ is superdifferentiable at the point $(x_1, x_2, z_1)'$ as well as the square of a superdifferentiable positive function (see book [8]).

b) In the case $h_1(x, z) = 0$ it is obvious that the function $h^2(x_1, x_2, z_1)$ is differentiable at the point $(x_1, x_2, z_1)'$ and its gradient vanishes at this point.

Now turn back to the functional $\varphi(x, z)$. At first give the formulas for calculating the superdifferential $\overline{\partial}h^2(x, z)$ at the point (x, z). At $i = \overline{1, n}$ one has

$$\overline{\partial}\left(\frac{1}{2}h_i^2(x, z)\right) = h_i(x, z)\left(\psi_i^*\mathbf{e}_{i+n} + \sum_{j=1}^{r}\overline{\partial}\left(-\max\left\{f_{i,j_1}(x)\psi_i^*, \ldots, f_{i,j_{k(j)}}(x)\psi_i^*\right\}\right)\right)$$

$$(12)$$

where at $j = \overline{1,r}$ and $j_p \in R_{ij}(x)$ we have

$$\overline{\partial}\left(-\max\left\{f_{i,j_1}(x)\psi_i^*, \ldots, f_{i,j_{k(j)}}(x)\psi_i^*\right\}\right) = \operatorname{co}\left\{\left[\psi_i^* \frac{\partial f_{i,j_p}(x)}{\partial x}, \mathbf{0}_n\right]\right\}, \quad (13)$$

$$R_{ij}(x) = \left\{j_p \in \{j_1, \ldots, j_{k(j)}\} \mid f_{i,j_p}(x)\psi_i^* = \max\left\{f_{i,j_1}(x)\psi_i^*, \ldots, f_{i,j_{k(j)}}(x)\psi_i^*\right\}\right\}.$$

Calculate the derivative of the function $\frac{1}{2}h_i^2(x, z)$, $i = \overline{1,n}$, in the direction $g = [g_1, g_2] \in \mathcal{R}^n \times \mathcal{R}^n$. By virtue of (12), (13) and superdifferential calculus rules [10] at $i = \overline{1,n}$, we have

$$\frac{\partial\left(\frac{1}{2}h_i^2(x, z)\right)}{\partial g} = h_i(x, z)\left(\psi_i^* g_{2i} + \sum_{j=1}^{r} \min_{j_p \in R_{ij}(x)}\left\langle -\psi_i^* \frac{\partial f_{i,j_p}(x)}{\partial x}, g_1\right\rangle\right). \quad (14)$$

Show that the functional $\varphi(x, z)$ is superdifferentiable and that its superdifferential is determined by the corresponding integrand superdifferential.

Theorem 1. *Let the interval $[0, T]$ may be divided into a finite number of intervals in every of which one (several) of the functions $\{f_{i,j_1}(x)\psi_i, \ldots, f_{i,j_{k(j)}}(x)\psi_i\}$, $i = \overline{1,n}$, $j = \overline{1,r}$, is (are) active. Then the functional $\varphi(x, z)$ is superdifferentiable, i.e.*

$$\frac{\partial\varphi(x, z)}{\partial g} = \lim_{\alpha\downarrow 0} \frac{1}{\alpha}\left(\varphi(x + \alpha g_1, z + \alpha g_2) - \varphi(x, z)\right) = \min_{w\in\overline{\partial}\varphi(x,z)} \int_0^T \langle w(t), g(t)\rangle dt$$

$$(15)$$

where $g = [g_1, g_2] \in C_n[0, T] \times P_n[0, T]$ and the set $\overline{\partial}\varphi(x, z)$ is defined as follows

$$\overline{\partial}\varphi(x, z) = \left\{w = [w_1, w_2] \in L_n^\infty[0, T] \times L_n^\infty[0, T] \mid \right. \quad (16)$$

$$\left. [w_1(t), w_2(t)] \in \overline{\partial}\left(\frac{1}{2}h^2(x(t), z(t))\right) \quad \text{for a.e. } t \in [0, T]\right\}.$$

Proof. In accordance with definition (2) of a superdifferentiable functional, in order to prove the theorem, one has to check that

1) the derivative of the functional $\varphi(x, z)$ in the direction g is actually of form (15),
2) herewith, the set $\overline{\partial}\varphi(x, z)$ is convex and weakly* compact subset of the space $\left(C_n[0, T] \times P_n[0, T], \|\cdot\|_{L_n^2[0,T]\times L_n^2[0,T]}\right)^*$.

Let us give just the scheme of proof for brevity:

1) With the use of the superdifferential definition (see formula (1)), Lebesgue's dominated convergence theorem (making use of the mean value theorem [11] applied to the superdifferential) and Filippov lemma, we first show that the direction derivative of the functional $\varphi(x, z)$ is of form (15).

2) The corresponding proof for the analogous functional may be found in [6].

Using formulas (8), (9), (10), (11), (16) obtained and superdifferential calculus rules [12] we have the following final formula for calculating the superdifferential of the functional $I(x, z)$ at the point (x, z):

$$\overline{\partial}I(x, z) = \overline{\partial}\varphi(x, z) + \overline{\partial}\overline{\chi}(x, z) + \overline{\partial}\overline{\omega}(x, z) + \overline{\partial}\upsilon(x, z), \tag{17}$$

where formally $\overline{\chi}(x, z) := \chi(z)$, $\overline{\omega}(x, z) := \omega(x)$.

Using the known minimum condition [12] of the functional $I(x, z)$ at the point (x^*, z^*) in terms of superdifferential, we obtain the following theorem.

Theorem 2. *Let the interval* $[0, T]$ *may be divided into a finite number of intervals, in every of which one (several) of the functions* $\{f_{i,j_1}(x)\psi_i, \ldots, f_{i,j_{k(j)}}(x)\psi_i\}$, $i = \overline{1, n}$, $j = \overline{1, r}$, *is (are) active. In order for the point* (x^*, z^*) *to minimize the functional* $I(x, z)$, *it is necessary to have*

$$\overline{\partial}I(x^*(t), z^*(t)) = \{0_{2n}\} \tag{18}$$

at each $t \in [0, T]$ *where the expression for the superdifferential* $\overline{\partial}I(x, z)$ *is given by formula (17). If one has* $I(x^*, z^*) = 0$, *then condition (18) is also sufficient.*

6   Constructing the Superdifferential Descent Direction of the Functional $I(x, z)$

In this section we consider only the points (x, z) which do not satisfy minimum condition in Theorem 2. Our aim here is to find the superdifferential (or the steepest) descent direction of the functional $I(x, z)$ at the point (x, z). Denote this direction by $G(x, z)$. Herewith, $G = [G_1, G_2] \in L_n^2[0, T] \times L_n^2[0, T]$. In order to construct the vector-function $G(x, z)$, consider the problem

$$\max_{w \in \overline{\partial}I(x,z)} \|w\|_{L_n^2[0,T] \times L_n^2[0,T]}^2 = \max_{w \in \overline{\partial}I(x,z)} \int_0^T w^2(t)dt. \tag{19}$$

Denote by $\overline{w}$ the solution of this problem (below we will see that such a solution exists). The vector-function $\overline{w}$, of course, depends on the point (x, z) but we omit this dependence in the notation for brevity. Then one can check that the vector-function

$$G(x(t), z(t), t) = -\frac{\overline{w}(x(t), z(t), t)}{\|\overline{w}\|_{L_{2n}^2[0,T]}} \tag{20}$$

is a superdifferential descent direction of the functional $I(x, z)$ at the point (x, z) (cf. (25), (26) below). Recall that we are seeking the direction $G(x, z)$ in the case when the point (x, z) does not satisfy minimum condition (18) so $\|\overline{w}\|_{L_{2n}^2[0,T]} > 0$.

It is easy to check that in this case the solution of problem (19) is such a selector of the multivalued mapping $t \to \overline{\partial}I(x(t), z(t), t)$ that maximizes the distance from zero to the points of the set $\overline{\partial}I(x(t), z(t), t)$ at each time moment $t \in [0, T]$. In other words, to solve problem (19) means to solve the problem

$$\max_{w(t) \in \overline{\partial}I(x(t), z(t), t)} w^2(t) \qquad (21)$$

for each $t \in [0, T]$ (see the justification of an analogous statement in paper [6]).

From formula (17) (see also (12)) we see that the superdifferential $\overline{\partial}I(x(t), z(t))$ is a convex polyhedron $P(t) \subset \mathcal{R}^{2n}$. Herewith, of course, the set $P(t)$ depends on the point (x, z). We will omit this dependence in the notation of this paragraph for simplicity. Thus, problem (21) at each fixed $t \in [0, T]$ is a finite-dimensional problem of finding the maximal distance from zero to the convex polyhedron points. It is clear that in this case it is sufficient to go over all the vertexes $p_j(t)$, $j = \overline{1, s}$ (here s is a number of vertexes of the polyhedron $P(t)$) and choose among the values $\|p_j(t)\|_{\mathcal{R}^{2n}}$ the largest one. Denote the corresponding vertex by $p_{\overline{j}}(t)$ ($\overline{j} \in \{1, \ldots, s\}$), and if there are several vertexes on which the maximal norm-value is achieved, then choose any of them. Finally, put $\overline{w}(t) = p_{\overline{j}}(t)$.

Further, one makes a (uniform) splitting of the interval $[0, T]$, and the problem is being solved for every point of the splitting (as it is described in the previous paragraph), i.e. one has to calculate $G(x(t_i), z(t_i), t_i)$ where $t_i \in [0, T]$, $i = \overline{1, N}$, are the points of discretization (see notation in Lemma 1 below). Under some natural additional assumption Lemma 1 (see [6]) below guarantees that the vector-function obtained with the help of piecewise linear interpolation of the superdifferential descent directions evaluated at every point of such a splitting of the interval $[0, T]$ converges in the space $L_{2n}^2[0, T]$ to the vector-function $G(x(t), z(t), t)$ sought when the discretization rank tends to infinity.

Lemma 1. *Let the function $v \in L_1^\infty[0, T]$ satisfy the following condition: for every $\overline{\delta} > 0$ the function $v(t)$ is piecewise continuous on the set $[0, T]$ with the exception of only the finite number of the intervals $(\overline{t}_1(\overline{\delta}), \overline{t}_2(\overline{\delta})), \ldots, (\overline{t}_r(\overline{\delta}), \overline{t}_{r+1}(\overline{\delta}))$ whose union length does not exceed the value $\overline{\delta}$.*

Choose the (uniform) finite splitting $t_1 = 0, t_2, \ldots, t_{N-1}, t_N = T$ of the interval $[0, T]$ and calculate the values $v(t_i)$, $i = \overline{1, N}$, at these points. Let $L(t)$ be the function obtained via piecewise linear interpolation with the nodes $(t_i, v(t_i))$, $i = \overline{1, N}$. Then for every $\varepsilon > 0$ there exists such a number $\overline{N}(\varepsilon)$ that for every $N > \overline{N}(\varepsilon)$ one has $\|L - v\|_{L_1^2[0, T]}^2 \le \varepsilon$.

At a qualitative level Lemma 1 condition means that the sought function does not have "too many" points of discontinuity on the interval $[0, T]$. If this condition is satisfied for the vector-function $\overline{w}(t)$ (what it natural for applications), then this lemma justifies the approximation of the vector-function $\overline{w}(t)$ and hence, the approximation of the vector-function $\overline{G}(x(t), z(t), t)$, by the values $\overline{w}(t_i)$, $i = \overline{1, N}$, at the separate points of discretization implemented as described above. Lemma 1 also means that during the algorithm realization we approximate the steepest descent direction sought with the polygonal function.

7 On a Method for Finding the Stationary Points of the Functional $I(x, z)$

The simplest steepest (superdifferential) descent algorithm is used for numerical simulations of the paper. In order to convey the main ideas, we firstly consider the convergence of this method for an analogous problem in a finite-dimensional case (which is of an independent interest); and then turn to a more general problem considered in this paper.

First consider the minimization problem of a function which is a minimum of the finite number of continuously differentiable functions. So let

$$\varphi(x) = \min_{i=\overline{1,M}} f_i(x)$$

where $f_i(x)$, $i = \overline{1, M}$, are continuously differentiable functions on $\mathcal{R}^n$.

It is known [10] that the function $\varphi(x)$ is differentiable at every point $x_0 \in \mathcal{R}^n$ in any direction $g \in \mathcal{R}^n$, $||g||_{\mathcal{R}^n} = 1$, and

$$\frac{\partial \varphi(x_0)}{\partial g} = \min_{i \in R(x_0)} \left\langle \frac{\partial f_i(x_0)}{\partial x}, g \right\rangle = \min_{w \in \overline{\partial} \varphi(x_0)} \langle w, g \rangle, \qquad (22)$$

$$R(x_0) = \{ i \in \{1, \dots, M\} \mid f_i(x_0) = \varphi(x_0) \},$$

$$\overline{\partial} \varphi(x_0) = \text{co} \left\{ \frac{\partial f_i(x_0)}{\partial x}, \ i \in R(x_0) \right\}.$$

It is also known [10] that in order for the point $x^* \in \mathcal{R}^n$ to minimize the function $\varphi(x)$, it is necessary to have $\Psi(x^*) \geq 0$ where

$$\Psi(x) = \min_{||g||_{\mathcal{R}^n} = 1} \min_{i \in R(x)} \left\langle \frac{\partial f_i(x)}{\partial x}, g \right\rangle. \qquad (23)$$

From formulas (22), (23) we see that by definition the vector $g_0 \in \mathcal{R}^n$, $||g_0||_{\mathcal{R}^n} = 1$, is the steepest descent direction of the function $\varphi(x)$ at the point x^0 if we have

$$\min_{i \in R(x_0)} \left\langle \frac{\partial f_i(x_0)}{\partial x}, g_0 \right\rangle = \Psi(x_0). \qquad (24)$$

In terms of superdifferential (see (22)) a steepest descent direction [8] is the vector

$$g_0 = -\frac{w_0}{||w_0||_{\mathcal{R}^n}} \qquad (25)$$

where w_0 is a solution of the problem

$$\max_{w \in \overline{\partial} \varphi(x_0)} ||w|| = ||w_0||. \qquad (26)$$

Apply the superdifferential (the steepest) descent method to minimizing the function $\varphi(x)$. Describe the algorithm as applied to the function and the space under consideration. Fix an arbitrary initial point $x_1 \in \mathcal{R}^n$. Suppose that the set

$$lev_{\varphi(x_1)}\varphi(\cdot) = \{x \in \mathcal{R}^n \mid \varphi(x) \leq \varphi(x_1)\}$$

is bounded (due to the arbitrariness of the initial point, in fact, one must assume that the set $lev_{\varphi(x_1)}\varphi(\cdot)$ is bounded for every initial point taken). Due to the function $\varphi(x)$ continuity [10] the set $lev_{\varphi(x_1)}\varphi(\cdot)$ is also closed. Let the point $x_k \in \mathcal{R}^n$ be already constructed. If $\Psi(x_k) \geq 0$ (in practice we check that this condition is satisfied only with some fixed accuracy $\bar{\varepsilon}$, i.e. $\Psi(x_k) \geq -\bar{\varepsilon}$), then the point x_k is a stationary point of the function $\varphi(x)$ and the process terminates. Otherwise, construct the next point according to the rule

$$x_{k+1} = x_k + \alpha_k g_k$$

where the vector g_k is a steepest (superdifferential) descent direction of the function $\varphi(x)$ at the point x_k (see (25), (26)) and the value α_k is a solution of the following one-dimensional minimization problem

$$\min_{\alpha \geq 0} \varphi(x_k + \alpha g_k) = \varphi(x_k + \alpha_k g_k).$$

Then $\varphi(x_{k+1}) \leq \varphi(x_k)$.

Theorem 3. *Under the assumptions made one has the inequality*

$$\varliminf_{k \to \infty} \Psi(x_k) \geq 0 \tag{27}$$

for the sequence built according to the algorithm above.

Proof. The following proof is partially based on the ideas of an analogous one in book [10].

Assume the contrary. Then there exist such a subsequence $\{x_{k_j}\}_{j=1}^{\infty}$ and such a number $b > 0$ that for each $j \in \mathcal{N}$ we have the inequality

$$\Psi(x_{k_j}) \leq -b. \tag{28}$$

Note that for every $j \in \mathcal{N}$ the set $R(x_{k_j})$ is nonempty.

From the steepest descent direction definition (see (24)) it follows that for every $j \in \mathcal{N}$ there exists an index $\bar{i} \in R(x_{k_j})$ such that for each $\alpha > 0$ the relation

$$f_{\bar{i}}(x_{k_j} + \alpha g_{k_j}) = f_{\bar{i}}(x_{k_j}) + \alpha \Psi(x_{k_j}) + o_{\bar{i}}(x_{k_j}, g_{k_j}, \alpha) \tag{29}$$

holds true.

Recall that $f_{\bar{i}}(x_{k_j}) = \varphi(x_{k_j})$ for each index $\bar{i}$ from the set $R(x_{k_j})$ by this index set definition. Then from (28) since $f_{\bar{i}}(x)$ is a continuously differentiable function by assumption, there exists [10] such $\bar{\alpha} > 0$ (which does not depend on the number k_j) such that for $\bar{i} \in R(x_{k_j})$ satisfying (29) and for each $\alpha \in (0, \bar{\alpha}]$ one has

$$f_{\bar{i}}(x_{k_j} + \alpha g_{k_j}) \leq \varphi(x_{k_j}) - \frac{1}{2}\alpha b.$$

Using the fact that $R(x_{k_j}) \neq \emptyset$ for all $j \in \mathcal{N}$, we finally have

$$\varphi(x_{k_j} + \overline{\alpha} g_{k_j}) \leq \varphi(x_{k_j}) - \frac{1}{2}\overline{\alpha}b = \varphi(x_{k_j}) - \frac{1}{2}\beta \qquad (30)$$

uniformly in $j \in \mathcal{N}$.

This inequality leads to a contradiction. Indeed, the sequence $\{\varphi(x_k)\}_{k=1}^{\infty}$ is monotonically decreasing and bounded below by the number $\min\limits_{x \in lev_{\varphi(x_{(1)})}\varphi(\cdot)} \varphi(x)$, hence it has a limit:

$$\{\varphi(x_k)\} \to \varphi^* \text{ at } k \to \infty,$$

herewith, at each $k \in \mathcal{N}$ one has

$$\varphi(x_k) \geq \varphi^*. \qquad (31)$$

Now choose such a large number $\overline{j}$ that $\varphi(x_{k_{\overline{j}}}) < \varphi^* + \frac{1}{2}\beta$. Due to (30) we have $\varphi(x_{k_{\overline{j}}+1}) \leq \varphi(x_{k_{\overline{j}}} + \overline{\alpha} g_{k_{\overline{j}}}) \leq \varphi^* - \frac{1}{2}\beta$ what contradicts (31).

Now turn back to the problem of functional $I(x, z)$ minimizing. Denote (see formulas (19), (20))

$$\Psi(x, z) = \min_{\|g\|_{L_n^2[0,T] \times L_n^2[0,T]}=1} \frac{\partial I(x, z)}{\partial g} = \frac{\partial I(x, z)}{\partial G}.$$

Then necessary minimum condition of the functional $I(x, z)$ at the point (x^*, z^*) may be written [12] as the inequality $\Psi(x^*, z^*) \geq 0$.

Apply the superdifferential (the steepest) descent method to the functional $I(x, z)$ minimization problem. Describe the algorithm as applied to the functional and the space under consideration. Fix an arbitrary initial point $(x_{(1)}, z_{(1)}) \in C_n[0, T] \times P_n[0, T]$. Suppose that the set

$$lev_{I(x_{(1)}, z_{(1)})} I(\cdot, \cdot) = \{(x, z) \in C_n[0, T] \times P_n[0, T] \mid I(x, z) \leq I(x_{(1)}, z_{(1)})\}$$

is bounded in $L_n^2[0, T] \times L_n^2[0, T]$-norm (due to the arbitrariness of the initial point, in fact, one must assume that the set $lev_{I(x_{(1)}, z_{(1)})} I(\cdot, \cdot)$ is bounded for every initial point taken). Let the point $(x_{(k)}, z_{(k)}) \in C_n[0, T] \times P_n[0, T]$ be already constructed. If $\Psi(x_{(k)}, z_{(k)}) \geq 0$ (in practice we check that this condition is satisfied only with some fixed accuracy $\overline{\varepsilon}$, i.e. $\Psi(x_{(k)}, z_{(k)}) \geq -\overline{\varepsilon}$) (in other words, if minimum condition (18) is satisfied (in practice with some fixed accuracy $\overline{\varepsilon}$, i.e. $\|\overline{w}(x_k(t), z_k(t), t)\|_{L_{2n}^2[0,T]} \leq \overline{\varepsilon}$)), then the point $(x_{(k)}, z_{(k)})$ is a stationary point of the functional $I(x, z)$ and the process terminates. Otherwise, construct the next point according to the rule

$$(x_{(k+1)}, z_{(k+1)}) = (x_{(k)}, z_{(k)}) + \alpha_{(k)} G(x_{(k)}, z_{(k)})$$

where the vector-function $G(x_{(k)}, z_{(k)})$ is a superdifferential descent direction of the functional $I(x, z)$ at the point $(x_{(k)}, z_{(k)})$ (see (19), (20)), and the value $\alpha_{(k)}$ is a solution of the following one-dimensional minimization problem

$$\min_{\alpha \geq 0} I\Big((x_{(k)}, z_{(k)}) + \alpha G(x_{(k)}, z_{(k)})\Big) = I\Big((x_{(k)}, z_{(k)}) + \alpha_{(k)} G(x_{(k)}, z_{(k)})\Big).$$

Then $I\big(x_{(k+1)}, z_{(k+1)}\big) \leq I\big(x_{(k)}, z_{(k)}\big)$.

Introduce now the set family $\mathcal{I}$. At first let us define the functional I_q, $q = \overline{1, \left(\prod_{j=1}^{r} k(j)\right)^n}$, as follows. Its integrand is the same as the functional I one but the maximum function $\max\big\{f_{i,j_1}(x)\psi_i, \ldots, f_{i,j_{k(j)}}(x)\psi_i\big\}$, $i = \overline{1,n}$, $j = \overline{1,r}$, is substituted for each $i = \overline{1,n}$ by only the one of the functions $f_{i,j_1}\psi_i, \ldots, f_{i,j_{k(j)}}\psi_i$, $j \in \{1, \ldots r\}$. Let the family $\mathcal{I}$ consist of the sums of the integrals over the intervals of the time interval $[0, T]$ splitting for all such possible finite splittings. Herewith, the integrand of each summand in the sum taken is the same as some functional I_q one, $q \in \left\{1, \ldots, \left(\prod_{j=1}^{r} k(j)\right)^n\right\}$.

Let for every point constructed by the method described the following assumption is valid: the interval $[0, T]$ may be divided into a finite number of segments, in every of which for each $i = \overline{1, n}$ either $h_i(x_{(k)}, z_{(k)}) = 0$, or one (several) of the functions $\left\langle -\psi_i^* \frac{\partial f_{i,j_p}(x_{(k)})}{\partial x}, G_1(x_{(k)}, z_{(k)}) \right\rangle$, $j = \overline{1, r}$, $p = \overline{1, k(j)}$, is (are) active.

Let us illustrate this assumption by an example. Consider the following simplest functional (whose structure, however, preserves the basic features of the general case)

$$\int_0^1 \frac{1}{2}\min^2\{x(t) + 1, -x(t) + 1\}dt.$$

Consider the point $x_{(1)} = 0$ (i.e. $x_{(1)}(t) = 0$ for all $t \in [0, 1]$). In order to find the steepest (the superdifferential) descent direction of the functional at this point one has, according to the theory described, to minimize the directional derivative (calculated at this point), i.e. to find such a function $G \in L_1^2[0, T]$, $\int_0^1 G^2(t)dt = 1$, that minimizes the functional

$$\int_0^1 \min\{x_{(1)}(t) + 1, -x_{(1)}(t) + 1\}\min\{g(t), -g(t)\}dt.$$

Here $g \in L_1^2[0, T]$, $\int_0^1 g^2(t)dt = 1$. Take $G(t) = \{-1, \ t \in [0, 0.5], \ 1, \ t \in (0.5, 1]\}$ as one of obvious solutions. Herewith, $\Psi(x_{(1)}) = \int_0^1 -1dt = -1$. We see that the assumption made is satisfied. Take

$$\hat{I}(x) = \int_0^{0.5} \frac{1}{2}(x(t) + 1)^2 dt + \int_{0.5}^1 \frac{1}{2}(-x(t) + 1)^2 dt.$$

Then we have

$$\hat{I}\big(x_{(1)} + \alpha G(x_{(1)})\big) = \hat{I}(x_{(1)}) + \alpha \Psi(x_{(1)}) + o(x_{(1)}, G(x_{(1)}), \alpha)$$

since $\nabla \hat{I}(x_{(1)}) = \{1, \ t \in [0, 0.5], \ -1, \ t \in (0.5, 1]\}$, i.e. we have $\nabla \hat{I}(x_{(1)})G(x_{(1)})$ $= \int_0^1 -1dt = -1$. It is obvious that $\hat{I} \in \mathcal{I}$. Note that with $\alpha_{(1)} = 1$ one gets the point $x_{(2)}(t) = \{-1,\ t \in [0, 0.5],\ 1,\ t \in (0.5, 1],\}$ which delivers the global minimum to the functional considered.

For functionals from the family $\mathcal{I}$ we make the following additional assumption. Let there exist such a finite number L that for every $\hat{I} \in \mathcal{I}$ and for all $\overline{x}, \overline{z}, \overline{\overline{x}}, \overline{\overline{z}}$ from a ball with the center in the origin and with some finite radius $r' + \hat{\alpha}$ (here $r' > \sup\limits_{(x,z) \in lev_{I(x_{(1)}, z_{(1)})} I(\cdot, \cdot)} \|(x, z)\|_{L_n^2[0,T] \times L_n^2[0,T]}$ and $\hat{\alpha}$ is some positive number) one has

$$\|\nabla \hat{I}(\overline{x}, \overline{z}) - \nabla \hat{I}(\overline{\overline{x}}, \overline{\overline{z}})\|_{L_n^2[0,T] \times L_n^2[0,T]} \leq L \|(\overline{x}, \overline{z}) - (\overline{\overline{x}}, \overline{\overline{z}})\|_{L_n^2[0,T] \times L_n^2[0,T]}. \quad (32)$$

Remark 2. At first glance it may seem that the Lipschitz constant L existence for all $\hat{I} \in \mathcal{I}$ simultaneously in the assumption is too burdensome. However, if one remembers that on each of the interval $[0, T]$ segments the functional $\hat{I} \in \mathcal{I}$ integrand coincides with the functional I_q, $q \in \left\{ 1, \ldots, \left(\prod_{j=1}^r k(j) \right)^n \right\}$, one by construction (see the set $\mathcal{I}$ definition), then this assumption is natural if we suppose the Lipschitz-continuity of every of the gradients ∇I_q, $q = \overline{1, \left(\prod_{j=1}^r k(j) \right)^n}$; and the gradient Lipschitz-continuity condition is a common assumption for justifying classical optimization methods constructed for differentiable functionals.

Lemma 2. *Let condition (32) be satisfied. Then for each functional $I \in \mathcal{I}$ and for all $(x, z) \in lev_{I(x_{(1)}, z_{(1)})} I(\cdot, \cdot)$, $G \in C_n[0, T] \times P_n[0, T]$, $\|G\|_{L_n^2[0,T] \times L_n^2[0,T]} = 1$, $\alpha \in \mathcal{R}$, $0 < \alpha \leq \hat{\alpha}$, one has the inequality*

$$I\left((x, z) + \alpha G\right) \leq I(x, z) + \alpha \langle \nabla I(x, z), G \rangle + \alpha^2 \frac{L}{2}.$$

Proof. The proof can be carried out with obvious modifications in a similar way as for the analogous statement in [13].

We will also suppose that during the method realization for each $k \in \mathcal{N}$ one has $\alpha_{(k)} < \hat{\alpha}$ (where $\hat{\alpha}$ is a number from Lemma 2 (see also the assumption before Remark 2)).

Arguing in an analogous fashion as in Theorem 3 proof, employing the family $\mathcal{I}$ definition with considering directional derivative (14) and making use of Lemma 2 and of the assumptions made it is not difficult to check the validity of the following theorem.

Theorem 4. *Under the assumptions made one has the inequality*

$$\underline{\lim}_{k \to \infty} \Psi(x_k, z_k) \geq 0 \quad (33)$$

for the sequence built according to the algorithm above.

Remark 3. It is easy to show that, in fact, in formulas (27), (33) the lower limit can be substituted by the "ordinary" limit and the inequality can be substituted by the equality in the cases considered.

8   An Example

One can give a practical problem example leading to the type of systems considered in the paper. Let from some physical considerations the "velocity" $\dot{x}_1$ of an object lie in the range $[\min\{x_1, x_2, x_3\}, \max\{x_1, x_2, x_3\}]$ of the "coordinates" x_1, x_2, x_3. The segment given may be written down as $\mathrm{co}\{x_1, x_2, x_3\} = \mathrm{co}\bigcup_{i=1}^{3}\{x_i\}$. The support function of this set is [7] $\max\{x_1\psi_1, x_2\psi_1, x_3\psi_1\}$. The constraints on the "velocities" $\dot{x}_2$, $\dot{x}_3$ in the system below may be interpreted analogously.

So consider the differential inclusion

$$\dot{x}_1 \in \mathrm{co}\{x_1, x_2, x_3\}, \quad \dot{x}_2 \in \mathrm{co}\{-2x_1, x_3\}, \quad \dot{x}_3 = x_1 + x_2$$

on the time interval $[0, 1]$ with the boundary conditions

$$x_1(0) = -1,\; x_2(0) = -1,\; x_3(0) = 2.5, \quad x_1(1) = 1,\; x_2(1) = 1,\; x_3(1) = 2.5.$$

As the phase constraint surface put

$$s(x) = x_1 - x_2 = 0.$$

Take $(x_{(1)}, z_{(1)}) = (0, 0, 0, 0, 0, 0)'$ as the first approximation (i.e. all the functions considered are identically equal to zero on the interval $[0, 1]$), then $I(x_{(1)}, z_{(1)}) = 8.125$. At the end of the process the discretization step was equal to 10^{-1}. The picture illustrates the trajectories obtained. From the picture we see that the differential inclusion is satisfied. The trajectory practically lies on the surface given as well. The boundary values error doesn't exceed the magnitude 2×10^{-3}. In order to obtain such an accuracy, 66 iterations has been required. The functional value on the trajectory obtained is approximately 2×10^{-5} (Fig. 1).

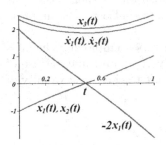 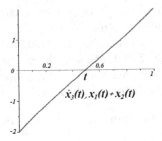

Fig. 1. Example

9   Conclusion

In the paper a differential inclusion with the special structure is studied. The support function of the set on its right-hand side is supposed to be subdifferentiable. The problem is reduced to minimizing a functional whose superdifferentiability is proved. The minimum conditions in terms of superdifferential are obtained. On the basis of these conditions the superdifferential descent method is described; the convergence of the method to a stationary point of the functional is proved. An example is given that illustrates the developed method.

References

1. Filippov, A.F.: Differential Equations with Discontinuous Righthand Sides. Springer, Dordrecht (1988). https://doi.org/10.1007/978-94-015-7793-9
2. Utkin, V.I.: Sliding Modes in Control and Optimization. Springer, Heidelberg (1992). https://doi.org/10.1007/978-3-642-84379-2
3. Levant, A.: Higher-order sliding modes, differentiation and output-feedback control. Int. J. Control **76**(9), 924–941 (2003)
4. Fominyh, A.V.: Open-loop control of a plant described by a system with nonsmooth right-hand side. Comput. Math. Math. Phys. **59**, 1639–1648 (2019)
5. Fominyh, A.V.: A numerical method for finding the optimal solution of a differential inclusion. Vestn. St. Petersbg. Univ. Math. **51**(4), 397–406 (2018)
6. Fominyh, A.V.: The subdifferential descent method in a nonsmooth variational problem. Optim. Lett. **17**, 675–698 (2023)
7. Blagodatskikh, V.I., Filippov, A.F.: Differential inclusions and optimal control. Proc. Steklov Inst. Math. **169**, 199–259 (1986)
8. Demyanov, V.F., Rubinov, A.M.: Basics of Nonsmooth Analysis and Quasidifferential Calculus. Nauka, Moscow (1990)
9. Bonnans, J.F., Shapiro, A.: Perturbation Analysis of Optimization Problems. Springer, New York (2000). https://doi.org/10.1007/978-1-4612-1394-9
10. Demyanov, V.F., Malozemov, V.N.: Introduction to Minimax. Dover Publications Inc., New York (1990)
11. Makela, M., Neittaanmaki, P.: Nonsmooth Optimization: Analysis and Algorithms with Applications to Optimal Control. World Scientific Publishing, London (1992)
12. Dolgopolik, M.V.: Nonsmooth problems of calculus of variations via codifferentiation. ESAIM Control Optim. Calc. Var. **20**(4), 1153–1180 (2014)
13. Kantorovich, L.V., Akilov, G.P.: Functional Analysis. Nauka Publication, Moscow (1977)

Approximate Solution of Small-Time Control Synthesis Problem Based on Linearization

Mikhail Gusev$^{(\boxtimes)}$ and Ivan Osipov

N.N. Krasovskii Institute of Mathematics and Mechanics, 16 S.Kovalevskaya str.,
620108 Yekaterinburg, Russia
{gmi,i.o.osipov}@imm.uran.ru

Abstract. We consider the problem of a feedback control design for a
nonlinear control-affine system. The aim of the control is to bring trajectories of the closed system to the origin of coordinates in a given time,
providing the minimal value of an integral functional. The object under
study is the nonlinear system, closed by a linear feedback controller. The
controller is obtained as a solution of the LQR problem for the linearized
system. We indicate sufficient conditions for this linear feedback to give
a local solution to the control synthesis problem under consideration. In
addition, we give some error estimates for the values of the functional.

Keywords: Control synthesis · Linear feedback · Linearization

1 Introduction

We propose here a method for solving the problem of a local control synthesis
for a control-affine system on a small time interval. This method is based on the
linearization of the original nonlinear system in the vicinity of the equilibrium
position. Linearization is often used in solving various control problems, such as
stabilization problems [1,2], stochastic and numerical control [3–6], MPC control
[7,8], etc.

In this article, we study the problem of control synthesis with an integral
quadratic cost. Note that the task is considered on a finite, and, moreover, small,
time interval. The goal of the control is to transfer the system to the origin in a
given time ensuring the minimum value of the cost. The linear feedback control
found for the linearized system is used as the input of the original non-linear
system. For a linear control system the optimal feedback is linear in state, with
gains increasing indefinitely when approaching the terminal time. The latter
makes it difficult to justify the applicability of the linearization method. The
restrictions on asymptotics of the controllability Gramian of the linearized system are needed in this case. Unlike, for example, the stabilization problem for

The work was performed as part of research conducted in the Ural Mathematical
Center with the financial support of the Ministry of Science and Higher Education of
the Russian Federation (Agreement number 075-02-2023-913).

M. Khachay et al. (Eds.): MOTOR 2023, LNCS 13930, pp. 362–377, 2023.
https://doi.org/10.1007/978-3-031-35305-5_25

which controllability (stabilizability) of the linearized system implies stabilizability of the nonlinear system. These restrictions coincide with the asymptotic equivalence conditions for reachable (null-controllable) sets of nonlinear and linearized systems. In [9] it was shown that, under these conditions the control in the form of a linear state feedback takes all trajectories starting from some neighborhood of zero to zero, if the control time interval is sufficiently small.

In this article, we generalize the main result of [9]. The proposed sufficient conditions have the form of an inequality for some improper integral. They depend on the smallest and the largest eigenvalues of the controllability Gramian of the linearized system and contain a scalar parameter. The choice of this parameter makes it possible to cover a wider class of control systems, the conditions from [9] are obtained here as a special case for a certain value of the parameter.

For a linearized system, the considered linear controller delivers the minimum value to the integral functional for any initial state. For a nonlinear system, this is not the case, so it is important to obtain an estimate of the resulting error. This was done in the second part of the article, where the relation between the values of the integral cost for the trajectories of the nonlinear and linearized systems was studied and the estimate for the relative error was given.

The article is structured as follows. The problem statement and some preliminary results are given in the second section. The third section contains the formulation and proof of the main results. Finally, we provide two illustrative examples in the fourth section.

2   Preliminary Results

2.1   Problem Statement

Let us consider the nonlinear control-affine system

$$\dot{z}(t) = f(z(t)) + Bu(t), \qquad 0 \leqslant t \leqslant T, \qquad z(0) = z_0. \tag{1}$$

where $x \in \mathbb{R}^n$ is a state vector, $u \in \mathbb{R}^r$ is a control vector, T is a positive number. We postulate that the function f has the following property.

Property 1. There exist $r > 0$, $k > 0$ such that for all $z \in B(0, r)$ the function $f(z)$ could be rewritten in the form $f(z) = Az + R(z)$, where $\|R(z)\| \leqslant k\|z\|^2$.

Here $B(0, r)$ is the ball of radius r centered at $0 \in \mathbb{R}^n$. This property holds if $f(0) = 0$, $\frac{\partial f}{\partial x}(0) = A$ and $f(z)$ is twice differentiable.

The space of square integrable scalar or vector functions on $[0, T]$ we will denote by $\mathbb{L}_2 = \mathbb{L}_2[0, T]$. The ball in the space $\mathbb{L}_2$ we denote as $B_{\mathbb{L}_2}(0, r)$. As the cost functional we consider the following

$$I(T, u) := \int_0^T u^\top(t)u(t)dt = \|u(\cdot)\|_{\mathbb{L}_2}^2. \tag{2}$$

The problem is to synthesize a feedback control $u(t) = u(t, z(t))$ that leads the trajectories of the closed system

$$\dot{z}(t) = f(z(t)) + Bu(t, z(t)), \qquad 0 \leqslant t \leqslant T, \qquad z(0) = z_0.$$

to the origin of coordinates at time T and to provide the minimum value of $I(T, u)$.

Consider the linear case $(R(z) = 0)$

$$\dot{z} = Az + Bu, \qquad 0 \leqslant t \leqslant T. \tag{3}$$

If system (3) is controllable the solution of the above problem is the linear in state feedback controller

$$u(t, z) = -B^\top Q_T(t)z \tag{4}$$

(see, for example, [9–11]). Here $Q_T(t) = W^{-1}(T - t)$ where $W(t)$ is the Controllability Gramian of system $\dot{x} = -Ax - Bu$:

$$W(t) = \int_0^t e^{-A\tau} BB^\top e^{-A^\top \tau} d\tau.$$

The Gramian $W(t)$ is positive definite for $t > 0$ iff system (3) is controllable. It may be shown that $Q_T(t)$ is the solution of differential equation

$$\dot{Q}_T = Q_T BB^\top Q_T - A^\top Q_T - Q_T A, \quad Q_T(0) = W^{-1}(T). \tag{5}$$

Thus, to find $Q_T(t)$ on $(0, T]$ one need first to calculate $W(T)$ and then integrate the system (5). Since $W(0) = 0$, $Q_T(t)$ is determined for $t < T$ and $\|Q_T(t)\| \to \infty$ as $t \to T$.

The following is true [9–11].

Assertion 1. *Any trajectory $z(t)$ of the system (3) with the control (4) starting from the point z_0 reaches the origin at time T. The integral cost $I(T, u)$ takes the minimum possible value $z_0^\top Q_T(0)z_0$ for each z_0.*

Further we are going to investigate the trajectories behavior of system (1) closed by linear feedback $u(t, z) = -B^\top Q_T(t)z$ assuming that T is small enough. Is it true that all trajectories starting in some neighborhood of the origin reach it? And what is the value of the cost functional?

2.2   Asymptotic Equality of Reachable Sets

In what follows, we use a notion of an asymptotic equality of reachable sets. Consider a control system whose equations are obtained from (1) by time reversal. Setting $\tau = T - t$ we have

$$\dot{x}(\tau) = -f(x(\tau)) - Bv(\tau), \qquad 0 \leqslant \tau \leqslant T; \tag{6}$$

here $x(\tau) = z(T - \tau)$, $v(\tau) = u(T - \tau)$. For a given $\mu > 0$ denote as $G_-(T, \mu)$ a reachable set of the system (6) under quadratic integral constraints on the cost functional, $G_-(T, \mu) = \{x \in \mathbb{R}^n : \exists v(\cdot) \in B_{\mathbb{L}_2}(0, \mu), \ x = x(T, v(\cdot)))\}$.

Here $x(\tau, v(\cdot)))$ denotes the solution of (6) with zero initial state. Properties of reachable sets of nonlinear systems with integral constraints on control have been studied in many papers (see, for example [12–14]). Consider also the linear system

$$\dot{x}(\tau) = -Ax(\tau) - Bv(\tau), \qquad 0 \leqslant \tau \leqslant T; \tag{7}$$

this system is the linearization of the system (6) at the origin. It's reachable set we denote by $G_-^0(T, \mu)$. This set is an ellipsoid in $\mathbb{R}^n$ described by inequality $G_-^0(T, \mu) = \{x \in \mathbb{R}^n : x^\top W^{-1}(T)x \leqslant \mu^2\}$.

Let $X, Y \subset \mathbb{R}^n$ be convex compact sets such that the origin is an interior point of both the sets.

Definition 1 (see, for example, [15,16]). *The Banach-Mazur distance between X and Y is defined as $\rho(X, Y) := \log\left(r(X, Y) \cdot r(Y, X)\right)$, where $r(X, Y) = \inf\{t \geqslant 1 : tX \supset Y\}$.*

From the definition it follows that for any $c > 0$, $\rho(cX, cY) = \rho(X, Y)$ and two inclusions are valid: $X \subset \exp(\rho(X, Y))Y$ and $Y \subset \exp(\rho(X, Y))X$.

Assume that X, Y depend on a small positive parameter τ, $0 < \tau \leqslant \tau_0$ and set-valued mappings $X(\tau), Y(\tau)$ are bounded.

Definition 2 ([16]). *The sets $X(\tau), Y(\tau)$ are called asymptotically equal under $\tau \to 0$ if $\rho(X(\tau), Y(\tau)) \to 0$, $\tau \to 0$.*

Denote by $\nu(\tau), \eta(\tau)$ the smallest and the largest eigenvalues of $W(\tau)$ respectively. From the results of [17–20] it follows that the reachable sets $G_-(\tau, \mu)$ and $G_-^0(\tau, \mu)$ are asymptotically equal under $\tau \to 0$ if the pair (A, B) is controllable and there exist $l > 0$, $\tau_0 > 0$ and $\alpha > 0$ such that for any $0 < \tau \leqslant \tau_0$

$$\nu(\tau) \geqslant l\tau^{4-\alpha}. \tag{8}$$

Remark 1. The reachable set $G_-(T, \mu)$ of system (6) coincides with the null-controllable set of system (1), i.e. the set of initial conditions from which the system can be led to the origin by controls from $B_{\mathrm{L}_2}(0, \mu)$ at time T. The same is true for systems (7) and (7) and their corresponding set $G_-^0(T, \mu)$.

3   The Control Synthesis Problem. Main Results

Everywhere below, we assume that the pair A, B is controllable without specifying this separately.

3.1   Asymptotics of the Trajectories

In this section we investigate asymptotic behavior of trajectories of system (1) closed by the linear feedback $u(t, z) = -B^\top Q_T(t)z$:

$$\dot{z} = f(z) - BB^\top Q_T(t)z, \qquad 0 \leqslant t \leqslant T, \qquad z(0) = z_0. \tag{9}$$

Recall that this control takes the trajectories of the linear system $\dot{z} = Az + Bz$ to the origin at time T, and provides the minimum to the cost. This minimum equals to $J_0(T, z_0) = z_0^T Q_T(0) z_0$.

To analyze the trajectories of the system (9) we implement the following lemma

Lemma 1. *Let $C \in \mathbb{R}^{n \times n}$, $D \in \mathbb{R}^{n \times n}$ be symmetric positive definite matrices, $C = D^{-1}$. Then for $\forall z \in \mathbb{R}^n$*

$$\frac{1}{\lambda_{max}(D)} \|z\|^2 \leqslant z^T C z \leqslant \frac{1}{\lambda_{min}(D)} \|z\|^2, \tag{10}$$

where $\lambda_{max}(D)$ and $\lambda_{min}(D)$ are the largest and the smallest eigenvalues of D.

Proof. Follows from the fact that the largest and the smallest eigenvalue of matrix C are $1/\lambda_{min}(D)$ and $1/\lambda_{max}(D)$, respectively.

If $C = Q_T(t) = W^{-1}(T - t)$ then $D = W(T - t)$ and inequalities (10) take the form

$$\frac{1}{\eta(T - t)} \|z\|^2 \leqslant z^T Q_T(t) z \leqslant \frac{1}{\nu(T - t)} \|z\|^2, \quad 0 \leqslant t < T.$$

Assumption 1. *There exist $\overline{T} > 0$ and a continuous positive function $\varphi(\tau) :$ $(0, \overline{T}] \to \mathbb{R}$ such that*

$$0 < \frac{\eta(\tau)}{\sqrt{\nu(\tau)}} \leqslant \varphi(\tau), \quad 0 < \tau \leqslant \overline{T}, \quad \int_0^{\overline{T}} \varphi(\tau) d\tau < \infty.$$

Define the function $\Phi(T) : [0, \overline{T}] \to \mathbb{R}$

$$\Phi(T) = \int_0^T \varphi(\tau) d\tau, \quad 0 < T \leqslant \overline{T}, \quad \Phi(0) = 0.$$

Recall that $\eta(\tau)$ and $\nu(\tau)$ the smallest and the greatest eigenvalues of $W(\tau)$. Further we will consider the system (7) to be completely controllable, therefore $\eta(\tau) \geqslant \nu(\tau) \geqslant 0$ with $\tau \geq 0$.

Since $\varphi(\tau)$ is not necessarily bounded at the zero, $\Phi(T)$ can take values equal to $+\infty$.

Lemma 2. *The following properties of $\Phi(T)$ are valid:*

1. *If $\Phi(T) < \infty$ for at least one T, then $\Phi(T) < \infty$ for all $T \in (0, \overline{T}]$.*
2. *If $\Phi(T) < \infty$ then $\Phi(T)$ is continuous and increasing on $[0, \overline{T}]$.*

Proof. Follows from properties of improper integrals.

Assumption 2 *There exists $0 < \beta \leqslant 1$ such that $\frac{\sqrt{\eta(T)}}{\Phi^\beta(T)} \to 0$ as $T \to 0$.*

If $\Phi(T)$ is finite there is at most one root of the equation $\Phi(T) = 1$ on $(0, \overline{T}]$, denote it by T^*. If $\Phi(T) < 1$ $T \in (0, \overline{T}]$ put $T^* = \overline{T}$. Clearly, for any $0 < \beta \leqslant 1$ $\Phi^\beta(T) \geqslant \Phi(T)$ if $T \leqslant T^*$.

For a given $T \in (0, \overline{T}]$ consider a quadratic form $V_T(t, z) = z^\top Q_T(t)z$.

Lemma 3. *Let Assumption 1 be fulfilled. Let $T \leqslant T^*$ and $z(t)$ be a trajectory of system(9) such that $z(t) \in B(0, r)$ $0 < t \leqslant T$ and $V_T(0, z(0)) \leqslant 1/(4k^2\Phi^{2\beta}(T))$ for some $0 < \beta \leqslant 1$. Then*

$$V_T(t, z(t)) \leqslant \frac{1}{k^2\Phi^{2\beta}(T)}, \qquad 0 \leqslant t \leqslant T.$$

Proof. Differentiating V_T along the solution $z(t)$ of system (1) on the time interval $[0, T]$ we obtain

$$\frac{d}{dt}V_T(t, z) = \frac{d}{dt}z^\top Q_T z = \dot{z}^\top Q_T z + z^\top \dot{Q}_T z + z^\top Q_T \dot{z}$$

$$= \left(z^\top A^\top + R^\top(z) - z^\top Q_T BB^\top\right) Q_T z$$

$$+ z^\top Q_T \left(A + R(z) - BB^\top Q_T\right) z + z^\top \left(QBB^\top Q_T - A^\top Q - Q_T A\right) z$$

$$= R^\top(z)Q_T z + z^\top Q_T R(z) - z^\top (Q_T BB^\top Q_T)z$$

$$= 2\left(R(z), Q_T z\right) - z^\top Q_T BB^\top Q_T z$$

While z and Q_T depend on t here, for brevity, we omit the explicit dependence in the notation. Since $z^\top Q_T BB^\top Q_T z \geqslant 0$, it follows that

$$\frac{d}{dt}V_T(t, z) \leqslant 2\left(R(z), Q_T z\right) = 2(R(z), z)_{Q_T} \leqslant 2\|R(z)\|_{Q_T}\|z\|_{Q_T}, \qquad (11)$$

where for $x, y \in \mathbb{R}^n$ we denote $(x, y)_{Q_T} = x^\top Q_T y$, $\|x\|_{Q_T} = \sqrt{(x, x)_{Q_T}}$. As $z = z(t) \in B(0, r)$ then, taking into account that $\|R(z)\| \leqslant k\|z\|^2$ and applying Lemma 1, we find that

$$\|R(z)\|_{Q_T} \leqslant \frac{1}{\sqrt{\nu(T-t)}}\|R(z)\| \leqslant \frac{k}{\sqrt{\nu(T-t)}}\|z\|^2 \leqslant k\frac{\eta(T-t)}{\sqrt{\nu(T-t)}}V_T. \qquad (12)$$

Recall, that $Q_T^{-1}(t) = W(T-t)$. Substituting the above estimate into (11) we obtain

$$\frac{d}{dt}V_T \leqslant 2k\frac{\eta(T-t)}{\sqrt{\nu(T-t)}}V_T^{3/2} \leqslant 2k\varphi(T-t)V_T^{3/2} \qquad (13)$$

Let's introduce the system

$$\dot{\psi} = 2k\varphi(T-t)\psi \qquad (14)$$

called a comparison system for (13). Integrating this system we have

$$d\psi^{-1/2} = -k\varphi(T-t)dt, \quad \psi^{-1/2}(t) = -k\int_0^t \varphi(T-\zeta)d\zeta + C,$$

where

$$0 < \int_0^t \varphi(T-\zeta)d\zeta \leqslant \int_0^T \varphi(T-\zeta)d\zeta = \int_0^T \varphi(\tau)d\tau = \Phi(T)$$

Choose $C = 2k(\Phi(T))^\beta$ then $\psi^{-1/2} \geqslant 2k(\Phi(T))^\beta \quad k\Phi(T) = k(2\Phi^\beta - \Phi) \geqslant k\Phi^\beta$.
Therefore, $\psi(t) \leqslant \left(k^2\Phi^{2\beta}(T)\right)^{-1}$ for all $0 \leqslant t \leqslant T^*$ and $\psi(0) = \left(4k^2\Phi^{2\beta}(T)\right)^{-1}$.
Thus, $V_T(0, z(0)) \leqslant \psi(0)$ and the comparison theorem [21] applied to (13), (14)
implies inequality $V_T(t, z(t)) \leqslant \psi(t)$. This completes the proof.

Theorem 1. *Let Assumptions 1, 2 be fulfilled. Then there exists $T_1 \leqslant T^*$ such that for any $T \leqslant T_1$, there exist $r_1(T)$ such that the trajectories of (9) with $z(0) = z_0 \in B(0, r_1(T))$ tends to 0 as $t \to T$.*

Proof. Since $\frac{\sqrt{\eta(T)}}{\Phi^\beta(T)}$ tends to 0 as T tends to 0, there exists $T_1 \leqslant T^*$, such that the following inequality holds $\sqrt{\eta(T)}/(k\Phi^\beta(T)) \leqslant r/2, \forall T \in [0; T_1]$.
    Define $r_1(T)$ by the equality

$$r_1(T) = \min\left\{\frac{r}{4}, \frac{\sqrt{\nu(T)}}{2k\Phi^\beta(T)}\right\}. \tag{15}$$

    Here we need to prove that the entire trajectory $z(t)$ lies in the sphere $B(0, r)$ in order to use the estimate (16) from the previous section. According to the theorem, $z_0 \in B(0, r_1(T)) \subset B(0, r)$, since $r_1(T) < r$. Furthermore, the continuity of the trajectory $z(t)$ implies that for values of t close to zero the condition $z(t) \in B(0, r)$ is also satisfied.
    Denote $t^* = \sup\{t : z(t) \in B(0, 0.5r)\}$. Suppose, that $t^* < T$, this means that $z(t) \notin B(0, 0.5r)$, for $t > t^*$. Thus, we can choose a positive ε such that $z(t) \in B(0, r)$ for $0 \leqslant t \leqslant t^* + \varepsilon$.
    Since (15), for $0 \leqslant t \leqslant t^* + \varepsilon$ the following condition is met

$$V_T(0, z_0) \leqslant \frac{1}{\nu(T)}\|z_0\|^2 \leqslant \frac{1}{\nu(T)}r_1^2 \leqslant \frac{1}{4k^2\Phi^{2\beta}(T)}.$$

From Lemma 3 it follows that $V_T(t, z(t)) \leqslant \psi(t)$ and

$$\|z(t)\|^2 \leqslant \eta(T-t)V_T(t, z(t)) \leqslant \eta(T-t)\psi(t) \leqslant \frac{\eta(T)}{k^2\Phi^{2\beta}(T)}, \tag{16}$$

so, $\|z(t)\| \leqslant \frac{\sqrt{\eta(T)}}{k\Phi^\beta(T)} \leqslant r/2$ for $0 \leqslant t \leqslant t^* + \varepsilon$, which is contrary to the definition of t^*. Therefore, $z(t) \in B(0, r)$ and inequality $V_T(t, z(t) \leqslant \psi(t)$ takes place for all $t \in [0; T]$

Conclusively, $\|z(t)\|^2 \leqslant \eta(T-t)\psi(t) \leqslant \eta(T-t)(k^2\Phi^{2\beta}(T))^{-1}$, where $\eta(T-t) \to 0$ as $t \to T$, which means, that $\|z(t)\|$ also tends to zero.

Corollary 1. *Let there exist $l > 0$, $\tau_0 > 0$ and $\alpha > 0$ such that for any $0 < \tau \leqslant \tau_0$*

$$\nu(\tau) \geqslant l\tau^{4-\alpha}. \tag{17}$$

Then Assumptions 1 and 2 are satisfied and hence the assertion of Theorem 1 is valid.

Proof. Let us put $\overline{T} = \tau_0$. There esists $m > 0$ such that $\eta(\tau) \leqslant m\tau$, $0 < \tau \leqslant \overline{T}$ (see, for example, [9]). Hence, we have $\eta(\tau)/\sqrt{\nu(\tau)} \leqslant ml^{-1/2}\tau^{-1+\alpha/2}$, and we can take $\varphi(\tau) := ml^{-1/2}\tau^{-1+\alpha/2}$. In this case $\Phi(T) = 2mT^{\alpha/2}/(l^{-1/2}\alpha)$, and Assumption 1 is, clearly, fulfilled. Since $\sqrt{\eta(T)}/\Phi^\beta(T) \leqslant c_0 T^{(1-\alpha\beta)/2}$, where c_0 is a constant, to fulfill Assumption 2, it is enough to take $\beta < 1/\alpha$.

Note that inequality (17) coincides with the condition that implies Theorem 1 from [9].

3.2   Error Estimates for the Value of Cost Functional

In this part of the paper, we will focus on the value of integral functional (2), when applying linear feedback (4) to a non-linear system (1). Recall that in the linear case of system (3), this functional takes the minimum value on the control (4). We previously denote it by $J_0(T, z_0)$. The designation $J(T, z_0)$ we use for the value of the functional on the trajectory of system (9). In order to obtain an expression for $J(T, z_0)$ we need to integrate (11) from 0 to t:

$$z^\top(t)Q_T(t)z(t) = z_0^\top Q_T(0)z_0 - \int_0^t u^\top(\xi)u(\xi)\,d\xi + 2\int_0^t R^\top(z(\xi))Q(\xi)z(\xi)d\xi, \tag{18}$$

where $u(\xi) = -B^\top Q_T(\xi)z(\xi)$ is a control at time ξ. In the linear case, $R(z) \equiv 0$ and $z^\top(t)Q_T(t)z(t) \to 0$ as $t \to T$ so

$$J_0(T, z_0) = z_0^\top Q_T(0)z_0 = \int_0^t u^\top(\xi)u(\xi)\,d\xi.$$

Further, we are going to investigate the behaviour of the quadratic form $z^\top(t)Q_T(t)z(t)$ and the residual term in (18), which we denote by

$$\gamma(t, z_0) = 2\int_0^t R^\top(z(\xi, z_0))Q(\xi)z(\xi, z_0)d\xi.$$

Theorem 2. *Let Assumption 1 be fulfilled. Let $x(t)$ be a trajectory of system (9) such that $x(t) \in B(0, r)$, $0 \leqslant t \leqslant \tilde{T} \leqslant \overline{T}$ and $V_{\tilde{T}}(t, x(t)) \to 0$ as $t \to \tilde{T}$. Then there exists $T_2 \leqslant \tilde{T}$ such that for all $0 < T < T_2$ the following estimate holds*

$$\left| \frac{J(T) - J_0(T)}{J_0(T)} \right| \leqslant 16k\Phi(T)J_0^{1/2}(T). \qquad (19)$$

Here $J(T) = J(T, x(\tilde{T} - T))$, $J_0(T) = J_0(T, x(\tilde{T} - T))$ are the values of functional $I(T, u(\cdot))$ on the trajectories of nonlinear and linearized systems accordingly.

Proof. Let $T \leqslant \tilde{T}$. Denote by $z(t)$ the trajectory of system (9) with initial state $z(0) = x(\tilde{T} - T)$ on the interval $[0, T)$. Then we have

$$Q_T(t) = W^{-1}(T - t) = W^{-1}(\tilde{T} - (\tilde{T} - T + t)) = Q_{\tilde{I}}(\tilde{T} - T + t) = Q(\tau),$$
$$V_T(t, z(t)) = z^{\top}(t)Q_T(t)z(t) = V_{\tilde{T}}(\tau, x(\tau)),$$

where $\tau = \tilde{T} - T + t$. $V_T(0, z(0)) = V_{\tilde{T}}(\tilde{T} - T, x(\tilde{T} - T))$, $V_{\tilde{T}}(t, x(t)) \to 0$ as $t \to \tilde{T}$. Since $V_{\tilde{T}}(t, x(t)) \to 0$, $t \to \tilde{T}$ we have that $V_T(0, z(0)) = J_0(T)$ tends to zero as $T \to 0$. Clearly, $V_T(t, z(t)) = z^{\top}(t)Q_T(t)z(t) = V_{\tilde{T}}(\tau, x(\tau))$ tends to zero as $t \to T$. Using this, let us rewrite (18)

$$\int_0^T u^{\top}(\xi)u(\xi)\,d\xi - z_0^{\top}Q_T(0)z_0 = 2\int_0^T R^{\top}(z(\xi))Q(\xi)z(\xi)\,d\xi = \gamma(T, z(0)), \quad (20)$$

and take a closer look at the integrand $(R(z), Q_T z) = (R(z), z)_{Q_T} \leqslant \|R(z)\|_{Q_T}\|z\|_{Q_T}$. Repeating the steps (12),(13) from the proof of the Lemma 3, we obtain an upper bound

$$2(R(z), Q_T z) \leqslant 2k\frac{\eta(T - t)}{\sqrt{\nu(T - t)}}V_T^{3/2} \leqslant 2k\varphi(T - t)V_T^{3/2}, \qquad (21)$$

which is identical to (13). However, further steps with the comparison system are slightly modified here. Integrating the comparison system $\dot{\psi} = 2k\varphi(T - t)\psi$ we have

$$d\psi^{-1/2} = -k\varphi(T - t)dt, \quad \psi^{-1/2}(t) = -k\int_0^t \varphi(T - \zeta)d\zeta + C.$$

Since $V_T(0, z(0)) \to 0$ and $\Phi(T) \to 0$ as $T \to 0$ there exists T_2 such that for $T \leqslant T_2$ we have $\Phi(T)\sqrt{V_T(0, z(0))} \leqslant 1/2k$, this implies the inequality

$$\frac{1}{2\sqrt{V_T(0, z(0))}} \geqslant k\Phi(T).$$

Let us take $C = 1/\sqrt{V_T(0, z(0))}$, then

$$\psi^{-1/2}(t) = -k\int_0^t \varphi(T - \zeta)d\zeta + C \geqslant -k\Phi(T) + C \geqslant \frac{1}{2\sqrt{V_T(0, z(0))}},$$

hence $\psi(t) \leqslant 4V_T(0, z(0)) = 4J_0(T)$. Since $\psi(0) = V_T(0, z(0))$, by the comparison theorem we obtain that $V_T(t, z(t)) \leqslant \psi(t) \leqslant 4J_0(T)$, for $\quad 0 \leqslant t < T$.

Substituting this estimate into (21) we get $(R(z), Q_T z) \leqslant k\varphi(T - t)(4J_0(T))^{3/2}$. Now we only need to integrate this expression to use it in (20),

$$J(T) - J_0(T) = \gamma(T, z(0)) = 2 \int_0^T R^\top(z(\xi))Q_T(\xi)z(\xi)d\xi \leqslant 2k\Phi(T)(4J_0(T))^{3/2}$$

that implies (19).

Since under the assumptions of Theorem 2 $\Phi(T)$ and $J_0(T)$ tends to zero, the right hand side of (19) also tends to zero as $T \to 0$.

In Theorem 1 we prove that the trajectory $z(t)$ of system (9) tends to zero as $t \to T$ and $V_T(t, z(t))$ is bounded in the neighborhood of T. The following theorem gives conditions under which $V_T(t, z(t)) \to 0$ as $t \to T$.

Theorem 3. *Let the inequality (8) holds. Let $T \leqslant \overline{T}$ and the trajectory $z(t)$ of system(9) tends to zero as $t \to T$. Then $V_T(t, z(t)) = z^\top(t)Q(t)z(t) \to 0$ as $t \to T$.*

Proof. Note that the function $u(\xi) = -B^\top Q(\xi)z(\xi)$ is continuous at $0 \leqslant \xi < T$.

Also note that the function $V_T(t, z(t))$ can only be infinite in the neighborhood of $t = T$. However, taking into account that the fulfillment of the inequality (8) implies the fulfillment of the assumption, one can see that it follows from $z(t) \to 0$ that the conditions of Lemma 3 will be met for t from the neighborhood of T. Hence, $V_T(t, z(t))$ is bounded along the whole trajectory $z(t)$. It is now clear from relation (18) that the integral cost $I(t, u) = \int_0^t u^\top(\xi)u(\xi)d\xi$ is uniformly bounded with respect to $t \in [0, T]$. Hence, $u(\cdot)$ belongs to the space $\mathbb{L}_2[0, T]$.

Let us assume, that the form $z^\top(t)Q_T(t)z(t)$ doesn't tends to zero as t tend to T. It means, that there exists sequence $t_k \to T$ and $\delta > 0$ such that

$$z^\top(t_k)Q(t_k)z(t_k) \geqslant \delta^2. \tag{22}$$

The inequality $\int_{t_k}^T u^\top(\xi)u(\xi)d\xi \leqslant \|u(\cdot)\|\sqrt{T - t_k}$ implies that there exists k_0 such that for all $k > k_0$ both of the following conditions are satisfied

$$\int_{t_k}^T u^\top(\xi)u(\xi)d\xi \leqslant \frac{\delta^2}{4}, \qquad T - t_k \leqslant \tau_0.$$

By the theorem, $z(t) \to 0$ as $t \to T$. Let us make a time change, then $y(\tau) = z(T - \tau)$ is the solution of (6) with initial condition $y(0) = 0$ and control $v(\tau) = u(T - \tau)$. Denote $\tau_k = T - t_k$, τ_k converges to zero. Obviously,

$$\int_0^{\tau_k} v^\top(\xi)v(\xi)d\xi \leqslant \frac{\delta^2}{4},$$

Therefore, $z(t_k) = y(\tau_k)$ belongs to the reachable set of system (6), i.e. the inclusion $z(t_k) \in G_-(T - t_k, \delta/2)$ holds.

From asymptotic equality of the reachable sets G_-^0, G_- [18] and properties of the Banach-Mazur distance it follows that

$$z(t_k) \in \exp(\rho(T - t_k))G_-^0(T - t_k, \frac{\delta}{2}) = G_-^0(T - t_k, \frac{\delta}{2}\exp(\rho(T - t_k))),$$

where $\rho(T - t_k) = \rho(G_-(T - t_k, \frac{\delta}{2}), G_-^0(T - t_k, \frac{\delta}{2}))$.

Since $t_k \to T$, $\rho(T - t_k) \to 0$, $\exp(\rho(T - t_k)) \to 1$ there exists k_1 such that for $k > k_1$ we have $\exp(\rho(T - t_k))\delta/2 \leqslant 2\delta/3$. Using the formula

$$G_-^0(T - t_k, \delta/2) = \left\{ x \in \mathbb{R}^n : xW^{-1}(T - t_k)x \leqslant \frac{4\delta^2}{9} \right\},$$

we obtain $z^\top(t_k)W^{-1}(T - t_k)z(t_k) \leqslant 4\delta^2/9$.

As the $W^{-1}(\tau_k) = Q(t_k)$, the inequality above contradicts the inequality (22). This means, that $V_T(t, z(t)) = z^\top(t)Q(t)z(t) \to 0$ as $t \to T$.

4   Examples

In this section, we present the results of numerical experiments that are intended to illustrate the application of the Theorems 1 and 2. Here we deal with the Duffing oscillator, whose equations

$$\dot{z}_1 = z_2, \qquad \dot{z}_2 = -z_1 - 10z_1^3 + u, \qquad 0 \leqslant t \leqslant T \qquad (23)$$

describe the motion of a non-linear elastic spring under the influence of an external force u. The desired final state is $z_1(T) = z_2(T) = 0$. This state is also a state of equilibrium.

Now let us check whether the right-hand side of the system (23) has a property 1. It is not difficult to see that it is fulfilled with $k = 10$, $r = 1$: for all z_1, z_2 such that $z_1^2 + z_2^2 \leqslant 1$, $\|R(z)\| = 10|z_1|^3 < 10(z_1^2 + z_2^2)$.

Linearization of the system (23) at the origin results in a system described by the following pair of matrices

$$A = \begin{pmatrix} 0 & 1 \\ -1 & 0 \end{pmatrix}, \qquad B = \begin{pmatrix} 0 \\ 1 \end{pmatrix}. \qquad (24)$$

To select the function $\varphi(\tau)$, we have to write out the controllability Gramian's eigenvalues of the system (24) $\nu(t) = \frac{t^3}{12} + O(t^5)$, $\eta(t) = t - \frac{t^3}{12} + O(t^5)$. These eigenvalues allow us to choose $\varphi(t) = \frac{4}{\sqrt{t}}$. In this case, $\Phi(T) = 8\sqrt{T}$ and $\overline{T}$ could be as large as one wants. We set $\beta = 0.5$ to have

$$\frac{\sqrt{\eta(T)}}{\Phi^\beta(T)} = \frac{\sqrt{30}\, t^{0.25}\, \sqrt{t^4 - 20\, t^2 + 240}}{240} \to 0 \text{ as } T \to 0.$$

In the first series of experiments, the initial conditions $z_0 = (-0.0108; 0.2722)$ are fixed, and only the length of the time interval T is varied. The point z_0 is chosen such that it lies inside the null-controllable set of the system (23) at $T = 0.075$ and $\mu = 1$, that is, $z_0 \in G_-(0.075, 1)$.

Table 1. Results of experiments with variable T

№	T	$z_0^\top Q_T(0) z_0$	$J(T, z_0)$	Δ_J
1	1.500	0.159686	0.159136	0.0034435
2	1.250	0.197308	0.197031	0.0013990
3	1.000	0.249346	0.249231	0.0004611
4	0.750	0.327395	0.327360	0.0001055
5	0.500	0.459094	0.459089	0.0000102
6	0.250	0.710789	0.710790	0.0000012
7	0.100	0.836541	0.836542	0.0000014
8	0.075	1.000000	1.000001	0.0000009

Now, by changing T, we will calculate $Q_T(0)$ and simulate the motion of the system (23), closed by the linear feedback $u(t) = -B^\top Q_T(t)x$. The simulation results are shown in Fig. 1 and in Table 1. The green areas denote the null-controllable set $G_-(T, 1)$ of the system (23) at $T = \{0.075, 0.1, 0.25, 0.5, 0.75, 1.0, 1.25, 1.5\}$, the solid lines show the trajectories of the system at the same T. The symbol "$\blacklozenge$" represents the point z_0, and "$\bullet$" — the target point located at the origin of the coordinates. The lower left part of the drawing shows the zoomed area around the origin, marked in red rectangle. The heading of the table uses the notation

$$\Delta_J = \frac{|J_0(T, z_0) - J(T, z_0)|}{J_0(T, z_0)}.$$

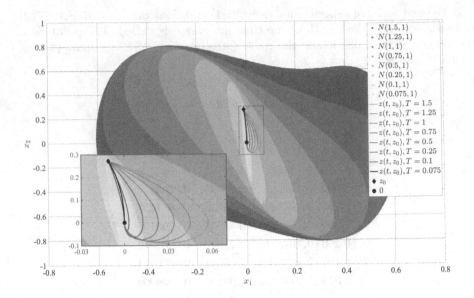

Fig. 1. Results of experiments with variable T.

Despite the fact that condition $z_0 \in B(0, r_1(T))$ of Theorem 1 is not satisfied with $r_1(T)$ used in the proof, the trajectories still tend to zero. This can be explained by a too strict choice of $r_1(T)$ and by the fact that Theorem 1 only formulates sufficient conditions for trajectories to tend to zero. One can observe that in the case of fixed initial conditions, as T decreases, the relative difference of the functionals Δ_J decreases, as follows from the estimate obtained in the Theorem 2.

Now we will change the conditions of the experiment quite a bit. We will change not only T but also the initial conditions z_0 so that, firstly, the equality $z_0^\top Q_T(0) z_0 = 1$, and second, so that the point z_0 is inside the corresponding null-controllable set $G_-(T, 1)$.

Table 2. Results of experiments with variable T and z_0

№	T	z_0	$\|z_0\|^2$	$z_0^\top Q_T(0) z_0$	$J(T, z_0)$	Δ_J
1	1.500	[−0.594; −0.057]	0.356680	0.999993	1.075833	0.0758407
2	1.250	[−0.578; 0.226]	0.385032	0.999997	1.110464	0.1104671
3	1.000	[−0.502; 0.508]	0.509514	0.999999	1.108159	0.1081607
4	0.750	[−0.354; 0.652]	0.550391	1.000000	1.048784	0.0487844
5	0.500	[−0.195; 0.638]	0.445513	1.000000	1.009848	0.0098475
6	0.250	[−0.069; 0.481]	0.236487	1.000000	1.000349	0.0003490

The results of this series of experiments are shown in Fig. 2 and in table 2. The notations are generally similar to Fig. 1: The green areas denote the null-controllable sets $G_-(T,1)$ of the system (23) at $T = \{0.25, 0.5, 0.75, 1.0, 1.25, 1.5\}$, the dashed lines show the boundaries of the null-controllable sets of the linearized system (24), the solid lines show the trajectories of the non-linear system at different T. The symbols "♦" in different colours represent initial conditions z_0, and "•" — the target point located at the origin.

The remark about not fulfilling the condition from the first part of the example is also relevant here. Table 2 shows that the values of Δ_J also decrease with decreasing T, but the decrease is not monotonic. This appears to be due to the fact that not only T but also z_0 is changing.

Also, in Fig. 2 we can observe that the null-controllable sets of the nonlinear and linearized systems are close in form at $T \leqslant 0.75$.

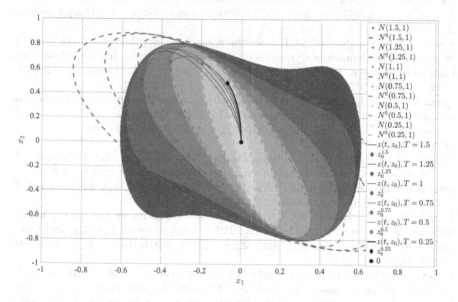

Fig. 2. Results of experiments with variable T and z_0.

5   Conclusion

It is shown that the linearization method may be applied to the problem of an optimal control synthesis on a finite time interval. The linear feedback controller, designed for the linearized system, also provides a local solution to synthesis problem for the nonlinear system if the interval is sufficiently small. This requires fairly strict restrictions on the controllability Gramian asymptotics, which coincide with sufficient conditions ensuring the asymptotic equivalence of reachable (null-controllable) sets. Under these conditions the estimate for the relative error values of an integral cost was given. We used an example of a nonlinear spring system under the influence of an external force to demonstrate the implementation of the described synthesis method.

References

1. Krasovskii, N.N.: Problems of stabilizing controlled movements, Editor's addendum to the book of Malkin I.G., Theory of motion stability. Nauka, Moscow (1966), pp. 475–514. (in Russian)
2. Khalil, K.H.: Nonlinear Systems, 3rd edn. Pearson, New Jersey (2001)
3. Roxin, E.O.: Linearization and approximation of control systems. In: Lakshmikantham, V. (ed.), Proceedings of the First World Congress of Nonlinear Analysts, Tampa, Florida, 19–26 August 1992. De Gruyter, Berlin, 2531–2540 (1996). https://doi.org/10.1515/9783110883237.2531
4. Li, L., Wang, T., Xia, Y., et al.: Trajectory tracking control for wheeled mobile robots based on nonlinear disturbance observer with extended Kalman filter. J. Franklin Inst. 357(13), 8491–8507 (2020). https://doi.org/10.1016/j.jfranklin.2020.04.043
5. van den Berg, J.: Iterated LQR smoothing for locally-optimal feedback control of systems with non-linear dynamics and non-quadratic cost. In: 2014 American Control Conference, Portland, OR, USA, 1912–1918 (2014). https://doi.org/10.1109/ACC.2014.6859404
6. Pang, Z.-H., Ma, B., Liu, G.-P., et al.: Data-driven adaptive control: an incremental triangular dynamic linearization approach. In: IEEE Transactions on Circuits and Systems II: Express Briefs, vol. 69, no. 12, pp. 4949–4953 (2022). https://doi.org/10.1109/TCSII.2022.3181232
7. Murillo, M.H., Limache, A.C., Rojas Fredini, P.S., Giovanini, L.L.: Generalized nonlinear optimal predictive control using iterative state-space trajectories: applications to autonomous flight of UAVs. Int. J. Control Autom. Syst. 13(2), 361–370 (2015). https://doi.org/10.1007/s12555-013-0416-y
8. Papadimitriou, D., Rosolia, U., Borrelli, F.: Control of unknown nonlinear systems with linear time-varying MPC. In: 59th IEEE Conference on Decision and Control (CDC) 2020, Jeju, Korea (South), 2258–2263 (2020). https://doi.org/10.1109/CDC42340.2020.9304441
9. Gusev, M.I., Osipov, I.O.: On a local synthesis problem for nonlinear systems with integral constraints. Vestnik Udmurtskogo Universiteta. Matematika. Mekhanika. Komp'yuternye Nauki 32(2), 171–186 (2022). https://doi.org/10.35634/vm220202
10. Abgaryan, K.A.: Matrix calculus with applications in the theory of dynamical systems. Fizzmatlit, Moscow (1994). (in Russian)
11. Kurzhanski, A.B., Varaiya, P.: Dynamics and control of trajectory tubes. theory and computation. Birkhauser (2014). https://people.eecs.berkeley.edu/varaiya/Download/KurzhanskiVaraiya.pdf
12. Guseinov, K.G.: Approximation of the attainable sets of the nonlinear control systems with integral constraint on controls. Nonlinear Anal. 71(1), 622–645 (2009). https://doi.org/10.1016/j.na.2008.10.097
13. Rousse, P., Alexandre dit Sandretto, J., Chapoutot, A., Garoche, P.-L.: Guaranteed simulation of dynamical systems with integral constraints and application on delayed dynamical systems. In: Chamberlain, R., Edin Grimheden, M., Taha, W. (eds.) CyPhy/WESE -2019. LNCS, vol. 11971, pp. 89–107. Springer, Cham (2020). https://doi.org/10.1007/978-3-030-41131-2_5

14. Gusev, M.I., Zykov, I.V.: On extremal properties of boundary points of reachable sets for a system with integrally constrained control. In: Proceedings of 20th World Congress International Federation of Automatic Control, vol. 50, pp. 4082–4087. Elsevier (2017). https://doi.org/10.1016/j.ifacol.2017.08.792
15. Lassak, M.: Banach–mazur distance from the parallelogram to the affine-regular hexagon and other affine-regular even-gons. RM **76**(2), 1–12 (2021). https://doi.org/10.1007/s00025-021-01368-8
16. Goncharova, E., Ovseevich, A.: Small-time reachable sets of linear systems with integral control constraints: birth of the shape of a reachable set. J. Optim. Theory Appl. **168**(2), 615–624 (2015). https://doi.org/10.1007/s10957-015-0754-4
17. Polyak, B.T.: Convexity of nonlinear image of a small ball with applications to optimization. Set-Valued Analysis **9**, 159–168 (2001). https://doi.org/10.1023/A:1011287523150
18. Gusev, M.I., Osipov, I.O.: Asymptotic Behavior of Reachable Setson Small Time Intervals. Proc. Steklov Inst. Math. **309**(1), S52–S64 (2020). https://doi.org/10.1134/S0081543820040070
19. Osipov, I.O.: On the convexity of the reachable set with respect to a part of coordinates at small time intervals. Vestn. Udmurtsk. Univ. Mat. Mekh. Komp. Nauki, 31(2), 210–225 (2021). https://doi.org/10.35634/vm210204
20. Gusev, M.: Estimates of the minimal eigenvalue of the controllability Gramian for a system containing a small parameter. In: Khachay, M., Kochetov, Y., Pardalos, P. (eds.) MOTOR 2019. LNCS, vol. 11548, pp. 461–473. Springer, Cham (2019). https://doi.org/10.1007/978-3-030-22629-9_32
21. Walter, W.: Differential and Integral Inequalities. Springer, Berlin (1970)

A Priori Estimates of the Objective Function in the Speed-in-Action Problem for a Linear Two-Dimensional Discrete-Time System

Danis N. Ibragimov[✉] [iD] and Sofya R. Guseva [iD]

Moscow Aviation Institute (National Research University), Volokolamskoe shosse, 4,
Moscow 125993, Russia
rikk.dan@gmail.com

Abstract. The two-dimensional linear system with discrete time and
bounded control is considered. It is assumed that the system matrix is
nondegenerate and diagonalizable by means of a similarity transformation and the set of admissible values of the control is convex and compact.
The speed-in-action problem is studied for a given system, in particular,
it is required to construct a priori estimates of the optimal value of the
objective function in the speed-in-action problem as a function of the
initial state and parameters of the system, which do not require an exact
construction of the class of 0-controllable sets. In the trivial case, when
the system matrix equals its real Jordan form, and the set of admissible
values of the control is either a ball or a rectangle, the optimal value of
the objective function in the speed-in-action problem is obtained explicitly. The method is proposed, which reduces case of an arbitrary control
system to a trivial case by means of the singular decomposition of the
system matrix and constructing internal and external approximations of
constraints on control actions. Numerical calculations are presented that
demonstrate the efficiency and accuracy of the developed technique.

Keywords: Linear Discrete-Time System · Speed-in-Action Problem ·
Optimal Control · A Priori Estimates of the Optimal Value of the
Objective Function

1 Introduction

The speed-in-action problem has been known for a long time as an optimal
control problem with a natural quality functional, that is the time, which system
spent to reach some given terminal state [2,14,15]. This problem does not have
any essential features in comparison with other problems of optimal control
theory, when we consider continuous-time systems. The solution obtained by
Pontryagin's maximum principle [15] guarantees the relay nature of control for
linear systems.

Supported by Russian Science Foundation № 23-21-00293.

At the same time, systems with discrete time have a number of fundamental differences from continuous-time systems in terms of constructing an optimal control [3,4,16]. Whereas in the continuous case the optimization problem is a problem of variational calculus, in discrete-time case it is a convex programming problem. This fact determines a fundamentally different set of tools for constructing optimal processes in the discrete case. But, despite the fact that by means of the discrete maximum principle [5,16] and the dynamic programming method [1] it is possible to solve most of the problems of the theory of optimal control of discrete-time systems, for solving the speed-in-action problem these principle and the method are inapplicable due to the irregularity of the extremum for almost all initial states, the non-uniqueness of the optimal trajectory, and the discrete nature of the control quality function, which is the number of steps required to reach a fixed terminal state from a given initial state [7,10]. For example, the use of well-known modern results on the optimal control of discrete-time systems [12,13,19,21] in relation to the speed-in-action problem are inefficient and incorrect.

In this regard, it is relevant to search for alternative approaches to solving the problem. At the moment, the method based on the apparatus of 0-controllable sets [9–11,18] has demonstrated its effectiveness, where each 0-controllable set is a set of those initial states from which it is possible to transfer the system to the origin by choosing an admissible control in a finite number of steps. At the same time, it is proved that the solution method depends on the type of control restrictions. In the case of strictly convex constraints, it is possible to modify the well-known maximum principle [7,10] to construct an optimal control. If the set of admissible control values is a polytope, then the solution of the original problem can be reduced to solving a number of linear programming problems due to the dynamic programming method [9,11]. Also [9] describes a method for reducing the case of arbitrary convex control constraints to the case of linear constraints by a polyhedral approximation of the set of admissible control values.

An essential drawback of the results of [9–11] is the lack of complexity estimates for the algorithms presented there. Moreover, the complexity of these methods for a given initial state directly depends on the number of 0-controllable sets that need to be constructed. And this number equal to the optimal value of the objective function in the considered problem. I.e. it is necessary to solve the speed-in-action problem to calculate its complexity. The purpose of this article is to construct estimates of the minimum number of steps required to transfer the system to the origin in the form of functions of the initial state and system parameters. The system is assumed to be two-dimensional, which is due to the possibility of obtaining the necessary estimates in an explicit form. On the other hand, in further studies it is possible to generalize the obtained results to the case of a system of arbitrary dimension by decomposition methods. The proofs of the stated Theorems and Lemmas are based on statements of convex analysis [17] and matrix analysis [6].

2   Problem Statement

We consider a two-dimensional discrete-time linear control system $(A, \mathcal{U})$:

$$x(k+1) = Ax(k) + u(k),$$
$$x(0) = x_0, \ u(k) \in \mathcal{U}, \ k \in \mathbb{N} \cup \{0\}, \tag{1}$$

where $x(k) \in \mathbb{R}^2$ is the system state vector, $u(k) \in \mathbb{R}^2$ is the control action at k-th step, $A \in \mathbb{R}^{2 \times 2}$ is the system matrix, $\mathcal{U} \subset \mathbb{R}^2$ is the set of admissible control values. It is assumed that $\mathcal{U}$ is a convex compact, $0 \in \mathrm{int}\,\mathcal{U}$, $\det A \neq 0$, the phase space $\mathbb{R}^2$ is normed with the norm defined by the following relation:

$$\|x\| = \sqrt{x_1^2 + x_2^2}.$$

For system (1) the speed-in-action problem is being solved, i.e. it is required to construct an admissible control $\{u^*(k)\}_{k=0}^{N_{\min}-1} \subset \mathcal{U}$, which we will call optimal, transferring the system from the initial state $x_0 \in \mathbb{R}^2$ to the origin in the minimum number of steps $N_{\min}$:

$$N_{\min} = \min\{N \in \mathbb{N} \cup \{0\} : \exists u(0), \dots, u(N-1) \in \mathcal{U} : x(N) = 0\}.$$

As demonstrated in [9,10], the solution of the speed-in-action problem is based on the apparatus of the 0-controllable sets $\{\mathcal{X}(N)\}_{N=0}^\infty$, where $\mathcal{X}(N) \subset \mathbb{R}^2$ is the set of those initial states from which system (1) can be transferred to the origin in N steps:

$$\mathcal{X}(N) = \begin{cases} \{x_0 \in \mathbb{R}^2 : \exists u(0), \dots, u(N-1) \in \mathcal{U} : x(N) = 0\}, & N \in \mathbb{N}, \\ \{0\}, & N = 0. \end{cases} \tag{2}$$

The solvability of the speed-in-action problem for each initial state is reduced to the construction of limit 0-controllable sets and is discussed in detail in [8,18]. In what follows, we will assume that the speed-in-action problem for a given initial state x_0 is solvable, i.e. the inclusion

$$x_0 \in \bigcup_{N=0}^{\infty} \mathcal{X}(N)$$

is correct.

The solution of the speed-in-action problem consists of two stages, which are calculation of the value $N_{\min}$ and construction of the optimal process $\{x^*(k), u^*(k-1), x_0\}_{k=1}^{N_{\min}}$. The second stage is described in detail in a series of papers [9–11]. The first stage is reduced to the sequential construction of 0-controllable sets:

$$N_{\min} = \min\{N \in \mathbb{N} \cup \{0\} : x_0 \in \mathcal{X}(N)\}. \tag{3}$$

But the process of constructing of the sequence $\{\mathcal{X}(N)\}_{N=0}^\infty$ is a very computationally difficult procedure, which follows from the following representation of 0-controllable sets.

Lemma 1 ([10]). *Let the sequence* $\{\mathcal{X}(N)\}_{N=0}^{\infty}$ *be defined by relations* (2), $\det A \neq 0$. *Then for all* $N \in \mathbb{N}$ *the following representation is correct:*

$$\mathcal{X}(N) = -\sum_{k=1}^{N} A^{-k}\mathcal{U}.$$

The operation of Minkowski sum of sets is often difficult to implement from a computational point of view. Even in the case when $\mathcal{U}$ is a polytope, and therefore every $\mathcal{X}(N)$ is also a polytope [17], their descriptive complexity (i.e. the number of vertices) grows exponentially with N [20].

On the other hand, there are cases when the exact value of $N_{\min}$ is not essential and it is sufficient to obtain only some estimates of this value. For example, such estimates are useful to calculate the complexity of the speed-in-action problem solving, which depends on the value of $N_{\min}$, or to compare two initial states. The aim of this work is to develop methods that will allow to calculate a priori estimate $N_{\min}$ for given matrix A and set $\mathcal{U}$ without the need to construct the sequence $\{\mathcal{X}(N)\}_{N=0}^{\infty}$ explicitly. I.e. it is required to determine the values $\overline{N_{\min}}, \underline{N_{\min}} \in \mathbb{N}$ as functions of $x_0 \in \mathbb{R}^2$, $A \in \mathbb{R}^{2 \times 2}$, $\mathcal{U} \subset \mathbb{R}^2$ so that the following inequality will be true:

$$\overline{N_{\min}} \leqslant N_{\min} \leqslant \underline{N_{\min}}.$$

3 The Exact Value of $N_{\min}$ in the Trivial Case

For a two-dimensional system (1), in some cases related to the structure of the system matrix A and the shape of the set $\mathcal{U}$, it is possible to construct an explicit description of the 0-controllable set $\mathcal{X}(N)$. Due to (3) this fact allows to explicitly describe $N_{\min}$ as a function of the initial state x_0.

The diagonalizability of the matrix A is essential. As is known [6], there are two fundamentally different cases of real and complex eigenvalues for a diagonalizable two-dimensional matrix. If $\lambda_1, \lambda_2 \in \mathbb{R}$ are real eigenvalues of A, then a non-degenerate maxtrix $S \in \mathbb{R}^{2 \times 2}$ exists such that the following representation is correct:

$$\Lambda = \begin{pmatrix} \lambda_1 & 0 \\ 0 & \lambda_2 \end{pmatrix} = S^{-1}AS. \tag{4}$$

If $\lambda_{1,2} = re^{\pm i\varphi} \in \mathbb{C}$ are complex eigenvalues of A, then a non-degenerate maxtrix $S \in \mathbb{R}^{2 \times 2}$ exists such that the following representation is correct:

$$\Lambda = r\begin{pmatrix} \cos(\varphi) & \sin(\varphi) \\ -\sin(\varphi) & \cos(\varphi) \end{pmatrix} = S^{-1}AS. \tag{5}$$

Next, we present the conditions that allow us to calculate $N_{\min}$ for the cases when A equals to one of its two possible real Jordan forms (4) or (5).

Lemma 2. *Let* $A = \mathrm{diag}(\lambda_1, \lambda_2)$, $\mathcal{U} = [-u_{1,\max}; u_{1,\max}] \times [-u_{2,\max}; u_{2,\max}]$, *for a given initial state* $x_0 \in \mathbb{R}^2$ *it is true that* $N_{\min} < \infty$. *Then the inclusion*

$x_0 \in \mathcal{X}(N)$ *is true if and only if for all* $i = \overline{1,2}$ *the following inequality is correct:*

$$
N \geqslant
\begin{cases}
\dfrac{|x_{0,i}|}{u_{i,max}}, & |\lambda_i| = 1, \\[3mm]
-\dfrac{\ln\left(1 - \dfrac{|x_{0,i}|}{u_{i,max}}(|\lambda_i| - 1)\right)}{\ln|\lambda_i|}, & |\lambda_i| \neq 1.
\end{cases}
$$

Proof. Due to Lemma 1 for any $N \in \mathbb{N}$ the following representation is valid:

$$
\mathcal{X}(N) = -\sum_{k=1}^{N}
\begin{pmatrix} \lambda_1 & 0 \\ 0 & \lambda_2 \end{pmatrix}^{-k}
[-u_{1,max}; u_{1,max}] \times [-u_{2,max}; u_{2,max}] =
$$

$$
\sum_{k=1}^{N} \left[-u_{1,max}|\lambda_1|^{-k}; u_{1,max}|\lambda_1|^{-k}\right] \times \left[-u_{2,max}|\lambda_2|^{-k}; u_{2,max}|\lambda_2|^{-k}\right] =
$$

$$
\left[-u_{1,max}\sum_{k=1}^{N}|\lambda_1|^{-k}; u_{1,max}\sum_{k=1}^{N}|\lambda_1|^{-k}\right] \times \left[-u_{2,max}\sum_{k=1}^{N}|\lambda_2|^{-k}; u_{2,max}\sum_{k=1}^{N}|\lambda_2|^{-k}\right].
$$

Then the inclusion $x_0 \in \mathcal{X}(N)$ is valid if and only if for each $i = \overline{1,2}$ the inequalities

$$
-u_{i,max}\sum_{k=1}^{N}|\lambda_i|^{-k} \leqslant x_{0,i} \leqslant u_{i,max}\sum_{k=1}^{N}|\lambda_i|^{-k},
$$

$$
|x_{0,i}| \leqslant u_{i,max}\sum_{k=1}^{N}|\lambda_i|^{-k}
$$

are true. Since $N_{min} < \infty$, the last inequality will be satisfied for at least one $N \in \mathbb{N}$. Moreover, for the case $|\lambda_i| = 1$ the corresponding condition is equivalent to the inequalities

$$
|x_{0,i}| \leqslant u_{i,max}N, \quad N \geqslant \frac{|x_{0,i}|}{u_{i,max}}.
$$

For the $|\lambda_i| \neq 1$ the corresponding condition is equivalent to the inequalities

$$
|x_{0,i}| \leqslant u_{i,max}\frac{1 - |\lambda_i|^{-N}}{|\lambda_i| - 1}, \quad N \geqslant -\frac{\ln\left(1 - \frac{|x_{0,i}|}{u_{i,max}}(|\lambda_i| - 1)\right)}{\ln|\lambda_i|}.
$$

$\square$

The inequalities for N in the statements of Lemma 2 can be replaced by an equivalent condition of a more general form:

$$
N \geqslant \lim_{\alpha \to |\lambda_i|} -\frac{\ln\left(1 - \frac{|x_{0,i}|}{u_{i,max}}(\alpha - 1)\right)}{\ln\alpha}.
$$

Lemma 3. *Let* $A = r \begin{pmatrix} \cos(\varphi) & \sin(\varphi) \\ -\sin(\varphi) & \cos(\varphi) \end{pmatrix}$, $\mathcal{U} = \mathcal{B}_{R_{\max}}(0) = \{u \in \mathbb{R}^2 : \|u\| \leq R_{\max}\}$, *for a given initial state* $x_0 \in \mathbb{R}^2$ *it is true that* $N_{\min} < \infty$. *Then the inclusion* $x_0 \in \mathcal{X}(N)$ *is true if and only if the following inequality is correct:*

$$N \geq \begin{cases} \dfrac{\|x_0\|}{R_{\max}}, & r = 1, \\[3mm] -\dfrac{\ln\left(1 - \frac{\|x_0\|}{R_{\max}}(r-1)\right)}{\ln r}, & r \neq 1. \end{cases}$$

Proof. Due to Lemma 1 for any $N \in \mathbb{N}$ the following representation is valid:

$$\mathcal{X}(N) = -\sum_{k=1}^{N} r^{-k} \begin{pmatrix} \cos(\varphi) & \sin(\varphi) \\ -\sin(\varphi) & \cos(\varphi) \end{pmatrix}^{-k} \mathcal{B}_{R_{\max}}(0) =$$

$$\sum_{k=1}^{N} r^{-k} \mathcal{B}_{R_{\max}}(0) = \sum_{k=1}^{N} \mathcal{B}_{R_{\max} r^{-k}}(0) = \mathcal{B}_{R_{\max} \sum_{k=1}^{N} r^{-k}}(0).$$

Then the inclusion $x_0 \in \mathcal{X}(N)$ is valid if and only if the inequality

$$\|x_0\| \leq R_{\max} \sum_{k=1}^{N} r^{-k}$$

is true. Since $N_{\min} < \infty$, the inequality will be satisfied for at least one $N \in \mathbb{N}$. Moreover, for the case $r = 1$ the corresponding condition is equivalent to the inequalities

$$\|x_0\| \leq R_{\max} N, \quad N \geq \frac{\|x_0\|}{R_{\max}}.$$

For the $r \neq 1$ the corresponding condition is equivalent to the inequalities

$$\|x_0\| \leq R_{\max} \frac{1 - r^{-N}}{r - 1}, \quad N \geq -\frac{\ln\left(1 - \frac{\|x_0\|}{R_{\max}}(r-1)\right)}{\ln r}.$$

$\square$

The inequalities for N in the statements of Lemma 3 can be replaced by an equivalent condition of a more general form:

$$N \geq \lim_{\alpha \to r} -\frac{\ln\left(1 - \frac{\|x_0\|}{R_{\max}}(\alpha - 1)\right)}{\ln \alpha}.$$

If we denote for an arbitrary $\alpha \in \mathbb{R}$ by $\lceil \alpha \rceil$ the minimum integer not less than α:

$$\lceil \alpha \rceil = \min\{k \in \mathbb{Z} : \alpha \leq k\},$$

then on the basis of Lemmas 2 and 3 we can obtain an explicit expression for $N_{\min}$. Under the conditions of Lemma 2 it is true that

$$
N_{\min} = \max_{i=\overline{1,2}} \left\{ \left[\lim_{\alpha \to |\lambda_i|} -\frac{\ln\left(1 - \frac{|x_{0,i}|}{u_{i,\max}}(\alpha - 1)\right)}{\ln \alpha} \right] \right\}. \tag{6}
$$

Under the conditions of Lemma 3 it is true that

$$
N_{\min} = \left[\lim_{\alpha \to r} -\frac{\ln\left(1 - \frac{\|x_0\|}{R_{\max}}(\alpha - 1)\right)}{\ln \alpha} \right]. \tag{7}
$$

4   A Priori Estimates of $N_{\min}$ in the General Case

The results of Sect. 3 can only be used for a small class of $(A, \mathcal{U})$ systems. But we can generalize them to the case of an arbitrary diagonalizable matrix A and a set $\mathcal{U}$ by introducing auxiliary systems that satisfy the conditions of Lemma 2 or Lemma 3, for which the optimal value of the objective function in the speed-in-action problem is be an upper or lower estimate of the exact value of $N_{\min}$.

Lemma 4. *Let the inclusion $\underline{\mathcal{U}} \subset \mathcal{U} \subset \overline{\mathcal{U}}$ be correct, where $\underline{\mathcal{U}}, \mathcal{U}, \overline{\mathcal{U}} \subset \mathbb{R}^2$ are convex and compact sets containing 0, the speed-in-action problem for $x_0 \in \mathbb{R}^2$ is solvable for systems $(A, \underline{\mathcal{U}})$, $(A, \mathcal{U})$, $(A, \overline{\mathcal{U}})$, and $\underline{N_{\min}}$, $N_{\min}$, $\overline{N_{\min}}$ are the optimal values of the objective function in the speed-in-action problem for these systems respectively.*
    Then

$$
\overline{N_{\min}} \leqslant N_{\min} \leqslant \underline{N_{\min}}.
$$

Proof. Due to condition (2) for any $N \in \mathbb{N} \cup \{0\}$ the inclusion

$$
\underline{\mathcal{X}}(N) \subset \mathcal{X}(N) \subset \overline{\mathcal{X}}(N)
$$

is true, where $\underline{\mathcal{X}}(N), \mathcal{X}(N), \overline{\mathcal{X}}(N)$ are 0-controllable sets in N steps of systems $(A, \underline{\mathcal{U}})$, $(A, \mathcal{U})$, $(A, \underline{\mathcal{U}})$ respectively. Then the proof of Lemma 4 follows directly from representation (3). $\qquad\square$

The problem of constructing a real Jordan form of the matrix A of type (4) or (5) is related to the calculation of eigenvectors and eigenvalues and the transformation to a real Jordan basis. I.e. any system $(A, \mathcal{U})$ in which A is diagonalizable can be reduced to the equivalent system $(\Lambda, \tilde{\mathcal{U}})$ by a nondegenerate linear transformation of the coordinate system. Lemma 4 allows us to construct upper and lower bounds for $N_{\min}$ by considering auxiliary systems $(A, \underline{\mathcal{U}})$, $(A, \overline{\mathcal{U}})$, where the sets $\underline{\mathcal{U}}, \overline{\mathcal{U}} \subset \mathbb{R}^2$ are chosen in accordance with the conditions of Lemmas 2 and 3 and depend on the type of real Jordan form Λ. We formulate this fact in the form of a Theorem.

Theorem 1. *Let $A \in \mathbb{R}^{2 \times 2}$ have two linearly independent eigenvectors, $\det A \neq 0$, $S \in \mathbb{R}^{2 \times 2}$ be the transformation matrix to the real Jordan basis of the matrix A, for a given initial state $x_0 \in \mathbb{R}^2$ it is true that $N_{\min} < \infty$.*
   Then following statements are correct.

1. if $\lambda_1, \lambda_2 \in \mathbb{R}$ are eigenvalues of the matrix A, then inequalities

$$\max_{i=\overline{1,2}}\left\{\left[\lim_{\alpha \to |\lambda_i|} -\frac{\ln\left(1 - \frac{|y_{0,i}|}{u''_{i,\max}}(\alpha - 1)\right)}{\ln \alpha}\right]\right\} \leqslant N_{\min} \leqslant$$

$$\max_{i=\overline{1,2}}\left\{\left[\lim_{\alpha \to |\lambda_i|} -\frac{\ln\left(1 - \frac{|y_{0,i}|}{u'_{i,\max}}(\alpha - 1)\right)}{\ln \alpha}\right]\right\}$$

are true, where $y_0 = S^{-1}x_0$ and the numbers $u'_{1,\max}, u'_{2,\max}, u''_{1,\max}, u''_{2,\max} > 0$ are determined from the condition

$$[-u'_{1,\max}; u'_{1,\max}] \times [-u'_{2,\max}; u'_{2,\max}] \subset S^{-1}\mathcal{U} \subset [-u''_{1,\max}; u''_{1,\max}] \times [-u''_{2,\max}; u''_{2,\max}].$$

2. if $\lambda_{1,2} = re^{\pm i\varphi} \in \mathbb{C}$ are eigenvalues of the matrix A, then inequalities

$$\left[\lim_{\alpha \to r} -\frac{\ln\left(1 - \frac{\|S^{-1}x_0\|}{R''_{\max}}(\alpha - 1)\right)}{\ln \alpha}\right] \leqslant N_{\min} \leqslant \left[\lim_{\alpha \to r} -\frac{\ln\left(1 - \frac{\|S^{-1}x_0\|}{R'_{\max}}(\alpha - 1)\right)}{\ln \alpha}\right]$$

are correct, where the numbers $R'_{\max}, R''_{\max} > 0$ are determined from the condition $\mathcal{B}_{R'_{\max}}(0) \subset S^{-1}\mathcal{U} \subset \mathcal{B}_{R''_{\max}}(0)$.

Proof. Denote by $\{\tilde{\mathcal{X}}(N)\}_{N=0}^{\infty}$ the class of 0-controllable sets of the system $(\Lambda, S^{-1}\mathcal{U})$. Due to Lemma 1 and formulas (4) and (5), the representation

$$\mathcal{X}(N) = -\sum_{k=1}^{N} (S\Lambda S^{-1})^{-k}\mathcal{U} = -\sum_{k=1}^{N} S\Lambda^{-k}S^{-1}\mathcal{U} = S\tilde{\mathcal{X}}(N).$$

is correct. It follows that the inclusion $x_0 \in \mathcal{X}(N)$ is true if and only if the inclusion $S^{-1}x_0 \in \tilde{\mathcal{X}}(N)$ is true. I.e. due to (3) the optimal values of the objective function in the speed-in-action problem for systems $(A, \mathcal{U})$ and $(\Lambda, S^{-1}\mathcal{U})$ are equal for initial states x_0 and $S^{-1}x_0$ respectively.

If $\lambda_1, \lambda_2 \in \mathbb{R}$ are eigenvalues of the matrix A, then due to Lemma 2 and expression (6) the optimal value of the objective function in the speed-in-action problem $N_{\min}$ for system $(\Lambda, [-u'_{1,\max}; u'_{1,\max}] \times [-u'_{2,\max}; u'_{2,\max}])$ and the initial state $S^{-1}x_0$ has the following form:

$$N_{\min} = \max_{i=\overline{1,2}}\left\{\left[\lim_{\alpha \to |\lambda_i|} -\frac{\ln\left(1 - \frac{|y_{0,i}|}{u'_{i,\max}}(\alpha - 1)\right)}{\ln \alpha}\right]\right\}.$$

And the optimal value of the objective function in the speed-in-action problem $\overline{N_{\min}}$ for the system $(\Lambda, [-u'_{1,\max}; u'_{1,\max}] \times [-u'_{2,\max}; u'_{2,\max}])$ and the initial state $S^{-1}x_0$ has the following form:

$$\overline{N_{\min}} = \max_{i=\overline{1,2}} \left\{ \left[\lim_{\alpha \to |\lambda_i|} - \frac{\ln\left(1 - \frac{|y_{0,i}|}{u''_{i,\max}}(\alpha - 1)\right)}{\ln \alpha} \right] \right\}.$$

Taking into account Lemma 4, item 1 of Theorem 1 follows from the obtained relations.

If $\lambda_{1,2} = re^{\pm i\varphi} \in \mathbb{C}$ are eigenvalues of the matrix A, then due to Lemma 3 and expression (7) the optimal value of the objective function in the speed-in-action problem $\underline{N_{\min}}$ for system $(\Lambda, \mathcal{B}_{R'_{\max}}(0))$ and the initial state $S^{-1}x_0$ has the following form:

$$\underline{N_{\min}} = \left[\lim_{\alpha \to r} - \frac{\ln\left(1 - \frac{\|S^{-1}x_0\|}{R'_{\max}}(\alpha - 1)\right)}{\ln \alpha} \right].$$

And the optimal value of the objective function in the speed-in-action problem $\overline{N_{\min}}$ for the system $(\Lambda, \mathcal{B}_{R''_{\max}}(0))$ and the initial state $S^{-1}x_0$ has the following form:

$$\overline{N_{\min}} = \left[\lim_{\alpha \to r} - \frac{\ln\left(1 - \frac{\|S^{-1}x_0\|}{R''_{\max}}(\alpha - 1)\right)}{\ln \alpha} \right].$$

Taking into account Lemma 4, item 2 of Theorem 1 follows from the obtained relations.

Theorem 1 is completely proved.   □

Theorem 1 allows us to obtain a priori estimates for $N_{\min}$ for an arbitrary initial state x_0 only if the system matrix A has two linearly independent eigenvectors. This fact is essential, because results similar to Lemmas 2 and 3 cannot be obtained if the real Jordan form of the matrix A is not a rotation matrix or a diagonal matrix. This relates to the complexity of constructing a class of bounded and convex sets invariant under transformations of the following form:

$$\Lambda = \begin{pmatrix} \lambda & 1 \\ 0 & \lambda \end{pmatrix}.$$

In case (4), this is the class of rectangles $[a; b] \times [c; d]$, on which Lemma 2 is based. In case (5), this is the class of circles $\mathcal{B}_R(0)$, on which Lemma 3 is based.

Also, the application of Theorem 1 in practice requires knowledge of the values $u'_{1,\max}, u'_{2,\max}, u''_{1,\max}, u''_{2,\max}$ for case (4) and $R'_{\max}, R''_{\max}$ for case (5). The formulation of Theorem 1 does not give instructions for their calculation, but according to Lemma 4 the highest accuracy of the obtained estimates will be achieved for the maximum possible values of $u'_{1,\max}, u'_{2,\max}, R'_{\max}$ and minimum possible values $u''_{1,\max}, u''_{2,\max}, R''_{\max}$. From where the corresponding parameters can be calculated by solving the following convex programming problems:

$$R'_{\max} = \min_{u \in \partial S^{-1}\mathcal{U}} \|u\| = \min_{u \in \partial \mathcal{U}} \|S^{-1}u\|, \tag{8}$$

$$R''_{\max} = \max_{u \in S^{-1}\mathcal{U}} \|u\| = \max_{u \in \mathcal{U}} \|S^{-1}u\|, \tag{9}$$

$$u''_{i,\max} = \max_{u \in S^{-1}\mathcal{U}} |u_i| = \min_{u \in \mathcal{U}} |(S^{-1}u)_i|, \quad i = \overline{1,2}. \tag{10}$$

Although the parameters $u'_{1,\max}, u'_{2,\max}$ are defined ambiguously, it is possible to construct a set of their admissible values based on the following conditions:

$$(\pm u'_{1,\max}, 0)^T \in S^{-1}\mathcal{U},$$

$$u'_{2,\max} = \max_{(\pm u'_{1,\max}, \pm u_2)^T \in S^{-1}\mathcal{U}} u_2. \tag{11}$$

Here problem (8) determines the radius of the largest inscribed circle in $S^{-1}\mathcal{U}$, and formula (9) determines the radius of the smallest escribed circle around $S^{-1}\mathcal{U}$. Similarly, problem (10) determines the smallest rectangle escribed around $S^{-1}\mathcal{U}$, and conditions (11) allow us to construct a rectangle inscribed in $S^{-1}\mathcal{U}$, for whose length or width cannot be increased while maintaining the nesting condition.

5   Numerical Simulation

Let us demonstrate the obtained in Sects. 3 and 4 results by numerical simulation. Numerical simulation will be performed only for systems (1), in which the eigenvalues of the matrix A do not exceed 1 in absolute value. This restriction is related to the necessity to take into account the initial assumption $N_{\min} < \infty$. It is shown [8] that only in this case the equality

$$\bigcup_{N=0}^{\infty} \mathcal{X}(N) = \mathbb{R}^2$$

is true, i.e. there is no need to take into the account any restrictions on the initial state, which makes it possible to carry out calculations without additional checks.

Example 1. Let

$$\mathcal{U} = \text{conv}\left\{ \begin{pmatrix} \cos\left(\frac{2\pi j}{5}\right) \\ \sin\left(\frac{2\pi j}{5}\right) \end{pmatrix} : j = \overline{1,5} \right\},$$

$$A_1 = \begin{pmatrix} \frac{4}{5}(\cos(1) + \sin(1)) & -\frac{8}{5}\sin(1) \\ \frac{4}{5}\sin(1) & \frac{4}{5}(\cos(1) - \sin(1)) \end{pmatrix}.$$

The choice of $\mathcal{U}$ in the form of a polytope is due to the possibility of calculating the exact value of $N_{\min}$ based on the results presented in [9]. The eigenvalues of the matrix A_1 have the form $\lambda_{1,2} = \frac{4}{5}e^{\pm i}$, and representation (5) is true, where S and Λ_1 are defined as follows:

$$S = \begin{pmatrix} 1 & 1 \\ 1 & 0 \end{pmatrix}, \quad \Lambda_1 = \frac{4}{5} \begin{pmatrix} \cos(1) & \sin(1) \\ -\sin(1) & \cos(1) \end{pmatrix}.$$

The values $R'_{\max}$ and $R''_{\max}$ according to expressions (8) and (9) can be calculated explicitly:

$$R'_{\max} = \frac{2\sin\left(\frac{2\pi}{5}\right)}{\sqrt{10 + 2\cos\left(\frac{\pi}{5}\right) - 4\sin\left(\frac{\pi}{5}\right) + 8\sin\left(\frac{2\pi}{5}\right) - 8\cos\left(\frac{2\pi}{5}\right)}},$$

$$R''_{\max} = \sqrt{1 + \sin\left(\frac{4\pi}{5}\right) + \sin^2\left(\frac{2\pi}{5}\right)}.$$

The sets $\mathcal{B}_{R'_{\max}}(0), \mathcal{B}_{R''_{\max}}(0), S^{-1}\mathcal{U}$ are represented graphically in Fig. 1.

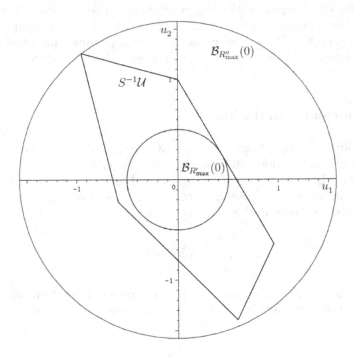

Fig. 1. Internal and external approximations of $S^{-1}\mathcal{U}$ by sets $\mathcal{B}_{R'_{\max}}(0)$, $\mathcal{B}_{R''_{\max}}(0)$, on the basis of which, according to Theorem 1, we construct estimates of $N_{\min}$.

The values of the $\overline{N_{\min}}$ and $\underline{N_{\min}}$ estimates according to Theorem 1 for fixed values of $R'_{\max}$, $R''_{\max}$, r are functions of $t = \|S^{-1}x_0\|$. Graphs of $\overline{N_{\min}}$ and $\underline{N_{\min}}$ for the system $(A_1, \mathcal{U})$ depending on t are shown in the Fig. 2. Due to

Lemma 4 as an error of $N_{\min}$ estimation we can consider the $\underline{N_{\min}} - \overline{N_{\min}}$ value, which is the difference between the values of the functions presented in Fig. 2. It can be seen that the estimation accuracy decreases as the true value of $N_{\min}$ increases, which is expressed in the increase in the distance between the graphs as t increases.

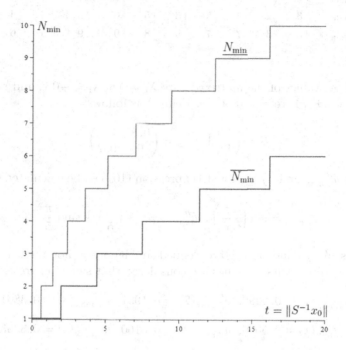

Fig. 2. Dependence of $\overline{N_{\min}}$ and $\underline{N_{\min}}$ estimates on $t = \|S^{-1}x_0\|$, obtained on the basis of Theorem 1.

Let us make calculations for various initial states $x_0^1, \ldots, x_0^{10} \in \mathbb{R}^2$. The values $x_0^1, \ldots, x_0^{10}$ are determined randomly inside the square $[-10; 10] \times [-10; 10]$:

$$(x_0^1, \ldots, x_0^{10}) = \begin{pmatrix} -3,19 & 0,59 & 4,32 & 9,76 & 4,40 & 8,25 & 0,10 & 1,16 & 0,06 & -0,75 \\ 0,93 & -1,04 & 7,08 & 2,08 & -0,02 & 9,59 & -9,31 & 9,54 & -2,73 & 3,59 \end{pmatrix}.$$

The value of $N_{\min}$ for each initial state is determined by exact construction of 0-controllable sets (2), based on the methods described in [9]. The results of numerical calculations are presented in Table 1.

Example 2. Let the set $\mathcal{U}$ be equal to the set considered in Example 1. Consider the system matrix with real eigenvalues:

$$A_2 = \begin{pmatrix} -0,7 & 1,5 \\ 0 & 0,8 \end{pmatrix}.$$

Table 1. Exact values and a priori estimates of N_{min} for various initial states for the case of complex eigenvalues of the system matrix.

j	1	2	3	4	5	6	7	8	9	10
$\|S^{-1}x_0^j\|$	4,23	1,95	7,60	7,96	4,43	9,69	13,24	12,69	3,91	5,63
$\overline{N_{min}}$	2	1	4	4	2	4	5	5	2	3
N_{min}	3	2	4	5	3	5	6	6	3	4
$\underline{N_{min}}$	5	3	7	7	5	8	9	9	5	6

Then the eigenvalues of the matrix A_2 are $\lambda_1 = 0.8$, $\lambda_1 = -0.7$, and representation (4) is valid, where S and Λ_2 are defined as follows:

$$S = \begin{pmatrix} 1 & 1 \\ 1 & 0 \end{pmatrix}, \quad \Lambda_2 = \begin{pmatrix} 0,8 & 0 \\ 0 & -0,7 \end{pmatrix}.$$

The values $u''_{1,\max}$ and $u''_{2,\max}$ due to expression (10) can be calculated explicitly:

$$u''_{1,\max} = \sin\left(\frac{2\pi}{5}\right), \quad u''_{2,\max} = \cos\left(\frac{\pi}{5}\right) + \sin\left(\frac{\pi}{5}\right).$$

The values $u'_{1,\max}$ and $u'_{2,\max}$ are defined ambiguously. For further numerical calculations, three pairs of values are considered that satisfy expression (11):

$$u'_{1,\max}(1) = 0,2896, \quad u'_{1,\max}(2) = 0,1931, \quad u'_{1,\max}(3) = 0,3861,$$

$$u'_{2,\max}(1) = 0,5000, \quad u'_{2,\max}(2) = 0,6160, \quad u'_{2,\max}(3) = 0,3333.$$

Sets $[-u''_{1,\max}; u''_{1,\max}] \times [-u''_{2,\max}; u''_{2,\max}]$, $S^{-1}\mathcal{U}$ and $[-u'_{1,\max}(k); u'_{1,\max}(k)] \times [-u'_{2,\max}(k); u'_{2,\max}(k)]$, $k = \overline{1,3}$ are represented graphically in Fig. 3.

The values of the estimates $\overline{N_{min}}$ and $\underline{N_{min}}(k)$ according to Theorem 1 for fixed values $u'_{1,\max}(k), u'_{2,\max}(k), u''_{1,\max}, u''_{2,\max}, \lambda_1, \lambda_2$ are functions of x_0. Graphs of $\overline{N_{min}}$ and $\underline{N_{min}}(k)$ for the system $(A_2, \mathcal{U})$ depending on x_0 are shown in Fig. 4.

In reality, all three values of $\underline{N_{min}}(k)$ are of no interest. It suffices to consider only the smallest value $\min_{k=\overline{1,3}} \{\underline{N_{min}}(k)\}$ for each x_0.

Let us make calculations for various initial states $x_0^1, \ldots, x_0^{10} \in \mathbb{R}^2$. The values $x_0^1, \ldots, x_0^{10}$ are determined randomly inside the square $[-10; 10] \times [-10; 10]$:

$$(x_0^1, \ldots, x_0^{10}) = \begin{pmatrix} 8,26 & 2,64 & -8,04 & -4,43 & 0,93 & 9,15 & 9,29 & -6,84 & 9,41 & 9,14 \\ -0,29 & 6,01 & -7,16 & -1,56 & 8,31 & 5,84 & 9,18 & 3,11 & -9,28 & 6,98 \end{pmatrix}.$$

The value of N_{min} for each initial state is determined by exact construction of 0-controllable sets (2), based on the methods described in [9]. The results of numerical calculations are presented in Table 2.

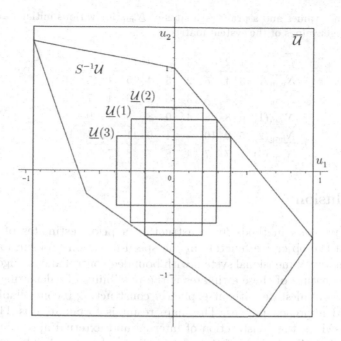

Fig. 3. Internal and external approximations of $S^{-1}\mathcal{U}$ by sets $\overline{\mathcal{U}} = [-u''_{1,\max}; u''_{1,\max}] \times [-u''_{2,\max}; u''_{2,\max}]$, $\underline{\mathcal{U}}(k) = [-u'_{1,\max}(k); u'_{1,\max}(k)] \times [-u'_{2,\max}(k); u'_{2,\max}(k)]$, $k = \overline{1,3}$, on the basis of which, according to Theorem 1, the estimates $N_{\min}$ are obtained.

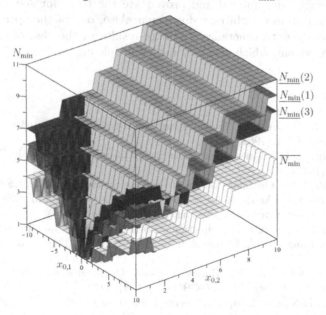

Fig. 4. Dependence of $\overline{N_{\min}}$ and $\underline{N_{\min}}(k)$, $k = \overline{1,3}$ estimates on x_0, obtained on the basis of Theorem 1.

Table 2. Exact values and a priori estimates of $N_{\min}$ for various initial states for the case of real eigenvalues of the system matrix.

j	1	2	3	4	5	6	7	8	9	10
$\overline{N}_{\min}$	3	4	5	2	5	4	5	4	5	5
$N_{\min}$	4	4	5	3	6	4	6	4	7	5
$N_{\min}(1)$	6	8	8	4	9	8	9	6	9	8
$N_{\min}(2)$	5	9	10	5	11	9	11	7	11	10
$N_{\min}(3)$	7	7	7	4	8	7	8	7	9	7

6   Conclusion

The paper presents methods for constructing a priori estimates of the optimal value of the objective function in the speed-in-action problem for a linear discrete-time two-dimensional system with bounded control and a diagonalizable matrix. The feature of these estimates is the possibility of calculating them for an arbitrary initial state without explicitly constructing 0-controllable sets, as the standard approach implies. The main result is formulated as Theorem 1, which is based on the construction of internal and external approximations of the set of admissible values of the control due to Lemma 4. The type of the approximation is determined on the basis of Lemmas 2 and 3. The accuracy of the estimates is studied in the numeraical simulation.

The statements formulated and proved are the basis for solving a similar problem for a system of arbitrary dimension. Also, one of the options for continuing the study is to generalize the main results to the case of an arbitrary matrix of the system, which also may have multiple eigenvalues.

References

1. Bellman, R., Dreyfus, S.: Applied dynamic programming. United states air force project rand, USA (1962)
2. Boltyanskii, V.G.: Mathematical Methods of Optimal Control. Holt, Rinehart and Winston (1971)
3. Boltyanskii, V.G.: Optimal Control of Discrete Systems. I.P.S.T, USA (1979)
4. Evtushenko, Y.G.: Methods for Solving Extremal Problems and their Application in Optimization systems. Nauka, Russia (1982). (in Russian)
5. Holtzman, J.M., Halkin, H.: Directional convexity and the maximum principle for discrete systems. J. SIAM Control 4(2), 263–275 (1966). https://doi.org/10.1137/0304023
6. Horn R.A., Johnson C.R.: Matrix Analysis. Cambridge University Press, USA (1985). https://doi.org/10.1017/CBO9780511810817
7. Ibragimov, D.N.: On the optimal speed problem for the class of linear autonomous infinite-dimensional discrete-time systems with bounded control and degenerate operator. Autom. Remote. Control. 80(3), 393–412 (2019). https://doi.org/10.1134/S0005117919030019

8. Berendakova A.V., Ibragimov D.N.: Method of constructing and estimating asymptotic controllable sets of two-dimensional linear discrete systems with limited control. Electronic Journal Trudy MAI 126. (in Russian). https://doi.org/10.34759/trd-2022-126-17

9. Ibragimov, D.N., Novozhilin, N.M., Portseva, E.Y.: On sufficient optimality conditions for a guaranteed control in the speed problem for a linear time-varying discrete-time system with bounded control. Autom. Remote. Control. 82(12), 2076–2096 (2021). https://doi.org/10.1134/S000511792112002X

10. Ibragimov, D.N., Sirotin, A.N.: On the problem of operation speed for the class of linear infinite-dimensional discrete-time systems with bounded control. Autom. Remote. Control. 78(10), 1731–1756 (2017). https://doi.org/10.1134/S0005117917100010

11. Ibragimov, D.N., Sirotin, A.N.: On the problem of optimal speed for the discrete linear system with bounded scalar control on the basis of 0-controllability sets. Autom. Remote. Control. 76(9), 1517–1540 (2015). https://doi.org/10.1134/S0005117915090015

12. Kurzhanskiy, A.F., Varaiya, P.: Theory and computational techniques for analysis of discrete-time control systems with disturbances. Optim. Method Sofw. 26(4–5), 719–746 (2011). https://doi.org/10.1080/10556788.2010.548867

13. Lin, X., Zhang, W.: A maximum principle for optimal control of discrete-time stochastic Systems with multiplicative noise. IEEE Trans. Automatic Control 60(4), 1121–1126 (2015). https://doi.org/10.1109/TAC.2014.2345243

14. Moiseev, N.N.: Elements of the Theory of Optimal Systems. Nauka, Russia (1975). (in Russian)

15. Pontryagin L.S.: Mathematical Theory of Optimal Processes. Routledge, London (1987). https://doi.org/10.1201/9780203749319

16. Propoi, A.I.: Elements of the Theory of Optimal Discrete Processes. Nauka, Russia (1973). (in Russian)

17. Rockafellar R.T.: Convex Analysis. Princeton University Press, USA (1970). https://doi.org/10.1515/9781400873173

18. Sirotin, A.N., Formal'skii, A.M.: Reachability and controllability of discrete-time systems under control actions bounded in magnitude and norm. Autom. Remote. Control. 64(3), 1844–1857 (2003). https://doi.org/10.1023/B:AURC.0000008423.93495.be

19. Wang, G., Yu, Z.: A Pontryagin's maximum principle for non-zero sum differential games of BSDEs with applications. IEEE Trans. Autom. Control 55(7), 1742–1747 (2010)

20. Weibel C.: Minkowski sums of polytopes: combinatorics and computation. EPFL, Suisse (2007). https://doi.org/10.5075/epfl-thesis-3883

21. Wu, Z.: A general maximum principle for optimal control of forward-backward stochastic systems. Automatica 49(5), 1473–1480 (2013). https://doi.org/10.1016/j.automatica.2013.02.005

An Approach to Solving Input Reconstruction Problems in Stochastic Differential Equations: Dynamic Algorithms and Tuning Their Parameters

Valeriy Rozenberg[1,2]($\boxtimes$) (iD)

[1] Krasovskii Institute of Mathematics and Mechanics UB RAS, S. Kovalevskoi str. 16, Yekaterinburg 620108, Russia
[2] Ural Federal University, Mira str. 19, Yekaterinburg 620002, Russia
rozen@imm.uran.ru

Abstract. Within the framework of the key approach from the theory of dynamic inversion, input reconstruction problems for stochastic differential equations are investigated. Different types of input information are used for the simultaneous reconstruction of disturbances in both the deterministic and stochastic terms of the equations. Feasible solving algorithms are designed; estimates of their convergence rates are derived. An empirical procedure adapting an algorithm to a specific system's dynamics to obtain best approximation results is discussed. An illustrative example for this technique is presented.

Keywords: Stochastic differential equation · Dynamic input reconstruction · Controlled model · Feedback control

1 Introduction

Input reconstruction problems for dynamical controlled systems provided inaccurate measurements of their phase states arise in both theoretical research and practice. They have been attracting a considerable attention since the 60's of the previous century, when key definitions of invertible dynamical systems and criteria of the unique solvability of inverse problems for systems described by ordinary differential equations (ODEs) were formulated [1,2].

Input reconstruction problems are rather often related to a special part of identification theory. They belong to the class of inverse problems of dynamics of controlled systems and, as a rule, are ill-posed and require the application of regularizing procedures. A specific solution approach oriented to the online reconstruction of unknown parameters was suggested and developed by Kryazhimskii, Osipov, and their colleagues, see [3–6]. This approach, well-known now as the method of dynamic inversion, is based on the combination of principles of the theory of positional control [7,8] and ideas of the theory of ill-posed problems [9,10]. Its essence is in the transformation of an input reconstruction problem to a feedback control problem for some auxiliary dynamical system called a model. Then, it is necessary to construct an appropriate control law for

M. Khachay et al. (Eds.): MOTOR 2023, LNCS 13930, pp. 394–408, 2023.
https://doi.org/10.1007/978-3-031-35305-5_27

this model in such a way that the adjustment of model controls to the current observation results provides an on-line approximation of the unknown inputs, for example, in decision making during some control process. In addition, this approximation should be as precise as one likes if the observations are sufficiently accurate.

In the beginning, the method of dynamic inversion was realized for systems described by ODEs [3–5], then, for equations and variational inequalities with distributed parameters [6], functional differential equations [11], fractional order systems [12]. The case of partially observed systems was also intensively studied [5,13]. For stochastic objects, the specific problem of modelling an unknown stochastic control in a system described by an ODE was first considered in [14]. Later, in [15], the stable approximation of an unknown input of a controlled system was constructed from inaccurate observations of phase states, assuming that the disturbances in the observation channel obey a probability distribution with a small expectation. At that, a probabilistic cost functional was first used within dynamic inversion theory. For the state of the art in the fields of identification theory close to the theory of dynamic inversion, see a detailed review [16] and references therein.

The present paper continues the dynamic inversion theory based studies of the problems of reconstructing unknown disturbances in linear/quasi-linear stochastic differential equations (SDEs), see [17–19]. In case of a linear SDE, the problem of dynamic reconstructing an input action entering the Ito integral and describing the amplitude of a random noise, with the use of the information on realizations of the whole phase vector, was studied in [17]. Then, the statement was supplemented by the possibility of measuring only a part of state coordinates [18]. The inverse problem for a quasi-linear system with diffusion depending on the phase state was investigated in [19], where the simultaneous reconstruction of disturbances both in the deterministic and stochastic terms under conditions of incomplete information was realized. As is shown in [17–19], the problems for SDEs under consideration can be transformed to inverse problems for systems of ODEs describing the mathematical expectation and covariance matrix of the process. So, the applicability of the input reconstruction algorithms elaborated earlier for systems of ODEs [3–5] was established, proper feasible modifications were designed, and estimates for their convergence rate with respect to the number of measurable realizations were derived in [17–19].

It is known that the usage of algorithms of such kind in practice is quite difficult due to the necessity of adapting the algorithm to a specific system. There is no a universal procedure for fitting parameters of reconstruction algorithms even in the case of ODEs; for some arguments, see [4,5]. Up to now, this important applied aspect of the theory of dynamic inversion remains vague. In [20], a first attempt to fill this gap was made based on the idea of the automatic tuning of algorithm's parameters to a given system using some test disturbances possessing key features typical for unknown real actions (maybe, subjecting to the known structural constraints). After performing some empirical procedure, the special set of parameters was considered as suitable for reconstructing other input actions.

The novelty of the present paper consists in considering a sufficiently general form of input information containing different variants of incomplete measurements and in testing the tuning procedure (actually developed earlier) for the new statement of the problem. Note that the parallelization of time-taking calculations can be involved to optimize the process of adapting the algorithm to a system.

The paper is organized as follows. The statement of the dynamic input reconstruction problem for a linear SDE is given in Sect. 2. The procedure transforming the problem to a problem for a system of ODEs is described in Sect. 3. Section 4 is devoted to a constructive algorithm for solving the problem. A procedure adapting the algorithm to a specific system in order to get best approximation results is discussed in Sect. 5. An illustrative example for the proposed technique is presented in Sect. 6. In Conclusion, the obtained results are resumed and perspective studies are outlined.

2 Problem Statement

We consider a linear SDE of the following form:

$$dx(t, \omega) = (A(t)x(t, \omega) + B_1(t)u_1(t) + f(t)) \, dt + B_2(t)U_2(t) \, d\xi(t, \omega), \quad x(0, \omega) = x_0. \quad (1)$$

Here, $t \in T = [0, \vartheta]$; $x = (x_1, x_2, \ldots, x_n) \in \mathbb{R}^n$; $\xi = (\xi_1, \xi_2, \ldots, \xi_k) \in \mathbb{R}^k$; x_0 is a known deterministic or random (normally distributed) vector of initial conditions; $\omega \in \Omega$, (Ω, F, P) is a probability space [21]; $\xi(t, \omega)$ is a standard Wiener process (i.e., a process starting from zero with zero mathematical expectation and covariance matrix equal to It, where I is the unit matrix from $\mathbb{R}^{k \times k}$); and $f(t) = \{f_i(t)\}$, $A(t) = \{a_{ij}(t)\}$, $B_1(t) = \{b_{1ij}(t)\}$, and $B_2(t) = \{b_{2ij}(t)\}$ are continuous matrix functions of dimension $n \times 1$, $n \times n$, $n \times r$, and $n \times k$, respectively. Two external disturbances act on the system: a vector $u_1(t) = (u_{11}(t), u_{12}(t), \ldots, u_{1r}(t)) \in \mathbb{R}^r$ and a diagonal matrix $U_2(t) = \{u_{21}(t), u_{22}(t), \ldots, u_{2k}(t)\} \in \mathbb{R}^{k \times k}$. The action u_1 enters the deterministic term and influences the mathematical expectation of the desired process. Since $U_2 d\xi = (u_{21}d\xi_1, u_{22}d\xi_2, \ldots, u_{2k}d\xi_k)$, we can assume that the vector $u_2 = (u_{21}, u_{22}, \ldots, u_{2k})$ characterizes the amplitude of random noises. Let the disturbances $u_1(\cdot) \in L_2(T; \mathbb{R}^r)$ and $u_2(\cdot) \in L_2(T; \mathbb{R}^k)$ take values from given convex compact sets S_{u_1} and S_{u_2}, respectively.

A solution of Eq. (1) is defined as a stochastic process satisfying the corresponding integral identity containing the stochastic Ito integral on the right-hand side for any t with probability 1. As is known, under the assumptions above, there exists a unique solution, which is a normal Markov process with continuous realizations [22], see Theorem 5.2.1.

The problem in question is as follows. At discrete, frequent enough, times $\tau_i \in T$, $\tau_i = i\delta$, $\delta = \vartheta/l$, $i \in [0 : (l-1)]$, the information on some number N of realizations of the stochastic process $x(\tau_i)$ is received; at that the signal is available:

$$y(\tau_i) = Qx(\tau_i), \quad (2)$$

where Q is a constant matrix of dimension $q \times n$. It is assumed that $l = l(N)$ and there exist an estimate m_i^N of the mathematical expectation $m(t) = Mx(t)$ and an estimate D_i^N of the covariance matrix $D(t) = M(x(t) - m(t))(x(t) - m(t))'$ (the prime stands for transposition) such that

$$P\left(\max_{i \in [1:(l(N)-1)]} \left\{ \left\| m_i^N - m(\tau_i) \right\|_{\mathbb{R}^n}, \left\| D_i^N - D(\tau_i) \right\|_{\mathbb{R}^{n \times n}} \right\} \leq h(N) \right) = 1 - g(N), \quad (3)$$

and $h(N)$, $g(N) \to 0$ as $N \to \infty$. By $\| \cdot \|$, we denote the Euclidean norm in the corresponding space. The standard statistical procedures [23] for designing the estimates m_i^N and D_i^N admit modifications providing the validity of (3).

It is required to elaborate an algorithm for the dynamic reconstruction of the unknown disturbances $u_1(t)$ and $u_2(t)$ generating the stochastic process $x(t)$ from available signal (2) on its realizations. The probability of an arbitrarily small deviation of approximations from the desired inputs in the metric of spaces $L_2(T; \mathbb{R}^r)$ and $L_2(T; \mathbb{R}^k)$, respectively, should be close to 1 for sufficiently large N and the time discretization step $\delta = \delta(N) = \vartheta / l(N)$ concordant with N in an appropriate way. In addition, a procedure tuning algorithm's parameters for the given system would be desirable in practical applications.

Note that linearized models similar to (1) are useful when studying, for example, the dynamics of a biological population in a stochastic medium or the processes of chaotic motion of one-type particles. In this context, the formulated inverse problem can be interpreted as the dynamic reconstruction of the external control action and of the amplitude of random noises under incomplete information, when the dynamics of the system admits simultaneous measurements of a sufficiently large number of trajectories (for example, of the motion of one-type particles).

Thus, the novelty of the problem statement in question, in comparison with [17–19], consists in the form of input signal (2) and in the consideration of applied aspects of algorithm's operation. Note that the general statement of inverse problem specified by the form and dimension of the matrix Q from (2) actually covers the problems considered earlier. If Q is the unit $n \times n$-matrix, since, obviously, all the coordinates of the random process $x(\tau_i)$ are available for measuring (Case 1). In such a formulation, the problem was solved in [17]. In the case when the matrix Q has the dimension $q \times n$ ($q < n$) and rank q is determined by the unit $q \times q$-submatrix started from the first column (all the rest elements of Q are zeros), we get measurements of the first q coordinates of the process (Case 2). The reconstruction problem of this type was investigated in [18]. The case of general form of the matrix Q (Case 3) is considered in the present paper.

In all the statements listed above, we use rather universal solving approach consisting in the reduction (by means of the method of moments [24]) of the problem formulated for an SDE to a problem for a system of ODEs describing the mathematical expectation and the covariance matrix of the desired process. In addition, we apply a tuning procedure approbated earlier for an SDE of other type [20]. In all cases, we organize the simultaneous reconstruction of disturbances in the deterministic and stochastic terms of the SDE based on the method of auxiliary controlled models, which is typical for problems of the theory of positional control [7,8] and of the theory of dynamic inversion [4,5].

Note that when the informativeness of input signal changes, we should adapt to this process by making changes in solvability conditions for the problem and solving algorithms. The less informative is the input signal, the more complicated is the algorithm and the worse is the estimate of its convergence rate.

The main goal of the present work is to get results in Case 3, which are similar to the ones obtained earlier in Cases 1 and 2.

3 Problem for ODEs

Following [17, 18], we reduce the reconstruction problem for a linear SDE to a problem for a system of ODEs. Let us introduce the notation: $m_0 = Mx_0$ and $D_0 = M(x_0 - m_0)(x_0 - m_0)'$. Since the original system is linear and the mathematical expectation of an Ito integral is zero, the value $m(t)$ depends only on $u_1(t)$; its dynamics is described by the equation

$$\dot{m}(t) = A(t)m(t) + B_1(t)u_1(t) + f(t), \quad m \in \mathbb{R}^n, \quad m(0) = m_0. \tag{4}$$

Note that, since N ($N > 1$) trajectories $x^r(\tau_i)$, $r \in [1 : N]$, of the original SDE are measured; then, according to the problem statement, we know values of q-dimensional signal (2) $y^r(\tau_i) = Qx^r(\tau_i)$.

The signal on the trajectory of Eq. (4) is denoted by $y_m(\tau_i) = Q_m m(\tau_i)$; its estimate formed by the information on $y^r(\tau_i)$, $r \in [1 : N]$, by $y_m^N(\tau_i)$. The latter is constructed from (2) as follows:

$$y_m^N(\tau_i) = \frac{1}{N}\sum_{r=1}^{N} y^r(\tau_i) = Q m_i^N, \quad m_i^N = \frac{1}{N}\sum_{r=1}^{N} x^r(\tau_i). \tag{5}$$

Evidently, $Q_m = Q$ and, for all the times $\tau_i \in T$, $i \in [1 : (l(N) - 1)]$, in view of relation (3), it holds that

$$y_m(\tau_i) = Q_m m(\tau_i), \quad P(\forall i \in [1 : (l(N) - 1)] \; \left\| y_m^N(\tau_i) - y_m(\tau_i) \right\|_{\mathbb{R}^q} \le C_1 h(N)) = 1 - g(N), \tag{6}$$

where the constant C_1 can be written explicitly.

The covariance matrix $D(t)$ does not depend on $u_1(t)$ explicitly; its dynamics is obtained by the method of moments applied in [17, 18]. As a result, we get the following equation for $D \in \mathbb{R}^{n \times n}$:

$$\dot{D}(t) = AD(t) + D(t)A' + B_2 U_2(t)U_2'(t)B_2', \quad D(0) = D_0. \tag{7}$$

Matrix Eq. (7) is rewritten in the form of a vector equation, which is more traditional for the problems in question. Due to the symmetry of the matrix $D(t)$, its dimension is defined as $n_d = (n^2 + n)/2$. We introduce the vector $d(t) = \{d_s(t)\}$, $s \in [1 : n_d]$, consisting of successively written and enumerated elements of the matrix $D(t)$ taken line by line starting with the element located at the main diagonal. The coordinates of this vector are found from the elements of the matrix $D(t) = \{d_{ij}(t)\}$, $i, j \in [1 : n]$:

$$d_s(t) = d_{ij}(t), \quad i \le j, \quad s = (n - i/2)(i - 1) + j. \tag{8}$$

Note that relations (8) between indices s and i, j are one-to-one due to the way of enumeration of elements of the matrix $D(t)$. The procedure described in [25] in detail allows us to transform Eq. (7) to the form

$$\dot{d}(t) = \bar{A}d(t) + \bar{B}\bar{u}(t), \quad d(0) = d_0, \tag{9}$$

where the matrix $\bar{A}$ of dimension $n_d \times n_d$ and the matrix $\bar{B}$ of dimension $n_d \times k$ are explicitly written, whereas the initial state d_0 is found from D_0 by formula (8). The product of the diagonal matrices $U_2(t)U_2'(t)$ leads to the appearance of the control vector $\bar{u}(t) = (u_{21}^2(t), u_{22}^2(t), \ldots, u_{2k}^2(t))$. It is evident that the vector $\bar{u}(\cdot) \in L_2(T; \mathbb{R}^k)$ takes values from some convex compact set $S_{\bar{u}} \in \mathbb{R}^k$, which is built from $S_{u_2} \in \mathbb{R}^k$ in natural way.

It should be noted that the existence, the uniqueness, and the form of a solution of Eq. (1) imply the existence and uniqueness of a solution of system (4), (9).

The signal on the trajectory of Eq. (9) is denoted by $y_d(\tau_i) = Q_d d(\tau_i)$; its estimate formed by the information on $y^r(\tau_i)$, $r \in [1 : N]$, by $y_d^N(\tau_i)$. The latter is constructed from original signal (2) and estimate (5) as follows:

$$\frac{1}{N-1} \sum_{r=1}^{N} (y^r(\tau_i) - y_m^N(\tau_i))(y^r(\tau_i) - y_m^N(\tau_i))'$$

$$= Q \frac{1}{N-1} \sum_{r=1}^{N} (x^r(\tau_i) - m_i^N)(x^r(\tau_i) - m_i^N)' Q' = Q D_i^N Q', \qquad (10)$$

where $D_i^N = \{d_{ij}^N(\tau_i)\}$, $i, j \in [1 : n]$ is the standard estimate of the covariance matrix $D(\tau_i)$ for an unknown (estimated by m_i^N) mathematical expectation $m(\tau_i)$. Symmetric $q \times q$-matrix (10) is represented in the form

$$QD_i^N Q' = \begin{pmatrix} \sum_{r=1}^{n} \sum_{p=1}^{n} q_{1r} q_{1p} d_{pr}^N & \sum_{r=1}^{n} \sum_{p=1}^{n} q_{2r} q_{1p} d_{pr}^N & \cdots & \sum_{r=1}^{n} \sum_{p=1}^{n} q_{qr} q_{1p} d_{pr}^N \\ \sum_{r=1}^{n} \sum_{p=1}^{n} q_{1r} q_{2p} d_{pr}^N & \sum_{r=1}^{n} \sum_{p=1}^{n} q_{2r} q_{2p} d_{pr}^N & \cdots & \sum_{r=1}^{n} \sum_{p=1}^{n} q_{qr} q_{2p} d_{pr}^N \\ \cdots & \cdots & \cdots & \cdots \\ \sum_{r=1}^{n} \sum_{p=1}^{n} q_{1r} q_{qp} d_{pr}^N & \sum_{r=1}^{n} \sum_{p=1}^{n} q_{2r} q_{qp} d_{pr}^N & \cdots & \sum_{r=1}^{n} \sum_{p=1}^{n} q_{qr} q_{qp} d_{pr}^N \end{pmatrix}.$$

Evidently, it is sufficient to consider $n_q = (q^2 + q)/2$ elements of this matrix. The element (i_1, j_1), $i_1 \leq j_1$, $i_1, j_1 \in [1 : q]$, namely $\sum_{r=1}^{n} \sum_{p=1}^{n} q_{i_1 r} q_{j_1 p} d_{pr}^N$, defines the s_1th row $(s_1 = (q - i_1/2)(i_1 - 1) + j_1$ by analogy with (8)) in the formed relation $y_d^N(\tau_i) = Q_d d_i^N$, where Q_d is a constant matrix of dimension $n_q \times n_d$, d_i^N is a vector of dimension n_d extracted from D_i^N by rule (8). Then, to write the element $Q_d[s_1, s_2]$, it is necessary to find, using s_1 and s_2, the indices i_1, j_1 $(i_1, j_1 \in [1 : q]$, $i_1 \leq j_1$, $(q-i_1/2)(i_1-1)+j_1 = s_1)$ and i_2, j_2 $(i_2, j_2 \in [1 : n]$, $i_2 \leq j_2$, $(n - i_2/2)(i_2 - 1) + j_2 = s_2)$, respectively. It is easy to verify that

$$Q_d[s_1, s_2] = \begin{cases} q_{i_1 i_2} q_{j_1 j_2}, & i_2 = j_2 \\ q_{i_1 i_2} q_{j_1 j_2} + q_{i_1 j_2} q_{j_1 i_2}, & i_2 \neq j_2 \end{cases}$$

The transformations described above are necessary for constructing, from known matrix (10), the estimate $y_d^N(\tau_i) = Q_d d_i^N$ of the signal $y_d(\tau_i) = Q_d d(\tau_i)$. Thus, for the finite set of times $\tau_i \in T$, $i \in [1 : (l(N) - 1)]$, taking into account (3), we obtain

$$y_d(\tau_i) = Q_d d(\tau_i), \quad P(\forall i \in [1 : (l(N) - 1)] \, \left\| y_d^N(\tau_i) - y_d(\tau_i) \right\|_{\mathbb{R}^{nq}} \leq C_2 h(N)) = 1 - g(N), \tag{11}$$

where the constant C_2 can be written explicitly.

Thus, the original dynamic reconstruction problem for SDE (1) can be reformulated in the form of a problem for system of ODEs (4), (9) as follows. Assume that during the process, at the discrete times $\tau_i \in T$, $\tau_i = i\delta$, $\delta = \vartheta/l(N)$, $i \in [0 : (l(N) - 1)]$, we receive the information allowing us to estimate the signals on the phase states of Eqs. (4) and (9), the vectors $y_m(\tau_i)$ and $y_d(\tau_i)$, respectively. The estimates found from (5), (10) satisfy relations (6), (11). It is required to design an algorithm for the dynamic reconstruction of the unknown disturbances $u_1(t)$ and $\bar{u}(t)$. The probability of an arbitrarily small deviation of approximations from the desired inputs in the metric of spaces $L_2(T; \mathbb{R}^r)$ and $L_2(T; \mathbb{R}^k)$, respectively, should be close to 1 for sufficiently large N and the time discretization step $\delta = \delta(N) = \vartheta/l(N)$ concordant with N in an appropriate way. In the general case, we reconstruct namely the vector $\bar{u}(t)$. Under additional, sufficiently natural, constraints on the real vector $u_2(t)$, its reconstruction is also possible.

In such a formulation, the problem corresponds to the problem for an ODE considered in [5] in the case of measuring a signal of form (2). Here, it is shown that the algorithm proposed in [5] can be applied to solving the problem obtained for system (4), (9), since this algorithm admits a constructive concordance of its parameters with the number of measurable realizations of the original stochastic process. Sufficient conditions for the solvability of the problem are formulated in terms of original Eq. (1) and can be easily verified.

4   Model and Solution Algorithm

In Case 3 under consideration, the solution algorithm is designed provided the following condition is fulfilled.

Condition 1. The dimension of the unknown vector function $u_1(\cdot)$ does not exceed the dimension of the signal $y_m(\cdot)$ ($r \le q$), and the matrix $Q_m B_1$ has rank r; i.e., it is a matrix of full rank. The dimension of the unknown vector function $u_2(\cdot)$ (and, respectively, $\bar{u}(\cdot)$) does not exceed the dimension of the signal $y_d(\cdot)$ ($k \le n_q$), and the matrix $Q_d \bar{B}$ has rank k; i.e., it is a matrix of full rank.

This condition guarantees the uniqueness of the functions $u_1(\cdot)$ and $\bar{u}(\cdot)$ generating the solution of system (4), (9) and, respectively, signals (6), (11). The proof of this statement is based on the properties of the pseudoinverse matrix for a matrix of full rank and, up to minor details, follows a similar argument from [5, 13]. Note that the number of rows q of the matrix Q (for example, the number of measured coordinates of original system (1)) should not be less than the dimension of the disturbance $u_1(\cdot)$ but can be less than the dimension of the disturbance $u_2(\cdot)$ (and $\bar{u}(\cdot)$).

The algorithm below (namely, the model dynamics and control law) is an application of the computational procedure from [5, §3.4] to system (4), (9) under measurements (6), (11). At the initial moment $\tau_0 = 0$, we fix a value N; calculate the values $l^N = l(N)$, $h^N = h(N)$, and $g^N = g(N)$, and construct the uniform partition of the interval T with the step $\delta^N = \vartheta/l^N$: $\tau_i \in T$, $\tau_i = i\delta^N$, $i \in [0 : l^N]$. We introduce a controlled model system containing two independent blocks related to Eq. (4), (6) and (9), (11). The phase vector of the model is denoted by $w(t)$; it consists of two triples: (i) a q-dimensional vector $w_{m0}(t)$, an n-dimensional vector $w_{m1}(t)$, and an n-dimensional vector $w_{m2}(t)$ and (ii) an n_q-dimensional vector $w_{d0}(t)$, an n_d-dimensional vector $w_{d1}(t)$, and an n_d-dimensional vector $w_{d2}(t)$. The dynamics of the model and its initial state are defined by the relations

$$\dot{w}_{m0}(t) = 0, \ \dot{w}_{m1}(t) = 0, \ \dot{w}_{m2}(t) = 0, \ t \in [0, \tau_1],$$
$$\dot{w}_{d0}(t) = 0, \ \dot{w}_{d1}(t) = 0, \ \dot{w}_{d2}(t) = 0, \ t \in [0, \tau_1],$$
$$\dot{w}_{m0}(t) = Q_m B_1 v_1^N(t), \ \dot{w}_{m1}(t) = A w_{m1}(t) + B_1 v_1^N(t), \ \dot{w}_{m2}(t) = w_{m1}(t), \ t \in (\tau_1, \vartheta], \quad (12)$$
$$\dot{w}_{d0}(t) = Q_d \bar{B} v_2^N(t), \ \dot{w}_{d1}(t) = \bar{A} w_{d1}(t) + \bar{B} v_2^N(t), \ \dot{w}_{d2}(t) = w_{d1}(t), \ t \in (\tau_1, \vartheta],$$
$$w_{m0}(0) = 0, \ w_{m1}(0) = 0, \ w_{m2}(0) = 0, \ w_{d0}(0) = 0, \ w_{d1}(0) = 0, \ w_{d2}(0) = 0.$$

Obviously, it is possible to rewrite a discrete analog of model (12). The model controls

$$v_1^N(t) = 0, \ v_2^N(t) = 0, \ t \in [0, \tau_1], \ v_1^N(t) = v_{1i}^N, \ v_2^N(t) = v_{2i}^N, \ t \in (\tau_i, \tau_{i+1}], \ i \in [1 : l^N - 1],$$

are calculated at the moment τ_i, $i \in [1 : l^N - 1]$, according to the feedback rules:

$$v_{1i}^N = \arg\min \left\{ 2\langle l_1^N(\tau_i), v_1 \rangle_r + \alpha_m^N \|v_1\|_{\mathbb{R}^r}^2 : v_1 \in S_{u1} \right\},$$
$$v_{2i}^N = \arg\min \left\{ 2\langle l_2^N(\tau_i), v_2 \rangle_k + \alpha_d^N \|v_2\|_{\mathbb{R}^k}^2 : v_2 \in S_{\bar{u}} \right\}, \quad (13)$$

where $\langle \cdot, \cdot \rangle$ is the scalar product in the corresponding space,

$$l_1^N(\tau_i) = (Q_m B_1)^+ \left[w_{m0}(\tau_i) + Q_m A w_{m2}(\tau_i) + Q_m m_0 - Q_m m_i^N + \int_0^{\tau_i} (Q_m A \bar{m}(\tau) + Q_m f(\tau)) \, d\tau \right],$$
$$l_2^N(\tau_i) = (Q_d \bar{B})^+ \left[w_{d0}(\tau_i) + Q_d \bar{A} w_{d2}(\tau_i) + Q_d d_0 - Q_d d_i^N + \int_0^{\tau_i} Q_d \bar{A} \bar{d}(\tau) \, d\tau \right],$$

$\bar{m}(\cdot)$ and $\bar{d}(\cdot)$ are solutions of the undisturbed equations, namely,

$$\dot{m}(t) = A m(t) + f(t), \ m(0) = m_0, \quad \dot{d}(t) = \bar{A} d(t), \ d(0) = d_0,$$

$\alpha_m^N = \alpha_m(h^N)$, $\alpha_d^N = \alpha_d(h^N)$ are regularization parameters.

Thus, the work of the algorithm is decomposed into l^N identical steps. At the ith step performed on the interval $(\tau_i, \tau_{i+1}]$, the input data for calculations are the estimates m_i^N, d_i^N and the model state $w(\tau_i)$ obtained by this moment. After calculating the model controls by formulas (13), the model state $w(\tau_{i+1})$ is computed according to the discrete analog of (12). The process stops at the terminal time ϑ.

Theorem 1. *Let the following conditions of concordance of the parameters hold:*

$$h^N \to 0, \quad g^N \to 0, \quad \delta^N \to 0, \quad \delta^N \geq h^N, \quad \alpha_m^N \to 0, \quad \alpha_d^N \to 0,$$

$$\frac{\delta^N + h^N}{\alpha_m^N} \to 0, \quad \frac{\delta^N + h^N}{\alpha_d^N} \to 0 \quad as \quad N \to \infty. \quad (14)$$

Then, for model controls $v_1^N(\cdot)$ and $v_2^N(\cdot)$ formed by (13), we have the convergence

$$P\left(\max\{ \|v_1^N(\cdot) - u_1(\cdot)\|_{L_2(T;\mathbb{R}^r)}, \ \|v_2^N(\cdot) - \bar{u}(\cdot)\|_{L_2(T;\mathbb{R}^k)} \} \to 0 \right) \to 1 \quad as \quad N \to \infty. \quad (15)$$

If the real disturbances $u_1(\cdot)$ and $u_2(\cdot)$ (and, consequently, $\bar{u}(\cdot)$) have a bounded variation on T, the following estimate for the accuracy of the algorithm with respect to the number of measurable realizations of the process is valid:

$$P\left(\max\{\,\|v_1^N(\cdot) - u_1(\cdot)\|_{L_2(T;\mathbb{R}^r)},\, \|v_2^N(\cdot) - \bar{u}(\cdot)\|_{L_2(T;\mathbb{R}^k)}\} \le C_1\left(\frac{1}{N}\right)^{2/73}\right) = 1 - C_2\left(\frac{1}{N}\right)^{2/73},$$

(16)

where C_1 and C_2 are some constants independent of N, $u_1(\cdot)$, and $u_2(\cdot)$.

Proof. Convergence (15) immediately follows from the application of results of [5] to system (4), (9) with signal (6), (11) and from relations (14). Since estimates (6) and (11) are fulfilled at all times $\tau_i \in T$ with probability $1 - g^N$, then, rewriting twice the convergence rate estimate of the algorithm above obtained in [5, Theorem 3.4.1] under the assumption on the boundedness of the variation of real disturbances, in the form

$$\|v_1^N(\cdot) - u_1(\cdot)\|_{L_2(T;\mathbb{R}^r)} \le \bar{C}_1\left(\frac{\delta^N}{\alpha_m^N} + (\delta^N + \alpha_m^N)^{1/2}\right)^{1/6},$$

$$\|v_2^N(\cdot) - \bar{u}(\cdot)\|_{L_2(T;\mathbb{R}^k)} \le \bar{C}_2\left(\frac{\delta^N}{\alpha_d^N} + (\delta^N + \alpha_d^N)^{1/2}\right)^{1/6},$$

and assuming $\delta^N = \bar{C}_3 h^N$, $\alpha_m^N = \bar{C}_4 (h^N)^{2/3}$, and $\alpha_d^N = \bar{C}_5 (h^N)^{2/3}$, we get

$$P\left(\max\{\,\|v_1^N(\cdot) - u_1(\cdot)\|_{L_2(T;\mathbb{R}^r)},\, \|v_2^N(\cdot) - \bar{u}(\cdot)\|_{L_2(T;\mathbb{R}^k)}\} \le \bar{C}_6 (h^N)^{1/18}\right) = 1 - g^N. \quad (17)$$

Here and below, we denote by $\bar{C}_i$ auxiliary constants, which are independent of the estimated values and can be written explicitly.

It is shown in [18] that the standard estimates m_i^N of the mathematical expectation $m(\tau_i)$ and D_i^N of the covariance matrix $D(\tau_i)$ constructed from N ($N > 1$) realizations $x^1(\tau_i), x^2(\tau_i), \ldots, x^N(\tau_i)$ of the random variables $x(\tau_i)$, $i \in [1 : (l^N - 1)]$, by the rules from (5) and (10), provide the validness of relations like (6) and (11). At that, the following explicit dependencies of their parameters on the number of measured realizations are obtained:

$$h^N = \bar{C}_7 \left(\frac{1}{N}\right)^{1/2 - \epsilon},\quad \delta^N = \bar{C}_8 \left(\frac{1}{N}\right)^{\min\{\alpha, \alpha(1/2 + 3\epsilon)\}},\quad g^N = \bar{C}_9 \left(\frac{1}{N}\right)^{\min\{1-\alpha, (1-\alpha)(1/2 + 3\epsilon)\}}, \quad (18)$$

where $0 < \epsilon < 1/2$ and $0 < \alpha < 1$.

Consider two cases.

1. Let $1/2 + 3\epsilon < 1$, $\epsilon < 1/6$. Then, relations (18) imply

$$h^N = \bar{C}_7 \left(\frac{1}{N}\right)^{1/2 - \epsilon},\quad \delta^N = \bar{C}_8 \left(\frac{1}{N}\right)^{\alpha(1/2 + 3\epsilon)},\quad g^N = \bar{C}_9 \left(\frac{1}{N}\right)^{(1-\alpha)(1/2 + 3\epsilon)}. \quad (19)$$

To obtain the equality $\delta^N = \bar{C}_3 h^N$, we can assume $1/2 - \epsilon = \alpha(1/2 + 3\epsilon)$, which implies $\epsilon = \dfrac{1 - \alpha}{2(3\alpha + 1)}$, $0 < \epsilon < 1/2$ for $0 < \alpha < 1$. For the found ϵ, the power exponent of the value $1/N$ in relations (19) is $\dfrac{2\alpha}{3\alpha + 1}$ for h^N and δ^N, and $\dfrac{2(1-\alpha)}{3\alpha + 1}$ for g^N. To obtain estimate (16), taking into account formula (17), we set $\dfrac{2\alpha}{18(3\alpha + 1)} = \dfrac{2(1-\alpha)}{3\alpha + 1}$, which implies $\alpha = 18/19$ and $\epsilon = 1/146$; consequently, the power exponents of the value $1/N$ in (16) are equal to $2/73$.

2. Let $1/2 + 3\epsilon \geq 1$, $\epsilon \geq 1/6$. Then,

$$ h^N = \bar{C}_7 \left(\frac{1}{N}\right)^{1/2-\epsilon}, \quad \delta^N = \bar{C}_8 \left(\frac{1}{N}\right)^{\alpha}, \quad g^N = \bar{C}_9 \left(\frac{1}{N}\right)^{1-\alpha}. $$

Evidently, the largest power exponent of the value $1/N$ in the approximation part of estimate (16) is $1/54$; this is worse than the value obtained in the first case. The theorem is proved.

Thus, to get estimate (16), it is necessary to set

$$ h^N = 1/N^{36/73}, \quad \delta^N = K_1/N^{36/73}, \quad \alpha_m^N = K_2/N^{24/73}, \quad \alpha_d^N = K_3/N^{24/73}, \tag{20} $$

where K_1, K_2, and K_3 are positive constants.

Note that the constants in (16) and (20) depend on the parameters of the given system, on its structural constraints a priori known, but are independent of the functions under reconstruction.

Varying ϵ and α from relations (18), we can obtain different values of the power exponents of the value $1/N$ in the approximating and probabilistic parts of estimates similar to (16). In the theorem above, we have got the same orders of smallness for the parts due to relations (20).

Remark 1. [19] If the vector u_2 is unique and the set S_{u_2} is such that $\forall u_2(\cdot) \in S_{u_2}$ $u_{2i}(t) \geq 0$ $\forall t \in T$ $\forall i \in [1:k]$, then algorithm (12)–(14) reconstructs $u_2(\cdot)$ in $L_2(T; \mathbb{R}^k)$-metric with an accuracy estimate of form (16).

5   Tuning of Algorithm's Parameters

Thus, we obtain the same system of ODEs for all three Cases of the inverse problem (measuring all the coordinates, a part of coordinates, and signal (2) in general form). It is reasonable to involve a universal procedure for the model calibration. The empirical procedure for tuning of algorithm's parameters was rather successfully applied to a problem for a quasi-linear SDE in [20]. Here, we use the same idea. Let us briefly describe the procedure from [20].

The number N of measured trajectories of SDE (1) influences both the accuracy h^N of estimates (3), (6), (11) and the parameters of the algorithm δ^N, α_m^N, α_d^N providing convergence (15), (16). The dependence of these parameters on N is specified via constants (20), which values should be concretized. Sensitivity of reconstruction algorithms to different constraints imposed on a given system was observed in various experiments [5,6,17]. Up to now, there is no a universal procedure for fitting parameters like δ^N, α_m^N, and α_d^N of reconstruction algorithms. We want to find constants in relations (20) providing the successful reconstruction of any possible disturbances in a specific system like (1). Toward this aim, we suggest to use some model disturbances possessing key features typical for unknown real actions and we believe that "optimal" relations between parameters can be found empirically through a solution of the next extremal problem:

$$(K_1^*, K_2^*, K_3^*) = \arg\min\left\{\sum_{N\in I^N} \beta^N \|u_{K_1,K_2,K_3}^N(\cdot) - u(\cdot)\|_{L_2(T)} : (K_1, K_2, K_3) \in S_K\right\}. \quad (21)$$

Here, $L_2(T) = L_2(T; \mathbb{R}^{r+k})$, $u_{K_1,K_2,K_3}^N(\cdot) = (v_1^N(\cdot), v_2^N(\cdot))$ is an algorithm's output depending on the constants from (20), $u(\cdot) = (u_1(\cdot), \bar{u}(\cdot))$; I^N is the set of chosen values of N, β^N are weighting coefficients, $\sum_{N\in I^N}\beta^N = 1$; S_K is some set of admissible values of the vector (K_1, K_2, K_3). We use all the constraints imposed on the admissible inputs, fix some test function $u(\cdot)$ satisfying these constraints, introduce a uniform grid with respect to K_1, K_2, K_3 and then solve problem (21) by exhaustive search. Then, we hope for the applicability of optimal values (K_1^*, K_2^*, K_3^*) in the process of reconstructing other admissible disturbances acting on the system. Numerous experiments (one of them is described below) demonstrate results in favor of our hypothesis.

The simulation of a large number of independent trajectories of SDE (1) in order to obtain estimates (3), (6), (11) for different N is performed by the Euler method with the substitution of a sequence of random impulses for the Wiener process [26]. For a linear SDE we consider, the mean square accuracy order of the method is $O(\delta_s)$, where δ_s is a simulation step, which is much smaller than the step δ^N between the times of measuring the realizations. Thus, two time grids are incorporated into the tuning procedure. The simulations with the smaller step are performed only at the stage of tuning; when solving a real problem, the input information should come from outside. Note that the tuning procedure is time-taking but admits an effective parallelization.

6   Numerical Example

Consider a linear system of SDEs of second order ($n = 2$) describing a stochastic process, which can be treated as a "spoiled" mean-reverting Ornstein–Uhlenbeck process [22]:

$$\begin{pmatrix} dx_1(t) \\ dx_2(t) \end{pmatrix} = \lambda \begin{pmatrix} -1 & a_{12} \\ a_{21} & -1 \end{pmatrix} \begin{pmatrix} x_1(t) \\ x_2(t) \end{pmatrix} dt + \begin{pmatrix} m_1 \\ m_2 \end{pmatrix} dt + \begin{pmatrix} b_{11} \\ b_{12} \end{pmatrix} u_1(t)dt + \begin{pmatrix} b_{21} \\ b_{22} \end{pmatrix} u_2(t)d\xi(t), \quad (22)$$

$$t \in T = [0, \vartheta], \; x_1(0) = x_{10}, \; x_2(0) = x_{20}.$$

Here, all the numerical parameters are fixed; $u_1(t)$ and $u_2(t)$ are unknown scalar bounded disturbances ($r = k = 1$), $\xi(t)$ is a standard scalar Wiener process. Equations like (22) are used, in particular, in some simplest models describing the dynamics of relatively stable biological populations. In this case, the values $x_1(t)$ and $x_2(t)$ represent the current sizes (in arbitrary units) of two interacting biological species; the parameter $0 < \lambda \leq 1$ provides the boundedness condition; the parameters m_1 and m_2 specify some averages, which are the aim of "subconscious" return of the populations. Such a return can be prevented by the influence of the neighbors defined by the coefficients $a_{ij}, i \neq j, |a_{ij}| < 1$, and by the action of outer forces presented by the disturbance $u_1(t)$ influencing the mathematical expectation of the process and $u_2(t)$ describing the amplitude of random noises. The unknown functions $u_1(t)$ and $u_2(t)$ should be reconstructed. The signal

$$y(\tau_i) = Qx(\tau_i) = x_1(\tau_i) + x_2(\tau_i)$$

is measurable at discrete times. Thus, the matrix Q is the row $(1, 1)$, $q = 1$, the total amount of two populations is measured.

In the computational experiment, the following values of parameters were chosen:

$$\lambda = 1, \ m_1 = 1, \ m_2 = 2, \ x_{10} = 5, \ x_{20} = 5, \ T = [0, 1],$$

$$a_{12} = 0.1, \ a_{21} = 0.2, \ b_{11} = 2, \ b_{12} = 1, \ b_{21} = 1, \ b_{22} = 2, \ -2 \leq u_1 \leq 2, \ 1 \leq u_2 \leq 4.$$

Reduced system of ODEs (4), (9) for the mathematical expectation and covariance matrix consists of two equations for $m(t)$ and three ones for $d(t)$:

$$\dot{m}_1(t) = -m_1(t) + 0.1m_2(t) + 1 + 2u_1(t), \ m_1(0) = 5,$$
$$\dot{m}_2(t) = 0.2m_1(t) - m_2(t) + 2 + u_1(t), \ m_2(0) = 5, \tag{23}$$

$$\dot{d}_1(t) = -2d_1(t) + 0.2d_2(t) + u_2^2(t), \ d_1(0) = 0,$$
$$\dot{d}_2(t) = 0.2d_1(t) - 2d_2(t) + 0.1d_3(t) + 2u_2^2(t), \ d_2(0) = 0, \tag{24}$$
$$\dot{d}_3(t) = 0.4d_2(t) - 2d_3(t) + 4u_2^2(t), \ d_3(0) = 0,$$

So, the matrices B_1 and $\bar{B}$, being important to verify Condition 1, take the form:

$$B_1 = \begin{pmatrix} 2 \\ 1 \end{pmatrix}, \quad \bar{B} = \begin{pmatrix} 1 \\ 2 \\ 4 \end{pmatrix}.$$

The calculation of the signals $y_m(\tau_i) = Q_m m(\tau_i)$ and $y_d(\tau_i) = Q_d d(\tau_i)$ on the trajectory according to (5) and (10) results in the following relations:

$$Q_m = (1, 1), \quad Q_d = (1, 2, 1).$$

Thus, Condition 1 is fulfilled: the dimension of the signal y_m is equal to the dimension of the unknown function u_1, the dimension of the signal y_d is equal to the dimension of the unknown function u_2, and the matrices $Q_m B_1 = 3$ and $Q_d \bar{B} = 9$ are of full rank. Then, for system (23), (24), in accordance with (12) and (13), a model system and its control law are rewritten in an explicit form.

The aim of calculations was the adjusting the algorithm to system (22)–(24) by means of appropriate relations between parameters δ^N, α_m^N, α_d^N, and N via K_1, K_2, K_3 (see (20)). In the first series, to solve problem (21), we chose the functions

$$u_1(t) = \sin 10t, \ u_2(t) = 2 + \sqrt{t}$$

and parameters $K_i \in [0.1, 10]$, $i = 1, 2, 3$, with $10 - 50$ grid nodes, $I^N = \{10^3, 10^5\}$, $\beta^N = \{1/2, 1/2\}$. Here, we took into account a priori information and the constraints imposed on real disturbances.

To obtain estimates (6) and (11), it is necessary to track N independent trajectories of system (22); toward this aim, we used the scheme described in the previous section. In this example, according to Remark 1, the function $u_2(t)$ can be reconstructed.

Optimal (in the sense of (21)) values of the constants from (20) were obtained: $K_1^* = 0.8$, $K_2^* = 2$, $K_3^* = 2.5$. Then, we specify the parameters δ^N, α_m^N, and α_d^N of the algorithm. The results of reconstructing the disturbances $u_1(t)$ and $u_2(t)$ for found K_i and different N are presented in Figs. 1 and 2, where $u_1(t)$ and $u_2(t)$ are shown by dashed line, and the results of their reconstruction, by solid line. The horizontal and vertical axes correspond to time t and to the values of disturbances, respectively. The results correspond to the essence of convergence (16): the more is N, the less is the reconstruction accuracy.

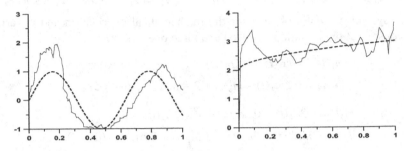

Fig. 1. Reconstruction of disturbances u_1 (left) and u_2. Variant 1: $N = 10^3$, $\delta^N = 0.025$, $\delta_s = \delta^N/10^4$, $\alpha_m^N = 0.2$, $\alpha_d^N = 0.25$, accuracy $\|u^N(\cdot) - u(\cdot)\|_{L_2(T)} = 0.944$.

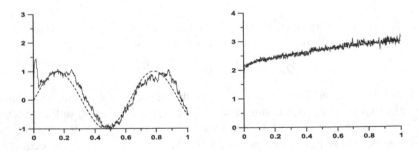

Fig. 2. Reconstruction of disturbances u_1 (left) and u_2. Variant 2: $N = 10^5$, $\delta^N = 0.0025$, $\delta_s = \delta^N/10^4$, $\alpha_m^N = 0.04$, $\alpha_d^N = 0.05$, accuracy $\|u^N(\cdot) - u(\cdot)\|_{L_2(T)} = 0.102$.

It seems that the output accuracy obtained in Variant 2 is satisfactory ($\|u^N(\cdot) - u(\cdot)\|_{L_2(T)} = 0.102$, whereas the measuring accuracy h^N of estimates (6), (18) does not exceed 0.003 (h^N is close to the order $1/N^{1/2}$, see (20), $N = 10^5$)).

Then, the relations obtained above were tested on other disturbances meeting the given geometrical constraints. Let us give the example (see Fig. 3, Variant 3) for $u_1(t) =$

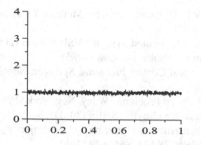 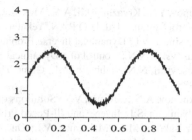

Fig. 3. Reconstruction of alternative disturbances u_1 (left) and u_2. Variant 3, parameters: $N = 10^5$, $\delta^N = 0.0025$, $\delta_s = \delta^N/10^4$, $\alpha_m^N = 0.04$, $\alpha_d^N = 0.05$, accuracy $\|u^N(\cdot) - u(\cdot)\|_{L_2(T)} = 0.094$.

1, $u_2(t) = 1.5 + \sin 10t$, where all the values of algorithm's parameters were the same as in Variant 2. The reconstruction accuracy is definitely comparable (even better) with the results in Variant 2. As to optimal values of the parameters for Variant 3 according to (21), they are not more 10% different from the optimal values for Variant 2. For other disturbances, the results are similar. This is a promising fact in favor of our hypothesis on the applicability of the procedure in question.

7 Conclusion

In the present paper, we continue to study inverse problems for linear SDEs consisting in the dynamic reconstruction of unknown nonrandom disturbances in the deterministic and stochastic terms of the equations. The novelty is in designing a solving algorithm in case of rather general form of input information containing different variants of incomplete measurements and in testing (for this new statement) the tuning procedure elaborated earlier to adapt algorithm's parameters to a specific system. As perspective directions of investigations, we mark out the theoretical substantiation of a tuning procedure (described above or similar) and the improvement of the algorithm from the viewpoint of its accuracy.

Acknowledgements. The work was performed as part of research conducted in the Ural Mathematical Center with the financial support of the Ministry of Science and Higher Education of the Russian Federation (Agreement number 075-02-2023-913).

References

1. Brockett, R.W., Mesarovich, M.P.: The reproducibility of multivariable systems. J. Math. Anal. Appl. **11**(1–3), 548–563 (1965)
2. Sain, M.K., Massey, J.L.: Invertibility of linear time-invariant dynamical systems. IEEE Trans. Automat. Contr. **14**, 141–149 (1969)
3. Kryazhimskii, A.V., Osipov, Y.S.: Modelling of a control in a dynamic system. Engrg. Cybernetics **21**(2), 38–47 (1984)
4. Osipov, Y.S., Kryazhimskii, A.V.: Inverse Problems for Ordinary Differential Equations: Dynamical Solutions. Gordon & Breach, London (1995)

5. Osipov, Y.S., Kryazhimskii, A.V., Maksimov, V.I.: Dynamic Recovery Methods for Inputs of Control Systems. Izd. UrO RAN, Yekaterinburg (2011)
6. Maksimov, V.I.: Dynamical Inverse Problems of Distributed Systems. VSP, Boston (2002)
7. Krasovskii, N.N.: Control of a Dynamical System. Nauka, Moscow (1985)
8. Krasovskii, N.N., Subbotin, A.I.: Game-Theoretical Control Problems. Springer, New York (1988)
9. Tikhonov, A.N., Arsenin, V.Y.: Solutions of Ill-Posed Problems. Wiley, New York (1981)
10. Kabanikhin, S.I.: Inverse and Ill-Posed Problems. De Gruyter, Berlin (2011)
11. Blizorukova, M.S., Maksimov, V.I.: On a reconstruction algorithm for the trajectory and control in a delay system. Proc. Steklov Inst. Math. **280**(1), S66–S79 (2013)
12. Surkov, P.G.: Dynamic right-hand side reconstruction problem for a system of fractional differential equations. Differ. Equ. **55**(6), 849–858 (2019)
13. Kryazhimskii, A.V., Osipov, Yu.S.: On a stable positional recovery of control from measurements of a part of coordinates. In: Some Problems of Control and Stability: Collection of Papers, pp. 33–47. Izd. AN SSSR, Sverdlovsk (1989)
14. Osipov, Yu.S., Kryazhimskii, A.V.: Positional modeling of a stochastic control in dynamical systems. In: Stochastic Optimization: Proceedings of the International Conference 1984, LNCS, vol. 81, pp. 696–704. Springer, Berlin (1986). https://doi.org/10.1007/BFb0007145
15. Kryazhimskii, A.V., Osipov, Y.S.: On dynamical regularization under random noise. Proc. Steklov Inst. Math. **271**(1), S125–S137 (2010)
16. Maksimov, V.I.: The methods of dynamical reconstruction of an input in a system of ordinary differential equations. J. Inverse Ill-Posed Probl. **29**(1), 125–156 (2021)
17. Rozenberg, V.L.: Dynamic restoration of the unknown function in the linear stochastic differential equation. Autom. Remote. Control. **68**(11), 1959–1969 (2007)
18. Rozenberg, V.L.: Reconstruction of random-disturbance amplitude in linear stochastic equations from measurements of some of the coordinates. Comput. Math. Math. Phys. **56**(3), 367–375 (2016). https://doi.org/10.1134/S0965542516030143
19. Rozenberg, V.L.: Reconstruction problem with incomplete information for a quasilinear stochastic differential equation. Comput. Math. Math. Phys. **62**(11), 1838–1848 (2022)
20. Melnikova, L.A., Rozenberg, V.L.: One dynamical input reconstruction problem: tuning of solving algorithm via numerical experiments. AIMS Math. **4**(3), 699–713 (2019)
21. Shiryaev, A.N.: Probability, Statistics, and Random Processes. Izd. Mosk. Gos. Univ, Moscow (1974)
22. Øksendal, B.: Stochastic Differential Equations: An Introduction with Applications. Springer, Berlin (1985). https://doi.org/10.1007/978-3-662-03620-4
23. Korolyuk, V.S., Portenko, N.I., Skorokhod, A.V., Turbin, A.F.: Handbook on Probability Theory and Mathematical Statistics. Nauka, Moscow (1985)
24. Chernous'ko, F.L., Kolmanovskii, V.B.: Optimal Control Under Random Perturbation. Nauka, Moscow (1978)
25. Rozenberg, V.L.: A control problem under incomplete information for a linear stochastic differential equation. Ural Math. J. **1**(1), 68–82 (2015)
26. Milshtein, G.N.: Numerical Integration of Stochastic Differential Equations. Izd. Ural. Gos. Univ, Sverdlovsk (1988)

Mathematical Modeling of the Household Behavior in the Labor Market

A. A. Shananin[1,2,3,4,5] and N. V. Trusov[1,2,3(✉)]

[1] Federal Research Center "Computer Science and Control" of RAS, Vavilova st. 40, Moscow 119333, Russia
[2] Moscow Center of Fundamental and Applied Mathematics, Leninskiye Gory, Moscow 119991, Russia
trunick.10.96@gmail.com
[3] Federal State Budgetary Institution "All-Russian Research Institute of Labor" of the Ministry of Labor and Social Protection of the Russian Federation, Parkovaya st. 29, Moscow 105043, Russia
[4] Moscow Institute of Physics and Technology, National Research University, 9 Institutsky pereulok, Dolgoprudny 141701, Moscow Region, Russia
[5] Peoples' Friendship University of Russia, RUDN University, Miklukho-Maklaya Street 6, 117198 Moscow, Russia

Abstract. We describe the economic behavior of the household. On the one hand, the household acts as a consumer that seeks to maximize the discounted consumptions in the imperfect lending and saving market. On the other hand, the household acts a worker in the labor market that receives salary and wants to enlarge its competencies to receive higher wages. In this work we present the model of the worker behavior that spends its salary on consumptions and on the investments in human capital. The investments in human capital helps to obtain new skills and increase the qualifications of the employee. This provides an opportunity to receive higher wages. The problem is formalized as an optimal control problem on the infinite time horizon. We introduce its solution in the form of the Pontryagin maximum principle, find the transversality conditions of the conjugate variables, and introduce the identification approach to reproduce the behavior of employees in different social layers based on the Russian Federation Household Budget Survey.

Keywords: Mathematical modeling · Optimal control · Infinite time horizon · Maximum principle · Identification problem

1 Introduction

The economic behavior of the household combines the rational consumer behavior and the employee behavior in the labor market. We use an approach of separation the problem by physical processes, and consider two independent

Supported by RSF (grant No.23-21-00281).

problems: the rational consumer behavior of the household and the employee behavior in the labor market.

The rational consumer model arises to the fundamental work by F. Ramsey [1], where the households operates in a perfect lending and saving market. It has an extension to the imperfect lending and saving market [2,3]. An imperfect lending and saving market takes place due to the fact that the interest rates on loans are higher than the interest rates on deposits (for instance, the interest rates on loans are 2.5–3 times higher than the interest rates on deposits over the past decade in Russia). This follows the non-smoothness of the right hand side of the differential equation (see [2]). The rational consumer model suppose that the household seeks to maximize the discounted consumptions, and has three opportunities: borrow consumer credit at the certain interest rate, save money in the form of deposits, or do not interact with the banking system at all. The crucial fact is that by the final moment of time the household has to pay off its credits. This follows to the phase restriction on the financial welfare of the household (see [2]). If this restriction incompletes, the situation turns into financial pyramid: the household borrows one consumer credit after another to pay off the active ones. Nevertheless, this Ramsey-type model has the synthesis in an almost analytical form. The synthesis leads to 7 different economic behavior types of the households that form a social layer [2,4]. This model is identified against Russian statistical data [5]. The model is used to analyze the state of consumer credit in Russia and its influence on the household economy under COVID-19 pandemic conditions and under the current conditions as a mechanism for social adaptation of the households [2–4,6,7].

In this work we present the household behavior model in the labor market as an employee. The employee receives the salary that it can spend on consumptions and on the investments in human capital. The employee seeks to maximize the discounted consumption as in the model of rational consumer behavior, and also enlarges its competencies in order to receive higher wages. The competencies of the household as an employee are described by the household's human capital. The bigger the human capital is, the higher wages the employee can pretend. In Sect. 2 we present the mathematical model of the household behavior as an employee in the labor market. We pose a theorem of an optimal control problem on the infinite time horizon and present the solution of the optimal control problem. In Sect. 3 we present the identification of the employee model in the different social layers of the households in Russia in different groups of regions that were previously determined in [4].

2   Modeling of Household Behavior in the Labor Market

The household as an employee seeks to maximize the discounted consumptions $C(t) \geq 0$ on the time horizon $t \in [0, T]$, where T is a random variable that describes the moment of receiving a higher wages. Denote by $P(t) \geq 0$ the human capital of the household. The household chooses the investments in human capital $X(t) \geq 0$ in order to maximize the functional

$$E\left\{\int_0^T e^{-\delta t} C^\alpha(t)dt + e^{-\delta T} \cdot \frac{\beta^\alpha(P(T))}{\delta}\right\} \rightarrow \max_{X \geq 0}, \qquad (1)$$

where $\delta > 0$ is a discount factor, $0 < 1 - \alpha < 1$ is a risk aversion, $\beta(P)$ is a non-negative monotonically increasing function that reflects the pretended salary of the employee depending on its qualification. We assume that the human capital is defined by the ordinary differential equation

$$\frac{dP}{dt} = X + \mu P, \qquad (2)$$

$$P(0) = 0. \qquad (3)$$

where $\mu < 0$ reflects the obsolescence of accumulated competencies of the employee.

In the model of the employee behavior in the labor market we suppose that the employee receives the salary S and spends it on the consumptions C and on the investments in human capital X. Thus,

$$S = C + X.$$

Denote by $F(t)$ the probability of receiving the new wage level $\beta(P)$, where

$$F(t) = 1 - e^{-\int_0^t H(P(\tau))d\tau}, \qquad (4)$$

where $H(P)$ is a non-negative monotonically increasing function.

Lemma 1. *The functional (1) can be written in the following form*

$$E\left\{\int_0^T e^{-\delta t} C^\alpha(t)dt + e^{-\delta T} \cdot \frac{\beta^\alpha(P(T))}{\delta}\right\}$$

$$= \int_0^{+\infty} e^{-\int_0^t (\delta + H(P(\tau)))d\tau} \left(C^\alpha(t) + H(P(t)) \cdot \frac{\beta^\alpha(P(t))}{\delta}\right) dt.$$

Proof. By definition,

$$E\left\{\int_0^T e^{-\delta t} C^\alpha(t)dt + e^{-\delta T} \cdot \frac{\beta^\alpha(P(T))}{\delta}\right\}$$

$$= \int_0^{+\infty} \left\{\int_0^\tau e^{-\delta t} C^\alpha(t)dt + \frac{e^{-\delta \tau}}{\delta}\beta^\alpha(P(\tau))\right\} dF(\tau).$$

The last term can be presented as a sum of two terms in accordance with (4)

$$\int\limits_0^{+\infty} \left\{ \int\limits_0^{\tau} e^{-\delta t} C^{\alpha}(t) dt + \frac{e^{-\delta \tau}}{\delta} \beta^{\alpha}(P(\tau)) \right\} dF(\tau)$$

$$= \int\limits_0^{+\infty} \int\limits_0^{\tau} e^{-\delta t} C^{\alpha}(t) dt dF(\tau) + \int\limits_0^{+\infty} e^{-\int\limits_0^{\tau}(\delta + H(P(s))) ds} \cdot H(P(\tau)) \cdot \frac{\beta^{\alpha}(\tau)}{\delta} d\tau.$$

Integration by parts follows that

$$\int\limits_0^{+\infty} \int\limits_0^{\tau} e^{-\delta t} C^{\alpha}(t) dt dF(\tau) = F(\tau) \cdot \int\limits_0^{\tau} e^{-\delta t} C^{\alpha}(t) \bigg|_{\tau=0}^{\tau=+\infty} - \int\limits_0^{+\infty} e^{-\delta t} C^{\alpha}(t) \cdot F(t) dt.$$

Thus, we obtain that

$$F(\tau) \cdot \int\limits_0^{\tau} e^{-\delta t} C^{\alpha}(t) \bigg|_{\tau=0}^{\tau=+\infty} - \int\limits_0^{+\infty} e^{-\delta t} C^{\alpha}(t) \cdot F(t) dt$$

$$+ \int\limits_0^{+\infty} e^{-\int\limits_0^{\tau}(\delta + H(P(s))) ds} \cdot H(P(\tau)) \cdot \frac{\beta^{\alpha}(\tau)}{\delta} d\tau$$

$$= \int\limits_0^{+\infty} e^{-\int\limits_0^{t}(\delta + H(P(\tau))) d\tau} \cdot \left(C^{\alpha}(t) + H(P(t)) \cdot \frac{\beta^{\alpha}(P(t))}{\delta} \right) dt.$$

This completes the proof of lemma.

Denote by

$$\Lambda(t) = \int\limits_0^{t} (\delta + H(P(\tau))) \, d\tau.$$

Let $S\theta$ be the annual salary of the worker. Introducing the dimensionless variables $\hat{P} = \frac{P}{S\theta}$, $\tau = \frac{t}{\theta}$, $\hat{\mu} = \mu\theta$, $\hat{H}\left(\hat{P}\right) = \frac{1}{\delta} H\left(S\theta \hat{P}\right)$, $\hat{\beta}\left(\hat{P}\right) = \frac{1}{S}\beta\left(S\theta \hat{P}\right)$, in account with Lemma 1, we obtain an optimal control problem (5)–(9).

$$\int\limits_0^{+\infty} e^{-\Lambda}\left(u^{\alpha} + \hat{H}\left(\hat{P}\right) \cdot \hat{\beta}^{\alpha}\left(\hat{P}\right)\right) d\tau \to \max_{0 \le u \le 1}, \qquad (5)$$

$$\frac{d\hat{P}}{d\tau} = (1 - u) + \hat{\mu}\hat{P}, \qquad (6)$$

$$\frac{d\Lambda}{d\tau} = \delta\theta\left(1 + \hat{H}\left(\hat{P}\right)\right), \qquad (7)$$

$$\hat{P}(0) = 0, \qquad (8)$$

$$\Lambda(0) = 0. \qquad (9)$$

Remark 1. The control u in the functional (5) represents the part of spending salary S on the consumptions C. The value $(1-u)$ represents the part of spending salary S on the investments in human capital.

Theorem 1. *1. Suppose that there exist such $\hat{C} > 0$:*

$$\hat{H}\left(\hat{P}\right) \leq \hat{C}(\hat{P}+1), \forall \hat{P} \geq 0, \tag{10}$$

Moreover, suppose that $\hat{H}\left(\hat{P}\right) \in C^1([0,+\infty))$, $\hat{\beta}\left(\hat{P}\right) \in C^1([0,+\infty))$. Then the optimal control problem (5)–(9) has a solution.

2. If the functions $\hat{P}(\tau)$, $\Lambda(\tau)$ are the solution of the optimal control problem (5)–(9) and are absolutely continuous at $\tau \geq 0$, $u^*(\tau)$ is a measurable function, then there exist such absolutely continuous functions $\varphi_1(\tau) > 0$ and $\varphi_2(\tau) < 0$, $\tau \geq 0$:

$$\varphi_1(\tau) = \int_{\tau}^{+\infty} \left\{ e^{-\hat{\mu}(\tau-s)} \left[-\delta\theta \left(1 + \hat{H}\left(\hat{P}(s)\right)\right) \varphi_1(s) \right. \right.$$

$$+ \left(\frac{d\hat{H}\left(\hat{P}(s)\right)}{d\hat{P}(s)} \cdot \hat{\beta}^\alpha\left(\hat{P}(s)\right) + \alpha \frac{d\hat{H}\left(\hat{P}(s)\right)}{d\hat{P}(s)} \cdot \frac{d\hat{\beta}(\hat{P}(s))}{d\hat{P}(s)} \cdot \hat{\beta}^{\alpha-1}\left(\hat{P}(s)\right)\right) \Bigg]$$

$$- \delta\theta \left[\int_0^s \frac{d\hat{H}\left(\hat{P}(\xi)\right)}{d\hat{P}(\xi)} e^{-\hat{\mu}(\tau-\xi)} d\xi - \int_0^\tau \frac{d\hat{H}\left(\hat{P}(\xi)\right)}{d\hat{P}(\xi)} e^{-\hat{\mu}(\tau-\xi)} d\xi \right]$$

$$\cdot \left[\delta\theta \left(1 + \hat{H}\left(\hat{P}(s)\right)\right) \varphi_2(s) + \min\left\{1, \left(\frac{\varphi_1(s)}{\alpha}\right)^{\frac{\alpha}{\alpha-1}}\right\} \right.$$

$$\left. + \hat{H}\left(\hat{P}(s)\right) \cdot \hat{\beta}^\alpha\left(\hat{P}(s)\right) \right] \right\} ds, \tag{11}$$

$$\varphi_2(\tau) = -\int_{\tau}^{+\infty} \left[\delta\theta \left(1 + \hat{H}\left(\hat{P}(s)\right)\right) \varphi_2(s) + \min\left\{1, \left(\frac{\varphi_1(s)}{\alpha}\right)^{\frac{\alpha}{\alpha-1}}\right\} \right.$$

$$\left. + \hat{H}\left(\hat{P}(s)\right) \cdot \hat{\beta}^\alpha\left(\hat{P}(s)\right) \right] ds, \tag{12}$$

$$u^*(\tau) = \min\left\{1, \left(\frac{\varphi_1(\tau)}{\alpha}\right)^{\frac{1}{\alpha-1}}\right\}. \tag{13}$$

Moreover,

$$\hat{P}(\tau) = \int_0^\tau \left(1 - \left(\frac{\varphi_1(s)}{\alpha}\right)^{\frac{1}{\alpha-1}}\right)_+ e^{\hat{\mu}(\tau-s)} ds, \tag{14}$$

$$\Lambda(\tau) = \int_0^\tau \left(1 + \hat{H}\left(\hat{P}(s)\right)\right) ds. \tag{15}$$

Proof. 1. The optimal control problem (5)–(9) under assumption (10) satisfy the assumptions A1-A7 in [8] of the solution existence on the infinite time horizon that are easy to check.

2. The proof of the second part of the theorem is based on [9]. The conjugate system defines by the matrix differential equation

$$\frac{d\psi(\tau)}{d\tau} = -A^T(\tau)\psi(\tau) + b(\tau), \tag{16}$$

where $\psi(\tau) = (\psi_1(\tau), \psi_2(\tau))^T$, $b(\tau) = (b_1(\tau), b_2(\tau))^T$,

$$A(\tau) = \begin{pmatrix} \hat{\mu} & 0 \\ \delta\theta\frac{d\hat{H}(\hat{P}(\tau))}{d\hat{P}(\tau)} & 0 \end{pmatrix},$$

$$b(\tau) = \begin{pmatrix} -e^{-\Lambda(\tau)} \cdot \left(\frac{d\hat{H}(\hat{P}(\tau))}{d\hat{P}(\tau)} \cdot \hat{\beta}^\alpha(\hat{P}(\tau)) + \alpha\frac{d\hat{H}(\hat{P}(\tau))}{d\hat{P}(\tau)} \cdot \frac{d\hat{\beta}(\hat{P}(\tau))}{d\hat{P}(\tau)} \cdot \hat{\beta}^{\alpha-1}(\hat{P}(\tau)) \right) \\ e^{-\Lambda(\tau)} \cdot \left((u^*(\tau))^\alpha + \hat{H}(\hat{P}(\tau)) \cdot \hat{\beta}^\alpha(\hat{P}(\tau)) \right) \end{pmatrix}.$$

Denote by $\varphi = \psi e^\Lambda$. Thus, the conjugate system (16) transforms to

$$\frac{d\varphi(\tau)}{d\tau} = -A^T(\tau)\varphi(\tau) + \delta\theta\left(1 + \hat{H}(\hat{P}(\tau))\right)\varphi(\tau) + b(\tau)e^{\Lambda(\tau)}. \tag{17}$$

The solution of (17) can be written as

$$\varphi(\tau) = -\int_\tau^{+\infty} Z(\tau)Z^{-1}(s)\left[\delta\theta\left(1 + \hat{H}\left(\hat{P}(s)\right)\right)\varphi(s) + b(s)e^{\Lambda(s)}\right]ds, \tag{18}$$

where $Z(\tau)$ is the solution of the general homogeneous matrix differential equation

$$\frac{dZ(\tau)}{d\tau} = -A^T(\tau)Z(\tau) \tag{19}$$

with initial condition

$$Z(0) = I, \tag{20}$$

where I is identity matrix. The solution of the matrix differential Eq. (19)–(20) is

$$Z(\tau) = \begin{pmatrix} e^{-\hat{\mu}\tau} & -\delta\theta\int_0^\tau \frac{d\hat{H}(\hat{P}(s))}{d\hat{P}(s)}e^{-\hat{\mu}(\tau-s)}ds \\ 0 & 1 \end{pmatrix}.$$

The inverse matrix $Z^{-1}(s)$ defines as

$$Z^{-1}(s) = \begin{pmatrix} e^{\hat{\mu}s} & \delta\theta\int_0^s \frac{d\hat{H}(\hat{P}(\xi))}{d\hat{P}(\xi)}e^{\hat{\mu}\xi}d\xi \\ 0 & 1 \end{pmatrix}.$$

In accordance with (18), we obtain that

$$\varphi_1(\tau) = \int_{\tau}^{+\infty} \left\{ e^{-\hat{\mu}(\tau-s)} \left[-\delta\theta \left(1 + \hat{H}\left(\hat{P}(s)\right) \right) \varphi_1(s) \right. \right.$$

$$+ \left(\frac{d\hat{H}\left(\hat{P}(s)\right)}{d\hat{P}(s)} \cdot \hat{\beta}^{\alpha}\left(\hat{P}(s)\right) + \alpha \frac{d\hat{H}\left(\hat{P}(s)\right)}{d\hat{P}(s)} \cdot \frac{d\hat{\beta}(\hat{P}(s))}{d\hat{P}(s)} \cdot \hat{\beta}^{\alpha-1}\left(\hat{P}(s)\right) \right) \right]$$

$$-\delta\theta \left[\int_0^s \frac{d\hat{H}\left(\hat{P}(\xi)\right)}{d\hat{P}(\xi)} e^{-\hat{\mu}(\tau-\xi)} d\xi - \int_0^\tau \frac{d\hat{H}\left(\hat{P}(\xi)\right)}{d\hat{P}(\xi)} e^{-\hat{\mu}(\tau-\xi)} d\xi \right]$$

$$\left. \cdot \left[\delta\theta \left(1 + \hat{H}\left(\hat{P}(s)\right) \right) \varphi_2(s) + (u^*(\tau))^{\alpha} + \hat{H}\left(\hat{P}(s)\right) \cdot \hat{\beta}^{\alpha}\left(\hat{P}(s)\right) \right] \right\} ds, \quad (21)$$

$$\varphi_2(\tau) = -\int_{\tau}^{+\infty} \left[\delta\theta \left(1 + \hat{H}\left(\hat{P}(s)\right) \right) \varphi_2(s) \right.$$

$$\left. + (u^*(\tau))^{\alpha} + \hat{H}\left(\hat{P}(s)\right) \cdot \hat{\beta}^{\alpha}\left(\hat{P}(s)\right) \right] ds. \quad (22)$$

An optimal control $u^*(\tau)$ is determined by the Hamiltonian function

$$\mathcal{H}\left(\tau, \hat{P}, \Lambda, \varphi_1, \varphi_2\right) = \sup_{0 \le u \le 1} \left\{ e^{-\Lambda}\left(u^{\alpha} + \hat{H}\left(\hat{P}\right) \cdot \hat{\beta}^{\alpha}\left(\hat{P}\right) \right) \right.$$

$$\left. + \varphi_1 e^{-\Lambda} \cdot \left((1-u) + \hat{\mu}\hat{P} \right) + \varphi_2 e^{-\Lambda} \cdot \delta\theta \left(1 + \hat{H}\left(\hat{P}\right) \right) \right\},$$

thus,

$$u^*(\tau) = \min\left\{ 1, \left(\frac{\varphi_1(\tau)}{\alpha} \right)^{\frac{1}{\alpha-1}} \right\}. \quad (23)$$

Substituting (23) to (21), (22), we obtain the formulae (11), (12). Integrating (6) with a help of variation of constants method in accordance with (23), we obtain the formula (14), and integrating (15), we obtain the formula (15). This completes the proof of the theorem.

Corollary 1. *Suppose that there exist such moment of time $\hat{\tau} \ge 0$ that the function $\hat{P}(\tau)$ becomes monotonically increasing for any $\tau \ge \hat{\tau}$, $\hat{P} < -\frac{1}{\hat{\mu}}$. Then,* $\lim_{\tau \to +\infty} \hat{P}(\tau) = \hat{P}^*$, $\lim_{\tau \to +\infty} \varphi_1(\tau) = \varphi_1^*$, $\lim_{\tau \to +\infty} \varphi_2(\tau) = \varphi_2^*$, *where*

$$\varphi_1^* = \alpha \left(1 + \hat{\mu}\hat{P}^* \right)^{\alpha-1}, \quad (24)$$

$$\varphi_2^* = -\frac{1}{\delta\theta \left(1 + \hat{H}\left(\hat{P}^*\right) \right)} \left[\left(1 + \hat{\mu}\hat{P}^* \right)^{\alpha} + \hat{H}\left(\hat{P}^*\right) \cdot \hat{\beta}^{\alpha}\left(\hat{P}^*\right) \right]. \quad (25)$$

$$-\frac{d\hat{H}\left(\hat{P}^*\right)}{d\hat{P}*}\cdot\hat{\beta}^\alpha\left(\hat{P}^*\right)-\alpha\hat{H}\left(\hat{P}^*\right)\cdot\hat{\beta}^{\alpha-1}(\hat{P}^*)\cdot\frac{d\hat{\beta}\left(\hat{P}^*\right)}{d\hat{P}*}$$

$$-\alpha\hat{\mu}\left(1+\hat{\mu}\hat{P}^*\right)^{\alpha-1}+\alpha\delta\theta\left(1+\hat{\mu}\hat{P}^*\right)^{\alpha-1}\cdot\left(1+\hat{H}\left(\hat{P}^*\right)\right)$$

$$+\frac{1}{1+\hat{H}\left(\hat{P}^*\right)}\cdot\left[\left(1+\hat{\mu}\hat{P}^*\right)^\alpha+\hat{H}\left(\hat{P}^*\right)\cdot\hat{\beta}^\alpha\left(\hat{P}^*\right)\right]\cdot\frac{d\hat{H}\left(\hat{P}^*\right)}{d\hat{P}*}=0. \quad (26)$$

Proof. Since the function $\hat{P}(\tau)$, $\tau \geq \hat{\tau}$ monotonically increasing and bounded, then there exist the limit $\lim_{\tau\to+\infty}\hat{P}(\tau)=\hat{P}^*$. As $\hat{P}^*$ is a stationary point, then $\frac{d\hat{P}}{d\tau}\to0$, $\tau\to+\infty$. Since

$$\frac{d\hat{P}}{d\tau}=1-\min\left\{1,\left(\frac{\varphi_1(\tau)}{\alpha}\right)^{\frac{1}{\alpha-1}}\right\}+\hat{\mu}\hat{P},$$

in order to satisfy the condition $\frac{d\hat{P}}{d\tau}\to0$, $\tau\to+\infty$, the function $\varphi_1(\tau)$ has the limit

$$\lim_{\tau\to+\infty}\varphi_1(\tau)=\alpha\left(1+\hat{\mu}\hat{P}^*\right)^{\alpha-1}. \quad (27)$$

Now we are going to prove that $\varphi_2(\tau)$ has limit when $\tau\to+\infty$, and this limit is (25). Denote by

$$h_1(s)=\delta\theta\left(1+\hat{H}\left(\hat{P}(s)\right)\right), \quad (28)$$

$$h_2(s)=\min\left\{1,\left(\frac{\varphi_1(s)}{\alpha}\right)^{\frac{\alpha}{\alpha-1}}\right\}+\hat{H}\left(\hat{P}(s)\right)\cdot\hat{\beta}^\alpha\left(\hat{P}(s)\right). \quad (29)$$

Then the formula (12) for φ_2 can be written in the following form

$$\varphi_2(\tau)=-\int_\tau^{+\infty}[h_1(s)\varphi_2(s)+h_2(s)]\,ds. \quad (30)$$

To solve the integral Eq. (30), at first we differentiate it

$$\frac{d\varphi_2(\tau)}{d\tau}=h_1(\tau)\varphi_2(\tau)+h_2(\tau). \quad (31)$$

With a help of variation of constants method, it is easy to obtain the solution of the ordinary differential Eq. (31)

$$\varphi_2(\tau)=\left(C_0+\int_0^\tau h_2(\eta)e^{-\int_0^\eta h_1(\xi)d\xi}\,d\eta\right)e^{\int_0^\tau h_1(\eta)d\eta}, \quad (32)$$

where $C_0 \in \mathcal{R}$. Since $h_1(\tau) > 0$, $\forall \tau \geq 0$, the only way the function $\varphi_2(\tau)$ has limit when $\tau \to +\infty$ is to set

$$C_0 = -\int_0^{+\infty} h_2(\eta) e^{-\int_0^\eta h_1(\xi) d\xi} d\eta.$$

This follows that

$$\varphi_2(\tau) = -\int_\tau^{+\infty} h_2(\eta) e^{\int_\tau^\eta h_1(\xi) d\xi} d\eta. \tag{33}$$

The exponent in (33) has the negative power since $\eta > \tau$. Thus, the integral in (33) converges, and $\lim_{\tau \to +\infty} \varphi_2(\tau) = \varphi_2^*$.

We are now in the position to show that φ_2^* satisfies (25). From (28), (29) follows that

$$h_1(\eta) = \hat{h}_1 + \varepsilon_1(\eta), \tag{34}$$

$$h_2(\eta) = \hat{h}_2 + \varepsilon_2(\eta), \tag{35}$$

where $\hat{h}_1$, $\hat{h}_2$ are positive constants, $\varepsilon_1(\eta) \to 0$, $\varepsilon_2(\eta) \to 0$ when $\eta \to +\infty$. Thus, we obtain that

$$\varphi_2(\tau) = -\int_\tau^{+\infty} \hat{h}_2 e^{\int_\tau^\eta (\hat{h}_1 + \varepsilon_1(\eta)) d\xi} d\eta - \int_\tau^{+\infty} \varepsilon_2(\eta) e^{\int_\tau^\eta (\hat{h}_1 + \varepsilon_1(\eta)) d\xi} d\eta. \tag{36}$$

The second term of the right hand side in (36) tends to 0 when $\tau \to +\infty$ due to the fact that the power of exponent is negative ($h_1(\xi) > 0$, $\eta > \tau$). Denote by

$$\tilde{\varphi}_1(\tau) = \int_\tau^{+\infty} e^{\int_\tau^\eta (\hat{h}_1 + \varepsilon_1(\eta)) d\xi} d\eta. \tag{37}$$

The formula (37) can be written in the following form

$$\tilde{\varphi}_1(\tau) = \int_\tau^{+\infty} e^{\hat{h}_1(\tau - \eta)} \cdot e^{\int_\tau^\eta \varepsilon_1(\eta) d\xi} d\eta. \tag{38}$$

The integration by parts of (38) leads to

$$\tilde{\varphi}_1(\tau) = -\frac{1}{\hat{h}_1} \cdot e^{\hat{h}_1(\tau - \eta)} e^{\int_\tau^\eta \varepsilon_1(\xi) d\xi} \Big|_\tau^{+\infty} - \frac{1}{\hat{h}_1} \cdot \int_\tau^{+\infty} \varepsilon_1(\eta) \cdot e^{\hat{h}_1(\tau - \eta) + \int_\eta^\tau \varepsilon_1(\xi) d\xi} d\eta. \tag{39}$$

The first term of (39) is equal to $\frac{1}{\hat{h}_1}$, the second term of (39) tends to zero when $\tau \to +\infty$ due to the fact that the power of the exponent is negative and $\varepsilon_1(\eta) \to 0$, $\eta \to +\infty$. Thus, in accordance with (36), we obtain that

$$\lim_{\tau \to +\infty} \varphi_2(\tau) = -\frac{\hat{h}_2}{\hat{h}_1}.$$

According to (34), (35), (27), we obtain that φ_2^* satisfy the formula (25). The proof of corollary is completed.

2.1   Optimal Control Trajectory

To find the optimal control trajectory, we have to investigate the system (40)–(42) near the equilibrium point. Consider the optimal control $u < 1$ (otherwise the human capital function $\hat{P}(\tau)$ do not increase). Thus, we obtain the following system

$$
\frac{d\varphi_1}{d\tau} = -\frac{d\hat{H}\left(\hat{P}\right)}{d\hat{P}} \cdot \hat{\beta}^\alpha\left(\hat{P}\right) - \alpha\hat{H}\left(\hat{P}\right) \cdot \hat{\beta}^{\alpha-1}\left(\hat{P}\right) \cdot \frac{d\hat{\beta}\left(\hat{P}\right)}{d\hat{P}} - \hat{\mu}\varphi_1
$$
$$
-\delta\theta\varphi_2\frac{d\hat{H}\left(\hat{P}\right)}{d\hat{P}} + \delta\theta\varphi_1\left(1 + \hat{H}\left(\hat{P}\right)\right), \quad (40)
$$

$$
\frac{d\varphi_2}{d\tau} = \left(\frac{\varphi_1}{\alpha}\right)^{\frac{\alpha}{\alpha-1}} + \hat{H}\left(\hat{P}\right) \cdot \hat{\beta}^\alpha\left(\hat{P}\right) + \delta\theta\varphi_2\left(1 + \hat{H}\left(\hat{P}\right)\right), \quad (41)
$$

$$
\frac{d\hat{P}}{d\tau} = 1 - \left(\frac{\varphi_1}{\alpha}\right)^{\frac{1}{\alpha-1}} + \hat{\mu}\hat{P}. \quad (42)
$$

The equilibrium point $\left(\hat{P}^*, \varphi_1^*, \varphi_2^*\right)$ of the system of ordinary differential equations (40)–(42) defines by formulae (24)–(26). Suppose that the root P^* of the Eq. (26) exists.

An optimal control trajectory can be found in case when the linearized system (40)–(42) in the equilibrium point $\left(\hat{P}^*, \varphi_1^*, \varphi_2^*\right)$ has two real positive eigenvalues and one real negative eigenvalue. In this case, the eigenvector corresponding to the negative eigenvalue is a stable separatrix of a saddle point. Starting at this point, we can solve the system of ordinary differential equations (40)–(42) backwards in time and find the synthesis $u^*\left(\hat{P}\right)$. We next present the linearization of the system (40)–(42) in the equilibrium point. Denote by $D_1(\varphi_1, \varphi_2, \hat{P})$, $D_2(\varphi_1, \varphi_2, \hat{P})$, $D_3(\varphi_1, \varphi_2, \hat{P})$ the right hand sides of the ordinary differential Eqs. (40)–(42), i.e.

$$
D_1(\varphi_1, \varphi_2, \hat{P}) = -\frac{d\hat{H}\left(\hat{P}\right)}{d\hat{P}} \cdot \hat{\beta}^\alpha\left(\hat{P}\right) - \alpha\hat{H}\left(\hat{P}\right) \cdot \hat{\beta}^{\alpha-1}\left(\hat{P}\right) \cdot \frac{d\hat{\beta}\left(\hat{P}\right)}{d\hat{P}} - \hat{\mu}\varphi_1
$$
$$
-\delta\theta\varphi_2\frac{d\hat{H}\left(\hat{P}\right)}{d\hat{P}} + \delta\theta\varphi_1\left(1 + \hat{H}\left(\hat{P}\right)\right),
$$

$$
D_2(\varphi_1, \varphi_2, \hat{P}) = \left(\frac{\varphi_1}{\alpha}\right)^{\frac{\alpha}{\alpha-1}} + \hat{H}\left(\hat{P}\right) \cdot \hat{\beta}^\alpha\left(\hat{P}\right) + \delta\theta\varphi_2\left(1 + \hat{H}\left(\hat{P}\right)\right),
$$

$$
D_3(\varphi_1, \varphi_2, \hat{P}) = 1 - \left(\frac{\varphi_1}{\alpha}\right)^{\frac{1}{\alpha-1}} + \hat{\mu}\hat{P}.
$$

The Jacobi matrix of the linearized system has the following form

$$
A = \begin{pmatrix} \frac{\partial D_1}{\partial \varphi_1} & \frac{\partial D_1}{\partial \varphi_2} & \frac{\partial D_1}{\partial \hat{P}} \\ \frac{\partial D_2}{\partial \varphi_1} & \frac{\partial D_2}{\partial \varphi_2} & \frac{\partial D_2}{\partial \hat{P}} \\ \frac{\partial D_3}{\partial \varphi_1} & \frac{\partial D_3}{\partial \varphi_2} & \frac{\partial D_3}{\partial \hat{P}} \end{pmatrix} \Bigg|_{(\varphi_1^*, \varphi_2^*, P^*)} = \begin{pmatrix} a_{11} & a_{12} & a_{13} \\ a_{21} & a_{22} & a_{23} \\ a_{31} & a_{32} & a_{33} \end{pmatrix},
$$

where the coefficients are

$$
a_{11} = -\hat{\mu} + \delta\theta \left(1 + \hat{H} \left(\hat{P}^* \right) \right),
$$

$$
a_{12} = -\delta\theta \frac{d\hat{H} \left(\hat{P} \right)}{d\hat{P}} \Bigg|_{\hat{P}_*},
$$

$$
a_{13} = -\frac{d^2\hat{H} \left(\hat{P} \right)}{d\hat{P}^2} \Bigg|_{\hat{P}_*} \cdot \hat{\beta}^\alpha \left(\hat{P}^* \right) - \alpha \frac{d\hat{H} \left(\hat{P} \right)}{d\hat{P}} \Bigg|_{\hat{P}_*} \cdot \hat{\beta}^{\alpha-1}(\hat{P}^*) \cdot \frac{d\hat{\beta} \left(\hat{P} \right)}{d\hat{P}} \Bigg|_{\hat{P}_*}
$$

$$
-\alpha \frac{d\hat{H} \left(\hat{P} \right)}{d\hat{P}} \Bigg|_{\hat{P}_*} \cdot \hat{\beta}^{\alpha-1}(\hat{P}^*) \cdot \frac{d\hat{\beta} \left(\hat{P} \right)}{d\hat{P}} \Bigg|_{\hat{P}_*} - \alpha \hat{H} \left(\hat{P}^* \right)
$$

$$
\cdot \left((\alpha - 1)\hat{\beta}^{\alpha-2} \left(\hat{P} \right) \cdot \left(\frac{d\hat{\beta} \left(\hat{P} \right)}{d\hat{P}} \Bigg|_{\hat{P}_*} \right)^2 + \hat{\beta}^{\alpha-1}(\hat{P}^*) \cdot \frac{d^2\hat{\beta} \left(\hat{P} \right)}{d\hat{P}^2} \Bigg|_{\hat{P}_*} \right)
$$

$$
+\frac{1}{1 + \hat{H} \left(\hat{P}_* \right)} \left[\left(1 + \hat{\mu}\hat{P}^* \right)^\alpha + \hat{H} \left(\hat{P}^* \right) \cdot \hat{\beta}^\alpha \left(\hat{P}^* \right) \right] \frac{d^2\hat{H} \left(\hat{P} \right)}{d\hat{P}^2} \Bigg|_{\hat{P}_*}
$$

$$
+\alpha\delta\theta \left(1 + \hat{\mu}\hat{P}^* \right)^{\alpha-1} \frac{d\hat{H} \left(\hat{P} \right)}{d\hat{P}} \Bigg|_{\hat{P}_*},
$$

$$
a_{21} = -\frac{1}{1-\alpha} \cdot \left(1 + \hat{\mu}\hat{P}^* \right),
$$

$$
a_{22} = \delta\theta \left(1 + \hat{H} \left(\hat{P}^* \right) \right),
$$

$$
a_{23} = \frac{d\hat{H} \left(\hat{P} \right)}{d\hat{P}} \Bigg|_{\hat{P}_*} \cdot \hat{\beta}^\alpha \left(\hat{P}^* \right) + \alpha\hat{H} \left(\hat{P}^* \right) \cdot \hat{\beta}^{\alpha-1}(\hat{P}^*) \cdot \frac{d\hat{\beta} \left(\hat{P} \right)}{d\hat{P}} \Bigg|_{\hat{P}_*}
$$

$$
-\frac{1}{1 + \hat{H} \left(\hat{P}_* \right)} \left[\left(1 + \hat{\mu}\hat{P}^* \right)^\alpha + \hat{H} \left(\hat{P}^* \right) \cdot \hat{\beta}^\alpha \left(\hat{P}^* \right) \right] \cdot \frac{d\hat{H} \left(\hat{P} \right)}{d\hat{P}} \Bigg|_{\hat{P}_*},
$$

$$a_{31} = \frac{1}{\alpha(1-\alpha)} \cdot \left(1 + \hat{\mu}\hat{P}^*\right)^{2-\alpha},$$

$$a_{32} = 0,$$

$$a_{33} = \hat{\mu}.$$

The characteristic equation is defined as

$$\chi_A(\lambda) = -\lambda^3 + \lambda^2(a_{11} + a_{22} + a_{33}) + \lambda(a_{12}a_{21} - a_{11}a_{33} + a_{13}a_{31} - a_{22}a_{33})$$
$$+a_{11}a_{22}a_{33} - a_{12}a_{21}a_{33} + a_{12}a_{23}a_{31} - a_{13}a_{22}a_{31},$$

where λ refers to the eigenvalues of the matrix A.

3   Numerical Results

In this section we present a numerical results for the household behavior in the labor market. Previously, in [4,7] we considered the Russian Federation Household Budget Survey statistics [5] on the period 2015–2020. The Russian Federation Household Budget Survey statistics are collected quarterly by Rosstat and cover about 50,000 households from 82 regions of Russia. The households answer 295 questions, including: salaries, other income, consumptions, value judgments about their financial situation, etc. We have split the 82 regions into 3 groups: rich, middle and poor per capital consumption. We observed that about a half of the households live in the villages in poor regions, and another half in cities. This motivated us to split the poor regions into two more groups. In [4] we have identified the behavior of the representative low income and high income households that borrow consumer credit in different groups of regions with a help of software package [10]. These representative households describe the behavior of a social layer.

Let us consider these representative households from different social layers from the point of their behavior in the labor market. Consider the linear functions

$$\hat{H}\left(\hat{P}\right) = \xi_1 \hat{P},$$

$$\hat{\beta}\left(\hat{P}\right) = \zeta_0 + \zeta_1 \hat{P}.$$

For the sake of simplicity, we set $\xi_1 = 75$.

In Appendix A we present the identification parameters for all considered social layers. All the parameters presented in Appendix A form one equilibrium point in the system (40)–(42) with a stable separatrix. Note that the parameters α, δ and S are taken from the software package [10] and were used in [4,7]. The salary S in Appendix A is in thousand rubles, and is considered for a working part of the people in a household as we model the behavior of the employee in the labor market. In Appendix B we present the information about how the salary of the employee changes (the value $\hat{\beta}(\hat{P}^*)$ shows how much wages will

increase), how long will it take to enlarge the competencies (the value $\frac{1}{H(\hat{P}*)}$)
and what the new salary will be.

Let us consider the high income representative household from the middle group of region with the following parameters: $\zeta_0 = 1.05$, $\zeta_1 = 0.15$, $\hat{\mu} = -1$, $\alpha = 0.7768$, $\delta = 0.010835$, $S = 54.47$. We obtain that there is one equilibrium point in the system (40)–(42). The linearized matrix in this equilibrium point has two positive eigenvalues and one negative that forms the stable separatrix. Starting in ε-area of the equilibrium point we obtain the solution of the system (40)–(42) that is illustrated on Fig. 1.

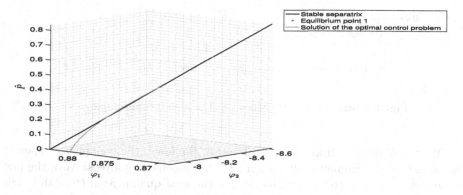

Fig. 1. Optimal trajectory among the stable separatrix.

On the Figs. 2 and 3 we present the optimal control synthesis of the employees in different social layers.

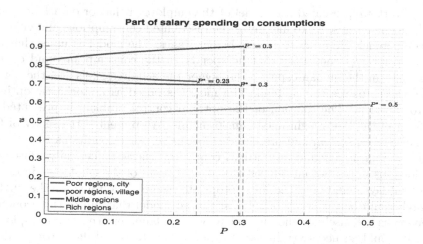

Fig. 2. Optimal control synthesis of the low income employees.

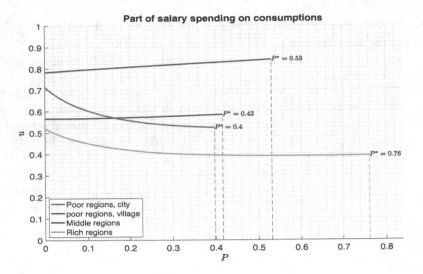

Fig. 3. Optimal control synthesis of the high income employees.

As we can observe from the Figs. 2 and 3, the low income employees spends less money on the human capital than the high income employees. Also, the low income employees spend more time to get the new qualification than the high income employees, and their salary does not increase as much as the salary of the high income employees (see Appendix B).

4   Conclusion

In this work we presented the model of the employee behavior on a labor market. The employee seeks to maximize the discounted consumptions and enlarges the competencies in order to receive higher wages. The model assumes that the employee receives the salary that he spends on the consumptions and on the investment in the human capital. The bigger the acquired skills are, the higher wages the employee can pretend. The model is posed as a stochastic optimal control problem. The solution of the optimal control problem is presented in the form of the Pontryagin maximum principle. We presented the study of the optimal control system and presented the transversality conditions to the conjugate system, and found an optimal control synthesis numerically. Based on the previous works, we introduced the identification of the employees in different social layers and obtained an appropriate results. The topic for the further research could be the study of the distributed model that unites the economic behavior as a household and the behavior on a labor market as an employee. Also, it would be necessary to use another set of statistical data to reproduce the behavior of the employees depending on their professions and competencies. Such models would help to study the structural changes in the behavior of agents under changes in external economic and social factors.

Appendix A

Monthly data actual at the end of 2021. Parameter $\theta = 12$.

Poor group of regions. City. Low income borrowers.
$\zeta_0 = 1.02$, $\zeta_1 = 0.15$, $\hat{\mu} = -1.2$, $\alpha = 0.924$, $\delta = 0.012101$, $S = 25.09$.

Poor group of regions. City. High income borrowers.
$\zeta_0 = 1.025$, $\zeta_1 = 0.155$, $\hat{\mu} = -1.2$, $\alpha = 0.8515$, $\delta = 0.011451$, $S = 43.56$.

Poor group of regions. Village. Low income borrowers.
$\zeta_0 = 1.01$, $\zeta_1 = 0.135$, $\hat{\mu} = -0.3$, $\alpha = 0.7989$, $\delta = 0.01096$, $S = 22.48$.

Poor group of regions. Village. High income borrowers.
$\zeta_0 = 1.01$, $\zeta_1 = 0.13$, $\hat{\mu} = -0.3$, $\alpha = 0.8562$, $\delta = 0.010459$, $S = 40.61$.

Middle group of regions. Low income borrowers.
$\zeta_0 = 1.025$, $\zeta_1 = 0.15$, $\hat{\mu} = -1$, $\alpha = 0.7988$, $\delta = 0.011446$, $S = 29.43$.

Middle group of regions. High income borrowers.
$\zeta_0 = 1.05$, $\zeta_1 = 0.15$, $\hat{\mu} = -1$, $\alpha = 0.7768$, $\delta = 0.010835$, $S = 54.47$.

Rich group of regions. Low income borrowers.
$\zeta_0 = 1.035$, $\zeta_1 = 0.16$, $\hat{\mu} = -0.8$, $\alpha = 0.8912$, $\delta = 0.012493$, $S = 43.46$.

Rich group of regions. High income borrowers.
$\zeta_0 = 1.06$, $\zeta_1 = 0.17$, $\hat{\mu} = -0.8$, $\alpha = 0.7479$, $\delta = 0.010618$, $S = 87.31$.

Appendix B

Poor group of regions. City. Low income borrowers.
The job search time is 4.7 months. The salary increases by 1.06 times and will be 26.48 thousand RUB.

Poor group of regions. City. High income borrowers.
The job search time is 2.9 months. The salary increases by 1.09 times and will be 47.34 thousand RUB.

Poor group of regions. Village. Low income borrowers.
The job search time is 4 months. The salary increases by 1.05 times and will be 23.63 thousand RUB.

Poor group of regions. Village. High income borrowers.
The job search time is 2.4 months. The salary increases by 1.08 times and will be 43.83 thousand RUB.

Middle group of regions. Low income borrowers.
The job search time is 3.9 months. The salary increases by 1.07 times and will be 31.49 thousand RUB.

Middle group of regions. High income borrowers.
The job search time is 2.9 months. The salary increases by 1.11 times and will be 60.61 thousand RUB.

Rich group of regions. Low income borrowers.
The job search time is 2.1 months. The salary increases by 1.12 times and will be 48.48 thousand RUB.

Rich group of regions. High income borrowers.
The job search time is 1.6 months. The salary increases by 1.19 times and will be 103.87 thousand RUB.

References

1. Ramsey, F.: A mathematical theory of savings. Econ. J. **152**(38), 543–559 (1928)
2. Shananin, A., Tarasenko, M., Trusov, N.: Consumer loan demand modeling of households in Russia. Comput. Math. Math. Phys. **61**, 1030–1051 (2021)
3. Shananin, A.A., Tarasenko, M.V., Trusov, N.V.: Consumer loan demand modeling. In: Strekalovsky, A., Kochetov, Y., Gruzdeva, T., Orlov, A. (eds.) MOTOR 2021. CCIS, vol. 1476, pp. 417–428. Springer, Cham (2021). https://doi.org/10.1007/978-3-030-86433-0_29
4. Shananin, A., Trusov, N.: Mathematical modeling of the consumer loan market in Russia under sanctions. Dokl. Math. **507**(6), 71–80 (2022)
5. Federal State Statistics Service of the Russian Federation. Russian Federation Household Budget Survey (HBS). Dataset downloaded https://obdx.gks.ru. Accessed Mar 2023
6. Shananin, A., Trusov, N.: The household behaviour modelling based on Mean Field Games approach. Lobachevskii J. Math. **42**, 1738–1752 (2021)
7. Trusov, N.: Identification of the household behavior modeling based on modified Ramsey model. Lobachevskii J. Math. **44**(1), 454–468 (2023)
8. Dmitruk, A., Kuz'kina, N.: Existence theorem in the optimal control problem on an infinite time interval. Math. Notes **78**(4), 466–480 (2005)
9. Aseev, S., Besov, K., Kryazhimskiy, A.: Infinite-horizon optimal control problems in economics. Russian Math. Surv. **67**(2), 195–253 (2012)
10. Certificate of state registration of the computer program No.2022619524. "Analysis of demand for consumer credit in the Russian Federation". Copyright holder: Trusov Nikolai Vsevolodovich. Application No.2022618580. Date of state registration in the Register of computer programs 23 May 2022

Visual Positioning of a Moving Object Using Multi-objective Control Algorithm

Margarita Sotnikova$^{(\boxtimes)}$ and Ruslan Sevostyanov

Saint-Petersburg State University, Universitetskaya 7-9, 199034 Saint-Petersburg, Russia
m.sotnikova@spbu.ru

Abstract. The paper is devoted to the problem of multi-objective control design for visual positioning of moving objects. This research area incorporates the methods of control theory and computer vision, and has a special importance for autonomous vehicles, where visual information from on board camera gives a rich data about the surrounding world. Visual information can be effectively used in feedback control, for example, in such applications as visual positioning, tracking a visually given line, moving in a changing environment with visual obstacles.

The main objective of this work is to develop the results obtained earlier and design a feedback control algorithm for visual positioning problem based on the multi-objective approach. This approach allows to take into account a set of requirements for closed-loop system performance in different regimes. These regimes, in particular, include the object motion under constant or random external disturbances. The multi-objective structure of the control law has adjustable elements that must be selected in accordance with the imposed requirements. It is convenient to formulate the problems of searching these elements as optimization tasks on the corresponding admissible sets.

The nonlinear mathematical model of object dynamics is considered. In addition, the nonlinear model of the dynamics in image plane of the camera is introduced. The control objective in visual positioning problem is to provide the desired projection of some three-dimensional object of the scene to the image plane. This projection is described by features vector. The main result of the work is the developed feedback control law, which is based on multi-objective structure and computer vision algorithms. Methods for searching adjustable elements of multi-objective structure are proposed. The efficiency of the approach is illustrated by a computational experiment in MATLAB environment.

Keywords: Visual Positioning · Computer Vision · Camera · Multi-Objective Control · External Disturbances · Moving Object

1 Introduction

Nowadays, essential attention is paid to the development of control systems for the operation of autonomous moving objects, their extensive implementation for solving practically important tasks without human participation, ensuring the safety and reliability of their work. In particular, in recent years, significant progress has been made

M. Khachay et al. (Eds.): MOTOR 2023, LNCS 13930, pp. 425–438, 2023.
https://doi.org/10.1007/978-3-031-35305-5_29

in the field of unmanned surface transportation. At the same time, the issues of saving natural resources, reducing energy consumption, decreasing harmful emissions into the atmosphere are acutely raised. Together, the requirements for the quality, reliability and safety of the functioning of autonomous systems continuously increase.

These circumstances determine the relevance of research aimed at developing methods for computer modeling, analysis and synthesis of automatic control systems for moving objects, ensuring the high quality of their functioning in a changing environment. In particular, purposeful processing of visual information obtained from onboard video cameras and used in the control loop plays an important role.

This paper is devoted to the problem of visual positioning of a moving object. In the framework of this problem the control goal is to provide the desired projection of some object in a three-dimensional scene to the image plane of the camera, which is mounted on the board of a moving object [1, 2]. The control design is performed on the basis of the introduced joint nonlinear mathematical model of the dynamics of a moving object and the dynamics of the mentioned projection on the image plane. This feature distinguishes proposed control design algorithm from traditional approaches, which are presented, for example, in papers [3–5]. These approaches include basic image-based visual servo control (IBVS), position based visual servo control (PBVS) and various hybrid schemes that integrate the elements of IBVS and PBVS. Typically, these approaches do not take into account the dynamics of a moving object and its inertial characteristics, which limit the possibilities of camera motion in three-dimensional space.

This paper develops the results obtained earlier [1, 2] and proposes a control design approach that is based on the multi-objective structure of control law [6, 7]. An essential feature of the multi-objective structure is that it allows to provide desired characteristics of closed-loop system performance in different regimes by means of a special adjustable elements. From practical reasons usually a special attention is paid to the following three regimes: basic motion without disturbances, motion under constant external disturbances and motion under random disturbances. The methods of searching the adjustable elements of multi-purpose structure in accordance with the requirements to the mentioned regimes are proposed. The developed approach is illustrated by an example of ship motion control.

The paper is organized as follows. Firstly, the problem statement is discussed. Secondly, the proposed control design approach is described. Thirdly, the practical example of ship motion control is given.

2 Problem Statement

Let introduce the mathematical model of moving object dynamics

$$\mathbf{M}\dot{\mathbf{v}} + \mathbf{D}\mathbf{v} = \boldsymbol{\tau} + \boldsymbol{\tau}_e,$$
$$\dot{\boldsymbol{\eta}} = \mathbf{J}(\boldsymbol{\eta})\mathbf{v}, \tag{1}$$

where $\mathbf{M} = \mathbf{M}^T$ is a positive definite inertia matrix, $\mathbf{D}$ is a positive definite damping matrix, $\mathbf{v} = \begin{pmatrix} u\ v\ w\ p\ q\ r \end{pmatrix}^T$ is a vector of linear $(u, v, w)^T$ and angular $(p, q, r)^T$ velocities, $\boldsymbol{\tau} \in E^6$ and $\boldsymbol{\tau}_e(t) \in E^6$ are control input and external disturbances respectively,

$\eta \in E^6$ is a vector that includes the position of the center of mass and the angles of rotation relative to the fixed frame, $\mathbf{J}(\eta)$ is the rotations matrix, such that $\det \mathbf{J}(\eta) \neq 0$, $\forall \eta$. Here and after the symbol E^n denotes Euclidean space of n-dimensional vectors with real components.

It is assumed that the moving object has onboard camera that can move and rotate only together with the object, that is, it does not have additional degrees of freedom. Let describe the dynamics in image plane. In the framework of visual positioning problem the control goal is to provide the desired projection of some object in three-dimensional (3D) scene to the image plane of the camera. Accordingly to computer vision algorithms [8, 9] and mathematical models, described in [3], the changes in projection of some object to image plane is related with the motion of a camera in space by following equation

$$\dot{\mathbf{s}} = \mathbf{L}_s(\mathbf{s}, \mathbf{Z}_c)\mathbf{v} + \mathbf{d}_c(t). \tag{2}$$

Here $\mathbf{s} \in E^{2N}$ is a vector representing the projection of the 3D object to the image plane. Its components are the projections (x_i, y_i), $i = \overline{1, N}$ to the image plane of a several points (X_i, Y_i, Z_i), $i = \overline{1, N}$ that belongs to the 3D object and are given by their coordinates in the camera coordinate frame [8, 9]. Vector $\mathbf{Z}_c \in E^N$ contains the applicate Z_i, $i = \overline{1, N}$ of each projected point, $\mathbf{d}_c(t)$ is an external disturbance, which describes the influence of the possible object motion on the projections to the image plane, $\mathbf{L}_s(\mathbf{s}, \mathbf{Z}_c)$ is an interconnection matrix whose components for each pair (x_i, y_i) are given by [3]:

$$\mathbf{L}_s(x_i, y_i, Z_i) = \begin{pmatrix} \frac{x_i}{Z_i} & -\frac{1}{Z_i} & 0 & y_i & x_i y_i & -(1 + x_i^2) \\ \frac{y_i}{Z_i} & 0 & -\frac{1}{Z_i} & -x_i & 1 + y_i^2 & -x_i y_i \end{pmatrix}$$

In addition to system (2), let also introduce the equations with respect to vector $\mathbf{Z}_c$:

$$\dot{\mathbf{Z}}_c = \mathbf{L}_z(\mathbf{s}, \mathbf{Z}_c)\mathbf{v} + \mathbf{d}_z(t), \tag{3}$$

where $\mathbf{d}_z(t)$ is an external disturbance, which describe the influence of the possible object motion on changes of coordinates Z_i, $i = \overline{1, N}$ in the camera frame, $\mathbf{L}_z(\mathbf{s}, \mathbf{Z}_c)$ is a matrix which row for each applicate Z_i and corresponding image point (x_i, y_i) is as follows

$$\mathbf{L}_z(x_i, y_i, Z_i) = \begin{pmatrix} 0 & 0 & -1 & -y_i Z_i & x_i Z_i & 0 \end{pmatrix}.$$

Thus, systems (1)–(3), considered together, forms the mathematical model of the object dynamics and dynamics in the image plane. It can be noted that Eqs. (1) play the role of differential constraints with respect to the camera motion. It is supposed that vectors η, $\mathbf{s}$ and $\mathbf{Z}_c$ can be measured. Value of the vector $\mathbf{Z}_c$ can be estimated from the images, for example, using the structure from motion approach.

In this paper we don't take into account the problem of visibility of the observed 3D points. It is known that the 3D point is in front of the camera if its applicate is positive. But the corresponding constraints are not considered in the mathematical model here and are the subject of further research.

The control goal in visual positioning problem is to provide the desired projection s_d of the 3D object on the image plane, that is

$$\lim_{t \to +\infty} s(t) = s_d. \tag{4}$$

In accordance with multi-objective approach, let us consider three main regimes of object (1) functioning, as well as requirements imposed on the performance of a closed-loop system at these regimes.

1) *Nominal (unperturbed) motion* is an object motion in the absence of external disturbances, determined only by the reference signal s_d. The control goal in this regime is described by equality (4), and the requirements for this regime are the stability of the equilibrium position (4) and the particular limits on the settling time and overshoot.
2) *Motion under constant disturbances* is an object motion induced by the external constant disturbances

$$\tau_e(t) \equiv \tau_{e0}, \quad d_c(t) \equiv d_{c0}, \quad d_z(t) \equiv d_{z0}.$$

This motion is realized with zero initial conditions and without reference signal. Thus, the control goal in this regime is to compensate the constant disturbances and provide an astatic property on the controlled output $s(t)$ for a closed-loop system. At the same time the following requirement is introduced: the maximum deviations of the controlled variables $s(t)$ should not exceed the specified values.
3) *Motion under random disturbances* is an object motion induced by the random external disturbances represented, for example, by a random stationary process with zero mean and a given matrix of spectral power densities. The control system performance in this regime is determined by the characteristics of the stabilization accuracy and intensity of control. In particular, this paper deals with the problem of filtering external disturbances in the control channel, that is, the problem of saving energy resources by reducing the intensity of control.

As a result, let us state the problem of synthesis of such a control law, which ensures the fulfillment of the mentioned requirements for all three presented basic regimes of moving object operation. The considered problem includes the asymptotic observer design, because only part of the components of the state vector is measured.

Currently, existing methods for control system design in the visual positioning problem include such approaches as IBVS, PBVS and a lot of hybrid variants that integrate the features of IBVS and PBVS [3–5]. All these methods do not take into account the dynamic properties of moving object and imply that the camera has six degree of freedom and can move freely in space.

Earlier in paper [1], a multi-purpose approach for the visual positioning problem was proposed. The control design algorithm proposed in [1] consists of two parts, where the control synthesis is performed separately for the moving object dynamics and for the dynamics in image plane. This approach can be effectively used in practice for moving objects with low mass and powerful actuators.

Another approach, which is based on model predictive control and multi-objective structure, was discussed in paper [2]. This approach can be used with nonlinear models, but its disadvantages are that it requires large computational capabilities and its performance depends significantly on the accuracy of the predictive model.

This paper develops the mentioned ideas and proposes a multi-objective approach to the control system design based on the nonlinear mathematical model (1)–(3), taking into account the requirements to closed-loop system performance in different regimes. The corresponding problems of the choice of adjustable elements are discussed.

3 Control System Design Based on Multi-objective Approach

The proposed approach to control system design for visual positioning problem is based on a special multi-objective structure of the control law. The theoretical basis of this approach is discussed in details in [6, 7]. Taking into account nonlinear mathematical model (1)–(3), let us now introduce the multi-objective structure that includes the following elements:

1) nonlinear asymptotic observer

$$
\begin{aligned}
\mathbf{M}\dot{\mathbf{z}}_v &= -\mathbf{D}\mathbf{z}_v + \boldsymbol{\tau} + \mathbf{J}^{\mathrm{T}}(\boldsymbol{\eta})\mathbf{K}_1\big(\boldsymbol{\eta} - \mathbf{z}_\eta\big), \\
\dot{\mathbf{z}}_\eta &= \mathbf{J}(\boldsymbol{\eta})\mathbf{z}_v + \mathbf{K}_2\big(\boldsymbol{\eta} - \mathbf{z}_\eta\big), \\
\dot{\mathbf{z}}_e &= \mathbf{L}_s(\mathbf{e}, \mathbf{Z}_c)\mathbf{z}_v + \mathbf{H}_e(\mathbf{e} - \mathbf{z}_e),
\end{aligned}
\tag{5}
$$

2) dynamical corrector

$$
\begin{aligned}
\dot{\mathbf{p}} &= \boldsymbol{\alpha}\mathbf{p} + \boldsymbol{\beta}_\eta(\boldsymbol{\eta} - \mathbf{z}_\eta) + \boldsymbol{\beta}_e(\mathbf{e} - \mathbf{z}_e), \\
\boldsymbol{\zeta} &= \boldsymbol{\gamma}\mathbf{p} + \boldsymbol{\mu}_\eta(\boldsymbol{\eta} - \mathbf{z}_\eta) + \boldsymbol{\mu}_e(\mathbf{e} - \mathbf{z}_e);
\end{aligned}
\tag{6}
$$

3) equation for control signal calculation

$$
\boldsymbol{\tau} = -\mathbf{K}_v\mathbf{z}_v - \mathbf{K}_e\mathbf{z}_e + \boldsymbol{\zeta}.
\tag{7}
$$

Here $\mathbf{z}_v \in E^6$, $\mathbf{z}_\eta \in E^6$ and $\mathbf{z}_e \in E^{2N}$ are the state vectors of asymptotic observer; $\mathbf{p} \in E^{n_p}$ is a state vector of the dynamical corrector; $\mathbf{e} = \mathbf{s} - \mathbf{s}_d$ is an error vector; $\boldsymbol{\xi} \in E^6$ is an output vector of a dynamical corrector. Adjustable elements of multi-objective structure (5)–(7) consists of:

a) matrices $\mathbf{K}_v$ and $\mathbf{K}_e$ of the basic control law;
b) matrices $\mathbf{K}_1$, $\mathbf{K}_2$ and $\mathbf{H}_e$ of the asymptotic observer;
c) matrices $\boldsymbol{\alpha}$, $\boldsymbol{\beta}_\eta$, $\boldsymbol{\beta}_e$, $\boldsymbol{\gamma}$, $\boldsymbol{\mu}_\eta$, $\boldsymbol{\mu}_e$ of the dynamical corrector.

The search for these adjustable elements, based on the requirements for the dynamics in the corresponding regimes of motion, is the essence of the problem of multi-objective synthesis. The important feature of a multi-objective structure is that its adjustable elements can be searched sequentially. This feature is proved by the corresponding statements [6]. It is often convenient to formalize the search for adjustable elements as optimization problems on the corresponding admissible sets.

Let us consider sequentially the problems of selecting adjustable elements of a multi-objective structure (5)–(7) in accordance with the requirements, stated above, for the three regimes of motion.

3.1 Synthesis of the Basic Control Law

Consider the problem of searching the matrices $\mathbf{K}_v$ and $\mathbf{K}_e$ of the basic control law

$$\boldsymbol{\tau} = -\mathbf{K}_v \boldsymbol{v} - \mathbf{K}_e \mathbf{e}. \tag{8}$$

These matrices must provide the desired performance of the closed-loop system in the nominal motion. Based on the definition given above and using models (1) and (2), we can obtain a system of equations that determine the dynamics in the nominal motion:

$$\mathbf{M}\dot{\boldsymbol{v}} + \mathbf{D}\boldsymbol{v} = \boldsymbol{\tau},$$
$$\dot{\mathbf{e}} = \mathbf{L}_s(\mathbf{e}, \mathbf{Z}_c)\boldsymbol{v}. \tag{9}$$

Let introduce a positive definite quadratic form

$$V = \frac{1}{2}\mathbf{e}^T\mathbf{e} + \frac{1}{2}\boldsymbol{v}^T\mathbf{M}\boldsymbol{v}.$$

Calculating its derivative for a closed-loop system (8), (9), we get

$$\dot{V}\big|_{(8),(9)} = -\boldsymbol{v}^T\mathbf{D}\boldsymbol{v} + \boldsymbol{v}^T(\mathbf{L}_s^T\mathbf{e} + \boldsymbol{\tau}).$$

Hence, the control law

$$\boldsymbol{\tau} = -\mathbf{L}_s^T\mathbf{e} - \mathbf{K}_v\boldsymbol{v}, \tag{10}$$

where $\mathbf{K}_v \succ 0$ is an arbitrary positive definite matrix, ensures the stability of the zero equilibrium position $\boldsymbol{v}_0 = \mathbf{0}$, $\mathbf{e}_0 = \mathbf{0}$ of the closed-loop system (8), (9), since in this case $\dot{V}\big|_{(8),(9)} = -\boldsymbol{v}^T\mathbf{D}\boldsymbol{v} - \boldsymbol{v}^T\mathbf{K}_v\boldsymbol{v} \leq 0$. If $\mathbf{e}_0 = \mathbf{0}$ then $\mathbf{s}_0 = \mathbf{s}_d$ and the control goal (4) achieved. Thus, taking into account (10), the matrices $\mathbf{K}_v$ and $\mathbf{K}_e$ of the basic control law (8) are equal to $\mathbf{K}_e = \mathbf{L}_s^T$, $\mathbf{K}_v \succ 0$. It is important to note that the only requirement to the matrix $\mathbf{K}_v$ is its positive definiteness. This can be used to meet the imposed limits on the settling time and overshoot for the closed-loop system.

3.2 Synthesis of the Asymptotic Observer

Consider the problem of searching the matrices $\mathbf{K}_1$, $\mathbf{K}_2$ and $\mathbf{H}_e$ of the asymptotic observer. Let us introduce the vectors of estimation errors: $\boldsymbol{\varepsilon}_v = \boldsymbol{v} - \mathbf{z}_v$, $\boldsymbol{\varepsilon}_\eta = \boldsymbol{\eta} - \mathbf{z}_\eta$, $\mathbf{e}_s = \mathbf{e} - \mathbf{z}_e$. In accordance with (5), these vectors satisfy the system of equations:

$$\mathbf{M}\dot{\boldsymbol{\varepsilon}}_v = -\mathbf{D}\boldsymbol{\varepsilon}_v - \mathbf{J}^T(\boldsymbol{\eta})\mathbf{K}_1\boldsymbol{\varepsilon}_\eta + \boldsymbol{\tau}_e,$$
$$\dot{\boldsymbol{\varepsilon}}_\eta = \mathbf{J}(\boldsymbol{\eta})\boldsymbol{\varepsilon}_v - \mathbf{K}_2\boldsymbol{\varepsilon}_\eta,$$
$$\dot{\mathbf{e}}_s = \mathbf{L}_s(\mathbf{e}, \mathbf{Z}_c)\boldsymbol{\varepsilon}_v - \mathbf{H}_e\mathbf{e}_s + \mathbf{d}_c(t). \tag{11}$$

If there are no external disturbances, then system (11) has a zero equilibrium position $\boldsymbol{\varepsilon}_v = \mathbf{0}$, $\boldsymbol{\varepsilon}_\eta = \mathbf{0}$, $\mathbf{e}_s = \mathbf{0}$. The first two equations in (11) are independent of the third one and can be studied separately. As shown in [10, 11], the global asymptotic stability (GAS) of the zero equilibrium position $\boldsymbol{\varepsilon}_v = \mathbf{0}$, $\boldsymbol{\varepsilon}_\eta = \mathbf{0}$ is guaranteed if the matrices $\mathbf{K}_1$ and $\mathbf{K}_2$ are positive definite and have a diagonal structure.

The third equation in (11) can be rewritten as

$$\dot{\mathbf{e}}_s = -\mathbf{H}_e \mathbf{e}_s + \mathbf{f}(t), \tag{12}$$

where $\mathbf{f}(t) = \mathbf{L}_s(\mathbf{e}, \mathbf{Z}_c)\boldsymbol{\varepsilon}_v$ can be viewed as a perturbation, such that $\mathbf{f}(t) \rightarrow \mathbf{0}$ as $t \rightarrow +\infty$. This property is determined by the GAS of the equilibrium position $\boldsymbol{\varepsilon}_v = \mathbf{0}$, $\boldsymbol{\varepsilon}_\eta = \mathbf{0}$, hence $\boldsymbol{\varepsilon}_v \rightarrow \mathbf{0}$ as $t \rightarrow +\infty$, and by boundedness of the components of matrix $\mathbf{L}_s(\mathbf{e}, \mathbf{Z}_c)$. Here we can apply the theorem for cascaded systems, presented in [12]. In accordance with this theorem the equilibrium $\boldsymbol{\varepsilon}_v = \mathbf{0}$, $\boldsymbol{\varepsilon}_\eta = \mathbf{0}$, $\mathbf{e}_s = \mathbf{0}$ is GAS, if the matrix $\mathbf{H}_e \succ \mathbf{0}$ is positive definite.

As a result, the searched matrices $\mathbf{K}_1$, $\mathbf{K}_2$ and $\mathbf{H}_e$ of the asymptotic observer must be positive definite, and besides that, the matrices $\mathbf{K}_1$, $\mathbf{K}_2$ must have diagonal structure. Within these properties, matrices can be arbitrary. This allows us to select them taking into account the following requirements for the processes quality:

a) in the case of a nominal motion, the performance of a closed-loop system only for the basic control law (8), (9), and for the system with an asymptotic observer (5), where the control input is given by $\boldsymbol{\tau} = -\mathbf{K}_v \mathbf{z}_v - \mathbf{K}_e \mathbf{z}_e$, must be approximately the same. The quality of the processes for the system with basic control law can be slightly better;

b) in the case of a motion under constant disturbances with an astatic property on the controlled output $\mathbf{s}(t)$, the maximum deviations of the controlled variables $\mathbf{s}(t)$ should not exceed the specified values.

Suppose that the matrices $\mathbf{K}_1$, $\mathbf{K}_2$ and $\mathbf{H}_e$ are selected to meet these requirements. In particular, this can be done using an optimization approach on a corresponding admissible set.

3.3 Synthesis of the Dynamical Corrector

Consider a problem of searching the matrices $\boldsymbol{\alpha}$, $\boldsymbol{\beta}_\eta$, $\boldsymbol{\beta}_e$, $\boldsymbol{\gamma}$, $\boldsymbol{\mu}_\eta$, $\boldsymbol{\mu}_e$ of the dynamical corrector (6), which is an additive element of the multi-purpose structure of the control law and can be turned on/off in real-time. The main objective of the dynamical corrector is to provide the desired performance of the closed-loop system for regimes of motion under constant and random disturbances. In the case of constant disturbances, the corrector is used to compensate them and provide astatic property on the controlled output $\mathbf{s}(t)$. If the disturbances are random, the corrector can be used to reduce the intensity of control actions while maintaining the accuracy of stabilization. Let consider the problem of the design of a dynamical corrector that satisfy imposed requirements.

First, consider the case of a constant disturbances $\boldsymbol{\tau}_e(t) \equiv \boldsymbol{\tau}_{e0}$ and $\mathbf{d}_c(t) \equiv \mathbf{d}_{c0}$, where $\boldsymbol{\tau}_{e0}$, $\mathbf{d}_{c0}$ are constant vectors. In a subsequent discussion, we will also use the dynamical corrector Eqs. (6) in the tf form:

$$\boldsymbol{\zeta} = \mathbf{F}_1(s)(\boldsymbol{\eta} - \mathbf{z}_\eta) + \mathbf{F}_2(s)(\mathbf{e} - \mathbf{z}_e), \tag{13}$$

where $\mathbf{F}_1(s) = \boldsymbol{\gamma}(\mathbf{E}_{n_p}s - \boldsymbol{\alpha})^{-1}\boldsymbol{\beta}_\eta + \boldsymbol{\mu}_\eta$, $\mathbf{F}_2(s) = \boldsymbol{\gamma}(\mathbf{E}_{n_p}s - \boldsymbol{\alpha})^{-1}\boldsymbol{\beta}_e + \boldsymbol{\mu}_e$. Taking into account (7) and (13), we obtain

$$\boldsymbol{\tau} = -\mathbf{K}_v \mathbf{z}_v - \mathbf{K}_e \mathbf{z}_e + \mathbf{F}_1(s)(\boldsymbol{\eta} - \mathbf{z}_\eta) + \mathbf{F}_2(s)(\mathbf{e} - \mathbf{z}_e). \tag{14}$$

Let form the equations of the closed-loop system. Using (11), (5) and (14), we get

$$
\begin{aligned}
\mathbf{M}\dot{\boldsymbol{\varepsilon}}_v &= -\mathbf{D}\boldsymbol{\varepsilon}_v - \mathbf{J}^\mathrm{T}(\eta)\mathbf{K}_1\boldsymbol{\varepsilon}_\eta + \boldsymbol{\tau}_{e0}, \\
\dot{\boldsymbol{\varepsilon}}_\eta &= \mathbf{J}(\eta)\boldsymbol{\varepsilon}_v - \mathbf{K}_2\boldsymbol{\varepsilon}_\eta, \\
\dot{\mathbf{e}}_s &= \mathbf{L}_s(\mathbf{e}, \mathbf{Z}_c)\boldsymbol{\varepsilon}_v - \mathbf{H}_e\mathbf{e}_s + \mathbf{d}_{c0}, \\
\mathbf{M}\dot{\mathbf{z}}_v &= -\mathbf{D}\mathbf{z}_v + \mathbf{J}^\mathrm{T}(\eta)\mathbf{K}_1\boldsymbol{\varepsilon}_\eta - \mathbf{K}_v\mathbf{z}_v - \mathbf{K}_e\mathbf{z}_e + \mathbf{F}_1(s)\boldsymbol{\varepsilon}_\eta + \mathbf{F}_2(s)\mathbf{e}_s, \\
\dot{\mathbf{z}}_e &= \mathbf{L}_s(\mathbf{e}, \mathbf{Z}_c)\mathbf{z}_v + \mathbf{H}_e\mathbf{e}_s.
\end{aligned}
\tag{15}
$$

The equilibrium position of this system can be found by equating its right-hand parts to zero. Doing so, from the first three equations, we obtain

$$
\begin{aligned}
-\mathbf{D}\boldsymbol{\varepsilon}_{v0} - \mathbf{J}^\mathrm{T}(\eta_0)\mathbf{K}_1\boldsymbol{\varepsilon}_{\eta0} + \boldsymbol{\tau}_{e0} &= \mathbf{0}, \\
\mathbf{J}(\eta_0)\boldsymbol{\varepsilon}_{v0} - \mathbf{K}_2\boldsymbol{\varepsilon}_{\eta0} &= \mathbf{0}, \\
\mathbf{L}_{s0}\boldsymbol{\varepsilon}_{v0} - \mathbf{H}_e\mathbf{e}_{s0} + \mathbf{d}_{c0} &= \mathbf{0},
\end{aligned}
\tag{16}
$$

where $\boldsymbol{\varepsilon}_{v0}$, $\mathbf{e}_{s0}$, $\boldsymbol{\varepsilon}_{\eta0}$, η_0 are corresponding equilibrium positions for vectors $\boldsymbol{\varepsilon}_v$, $\mathbf{e}_s$, $\boldsymbol{\varepsilon}_\eta$, η and $\mathbf{L}_{s0} = \mathbf{L}_s(\mathbf{e}_0, \mathbf{Z}_{c0})$ is a constant matrix calculated for equilibrium values $\mathbf{e}_0$, $\mathbf{Z}_{c0}$ of vectors $\mathbf{e}$, $\mathbf{Z}_c$. From (16), we can see that there is a direct relation between the non-zero disturbance vectors $\boldsymbol{\tau}_{e0}$, $\mathbf{d}_{c0}$ and non-zero error vectors $\boldsymbol{\varepsilon}_{\eta0}$, $\mathbf{e}_{s0}$. Hence, the vectors $\boldsymbol{\varepsilon}_{\eta0}$, $\mathbf{e}_{s0}$ can be considered as a disturbances in the fourth and fifth equations of the system (15).

From fifth equation, we get

$$
\mathbf{z}_{v0} = -(\mathbf{L}_{s0}^\mathrm{T}\mathbf{L}_{s0})^{-1}\mathbf{L}_{s0}^\mathrm{T}\mathbf{H}_e\mathbf{e}_{s0} = \mathbf{T}\mathbf{e}_{s0},
\tag{17}
$$

where $\mathbf{T} = -(\mathbf{L}_{s0}^\mathrm{T}\mathbf{L}_{s0})^{-1}\mathbf{L}_{s0}^\mathrm{T}\mathbf{H}_e$ is an auxiliary matrix. Substituting (17) in the forth equation of (15), we find

$$
\mathbf{K}_e\mathbf{e}_0 = (\mathbf{K}_e - \mathbf{D}\mathbf{T} - \mathbf{K}_v\mathbf{T} + \mathbf{F}_2(0))\mathbf{e}_{s0} + (\mathbf{J}^\mathrm{T}(\eta_0)\mathbf{K}_1 + \mathbf{F}_1(0))\mathbf{e}_{\eta0}.
\tag{18}
$$

From here, we can conclude that the astatic property is satisfied for constant disturbances $\boldsymbol{\tau}_{e0}$, $\mathbf{d}_{c0}$ if the following conditions are hold

$$
\mathbf{F}_2(0) = -\mathbf{K}_e + \mathbf{D}\mathbf{T} + \mathbf{K}_v\mathbf{T}, \quad \mathbf{F}_1(0) = -\mathbf{J}^\mathrm{T}(\eta_0)\mathbf{K}_1.
\tag{19}
$$

As a result, the matrices $\boldsymbol{\alpha}$, $\boldsymbol{\beta}_\eta$, $\boldsymbol{\beta}_e$, $\boldsymbol{\gamma}$, $\boldsymbol{\mu}_\eta$, $\boldsymbol{\mu}_e$ of the dynamical corrector (6) must satisfy conditions (19) which provides the astaticism of the controlled output $\mathbf{s}(t)$ with respect to constant disturbances.

Let now consider a case of random disturbances with a given spectral power density. For simplicity, we assume that this spectral density has one dominant frequency ω_0 and thus, for the purpose of a dynamical corrector design, we will consider the disturbances $\tau_e(t)$ and $d_c(t)$ as a harmonic oscillations with frequency ω_0. In the subsequent discussion the additional conditions on the transfer matrices $F_1(s)$ and $F_2(s)$ are derived. These conditions allows to reduce the intensity of control actions for external disturbances with frequencies close to ω_0 while maintaining the accuracy of stabilization.

Let consider the filtering problem when the object is moving with a constant linear velocity and zero angular velocity, so that the coordinate transformation matrix $J(\eta)$ remains constant. From the equations of the asymptotic observer in system (15) we can express the values of vectors z_v и z_e. As a result, we obtain

$$z_v = T_{11}(s)\varepsilon_\eta + T_{12}(s)e_s,$$
$$z_e = T_{21}(s)\varepsilon_\eta + T_{22}(s)e_s, \tag{20}$$

where $T_{11}(s)$, $T_{12}(s)$, $T_{21}(s)$, $T_{22}(s)$ are the auxiliary transfer functions, expressions for which are determined by (15). Substituting (20) to (14), we have

$$\tau = (-K_v T_{11}(s) - K_e T_{21}(s) + F_1(s))\varepsilon_\eta + (-K_v T_{12}(s) - K_e T_{22}(s) + F_2(s))e_s. \tag{21}$$

As a result, the condition of filtering the external disturbances $\tau_e(t)$ and $d_c(t)$ at a given frequency ω_0 takes the form

$$F_1(j\omega_0) = K_v T_{11}(j\omega_0) + K_e T_{21}(j\omega_0),$$
$$F_2(j\omega_0) = K_v T_{12}(j\omega_0) + K_e T_{22}(j\omega_0). \tag{22}$$

So the matrices α, β_v, β_e, γ, μ_v, μ_e of the dynamical corrector must satisfy conditions (19), (22) to ensure the astatic property of the closed-loop system and to reduce the intensity of control actions in the case of random disturbances.

4 Numerical Example

Consider the 3DOF model of the marine vessel, which can be described by the equations [1]

$$\dot{v} = -v + \tau,$$
$$\dot{\eta} = J(\eta)v,$$

where $v = \begin{pmatrix} u & v & r \end{pmatrix}^T$ is a velocity vector, τ is a control signal, $\eta = \begin{pmatrix} X_c & Y_c & \psi \end{pmatrix}^T$ is a position vector, X_c and Y_c are the coordinates of the vessel's center of mass in the fixed frame, ψ is the course angle. Rotation matrix here is

$$J(\eta) = J(\psi) = \begin{pmatrix} \cos\psi & -\sin\psi & 0 \\ \sin\psi & \cos\psi & 0 \\ 0 & 0 & 1 \end{pmatrix}.$$

Interconnection matrix $\mathbf{L}_s$ for this case is simplified to

$$\mathbf{L}_s(x_i, y_i, Z_i) = \begin{pmatrix} \frac{x_i}{Z_i} & -\frac{1}{Z_i} & -(1+x_i^2) \\ \frac{y_i}{Z_i} & 0 & -x_i y_i \end{pmatrix}.$$

Let us assume that there is some box in the field of view of the camera mounted on the ship. The task here is to achieve the desired position of the box $\{(2\,2), (2\,-2), (-2\,-2), (-2\,2)\}$ in the image plane. Starting position of the box is $\{(0.7071\,1), (0.7071\,-1), (-0.2929\,-0.4142), (-0.2929\,0.4142)\}$. This setup is presented on the Fig. 1. The initial value of the vector $\mathbf{Z}_c$ is $\mathbf{Z}_c(0) = (-1\,-1\,-2.4142\,-2.4142)^{\mathrm{T}}$.

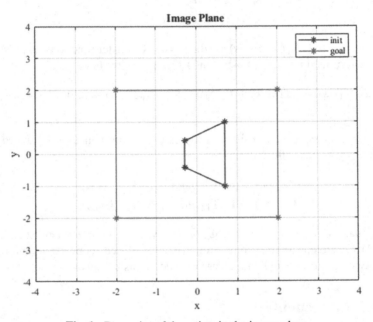

Fig. 1. Dynamics of the points in the image plane.

Let us assume that there are no external disturbances, so we can turn off the dynamic corrector and take the following values for the tunable parameters of the multipurpose controller (5)–(7):

$$\mathbf{K}_v = 8\mathbf{E}_{3\times3}, \mathbf{K}_1 = 10\mathbf{E}_{3\times3}, \mathbf{K}_2 = 5\mathbf{E}_{3\times3}, \mathbf{H} = 0.2\mathbf{E}_{8\times8},$$

where $\mathbf{E}$ is an identity matrix. Figure 2 demonstrates the dynamics of the points in the image plane and Fig. 3 shows the dynamics of the velocity vector. As we can see, the proposed regulator drives the ship to the goal smoothly and successfully.

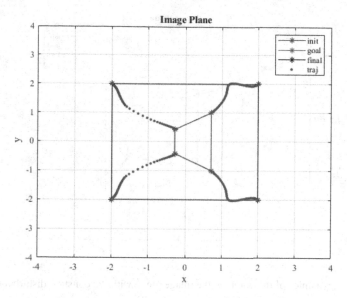

Fig. 2. Dynamics of the points in the image plane.

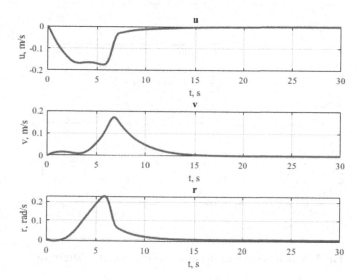

Fig. 3. Dynamics of the velocities.

Now let us consider another situation. The vessel is initially in the desired position, but there is constant disturbance $\tau_e = \begin{pmatrix} 0.3\ 0.3\ 0.1 \end{pmatrix}^T$ with the dynamic corrector still turned off. The initial value of the vector $\mathbf{Z}_c$ is $\mathbf{Z}_c(0) = \begin{pmatrix} -0.5\ -0.51\ -0.5\ -0.5 \end{pmatrix}^T$. As we can see from the Fig. 4 and Fig. 5, in this case the vessel can't fight the disturbance and moves away from the initial position.

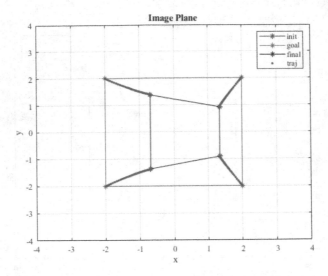

Fig. 4. Dynamics of the points in the image plane with the constant disturbance.

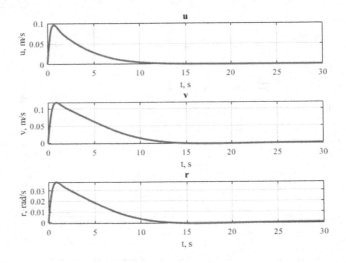

Fig. 5. Dynamics of the velocities with the constant disturbance.

Finally let us turn on the dynamic corrector that satisfies the conditions (19). Now the vessel actually manages to go back to the initial position after the action of the disturbance. Figure 6 and Fig. 7 demonstrate the dynamics for this situation.

The experiments with the random disturbances, related with the sea waves, have been carried out. In this case the corrector is designed taking into account conditions (22). The obtained results show the effectiveness of the proposed approach.

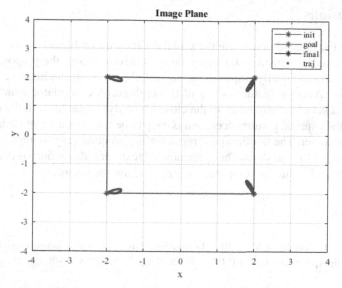

Fig. 6. Dynamics of the points in the image plane with the dynamic corrector.

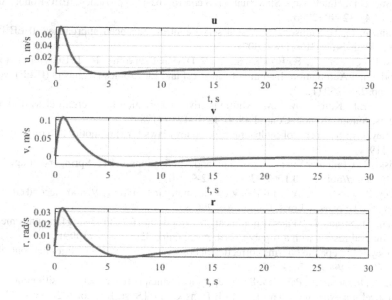

Fig. 7. Dynamics of the velocities with the dynamic corrector.

5 Conclusions

In this paper the method of solving the visual positioning problem based on the visual servoing approach is considered. One of the main advantages of the proposed solution is taking into account the dynamics of the control object which can improve the quality of the control. Another feature is use of the multipurpose regulator which can take into account a set of requirements to the closed loop dynamics (also in the case of the presence of the external disturbances) and is easy to be tuned as a set of easier subtasks in a parallel manner. The multipurpose regulator is also quite easy to implement on the digital devices on the real ships. The efficiency of the proposed approach is demonstrated on the experiments with the computer model of the marine vessel.

References

1. Veremey, E.I., Sotnikova, M.V.: Visual image based dynamical positioning using control laws with multipurpose structure. In: IFAC Proceedings Volumes, vol. 48, no. 16, pp. 184–189 (2015)
2. Sotnikova, M.V.: Control system design for visual positioning of a ship based on NMPC and multi-objective structure. IFAC-PapersOnLine 51(32), 445–450 (2018)
3. Chaumette, F., Hutchinson, S.: Visual servo control: basic approaches. IEEE Robot. Autom. Mag. 13(4), 82–90 (2006)
4. Chaumette, F., Hutchinson, S.: Visual servo control: advanced approaches. IEEE Robot. Autom. Mag. 14(1), 109–118 (2007)
5. Malis, E., Chaumette, F., Boudet, S.: 2 1/2 D visual servoing. In: IEEE Transactions on Robotics and Automation, Institute of Electrical and Electronics Engineers (IEEE), vol. 15, no. 2, pp. 238–250 (1999)
6. Veremei, E.I., Korchanov, V.M.: Multiobjective stabilization of a certain class of dynamic systems. Autom. Remote Control 49(9), 1210–1219 (1989)
7. Veremey, E.I.: Synthesis of multi-objective control laws for ship motion. Gyroscopy Navig. 1(2), 119–125 (2010)
8. Szeliski, R.: Computer Vision: Algorithms and Applications, p. 812. Springer-Verlag, London (2011). https://doi.org/10.1007/978-1-84882-935-0
9. Hartley, R., Zisserman, A.: Multiple View Geometry in Computer Vision, Second ed., p. 672. Cambridge University Press, Cambridge (2003)
10. Fossen, T.I., Strand, J.P.: Passive nonlinear observer design for ships using Lyapunov methods: experimental results with a supply vessel. Automatica 35(1), 3–16 (1999)
11. Veremey, E.I.: Separate filtering correction of observer-based marine positioning control laws. Int. J. Control 90(8), 1561–1575 (2017)
12. Loria, A., Fossen, T.I., Panteley, E.: A separation principle for dynamic positioning of ships: theoretical and experimental results. IEEE Trans. Control Syst. Technol. 8(2), 332–343 (2000)

Author Index

M. Khachay et al. (Eds.): MOTOR 2023, LNCS 13930, p. 439, 2023.
https://doi.org/10.1007/978-3-031-35305-5

Printed in the United States
by Baker & Taylor Publisher Services